ANALOG DEVICES AND CIRCUITS

AF731109

By

J.B. Gupta
B.Sc. (Elec. Engg.); C. Eng. (I);
F.I.E. (India); M.I.S.T.E.

Rajeev Manglik
B.E. (Elec.) Pb. University

Rohit Manglik
B.E. (Elec.) Univ. of Roorkee

S.K. KATARIA & SONS®
Publisher of Engineering Books
4885/109, Prakash Mahal, Dr. Subhash Bhargav Lane,
Opposite Delhi Medical Association, Daryaganj, New Delhi–110002 (INDIA)
Phone: +91-11-43551243; *Mobile:* +91-9871775858, 7678425209
e-mail: katariabook@gmail.com, katariabooks@yahoo.com
Website: www.skkatariaandsons.com
(e-books, subscription and bulk buying also available on our website)

Published by:

S.K. Kataria & Sons®

4885/109, Prakash Mahal, Dr. Subhash Bhargav Lane,
Opposite Delhi Medical Association, Daryaganj, New Delhi-110002 (INDIA)
Phone: +91-11-43551243; *Mobile:* +91-9871775858, 7678425209
e-mail: katariabook@gmail.com, katariabooks@yahoo.com
Website: www.skkatariaandsons.com

Edited By : **Shalini Manglik**

First Edition: July, 2012; Reprint: 2026

Price: ₹ 495/-

Laser Typesetting by :
AISA ENTERPRISES, Delhi-110092

Printed at : New A.S. Offset Printers, Delhi-110051

PREFACE

The primary aim of presenting this book is to meet the needs of 3rd Semester (ECE and EIE branches) B.Tech. students studying in Engineering colleges affiliated to Punjab Technical University. The subject matter has been presented in easy comprehensive manner, logical sequence using simple language and to provide a detailed exposition of the fundamentals.

The book consists of 9 chapters divided in 5 units providing a comprehensive coverage of courses as prescribed by Punjab Technical University for the subject of Analog Devices and Circuits. A large number of worked-out examples, theoretical exercises and numerical problems, all of degree standard, have been included in the text. The solution to illustrative examples covering almost all topics and subtopics makes the entire presentation easy to follow. Exercises and numerical problems with answers given at the end of each chapter help the students in evaluating their understanding of the subject. Short Answer Type Questions given at the end of each chapter will certainly help the students in making preparations for viva voce/practical examinations.

Every care has been taken to eliminate misprints, omissions and errors but it is too much to expect that no inaccuracy/misprint/error has crept in and the authors would be grateful to the readers for bringing to their notice any such errors/misprints/omissions they may come across while going through the book.

The authors will feel satisfied if the book meets the needs of the students for whom it is meant. Suggestions and criticism for the improvement of the book shall be welcomed.

The authors will like to thank to all the Scientists, Engineers, Authors and Publishers whose works and text have been the source of enlightenment, inspiration and guidance to us in the presentation.

Lastly, but not the least, the authors are thankful to their publisher for bringing out the book in such a short time and for pricing it moderately despite heavy cost of paper and printing.

July 11, 2012

Authors

OTHER USEFUL BOOKS

Contents

Unit-I

Semiconductor Diodes

CONTENTS OF THIS UNIT

UNIT-I : SEMICONDUCTOR DIODES : Theory of P-N Junction Diode; Band Structure of Open-Circuited P-N Junction; Volt-Ampere Characteristics; Temperature Dependence of PN Diode; LED; LCD; and Photodiode; Zener Diode As Voltage Regulator.

CHAPTER

1

JUNCTION DIODES

1.1. INTRODUCTION

INSIDE THIS CHAPTER

One of the first uses of semiconductor materials was to join a P-type material with an N-type to form a simple but important semiconductor device called a *diode.*

The term *diode* denotes a two-electrode device. A *semiconductor diode* is simply a P-N junction with connecting leads or terminals on the two sides of the P-N junction. P-N junction is also called the crystal diode because the P-N junction is grown out of crystal. A diode is a unidirectional device permitting the easy flow of current in one direction but restraining the flow in opposite direction. A major application of diodes is in *rectification i.e.,* in conversion of ac into dc. Semiconductor diode is becoming more and more popular these days due to its smaller size, cheapness, robustness and higher operating efficiency.

The P-type material of the diode is the positive electrode and is called the *anode* whereas the N-type material is the negative electrode and is called the *cathode.* The arrow symbol used for the anode represents the *direction of flow of conventional current.* The *electron current* flows in a direction *opposite* to the arrow.

The symbol for semiconductor diode is shown in Fig. 1.1. Diodes have *polarity* and, therefore, must be connected in circuits in the correct direction. Small diodes are marked with a band to indicate the cathode end. Larger diodes use the diode symbol to indicate their polarity. Diodes are available in many package configurations, depending on their ratings. Small diodes, used for low current application in circuits up to 1 A of average current, are generally packaged in a simple glass or plastic tube. Diodes for handling larger currents are packaged in metal containers and are often mounted by studs on metal plates to dissipate heat.

Diodes are rated in terms of their peak and average forward current-carrying capability and in terms of their reverse voltage or peak inverse voltage ratings. None of these values can be exceeded, otherwise the diode will get damaged. For the P-N diode junction, the reverse voltage rating applies to the instantaneous peak reverse voltage across the diode.

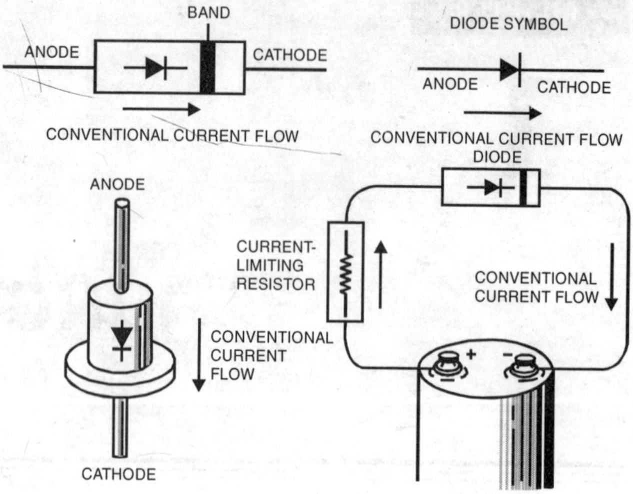

Fig. 1.1. *Semiconductor Diode*

The diode junction can be viewed as an automatic switch that is closed when the P-type material is relatively positive and open when the P-type material is relatively negative. Remember, it is the *voltage across the diode* that is important *not* the voltage to another part of the circuit. Because the depletion region is never completely absent, it takes a small voltage across a diode to make current flow, even when the polarity is correct for flow of larger current.

Most diodes in use today make use of silicon semiconductor material; however, for some applications, germanium is also used.

The silicon junction diode is similar in appearance to the germanium diode but differs in properties. Silicon diodes have, in general, higher peak inverse voltage (PIV) and current ratings and wider temperature ranges than germanium diode.

(*i*) The peak inverse voltage ratings for the silicon can be as high as 1,000 V whereas for germanium it is near about 400 V.

(*ii*) Silicon can be operated at temperatures up to about 200ºC compared with about 75–80ºC for germanium.

(*iii*) The forward voltage drop of silicon diode is approximately double of that of the corresponding germanium diode. It is typically of the order of 0.7 V for commercially available silicon diodes and 0.3 V for germanium diodes.

(*iv*) The reverse current at a given temperature and voltage is approximately a hundredth of that of the corresponding germanium diode, but there is little sign of current saturation as in case of germanium. In fact, the reverse current of a silicon diode is approximately proportional to the square root of the voltage until breakdown is approached.

(*v*) The reverse current of a silicon diode, for a given voltage, practically doubles for every 8ºC rise of temperature, compared with 10ºC for germanium.

1.2. BASIC STRUCTURE OF P-N JUNCTION

Most semiconductor devices employ one or more P-N junctions. The P-N junction is the control element for the performance of all semiconductor devices such as rectifiers, amplifiers, switching devices, linear and digital integrated circuits. The P-N junction is produced* by placing a layer of P-type semiconductor next to the layer of N-type semiconductor. The interface separating the N and P regions is referred to as the *metallurgical junction.*

* It is to be noted that a P-N junction is not formed by merging two separate P-type and N-type materials. If this were done, a polycrystalline structure would have resulted in, which would have not performed very well. The P-N junction found in the semiconductor devices is actually formed in a single monocrystalline structure by adding carefully controlled amounts of donor and acceptor impurities.

Figure 1.2 represents two blocks of semiconductor material, one P-type, and the other N-type. The P-type semiconductor block has mobile holes (shown by small circles) and the same number of fixed negative acceptor ions (shown by encircled minus sign). Similarly the N-type semiconductor block has mobile or free electrons (shown by dots) and the same number of fixed donor positive ions. Normally the holes, which are the majority charge carriers in P-type of material, are uniformly distributed throughout the volume of that material. Similarly the electrons, which are the majority charge carriers in N-type of material, are uniformly distributed throughout the volume of that material. Each region is electrically neutral because each of them carries equal positive and negative charges.

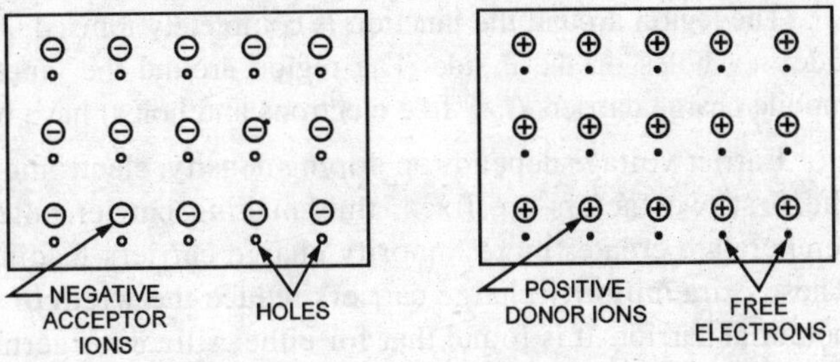

(a) P-Type Semiconductor Material *(b) N-Type Semiconductor Material*

Fig. 1.2

On the formation of P-N junction some of the holes from P-type material tend to diffuse across the boundary into N-type material and some of the free electrons similarly diffuse into the P-type material, as illustrated in Fig. 1.3. This happens due to density gradient (as concentration of holes is higher on P-side than that on N-side and concentration of electrons is higher on N-side than that on P-side). This process is known as *diffusion*. The potential distribution diagram is shown in Fig. 1.3. From Fig. 1.3, it is obvious that a potential barrier V_B or V_0 is developed which results in an electric field. This field prevents the respective majority carriers from crossing the barrier region.

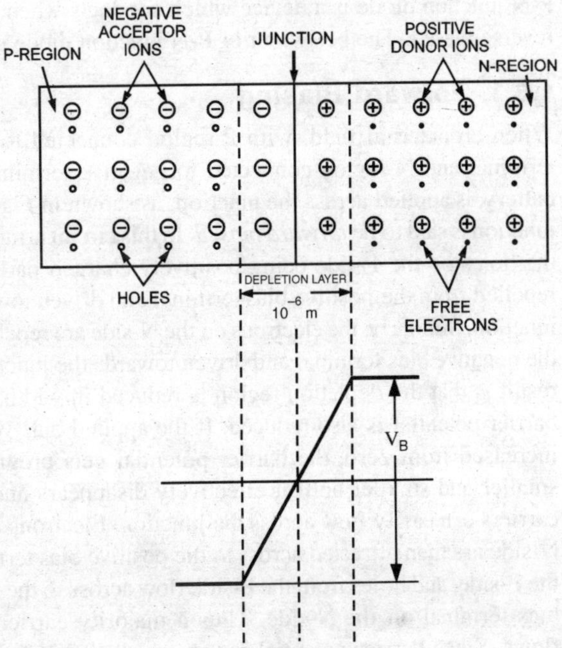

Fig. 1.3

The doping profile of an ideal uniformly doped P-N junction is depicted in Fig. 1.4.

It is evident that the free electrons crossing the junction create negative ions on the P-side by giving some atoms one more electron than their total number of protons. The electrons also leave positive ions (atoms with one fewer electron than the number of protons) behind them on the N-side. As negative ions are created on the P-side of the junction, the P-side acquires a negative potential. Similarly, the positive ions are created on the N-side and the N-side acquires a positive potential. The negative potential on the P-side prevents the migration of any more electrons from the N-type material to the P-type material. Similarly, the positive potential on the N-side prevents any further migration of holes across the boundary. Thus, the initial diffusion of charge carriers creates a *barrier potential* at the junction. The magnitude of barrier potential V_0 is of the order of few tenths of a volt (0.3 V in case of Ge and 0.7 V for Si).

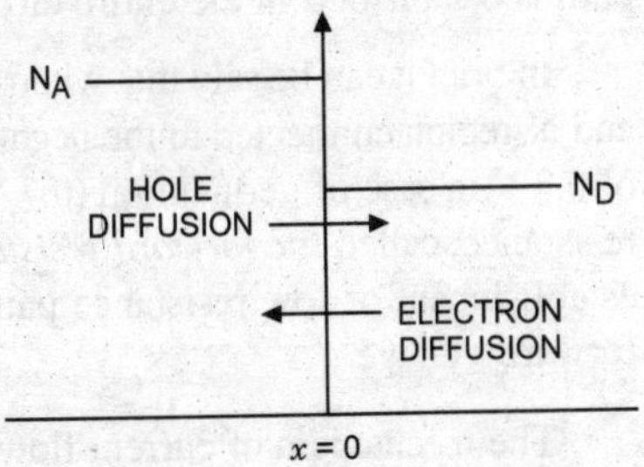

Fig. 1.4. ***Doping Profile of an Ideal Uniformly Doped P-N Junction***

The region around the junction is completely ionised. As a result, there are no free electrons on the N-side, nor holes on the P-side. The region around the junction is termed as *depletion region* because the mobile charge carriers (*i.e.,* free electrons and holes) have been emptied in this region.

Barrier voltage depends on doping density, electronic charge and temperature. For a given junction, the first two factors are fixed, thus making barrier potential dependent on temperature. Increase in temperature creates more minority charge carriers leading to their increased drift across the junction. These extra minority charge carriers reduce the width of the depletion layer equivalent to reduction in potential barrier. It is found that for either silicon or germanium diodes, the barrier potential decreases 2 mV for each Celsius degree rise.

1.3. P-N JUNCTION AS A DIODE

P-N junction diode is a device which conducts when forward biased and practically does not conduct when reverse biased. The behaviour of P-N junction diode under forward biasing will be discussed here.

1.3.1. Forward Biasing

When an external field, with P-region connected to positive terminal and N-region connected to negative terminal of the battery, is applied across the junction, as shown in Fig. 1.5, the junction is said to be *forward biased.* In this circuit arrangement, the holes on the P-side being positively charged particles are repelled from the positive bias terminal and driven towards the junction. Similarly, the electrons on the N-side are repelled from the negative bias terminal and driven towards the junction. The result is that the depletion region is reduced in width, and the barrier potential is also reduced. If the applied bias voltage is increased from zero, the barrier potential gets progressively smaller and smaller until it effectively disappears and charge carriers can easily flow across the junction. Electrons from the N-side are then attracted across to the positive bias terminal on the P-side, and holes from the P-side flow across to the negative bias terminal on the N-side. Thus a majority carrier current flows. Since barrier potential is very small (0.3 V for Ge and 0.7 V for Si), a small forward voltage is sufficient to eliminate the barrier completely. Once the barrier is eliminated by the application of forward voltage, junction resistance becomes almost zero and a low resistance path is established in the entire circuit. The current, called the *forward current,* flows in the circuit.

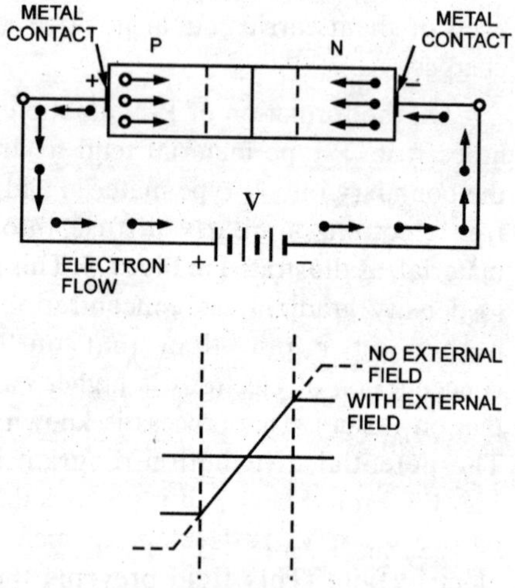

Fig. 1.5. ***Forward Biasing***

In brief it can be said that when the junction is forward biased (P-region connected to the positive terminal and N-region connected to the negative terminal of the battery), (*i*) barrier is reduced and at a forward voltage of 0.3 V in case of germanium (0.7 V in case of silicon) it is eliminated altogether, (*ii*) the junction offers low resistance, called the *forward resistance* r_f, to the flow of current and (*iii*) current flows in the circuit due to establishment of low resistance path and the magnitude of current depends upon the magnitude of applied forward voltage.

The mechanism of current flow in a forward biased P-N junction is summed up as follows :

(*i*) The electrons from the negative bias terminal continue to arrive into the N-region while the free electrons in the N-region move towards the P-N junction.

(*ii*) The electrons travel through the N-region as free electrons *i.e.,* the current in N-region is by free electrons.

(*iii*) These free electrons on reaching P-N junction, combine with holes and become valence electrons. Since a hole is in the covalent bond and, therefore, when a free electron combines with a hole, it becomes a valence electron.

(*iv*) The electrons travel through P-region as valence electrons and current in this P-region is, therefore, by holes.

(*v*) These valence electrons, on reaching the left end of the crystal, flow into the positive terminal of the battery.

Thus we see that the current in the N-type region is due to movement of electrons whereas in the P-type region it is carried by the holes. However, in the external circuit (*i.e.,* in connecting wires) the current is carried by electrons.

1.4. ZERO APPLIED BIAS

Here we will examine the properties of the step junction in thermal equilibrium, where no currents exist and no external excitation is applied. We will determine built-in potential barrier through the depletion or space-charge region, electric field and width of the space-charge region.

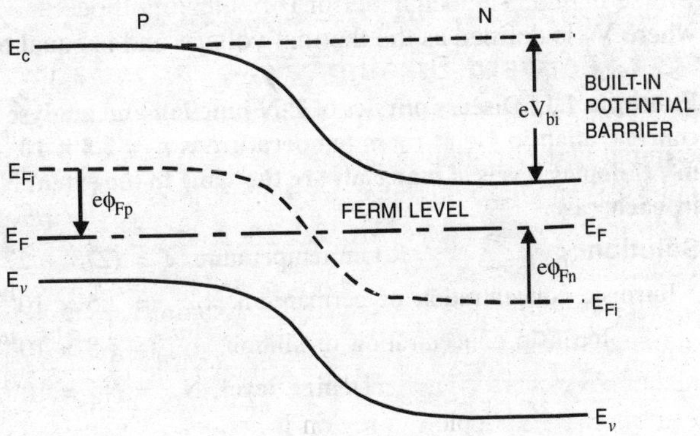

Fig. 1.6. ***Energy-Band Diagram of a P-N Junction in Thermal Equilibrium***

1.4.1. Built-in Potential Barrier

When no voltage is applied across the P-N junction (*i.e.,* with zero applied bias), the junction is in the thermal equilibrium—the Fermi energy level is constant throughout the entire specimen. The valence and conduction bands energies have to bend while moving through the spacc-charge region because the relative position of these bands with respect to the Fermi energy changes between P and N regions.

Electrons in the conduction band of the N-region face a potential barrier when moving into the conduction band of the P-region. This potential barrier is known as the *built-in potential barrier* and is denoted by V_{bi} or V_0. The built-in potential barrier maintains equilibrium between majority carrier electrons in the N-region and minority carrier electrons in the P-region. It also maintains equilibrium between majority carrier holes in the P-region and minority carrier holes in the N-region. This potential difference across the junction cannot be measured with a voltmeter as new potential barriers are being formed between the semiconductor and the probes which will cancel built-in potential barrier V_{bi}. As the condition pertains to equilibrium so no current is caused by this voltage.

The intrinsic Fermi level is equidistant from the conduction band edge through the junction, as illustrated in figure. The built-in potential barrier V_{bi} is the difference between the intrinsic Fermi levels of the N and P-regions.

$$V_{bi} = \phi_{Fn} + \phi_{Fp} \quad ...(1.1)$$

The electron concentration in the N-region conduction band,

$$n_0 = n_C e^{-(E_C - E_F)/k'T} = n_i e^{(E_F - E_{Fi})/k'T} = n_i e^{-e\phi_{Fn}/k'T} \quad ...(1.2)$$

where E_{Fi} and n_i are the intrinsic Fermi energy and intrinsic carrier concentration, respectively.

Potential ϕ_{Fn} in the N-region is defined as

$$e\,\phi_{Fn} = E_{Fi} - E_F \quad ...(1.3)$$

Taking natural log of both sides of Eq. (1.2), setting $n_o = N_D$ we have

$$\phi_{Fn} = \frac{-k'T}{e}\log_e \frac{N_D}{n_i} \quad ...(1.4)$$

Similarly for the P-region,

$$\phi_{Fp} = \frac{+k'T}{e}\log_e \frac{N_A}{n_i} \quad ...(1.5)$$

The built-in potential barrier for the step junction is then

$$V_{bi} = \frac{k'T}{e}\log_e \frac{N_A N_D}{n_i^2} = V_T \log_e \frac{N_A N_D}{n_i^2} \quad ...(1.6)$$

where V_T is defined as the thermal voltage and is equal to $\frac{k'T}{e}$.

Example 1.1. Discuss physics of P-N junction and analyse it when it is reverse and forward biased. The intrinsic concentration in Ge at room temperature is $n_i = 2.5 \times 10^{19}/m^3$ and of Si at room temperature is $n_i = 1.5 \times 10^{16}/m^3$. If doping levels of materials are the same to the extent $N_D = N_A = 10^{21}/m^3$, obtain the values of built-in voltage in each case.

[Pb. Technical Univ. Electronic Devices and Circuits, May 2007]

Solution : Room temperature, T = (273 + 27) = 300 K   Assuming room temperature to be 27°

Intrinsic concentration of germanium, $n_{i(Ge)} = 2.5 \times 10^{19}/m^3$

Intrinsic concentration of silicon, $n_{i(Si)} = 1.5 \times 10^{16}/m^3$

Doping level, $N_D = N_A = 10^{21}/m^3$

Built-in voltage for germanium, $V_{0(Ge)} = V_T \log_e \frac{N_D N_A}{n_i^2}$

$$= \frac{T}{11{,}600}\log_e \frac{N_D N_A}{n_i^2} \qquad \because V_T = \frac{T}{11{,}600}$$

$$= \frac{300}{11{,}600}\log_e \frac{(1\times 10^{21})^2}{(2.5\times 10^{19})^2} = 0.191 \text{ V} \quad \textbf{Ans.}$$

Built-in voltage of silicon, $V_{0(Si)} = \frac{300}{11{,}600}\log_e \frac{(1\times 10^{21})^2}{(1.5\times 10^{16})^2} = 0.5745 \text{ V}$ **Ans.**

Example 1.2. The resistivities of the P-region and N-region of a Ge diode are 6 Ω-cm and 4 Ω-cm respectively. Calculate the contact potential V_0. Given that $q = 1.6 \times 10^{-19}$ C, $n_i = 2.5 \times 10^{13}/cm^3$; $\mu_p = 1800$ cm²/V-s; $\mu_n = 3{,}800$ cm²/V-s and $V_T = 0.026$ V at 300 K.

[G.G.S.I.P. Univ. Analog Electronics December 2009]

Solution : Density of electrons, $n = N_D = \frac{\sigma_i \text{ (electrons)}}{e\,\mu_e} = \frac{1/\text{resistivity of N-region}}{e\,\mu_e}$

$$= \frac{1/4}{1.6\times 10^{-19}\times 3{,}800} = 4.1 \times 10^{14}/cm^3$$

Density of holes, $p = N_A = \frac{1/6}{1.6\times 10^{-19}\times 1{,}800} = 5.78 \times 10^{14}/cm^3$

Contact potential, $V = KT \log_e \frac{N_D N_A}{n_i^2} = V_T \log_e \frac{N_D N_A}{n_i^2} = 0.026 \times \log_e \frac{4.1\times 10^{14}\times 5.78\times 10^{14}}{(2.5\times 10^{13})^2}$

$$= 0.154 \text{ V} \quad \textbf{Ans.}$$

1.5. REVERSE APPLIED BIAS

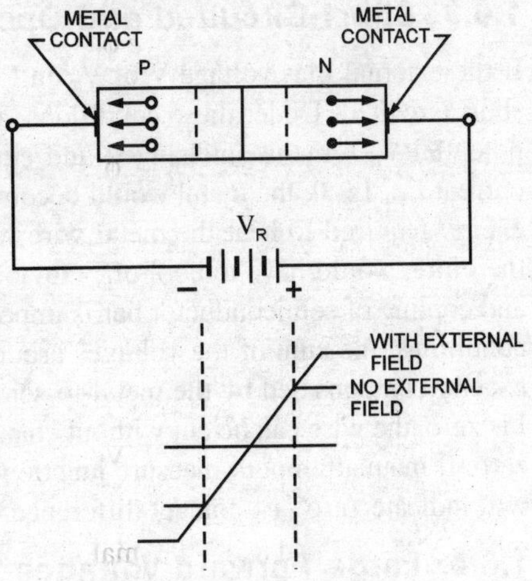

Fig. 1.7. ***Reverse Biasing***

If an external bias voltage is applied with positive terminal to N-side and negative terminal to the P-side of a P-N junction, as shown in Fig. 1.7, the junction is said to *be reverse biased.* In this arrangement, electrons from the N-side are attracted to the positive bias terminal, and holes from P-side are attracted to the negative bias terminal. Thus, as shown in Fig. 1.7, holes from the impurity atoms on the P-side of the junction are attracted away from the junction, and the electrons are attracted away from the atoms on the N-side of the junction. Thus, the depletion region is widened, and the barrier potential is increased by the magnitude of the applied bias. With the increased barrier potential, there is no possibility of majority carrier current flow across the junction. Thus the P-N junction is in a non-conductive state.

In brief it can be said that when the P-N junction is reverse biased (*i*) potential barrier is increased and width of depletion is also increased (*ii*) the junction offers high resistance, called the reverse resistance r_r, to the flow of current (*iii*) owing to establishment of high resistance path, very small current, called the *leakage current* or *reverse current*, flows through the circuit. The reverse or leakage current is not affected by barrier and depends on temperature, material and construction of the junction.

Thus the P-N junction allows flow of large current only when forward biased and only small reverse (or leakage) current when reverse biased.

1.6. METAL-SEMICONDUCTOR JUNCTION

Metal-semiconductor junctions are very common in all semiconductor devices and are very important. Depending upon the doping concentration, materials, and the characteristics of the interface, the metal semiconductor junctions can act as either an ohmic contact or as a Schottky barrier.

1.6.1. Structure of Metal-Semiconductor Junction

A metal-semiconductor junction, as its name indicates, consists of a metal in contact with a piece of semiconductor, as shown in Fig. 1.8. The active junction is the interface between the metal, which acts as an anode, and the semiconductor. The other interface between the semiconductor and the metal, which acts as a cathode, is an Ohmic contact and there is no potential drop at this junction.

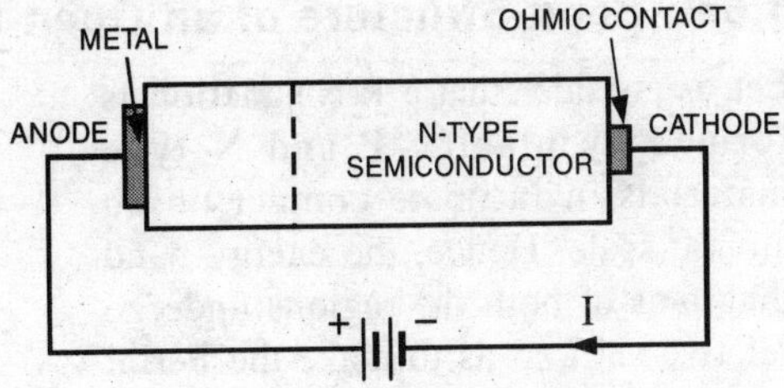

Fig. 1.8. ***Structure of Metal-Semiconductor Junction***

1.6.2. Ohmic Contacts

An ohmic contact may be defined as a metal-semiconductor contact that has a negligible contact resistance compared with the bulk or spreading resistance of the semiconductor. A satisfactory ohmic contact should not significantly perturb device performance, and it can supply the required current with a voltage drop that is considerably small compared with the drop across the active region of the device. Ideally, the current through the ohmic contact is a linear function of applied voltage, and the applied voltage should be very small.

1.6.3. Short-Circuited and Open-Circuited P-N Junction

If the external bias voltage V or V_R in Figs. 1.5 and 1.7 were set equal to zero, the P-N junction would be short circuited. Under these conditions, as explained below, no current can flow *i.e.,* I = 0 and electrostatic potential V_0 remains unchanged and equal to the value under open-circuited conditions. If there were a current *i.e.,* $I \neq 0$, the metal would become heated. Since there is no external source of energy available, the energy required to heat the metal wire would have to be supplied by the P-N bar. The semiconductor bar, therefore, would have to cool off. Obviously, under thermal equilibrium the simultaneous heating of metal and cooling of semiconductor bar is impossible, and so it is concluded that I = 0. Since under short-circuited conditions, the sum of the voltages around the closed loop must be zero the junction potential V_0 must be exactly compensated by the metal-to-semiconductor contact potentials at ohmic contacts. Since the current I is zero, the wire can be cut without changing the situation, and the voltage drop across the cut must remain zero. If in an attempt to measure junction potential V_0 a voltmeter is connected across the cut, the voltmeter will indicate zero *i.e.,* contact difference of potential cannot be measured directly with a voltmeter.

1.6.4. Large Forward Voltages

Suppose forward bias voltage V, in Fig. 1.5, is increased until V approaches junction potential V_0. If V were equal to V_0, the barrier would disappear and the current could be arbitrarily large, exceeding the diode rating. In practice, barrier cannot be reduced to zero because, as the current increases without limit, the bulk resistance of the crystal and the resistance of the ohmic contacts, will limit the current. Thus it is no longer possible to assume that all the voltage V appears as a change across the P-N junction. It is concluded that, as the forward bias voltage V becomes comparable with V_0, the current through a P-N junction diode will be governed by the ohmic contact resistance and the crystal bulk resistance. Thus the V-I characteristic becomes almost a straight line.

1.6.5. Bulk Resistance

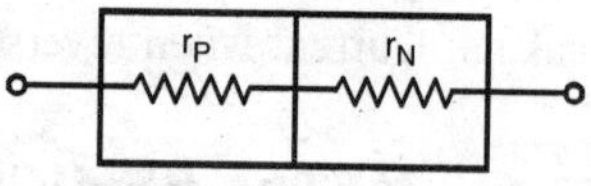

Fig. 1.9. ***Bulk Resistance***

After overcoming the barrier potential (or offset voltage), all that impedes the diode current is the resistance of the P and N regions, symbolized by the r_P and r_N in Fig. 1.9, if ohmic contact resistance is neglected. The sum of these resistances is called the bulk resistance r_B of the diode *i.e.,*

$$r_B = r_P + r_N \quad \text{...(1.7)}$$

The bulk resistance of the diode depends upon the doping level, and the size of the P and N regions. Typically, the bulk resistance of rectifier diode is less than 1 Ω.

1.6.6. Band Structure of an Open-Circuited P-N Junction

Let us consider that a P-N junction is formed by placing P and N type materials in intimate contact on an atomic scale. Hence, the energy band diagrams of both the regions undergo relative shift so as to make the Fermi level constant throughout the specimen. If this was not so, electrons on one side of the junction would have an average energy greater than those on the other side and as a result there would be transfer of electrons and energy until the Fermi levels on the two sides get equalised. However, because of this

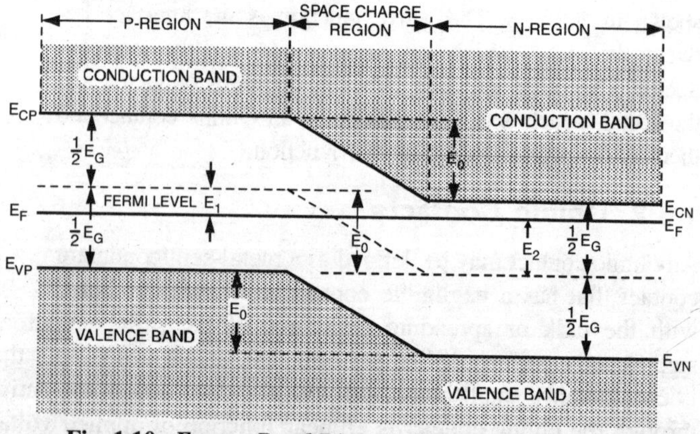

Fig. 1.10. ***Energy Band Diagram For a P-N Junction Under Open-Circuited Conditions***

shift, there is no disturbance in the relative positions of the conduction and valence bands and Fermi level in any region.

The energy band diagram for a P-N junction is given in Fig. 1.10, where Fermi level E_F is closer to the conduction band edge E_{CN} in the N-type material and closer to the valence band edge E_{VP} in the P-type material. Obviously, the conduction band edge E_{CP} in the P-type material cannot be at the same level as E_{CN}. Similarly, the valence band edge E_{VN} in the N-type material cannot line up with E_{VP} in the P-type material.

As shown in Fig. 1.10, E_1 and E_2 represent the shifts in the Fermi level from the intrinsic conditions in the P- and N-type materials respectively. Then, the total shift in the energy level E_0 is given as

$$E_0 = E_{CP} - E_{CN} + E_{VP} - E_{VN} = E_1 + E_2 \quad \text{...(1.8)}$$

The energy E_0 represents the potential energy of electrons, in eV, at the junction and is equal to eV_0 where V_0 is the contact potential or barrier potential in volts.

From Fig. 1.10, we have

$$E_F - E_{VP} = \frac{1}{2} E_G - E_1 \quad \text{...(1.9)}$$

$$\text{and } E_{CN} - E_F = \frac{1}{2} E_G - E_2 \quad \text{...(1.10)}$$

Adding Eqs. (1.9) and (1.10) we have

$$E_1 + E_2 = E_G - (E_{CN} - E_F) - (E_F - E_{VP})$$

$$\text{or } E_0 = E_G - (E_{CN} - E_F) - (E_F - E_{VP}) \quad \text{...(1.11)}$$

and we know that

$$n_i^2 = np = n_C n_V \, e^{-[(E_C - E_F) + (E_F - E_V)]/kT}$$

$$= n_C n_V \, e^{-(E_C - E_V)/kT} = n_C n_V \, e^{-E_G/kT}$$

$$\text{or} \quad E_G = kT \log_e \frac{n_C n_V}{n_i^2} \quad \text{...(1.12)}$$

Exact position of fermi level in an N-type and P-type extrinsic semiconductor is given by the following expressions

$$E_{CN} - E_F = kT \log_e \frac{n_C}{N_D} \quad \text{...(1.13)}$$

$$\text{and } E_F - E_{VP} = kT \log_e \frac{n_V}{N_A} \quad \text{...(1.14)}$$

Substituting values of E_G, $(E_{CN} - E_F)$ and $(E_F - E_{VP})$ from Eqs. (1.12), (1.13) and (1.14) in Eq. (1.11) we have

$$E_0 = kT \log_e \frac{n_C n_V}{n_i^2} - kT \log_e \frac{n_C}{N_D} - kT \log_e \frac{n_V}{N_A}$$

$$= kT \log_e \left[\frac{n_C n_V}{n_i^2} \cdot \frac{N_D}{n_C} \cdot \frac{N_A}{n_V} \right] = kT \log_e \frac{N_D N_A}{n_i^2} \quad \text{...(1.15)}$$

It is to be noted that in above equations, the E's are expressed in electron volts and k has the dimensions of electron volts per Kelvin. The contact difference of potential V_0 is expressed in volts and is numerically equal to E_0. It is obvious from Eq. (1.15) that E_0 depends upon the equilibrium concentrations and not upon the charge density in transition region.

Example 1.3. Consider a Ge diode with $N_D = N_A$ and with impurity concentration of $8 \times 10^{14}/cm^3$. Assume $n_i = 2 \times 10^{13}/cm^3$. At room temperature of 300 K calculate the height of the potential barrier under open-circuit conditions. Assume Boltzmann's constant k = 8.61×10^{-5} eV/K and electron charge, $e = 1.6 \times 10^{-19}$ C.

[U.P. Technical Univ. Fundamentals of Electronic Devices 2009-10]

Solution : Impurity concentration, $N_D = N_A = 8 \times 10^{14}/cm^3$

Intrinsic carrier concentration, $n_i = 2 \times 10^{13}/cm^3$

Boltzmann's constant, k = 8.61×10^{-5} eV/K

Temperature, T = 300 K

$$\text{Potential barrier, } V_0 = kT \log_e \frac{N_D N_A}{n_i^2} = 300 \times 8.61 \times 10^{-5} \log_e \frac{(8 \times 10^{14})^2}{(2 \times 10^{13})^2} = 0.19 \text{ V } \textbf{Ans.}$$

1.6.7. Volt-Ampere Characteristics of P-N Junction

The graph plotted between the potential difference across the P-N junction and the circuit current is known as volt-ampere characteristic of a P-N junction.

1. *Forward Characteristics.* When the external voltage is zero *i.e.,* when the circuit is open, the potential barrier at the junction does not allow the flow of current and, therefore, the circuit current is zero, as shown in Fig. 1.11.

With forward bias to the P-N junction (*i.e.,* P-region connected to +ve terminal and N-region connected to –ve terminal of the battery) very little current, called the *forward current,* flows until the *forward voltage exceeds* the junction barrier potential (0.3 V for Ge and 0.7 V for Si). The characteristics follow an exponential law. As the forward voltage is increased to the knee of the characteristic, the barrier potential is progressively reduced to zero, allowing more and more majority charge carriers to flow across the junction. Beyond the knee of the characteristic, the potential barrier is completely eliminated, forward current increases almost linearly with the increase in forward voltage and the P-N junction starts behaving as a resistor. If the forward voltage is increased beyond a certain value, extremely large current will flow and the P-N junction may get destroyed due to overheating.

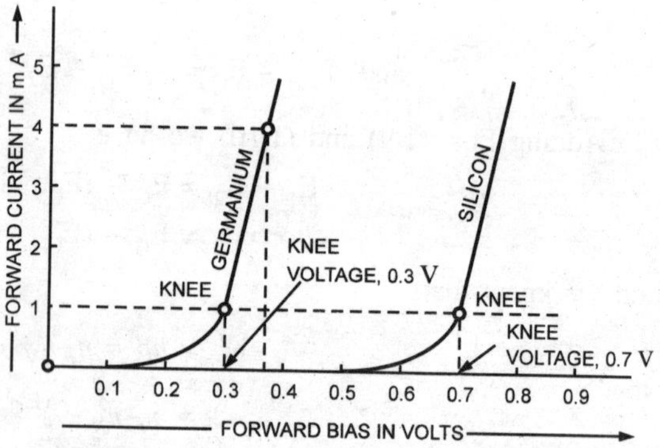

Fig. 1.11. ***Forward Characteristics of P-N Junction***

Because of the nonlinear knee of the curve, if a very small signal voltage is applied to the diode so that it only operates around the knee, the signal will be distorted. The signal must, therefore, be large enough to operate the diode over the linear part of the characteristic. Forward resistance of P-N junction is low (of the order of 75 Ω in case of Ge and 150 Ω in case of Si).

2. *Reverse Characteristics.* When a reverse bias is applied (*i.e.,* when P-region is connected to the negative bias terminal and N-region is connected to the positive bias terminal), as shown in Fig. 1.7, the potential barrier at junction is increased, therefore, the junction resistance becomes very high and there is no possibility of majority carriers flowing across a reverse biased junction. But minority carriers generated on each side can still cross the junction. Electrons on the P-side are attracted across the junction to the positive bias terminal on the N-side and holes on the N-side may travel across the junction to the negative bias terminal on the P-side. This results in a very small current, as illustrated in Fig. 1.12. This current is called *reverse current* and is due to minority carriers.

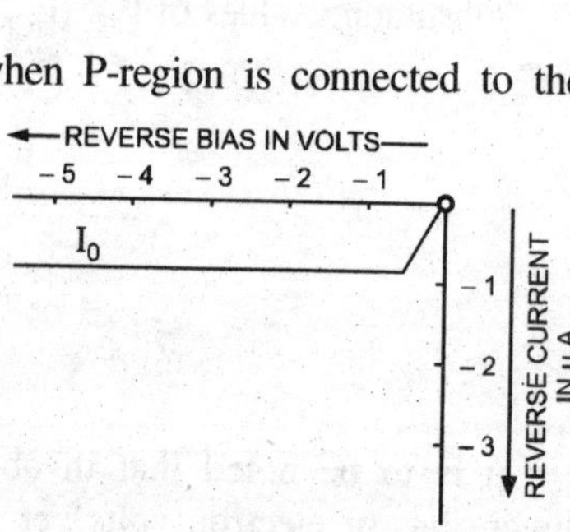

Fig. 1.12. ***Reverse Characteristics of P-N Junction***

For silicon the reverse current* is typically less than 1 μA and may be as low as 1 *n*A. In case of germanium, the reverse current may exceed 10 μA. Only a very small reverse voltage is necessary to direct all available minority carriers across the junction, and when all minority carriers are flowing across, further increase in bias voltage will not increase the current. This current is referred to as a *reverse saturation current* and is designated as I_0. Reverse resistance is very high (of the order of MΩ).

As discussed above, the reverse saturation current I_0 is made up of minority charge carriers crossing the junction. When the temperature of the semiconductor material is increased, the additional thermal energy causes more electrons to break away from atoms. This creates more hole-electron pairs and generates more minority charge carriers. So reverse saturation current I_0 increases with the increase in the junction temperature.* It has been found that I_0 approximately doubles for each 10ºC rise in temperature in case of germanium and for each 8ºC rise in temperature in case of silicon.

1.6.8. Important Terms Often Used in P-N Junction

1. *Breakdown Voltage.* Under normal reverse voltage, a very little reverse current flows through a P-N junction. However, on increasing the reverse voltage, a point may reach at which the junction breaks down with a sudden rise in reverse current. Breakdown voltage is defined as the reverse voltage at which P-N junction breaks down with a sudden rise in reverse current.

2. *Knee Voltage.* As already explained in Art. 1.6.7, with forward bias to P-N junction very little current flows until the forward voltage exceeds the junction barrier potential (0.3 V in case of Ge and 0.7 V in case of Si). The forward voltage, at which the current through the junction starts increasing rapidly, is called the *knee voltage* or the *cutin voltage*.

3. *Maximum Forward Current.* It is the highest instantaneous forward current that a P-N junction can conduct without damage to the junction. In case the forward current exceeds this rating, the junction will get destroyed due to overheating.

4. *Peak Inverse Voltage.* It is the maximum reverse voltage that can be applied to the P-N junction without damage to the junction. If the reverse voltage across the junction exceeds its peak inverse voltage (PIV), the junction may get destroyed owing to excessive heat.

5. *Maximum Power Rating.* It is the maximum power that can be dissipated at the junction without any damage to it.

Note : Every P-N junction has limiting values of maximum forward current, peak inverse voltage and maximum power rating. The P-N junction gives satisfactory performance when operated within these limiting values. The P-N junction may get destroyed due to excessive heat if any of these values is exceeded.

1.7. JUNCTION BREAKDOWN

When an ordinary P-N junction diode is reverse biased, normally only very small reverse saturation current flows. This current is due to movement of minority carriers. It is almost independent of the voltage applied. However, if the reverse bias is increased, a point is reached when the junction breaks down and the reverse current increases abruptly, as shown in Fig. 1.13. This current could be large enough to destroy the junction. If the reverse current is limited by means of a suitable series resistor, the power dissipation at the junction will not be excessive, and the device may be operated continuously in its breakdown region to its normal (reverse saturation) level. It is found that for a suitably designed diode, the breakdown voltage is very stable over a wide range of reverse currents. This quality gives the *breakdown diode* many useful applications as a *voltage reference source*.

* There are more minority charge carriers available in germanium than in silicon, since charge carriers are more easily detached from germanium atoms.

The critical value of the voltage, at which the breakdown of a P-N junction diode occurs is called the *breakdown voltage.* The breakdown voltage depends on the width of the depletion region, which, in turn, depends on the doping level. The junction offers almost zero resistance at the breakdown point.

There are two mechanisms by which breakdown can occur at a reverse biased P-N junction : *avalanche breakdown* and *zener breakdown.*

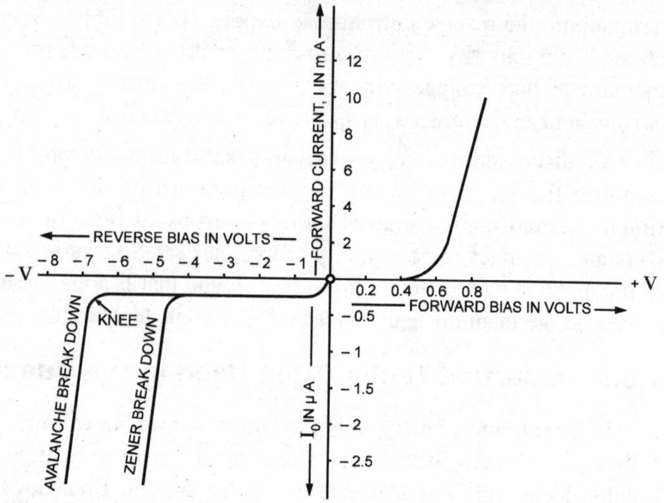

Fig. 1.13

The minority carriers, under reverse biased conditions, flowing through the junction acquire a kinetic energy which increases with the increase in reverse voltage. At a sufficiently high reverse voltage (say 5 V or more), the kinetic energy of minority carriers becomes so large that they knock out electrons from the covalent bonds of the semiconductor material. As a result of collision, the liberated electrons in turn liberate more electrons and the current becomes very large leading to the breakdown of the crystal structure itself. This phenomenon is called the *avalanche breakdown.* The breakdown region is the *'knee'* of the characteristic curve. Now the current is not controlled by the junction voltage but rather by the external circuit.

Under a very high reverse voltage, the depletion region expands and the potential barrier increases leading to a very high electric field across the junction. The electric field will break some of the covalent bonds of the semiconductor atoms leading to a large number of free minority carriers, which suddenly increase the reverse current. This is called the *zener effect.* The breakdown occurs at a particular and constant value of reverse voltage called the *breakdown voltage.* It is found that zener breakdown occurs at electric field intensity of about 3×10^7 V/m.*

Either of the two (zener breakdown or avalanche breakdown) may occur independently, or both of these may occur simultaneously. Diode junctions that break down below 5 V are caused by *zener effect.* Junctions that experience breakdown above 5 V are caused by *avalanche effect*. Junctions that break down around 5 V are usually caused by combination of two effects. The zener breakdown occurs in heavily doped junctions (P-type semiconductor moderately doped and N-type heavily doped), which produce narrow depletion layers. The avalanche breakdown occurs in lightly doped junctions, which produce wide depletion layers.

With the increase in junction temperature zener breakdown voltage is reduced while the avalanche breakdown voltage increases. The zener diodes have a negative temperature coefficient while avalanche diodes have a positive temperature coefficient. Diodes that have breakdown voltages around 5 V have zero temperature coefficient.

The breakdown phenomenon is reversible and harmless so long as the safe operating temperature is maintained.

---

* From basic circuit theory electric field intensity E is given as $\frac{V}{d}$, where V is the applied reverse voltage in volts and d is the width of depletion region in metres. The narrower the depletion region, the smaller the required bias to cause zener breakdown. A small reverse bias can produce a sufficiently strong electric field in a narrow depletion region.

Despite the fact that avalanche and zener constitute two types of breakdown diode, the name zener is commonly applied to both types of diodes.

1.8. IDEAL DIODE

It is a two terminal device that permits only *unidirectional conduction.* It conducts well in the forward direction and poorly in the reverse direction. It would have been ideal if it acted as a perfect conductor (with zero resistance or zero voltage drop across it) when forward biased and as a perfect insulator (with an infinite resistance or with no current through it) when reverse biased. The volt-ampere characteristics of such an ideal diode have been shown in Fig. 1.14 (*b*). An ideal diode acts like an *automatic switch.* The switch is closed when the diode is forward biased and is opened when reverse biased.

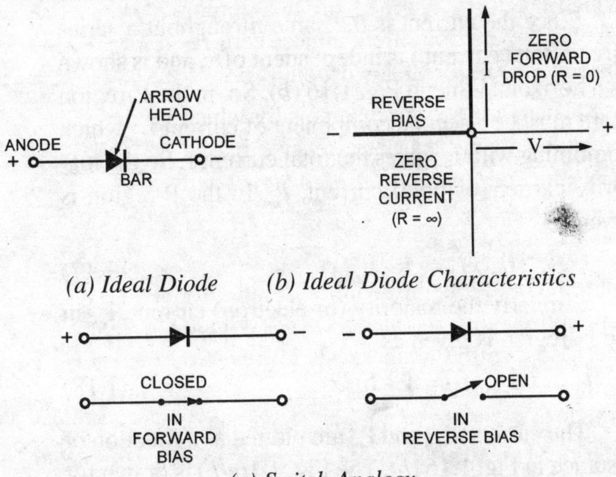

(*a*) *Ideal Diode* (*b*) *Ideal Diode Characteristics*

(*c*) *Switch Analogy*

Fig. 1.14

No diode can act as an ideal diode. An actual diode does not behave as a perfect conductor when forward biased and as a perfect insulator when reverse biased. Neither it offers zero resistance when forward biased nor infinite resistance when reverse biased. However, there are many applications in which diodes can be assumed to be nearly ideal devices, if the voltage drop across the diode when it is forward biased *i.e.,* V_D is taken into account. In cases when the circuit supply voltage V_S is much larger than the forward voltage drop V_D, V_D can be assumed constant without introducing any serious error. Also, the diode forward current I is usually so much larger than the reverse saturation current I_0 that the I_0 can just be ignored. These assumptions lead to a nearly ideal, or approximate, characteristics for germanium and silicon diodes, as illustrated in Fig. 1.15.

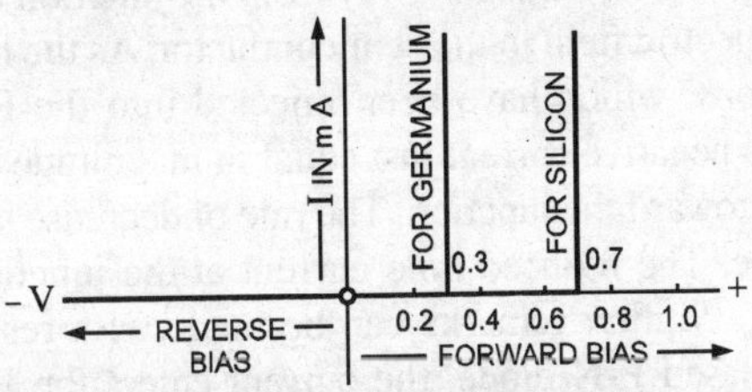

Fig. 1.15. ***Approximate Characteristics For Ge and Si Diodes***

1.9. CURRENT COMPONENTS IN A P-N DIODE

A P-N junction with a forward bias with an external voltage V is shown in Fig. 1.16 (*a*). Due to forward bias, there exists a potential gradient in P and N materials. Now the holes from P-region and electrons from N-region are injected into the N-region and P-region respectively. The holes injected into the N-region and electrons injected into the P-region are the minority carriers. The minority carriers diffuse away from the junction exponentially with the distance, as illustrated in Fig. 1.16 (*b*). Their concentration falls steadily due to recombination with electrons and holes respectively. Since the diffusion current due to minority carriers is proportional to the concentration gradient, this current must also vary exponentially with distance. There are two minority carrier (diffusion) currents, I_{ep} and I_{hn} as shown in Fig. 1.16 (*b*). Let the symbols I_{ep} (*x*) and I_{hn} (*x*) represent the electron current in P-region and hole current in N-region respectively.

Electrons crossing the junction at $x = 0$ from right to left constitute a current in the same direction as the holes crossing the junction from left to right. Thus the total current I at junction (*i.e.*, $x = 0$) is given as

$$I = I_{hn}(0) + I_{ep}(0) \qquad ...(1.16)$$

Since the current is the same throughout a series circuit, total current I is independent of x, and is shown as a horizontal line in Fig. 1.16 (*b*). So, in the P-region there must be a second component of current I_{hp} which combining with I_{ep} gives the total current I. So the majority carrier (or hole) current, I_{hp} in the P-region is given as

$$I_{hp}(x) = I - I_{ep}(x) \qquad ...(1.17)$$

Similarly the majority (or electron) current, I_{en} in the N-region is given as

$$I_{en}(x) = I - I_{hn}(x) \qquad ...(1.18)$$

The currents I_{hp} and I_{en} are plotted as a function of distance in Fig. 1.16 (*b*). The Fig. 1.16 (*b*) is drawn for unsymmetrically doped diode, so that $I_{ep} \neq I_{hn}$.

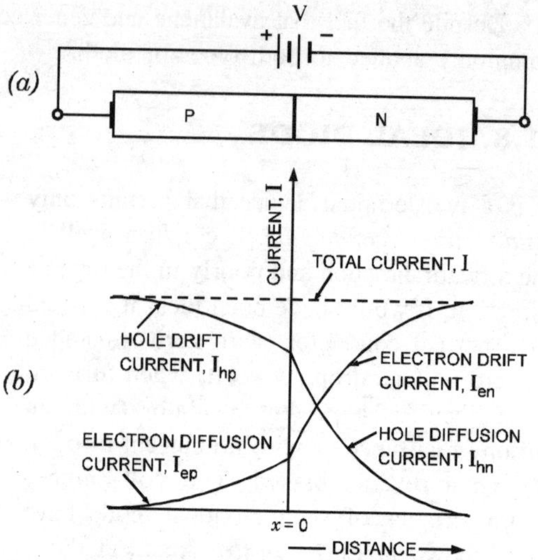

Fig. 1.16. ***Current Components in a Forward Biased P-N Junction Diode***

It may be noted that far away from the junction in P-region the current is drift current I_{hp} of holes sustained by small electric field in the semiconductor. As the holes approach the junction, some of them recombine with the electrons, which have been injected into the P-region from the N-region. Thus part of the current I_{hp} becomes a negative current just equal in magnitude to the diffusion current I_{ep}. The drift current I_{hp}, therefore, decreases toward the junction. The rate of decrease is such that the total current remains constant, independent of distance. The reduced hole current at the junction now enters the N-region and becomes hole diffusion current I_{hn}. Similar remarks can be made with respect to electron drift current in N-region, I_{en}. Thus in a forward-biased P-N diode, the current enters the P-region as a hole current and leaves the N-region as an electron current of the same magnitude.

From the above discussion it is concluded that the current in a P-N diode is bipolar in character (because it is made up of both positive and negative charge carriers) and the total current is constant throughout the device, but the proportion of current due to holes and that due to electrons varies with the distance, as illustrated in Fig. 1.16 (*b*).

1.10. QUANTITATIVE THEORY OF P-N DIODE CURRENTS

Here an expression for the total current will be derived as a function of the applied voltage with the assumption that depletion layer thickness is negligible (*i.e., barrier* width is zero).

When a P-N diode is forward biased, holes are injected from P-region into the N-region. The concentration p_n of the holes in the N-region is increased above its thermal equilibrium value p_{no}. The hole concentration in N-region is given as

$$p_n(x) = p_{no} + p'_n(0)\, e^{-x/L_h} \qquad ...(1.19)$$

where p_{no} is the hole concentration in thermal-equilibrium condition, L_h is the diffusion length for holes in N-type material and x is the distance from the junction where concentration is considered.

The injected, or excess, concentration of holes at $x = (0)$, $p'_n(0)$ is given as

$$p'_n(0) = p_n(0) - p_{no} \qquad ...(1.20)$$

These several hole concentration components are shown in Fig. 1.17, which shows that the concentration $p_n(x)$ falls exponentially with distance x into the N-type material.

The diffusion hole current in the N-region is given as

$$I_{hn} = -AeD_h \frac{dp_n}{dx} \quad ...(1.21)$$

Taking the derivative of Eq. (1.19) and substituting the value of $\frac{dp_n}{dx}$ in above Eq. (1.21) we have

$$I_{hn}(x) = \frac{AeD_h\, p'_n(0)}{L_h} e^{-x/L_h} \quad ...(1.22)$$

At junction *i.e.*, $x = 0$

$$I_{hn}(0) = \frac{AeD_h\, p'_n(0)}{L_h}$$

$$= \frac{AeD_h}{L_h}[p_n(0) - p_{no}] \quad ...(1.23)$$

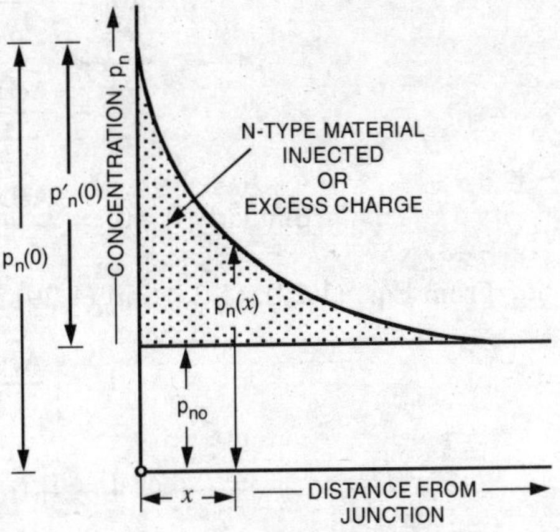

Fig. 1.17. ***Hole Concentration Components in N-Type Material***

where A is area of material in m^2, D_h is diffusion constant for holes in m^2/s, e is the magnitude of charge on holes, L_h is the diffusion length of holes in N-type material in metre, and $p'_n(0)$ is the excess of hole concentration at junction.

Using Boltzmann relationship of kinetic theory of gases, we have

$$p_n(0) = p_{no}\, e^{V/V_T} \quad ...(1.24)$$

where V is the voltage applied across the P-N diode and V_T is the volt-equivalent of temperature and it is given as

$$V_T = \frac{k'T}{e} = \frac{T}{11,600}$$ where k' is the Boltzmann constant.

The Eq. (1.24) is known as the *law of junction.*

1.10.1. Diffusion Length

Diffusion length is defined as the distance travelled by free charge carriers (electrons or holes) before recombination. It may also be defined as the average distance covered by an excess charge carrier during its lifetime τ.

The diffusion length L_h and lifetime τ_h are related to each other as :

$$L_h = \sqrt{D_h \tau_h} \quad ...(1.25)$$

Similarly in case of P type semiconductor

$$L_e = \sqrt{D_e \tau_e} \quad ...(1.26)$$

1.10.2. Forward Currents

The total diode current I at $x = 0$ is given as

$$I = I_{hn}(0) + I_{ep}(0) \quad ...(1.27)$$ Refer to Eq. (1.16)

where $I_{hn}(0)$ is the current caused by holes entering the N-region and $I_{ep}(0)$ is the current caused by electrons entering the P-region.

From Eqs. (1.23) and (1.24)

$$I_{hn}(0) = \frac{AeD_h}{L_h}[p_{no}\,e^{V/V_T} - p_{no}]$$

$$= \frac{AeD_h p_{no}}{L_h}[e^{V/V_T} - 1] \quad ...(1.28)$$

$$\text{Similarly } I_{ep}(0) = \frac{AeD_e n_{po}}{L_e}[e^{V/V_T} - 1] \quad ...(1.29)$$

From Eqs. (1.27), (1.28) and (1.29) we have total diode current

$$I = \left[\frac{AeD_h p_{no}}{L_h} + \frac{AeD_e n_{po}}{L_e}\right](e^{V/V_T} - 1)$$

$$\text{or } I = I_0(e^{V/V_T} - 1) \quad ...(1.30)$$

$$\text{where } I_0 = \frac{AeD_h p_{no}}{L_h} + \frac{AeD_e n_{po}}{L_e} \quad ...(1.31)$$

1.10.3. Reverse Saturation Current

In the foregoing discussion a positive value of V indicates a forward bias. Equation (1.30) is equally valid for reverse bias [*i.e.,* for negative values of applied voltage V]. For a reverse bias voltage exceeding V_T (*i.e.,* 26 mV) at room temperature (27 °C or 300 K), current $I \to -I_0$. Thus I_0 is called the *reverse saturation current.*

Concentration of holes in N-type region is given by the equation

$$p_{no} = \frac{n_i^2}{N_D}$$

Similarly, the concentration of electrons in P-type region is given by the equation

$$n_{po} = \frac{n_i^2}{N_A}$$

where N_D and N_A are the concentrations of donor and acceptor atoms respectively.

Substituting $p_{no} = \frac{n_i^2}{N_D}$ and $n_{po} = \frac{n_i^2}{N_A}$ in Eq. (1.31) we have

$$I_0 = Ae\left[\frac{D_h}{L_h N_D} + \frac{D_e}{L_e N_A}\right]n_i^2 \quad ...(1.32)$$

$$\text{Also, } n_i^2 = A_0 T^3 e^{-E_{GO}/kT} = A_0 T^3 e^{-V_{GO}/VT} \quad ...(1.33)$$

where V_{GO} is a voltage numerically equal to the forbidden energy gap E_{GO} in electron volts, and V_T is the volt equivalent of temperature and is equal to T/11,600.

For germanium the diffusion constants D_h and D_e vary approximately inversely proportional to temperature T. Hence I_0 may be given as

$$I_0 = K_1 T^2 e^{-V_{GO}/V_T} \quad ...(1.34)$$

where K_1 is a constant, independent of temperature T.

In the discussion made above, the effect of carrier generation and recombination in the space-charge region has been ignored. Such an assumption is valid for a germanium diode (not for a silicon diode). For the silicon diode, a diffusion current is negligible in comparison with the transition-layer charge-generation current, which is given approximately by

$$I = I_0 \left(e^{V/\eta V_T} - 1\right) \qquad ...(1.35)$$

where $\eta \approx 2$ for small (rated) currents and $\eta \approx 1$ for large currents. Also I_0 is found to be proportional to n_i instead of n_i^2 so

$$I_0 = K_2 T^{1.5} e^{-V_{GO}/2V_T} \qquad ...(1.36)$$

where K_2 is a constant, independent of temperature T.

1.11. TERMINAL CHARACTERISTICS OF JUNCTION DIODES

The volt-ampere characteristic of a silicon junction diode is depicted in Fig. 1.18. The same characteristic is depicted in Fig. 1.19 with some scales expanded and others compressed to reveal details—the changes in scale have resulted in the apparent discontinuity at the origin.

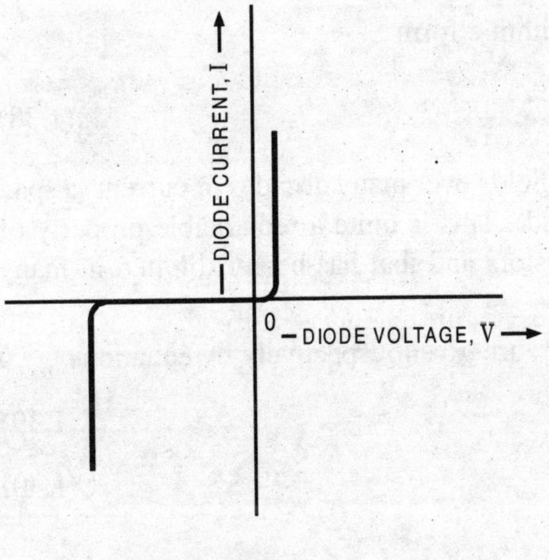

Fig. 1.18. ***Volt-Ampere Characteristic of a Silicon Junction Diode***

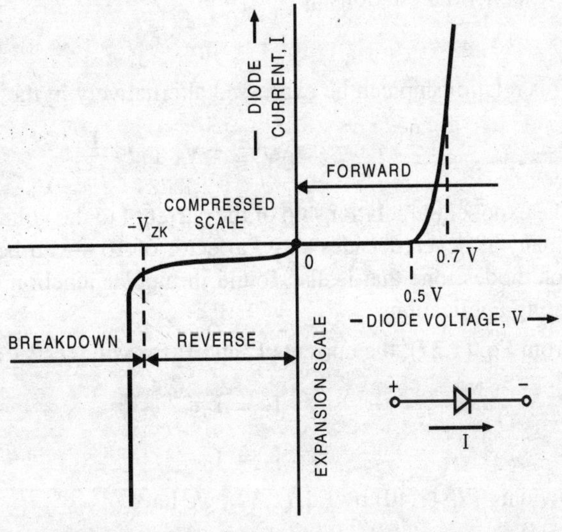

Fig. 1.19. ***V-I Characteristics of the Diode With Some Scales Expanded and Others Compressed In Order To Reveal Details***

As shown in the figure, the characteristic curve consists of three distinct regions :

1. The forward-bias region, determined by $V > 0$.
2. The reverse-bias region, determined by $V < 0$.
3. The breakdown region, determined by $V < -V_{ZK}$.

The above three regions of operation are described below :

1.11.1. Forward-Bias Region

A positive value of current I means the current flowing from the P-region to the N-region. The diode is forward biased if the applied voltage V is positive *i.e.,* P-side is connected to the positive terminal and N-side is connected to negative terminal as shown in Fig. 1.5.

The current flowing through the junction diode is given by Eq. (1.35).

In Eq. (1.35) I_0 is a constant for a given diode at a given temperature. A formula for I_0 in terms of the temperature will be given in Art. 1.12. The current I_0 is usually called the saturation current, and sometimes denoted by I_s. Another name for the saturation current is the *scale current.* This name arises from the fact that I_0 is directly proportional to the cross-sectional area of the diode. Thus doubling of junction area results in a diode with double the value of I_0 and, as the diode equation indicates, double the value of current I for a given forward voltage V. For "small-signal" diodes (small size diodes intended for low-power applications), I_0 is of the order of 10^{-15} A. The value of I_0 is, however, a very strong function of temperature. As a thumb rule, I_0 approximately doubles for every 5°C rise in temperature.

Voltage V_T in Eq. (1.35) is the volt equivalent of temperature and is equal to $\frac{T}{11,600}$. At room temperature V_T is taken equal to be 26 mV.*

In Eq. (1.35) the constant η has a value between 1 and 2, depending on the material and physical structure of the diode. Diodes made using the standard IC fabrication process exhibit $\eta = 1$ when operated under normal conditions. Diodes available as discrete two-terminal components generally exhibit $\eta = 2$.

For appreciable current I in the forward direction, specifically for $I >> I_0$, Eq. (1.35) can be approximated by the exponential relationship

$$I \simeq I_0 e^{V/\eta V_T} \quad ...(1.37)$$

This relationship can be expressed alternatively in the logarithmic form

$$V = \eta V_T \log_e \frac{I}{I_0} \quad ...(1.38)$$

The exponential relationship of the current I to the voltage V holds over many decades of current (a span of as many as seven decades—*i.e.*, a factor of 10^7—can be found). This is quite a remarkable property of junction diodes, one that is also found in bipolar junction transistors and that has been exploited in many interesting applications.

From Eq. (1.37), the currents I_1 and I_2 for voltages V_1 and V_2 are given respectively by equations

$$I_1 = I_0 e^{V_1/\eta V_T} \quad ...(1.39)$$

$$I_2 = I_0 e^{V_2/\eta V_T} \quad ...(1.40)$$

Dividing Eq. (1.40) by Eq. (1.39), we have

$$\frac{I_2}{I_1} = e^{(V_2 - V_1)/\eta V_T}$$

$$\text{or } V_2 - V_1 = \eta V_T \log_e \frac{I_2}{I_1}$$

$$\text{or } V_2 - V_1 = 2.3\eta V_T \log_{10} \frac{I_2}{I_1} \quad ...(1.41)$$

The above equation reveals that for a decade (factor of 10) change in diode current, the diode voltage drop changes by $2.3\eta V_T$, which is about 60 mV for $\eta = 1$ and 120 mV for $\eta = 2$. This also suggests that the diode volt-ampere characteristic is most conveniently drawn on semilog paper. Using the vertical, linear axis

* A slightly higher ambient temperature (25°C or so) is usually assumed for electronic equipment operating inside a cabinet. At this temperature, $V_T = 25.8$ mV $\simeq$ 26 mV. Nevertheless, for the sake of simplicity, and to promote rapid circuit analysis, sometimes more arithmetically convenient value of 25 mV is used.

for V and horizontal, log axis for current I, we have a straight line with a slope of $2.3\eta V_T$ per decade of current. Finally, it is noteworthy that not knowing the exact value of η, circuit designer use the convenient approximate number of 0.1 V/decade for the slope of the diode logarithmic characteristic.

Voltage-current characteristic in the forward region (Fig. 1.19) reveals that the current is negligibly small for voltage smaller than about 0.5 V. This value is usually referred to as the *cutin* voltage. It needs to be emphasized, however, that this apparent threshold in the characteristic is simply a consequence of the exponential relationship. Another consequence of this exponential relationship is the rapid increase of current I. Thus, for a "fully conducting" diode, the average drop lies in a narrow range, approximately 0.6 V to 0.8 V. This gives rise to a simple "model" for the diode where it is assumed that a conducting diode has approximately 0.7 V drop across the diode. Diodes with different current ratings (*i.e.,* different areas and correspondingly different I_0) will exhibit the 0.7 V drop at different currents.

1.11.2. Reverse-Bias Region

When the diode is reverse biased and V is several times V_T in magnitude, current, $I = -I_0$. The reverse current, is, therefore, constant, independent of the applied reverse bias, as shown in Fig. 1.19. Consequently, I_0 is referred to as the *reverse saturation current.*

Real diodes exhibit currents, that though quite small, are much larger than I_0. For instance, a small-signal diode whose current I_0 is of the order of 10^{-14} to 10^{-15} A could show a reverse current of the order of 1 nA. The reverse current also increases somewhat with the increase in magnitude of the reverse voltage. Because of the very small magnitude of the current, these details are not clearly evident on the diode V-I characteristic depicted in Fig. 1.19.

A large component of the reverse current is due to leakage effects. These leakage currents are proportional to junction area, just as I_0 is. Their dependence on temperature, however, is different from that of I_0. Thus, whereas I_0 doubles for every 5°C rise in temperature, the corresponding rule of thumb for the temperature dependence of the reverse current is that it doubles for every 10°C rise in temperature.

1.11.3. Breakdown Region

The third distinct region of diode operation is the breakdown region, which can be easily identified on the volt-ampere characteristic depicted in Fig. 1.19. The breakdown region is entered when the magnitude of the reverse voltage exceeds a threshold value that is specific to the particular diode, called the *breakdown voltage.* This is the voltage at the "knee" of the V-I characteristic curve in Fig. 1.19, and is denoted by V_{ZK}, where the subscript Z stands for zener (to be explained shortly) and K denotes knee.

As can be seen from Fig. 1.19, in the breakdown region the reverse current increases rapidly, with the associated increase in voltage drop being very small. Diode breakdown is normally not destructive provided that the power dissipated in diode is limited by external circuitry to a "safe" level. This safe value is normally specified on the data sheets of the device. It is, therefore, necessary to limit the reverse current in the breakdown region to a value consistent with the permissible power dissipation.

The fact that the diode V-I characteristic in breakdown region is almost a vertical line enables it to be used in voltage regulation.

1.12. TEMPERATURE DEPENDENCE OF P-N DIODES

Temperature is very important consideration in the analysis or design of electronic devices or systems. It affects virtually all of the characteristics of a semiconductor device.

The diode current at a given voltage depends upon the operating temperature. The variation may be determined from Eq. (1.35), where the temperature is contained implicitly in V_T and also in the reverse saturation current I_0.

From Eq. (1.35) $\quad I = I_0\,(e^{V/\eta V_T} - 1)$

At room temperature $V_T = 26\text{ mV} = 0.026\text{ V}$

For germanium $\eta = 1$ and for silicon $\eta = 2$

So substituting the above values in the expression for diode current we have

$$\text{For germanium, } I = I_0\,(e^{V/0.026} - 1) \simeq I_0\,e^{40\,V} \qquad ...(1.42)$$

$$\text{and for silicon, } I = I_0\,(e^{V/2\times 0.026} - 1) \simeq I_0\,e^{20\,V} \qquad ...(1.43)$$

The dependence of I_0 on temperature T is, from Eqs. (1.34) and (1.36), given approximately as

$$I_0 = K T^m\, e^{-V_{GO}/\eta V_T} \qquad ...(1.44)$$

where K is a constant and eV_{GO} is the forbidden energy gap in joules

For silicon : $\eta = 2;\ m = 1.5;\ V_{GO} = 1.21\text{ V}$

For germanium : $\eta = 1;\ m = 2;\ V_{GO} = 0.785\text{ V}$

Taking the logarithm on both sides of Eq. (1.44), we have

$$\log_e I_0 = mK \log_e T - \frac{V_{GO}}{\eta V_T}$$

Taking the derivative of the above equation we have

$$\frac{d(\log_e I_0)}{dT} = \frac{1}{I_0}\frac{dI_0}{dT} = \frac{m}{T} + \frac{V_{GO}}{\eta T V_T} \qquad ...(1.45)$$

At room temperature (300 K), from Eq. (1.45)

$$\frac{d(\log_e I_0)}{dT} = 0.08 \text{ per °C for silicon and } 0.11 \text{ per °C for germanium.}$$

The performance of commercial diodes is only consistent with above result. The reason for the discrepancy is that, in a practical diode, there is a component of the reverse saturation current due to leakage over the surface that has not been taken into account in Eq. (1.44). Since leakage compo-nent is independent of temperature, it may be expected to find a smaller rate of change of I_0 with temperature than that predicted above.

From experimental data it is found that reverse saturation current I_0 increases 7 per cent per °C for both silicon and germanium. Since $(1.07)^{10} \simeq 2.0$, the reverse saturation current approximately doubles for every 10°C rise in temperature.

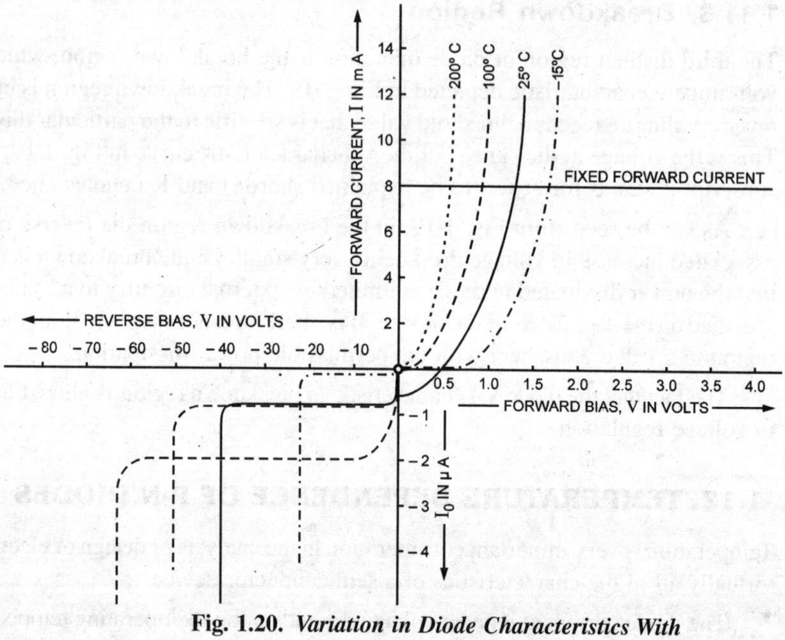

Fig. 1.20. ***Variation in Diode Characteristics With Variations in Temperature***

Germanium is more temperature dependent than silicon because its reverse saturation current is approximately 1,000 times larger.

The change in V-I characteristics of a semiconductor diode due to variations in temperature above and below room temperature is shown in Fig. 1.20.

From Fig. 1.20 we see that for a fixed level of forward voltage, forward current increases with the increase in temperature. If the forward current is measured, at the increased temperature, for several levels of forward voltage and the results are plotted, it is seen that current-voltage characteristic is moved to the left. The horizontal line on Fig. 1.20 shows that for a fixed forward current, forward voltage drops as the junction temperature increases *i.e.,* forward voltage has a negative temperature coefficient. It is found that the temperature coefficient for the forward voltage of a P-N junction is approximately – 1.8 mV/°C for Si and – 2.02 mV/ °C for Ge. Zener potential is also experiencing a pronounced change in level.

1.13. DIODE RESISTANCE

As already discussed, a forward-biased diode conducts easily whereas a reverse-biased diode restrains the flow of current, so a diode has a low forward resistance as compared to high blocking or reverse resistance.

Forward Resistance. The resistance offered by a diode in the circuit, when forward biased, is known as the *forward resistance.* This resistance is not the same for dc as well as changing current. Thus, forward resistance is of two types viz. dc or *static resistance* and ac or *dynamic resistance*

DC or **static resistance,** R is the resistance offered by a diode to the direct current. It is simply the ratio of the dc voltage across the diode to the direct current flowing through it. It can be determined by determining the ratio of voltage drop across the diode to the current flowing through it. At any point P on the V-I characteristic of the diode (Fig. 1.21), the voltage across the diode is OA and the corresponding current is OB.

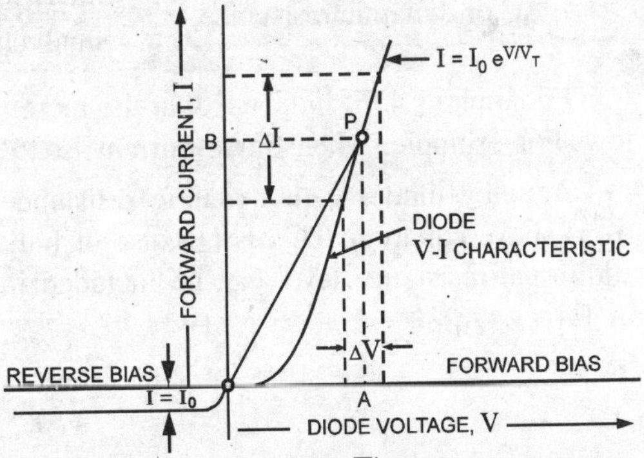

Fig. 1.21

So dc or static resistance, $R = \frac{V}{I} = \frac{OA}{OB}$

Thus at any point on the V-I characteristic of the diode, the dc or static resistance R is equal to the reciprocal of the slope of the line joining the operating point to the origin.

DC or static resistance is not constant but depends on the operating point on the V-I characteristic of the diode. It is different at different points, as illustrated in Fig. 1.21. From Fig. 1.21 it is obvious that the dc resistance of a diode in the forward-bias region decreases as the region of higher voltages and currents is approached.

In the reverse-bias region, the current $I = I_0$, constant in magnitude, the resistance of the ideal diode is essentially infinite (open-circuit equivalent).

Any way dc or static resistance does not form a useful parameter of the diode.

AC or **dynamic resistance** *r* is resistance offered by a diode to the changing forward current. It may also be defined as the reciprocal of the slope of the forward characteristic of the diode. It forms an important device parameter for small signal operation. However, it varies with the operating voltage.

From Eq. (1.35) $$r = \frac{dV}{dI} = \frac{\eta V_T}{I_0 e^{V/\eta V_T}} = \frac{\eta V_T}{I + I_0} \quad ...(1.46)$$

For a forward bias greater than a few tenths of a volt, $\left|\frac{V}{\eta V_T}\right|$ >> 1 and according to Eq. (1.35) $I >> I_0$, and dynamic resistance r is given approximately by

$$r = \frac{\eta V_T}{I} \quad \text{...(1.47a)}$$

From above equation it is obvious that for forward bias r varies inversely with the current I. At room temperature (27°C or 300 K) V_T = 0.026 V or 26 mV.

Substituting $\eta = 1$ and V_T = 26 mV in above Eq. (1.47*a*), we have

$$r = \frac{26 \text{ mV}}{I \text{ (mA)}} \quad \text{...(1.47b)}$$

The above equation implies that the dynamic resistance of a diode can be determined simply by substituting the quiescent value of the diode current into the equation. There is no need to have the characteristic available or to worry about sketching tangent lines as defined by Eq. (1.47*c*), given below.

The *ac* or *dynamic resistance* can be determined from the forward characteristic, as illustrated in Fig. 1.21.

$$\text{AC or dynamic resistance, } r = \frac{\text{Small change in forward voltage}}{\text{Small change in forward current}} = \frac{\Delta V}{\Delta I} \quad \text{...(1.47c)}$$

Equation (1.47*b*) indicates that the incremental forward resistance of the P-N junction diode is quite low. For example, when diode current is 26 mA, the diode resistance is 1 Ω.

Actually diode ac or dynamic resistance for forward bias is slightly different from that obtained from above equation. This is because of bulk and contact resistance of the semiconductor device. This additional resistance level can be included in Eq. (1.47*b*), by adding a factor denoted r_B as appearing in Eq. (1.47*d*)

$$r' = \frac{26 \text{ mV}}{I \text{ (mA)}} + r_B \quad \text{...(1.47d)}$$

For large values of forward current the first term $\frac{26 \text{ mV}}{I \text{ (mA)}}$ giving apparent forward or junction resistance r is negligible so $r' = r_B$. For small values of forward current, r_B is negligible so $r' = \frac{26 \text{ mV}}{I \text{ (mA)}}$.

The incremental backward resistance (*i.e.*, resistance with reverse bias), is ideally infinite because of the zero slope of the V-I characteristic in this region.

This conclusion also follows from Eq. (1.46), upon inserting $I = -I_0$ for negative values of bias voltage. Thus diode resistance for reverse bias

$$= \frac{\eta V_T}{I + I_0} = \frac{\eta V_T}{-I_0 + I_0} = \infty$$

However, the apparent backward resistance is not infinite because for any value of reverse bias there does exist a small current I_0. For germanium and silicon junction diodes this resistance is quite large compared to the apparent forward resistance. In germanium diodes the ratio of apparent backward to forward resistance is 400,000 : 1 while for silicon diodes this ratio is 1,000,000 : 1.

Example 1.8. Find the dynamic resistance of a P-N junction diode at a forward current of 2 mA. Assume $\frac{k'T}{e}$ = 25 mV.

Solution : Forward current, I = 2 mA or 0.002 A

Volt equivalent of temperature, $V_T = \frac{k'T}{e}$ = 25 mV or 0.025 V

From Eq. [1.47*a*], dynamic resistance is given approximately by $r = \frac{\eta V_T}{I}$

Substituting $\eta = 1$, $V_T = 0.025$ V and I = 0.002 A in above equation we have

$$\text{Dynamic resistance, } r = \frac{0.025}{0.002} = 12.5\ \Omega \quad \textbf{Ans.}$$

Example 1.9. Determine ac resistance for a germanium semiconductor diode having a forward bias of 200 mV and reverse saturation current of 1 μA at room temperature.

Solution : At room temperature (300 K), V_T = 26 mV = 0.026 V

Applied forward voltage, V = 200 mV = 0.2 V

Reverse saturation current, $I_0 = 1\ \mu A = 1 \times 10^{-6}$ A

From Eq. (1.46), ac resistance for the diode, $r = \frac{\eta V_T}{I_0 e^{V/\eta V_T}} = \frac{0.026}{1 \times 10^{-6} \times e^{0.2/0.026}} = 11.86\ \Omega$ **Ans.**

Taking $\eta = 1$ for Ge diode

Example 1.10. Explain the static and dynamic resistances in P-N junction diode. Determine dc resistance levels for the diode in following figure at (*i*) I_D = 2 mA, (*ii*) I_D = 20 mA, (*iii*) V_D = –10 V.

[*U.P. Technical Univ. Electronics Engineering Second Semester 2006-07*]

Solution : (*i*) When I_D = 2 mA, V_D = 0.5 V

So resistance, $R_1 = \frac{V_D}{I_D} = \frac{0.5}{2 \times 10^{-3}} = 250\ \Omega$ **Ans.**

(*ii*) When I_D = 20 mA = 0.02 A, V_D = 0.8 V

So resistance, $R_2 = \frac{0.8}{0.02} = 40\ \Omega$ **Ans.**

(*iii*) When V_D = –10 V,

$I_D = -1\ \mu A = -1 \times 10^{-6}$ A

So resistance, $R_3 = \frac{-10}{-1 \times 10^{-6}} = 10\ M\Omega$ **Ans.**

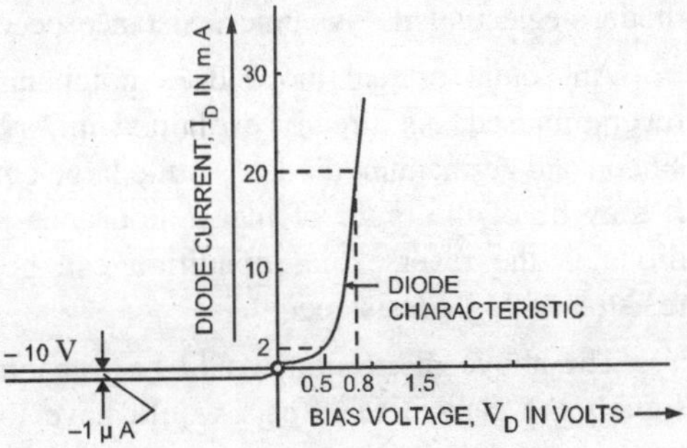

Fig. 1.22

1.14. REAL DIODE

As already mentioned in Art. 1.8, an actual diode neither behaves as a perfect conductor when forward biased nor as a perfect insulator when reverse biased.

When a diode is forward biased, it will not conduct till the potential barrier V_0 (0.7 V for Si and 0.3 V for Ge) at the junction is overcome. When the applied voltage exceeds the potential barrier V_0, the diode starts conducting. The forward current, I flowing through the diode causes a voltage drop in its internal resistance *r'*. Thus the voltage V applied across the diode is to overcome (*i*) the potential barrier V_0 and (*ii*) internal drop I*r'*

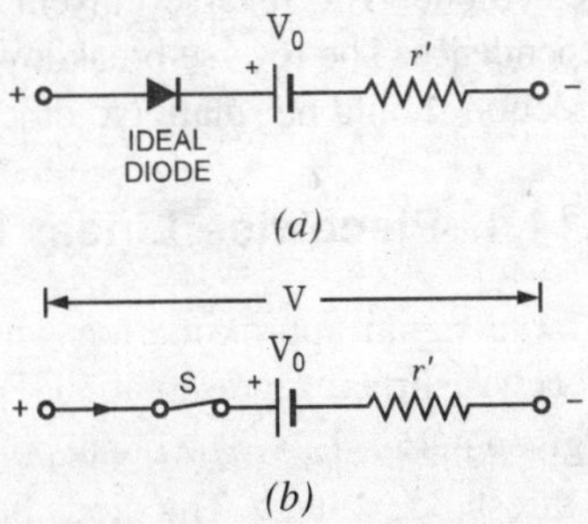

Fig. 1.23. ***Real Diode Representation Under Forward Bias Condition***

So $V = V_0 + Ir'$ ...(1.48)

Hence, a real diode may be considered as equivalent to an ideal diode in series with a barrier potential of V_0 volts and a forward bias ac resistance r', as illustrated in Fig. 1.23 (*a*). The equivalent circuit of a real diode under forward bias condition is shown in Fig. 1.23 (*b*). This circuit shows that a real diode still acts as a switch when forward biased but the voltage required to operate this switch is V given by Eq. (1.48).

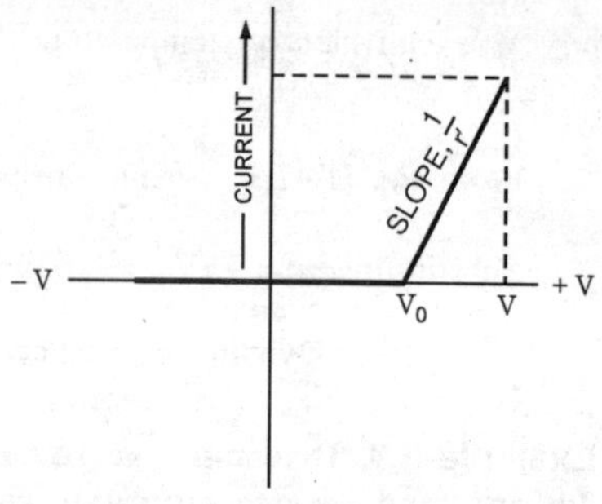

Fig. 1.24

Voltage-current characteristic of a real diode under forward bias condition is shown in Fig. 1.24 which follows the Eq. (1.48). Here, the reciprocal of the slope of this characteristic represents the ac or dynamic resistance r' of the diode.

For *large signal sources* whose voltage is much greater than the diode barrier potential V_0 (or knee voltage V_k), forward current I would be large and the ac or dynamic resistance r' of the diode will be equal to bulk resistance r_B of the diode, as already explained in Art. 1.13.

For *small signal sources* whose voltage is much smaller than V_0 or V_k, forward current I would be small and the ac or dynamic resistance of r' of the diode will be equal to apparent forward or junction resistance r given by Eq. (1.47*b*) because for smaller values of forward current r_B is negligible.

For most applications, the ac or dynamic resistance r' of a diode can be neglected, being very small as compared to other elements and the real diode can be represented as shown in Fig. 1.25. The V-I characteristics for a real diode, neglecting its dynamic resistance become as illustrated in Fig. 1.15.

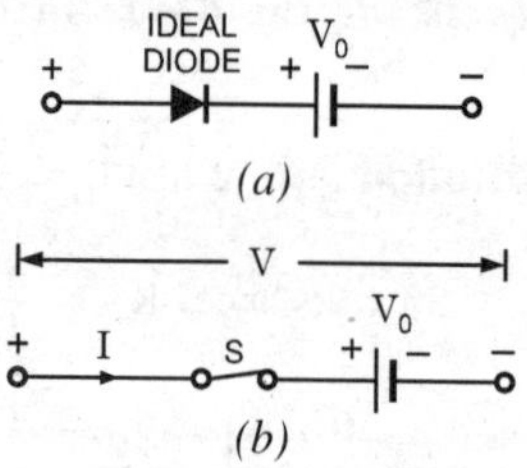

Fig. 1.25

An actual or real diode does not behave as a perfect insulator when reverse biased. As already explained in Art. 1.13, the reverse resistance of silicon and germanium diodes is quite large compared to the forward resistance. It may be of the order of many thousands of mega ohms. Hence an actual diode in the reverse bias condition can be thought of as equivalent to a resistor of high resistance.

The above discussion would be true only in case of small frequency signals. For large frequency signals the capacitive effects would have to be taken into account.

Unilateral diodes can be used for rectification of alternating currents, waveshaping of ac signals and many other applications. Waveshaping involves alteration of ac signals by passing only certain portions of the signal while blocking other portions.

Junction diodes designed for use as rectifiers must have volt-ampere characteristics close to those of an ideal diode. The reverse current should be negligible and forward current should exhibit little voltage dependence. The reverse breakdown voltage should be large and the offset or barrier potential V_0 in forward direction should be small. No diode can offer all these properties.

1.14.1. Piecewise Linear Diode Characteristic

A large-signal approximation which usually leads to a sufficiently accurate engineering solution is the **piecewise linear** representation. For instance, the piecewise linear approximation for a semiconductor diode is given in Fig. 1.24. The break point is not at the origin, and therefore, V_0 (or V_γ) is also known as the offset, or threshold, voltage. The diode behaves like an open circuit when the applied voltage is less than potential barrier V_0, and has a constant incremental resistance $r' = \dfrac{dV}{dI}$ when applied voltage exceeds potential barrier or threshold voltage. The equivalent circuit for this linear approximation is a constant voltage source in series with a resistor [Fig. 1.23 (*b*)]. This method models the diode with segments of straight lines : thus the name *piecewise linear model.*

1.15. INTERCONNECTION OF DIODES

Diodes can be connected in several ways such as series, parallel or series-parallel. In series arrangement, the diodes may be connected with polarities in the same direction (in forward-biased mode or in reverse-biased mode) or in opposite directions.

1.15.1. Series Diode Configurations

The approximate models discussed in preceding section are used for investigation of a number of series diode configurations with dc inputs. This establishes a foundation in diode analysis that will carry over into the sections and chapters to follow. The procedure described can, in fact, be applied to networks with any number of diodes in a variety of configurations.

For each configuration the state of each diode is first determined which diodes are "on" and which are "off"? Once determined, the appropriate equivalent can be substituted and the remaining parameters of the network determined.

In general, a diode is in the "on" state if the current caused by the applied sources is such that its direction matches that of the arrow in the diode symbol, and the diode voltage V_D exceeds V_0 (0.7 V in case of silicon and 0.3 V for germanium diode).

For each configuration, mentally replace the diodes with resistive elements and note the resulting current direction. If the direction of resulting current "matches" with the arrow in the diode symbol, the diode is in conducting or "on" state. If the diode is in the "on state", diode can be replaced by a 0.7 V drop across the element or the network is redrawn with the V_0 or V_K equivalent circuit. It is preferable to include the 0.7 V drop across each "on" diode and draw a line through each diode in the "off" or open state.

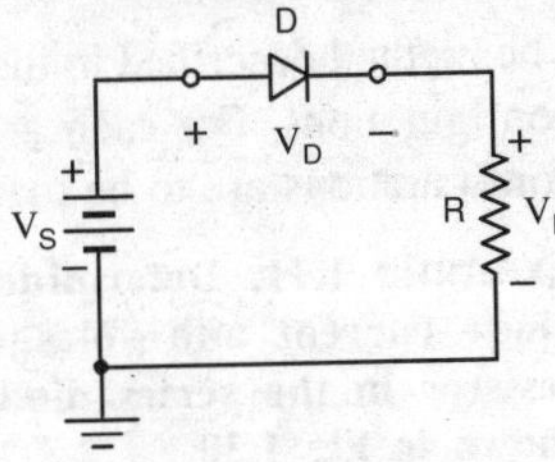

Fig. 1.26. ***Series Diode Configuration***

Let us consider a series circuit, shown in Fig. 1.26, to understand the approach described above.

The state of the diode is first determined by mentally replacing the diode with a resistive element as shown in Fig. 1.27 (*a*). The resulting direction of current I matches with the arrow in the diode symbol, and since V_S exceeds V_0 or V_K, the diode is in the conducting (or "on") state. The network is redrawn, as depicted in Fig. 1.27 (*b*), with the appropriate equivalent model for the forward-biased semiconductor diode. Note for future reference that the polarity of V_D is the same as would result if in fact the diode were a resistive element. The various relations between different voltages and current are given below :

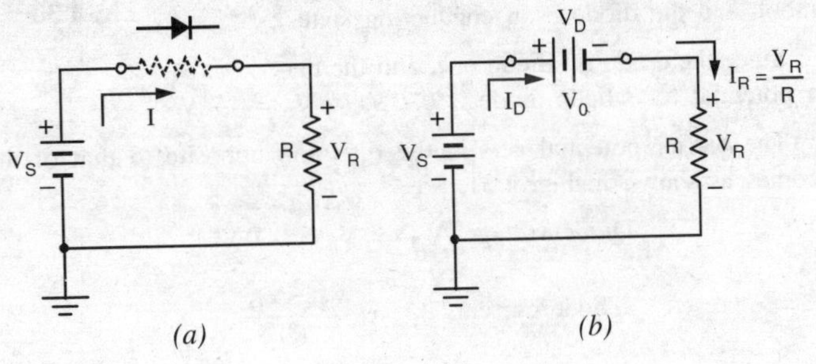

Fig. 1.27

$$\text{Voltage drop across diode, } V_D = V_0 \text{ (or } V_K) \quad \text{...(1.49)}$$

$$\text{Voltage drop across resistor R, } V_R = V_S - V_D = V_S - V_0 \quad \text{...(1.50)}$$

$$\text{Diode current, } I_D = I_R = \frac{V_R}{R} \quad \text{...(1.51)}$$

Now let us consider the diode connected in reverse-biased mode, as shown in Fig. 1.28. If the diode is replaced mentally with a resistive element as depicted in Fig. 1.29 (*a*), it is revealed that the direction of flow of resulting current does not match the arrow in the diode symbol. It means that the diode is in "off" state and the resultant equivalent circuit will be as shown in Fig. 1.29 (*b*). Open circuit will not allow the current to flow in the circuit *i.e.*, $I_D = I_R = 0$.

Voltage drop across resistor, $V_R = I_R R = 0 \times R = 0$ V ...(1.52)

and Voltage drop across diode, $V_D = V_S$ volts ...(1.53)

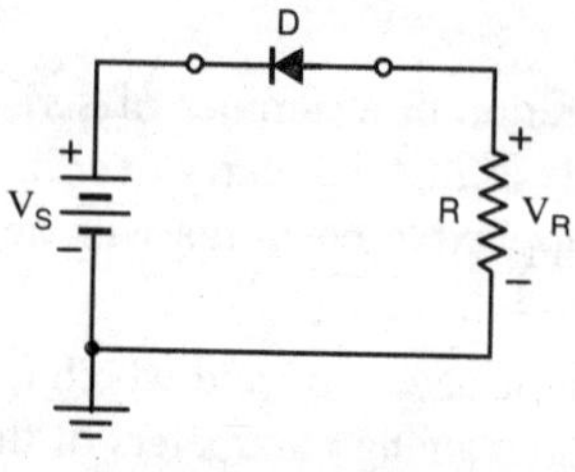

Fig. 1.28

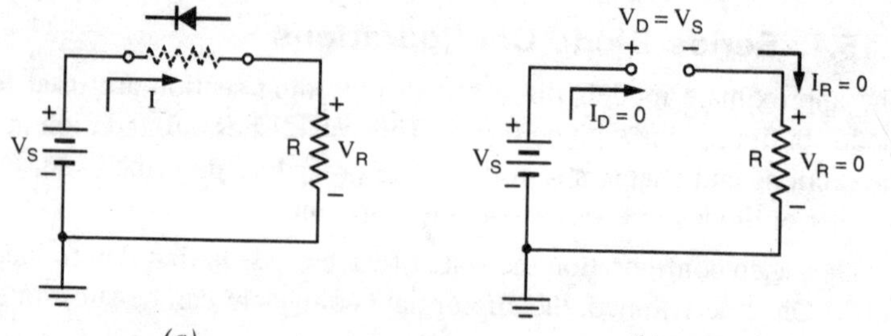

(*a*) (*b*)

Fig. 1.29

1.15.2. Parallel and Series-Parallel Configurations

The method described in the preceding section can be extended to the analysis of parallel and series-parallel configurations. For each area of application, simply the sequential series of steps applied to series diode configurations are to be matched.

Example 1.11. Determine diode voltage, diode current and voltage across 100 Ω resistor in the series diode configuration shown in Fig. 1.30.

Solution : Since the applied voltage establishes the current to flow in clockwise direction in the given circuit, it matches the arrow of the diode symbol, and the diode is in conducting state.

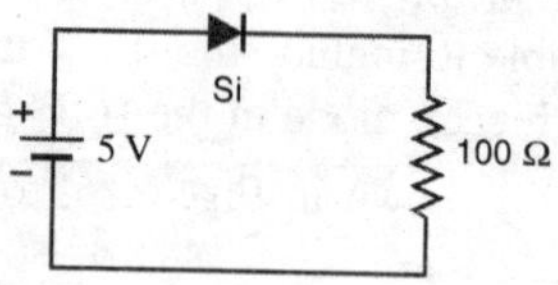

Fig. 1.30

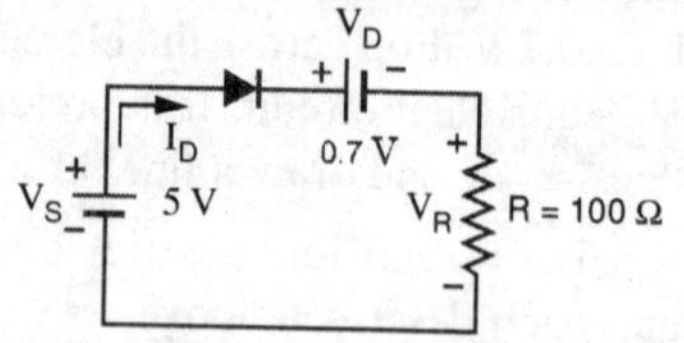

Fig. 1.31

Since the diode is silicon one, and the barrier potential for silicon diode is 0.7 V, so $V_0 = 0.7$ V.

The barrier potential acts in the direction opposite to that of the applied voltage and the equivalent circuit becomes as shown in Fig. 1.31. So

Diode voltage, $V_D = V_0 = 0.7$ V

Diode current, $I_D = \dfrac{V_S - V_D}{R}$ neglecting diode forward resistance

$= \dfrac{5 - 0.7}{100} = 0.043$ A or 43 mA **Ans.**

Voltage across resistor, $V_R = V_S - V_D = 5 - 0.7 = 4.3$ V **Ans.**

Example 1.12. Calculate forward current I_F for silicon diode with dynamic resistance $r_d = 0.25\ \Omega$ used in the following circuit (Fig. 1.32).

Solution : Supply voltage, E = 1.5 V

Dynamic resistance, $r_d = 0.25\ \Omega$

Resistance, $R_1 = 10\ \Omega$

Potential barrier of silicon diode, $V_0 = 0.7$ V

The equivalent circuit is shown in Fig. 1.33.

Forward current, $I_F = \dfrac{E - V_0}{R_1 + r_d}$

$= \dfrac{1.5 - 0.7}{10 + 0.25}$

$= 0.078$ A or 78 mA **Ans.**

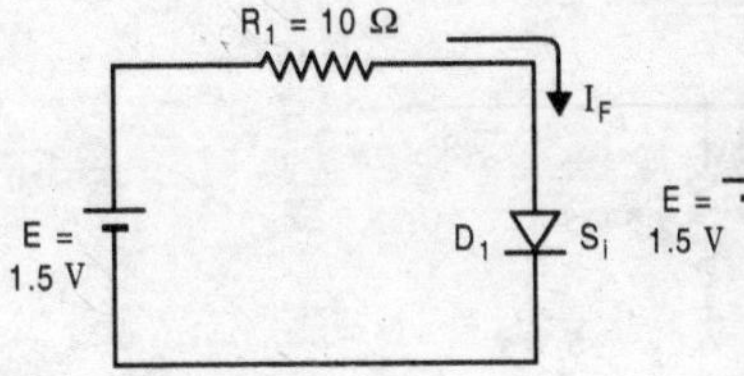

Fig. 1.32

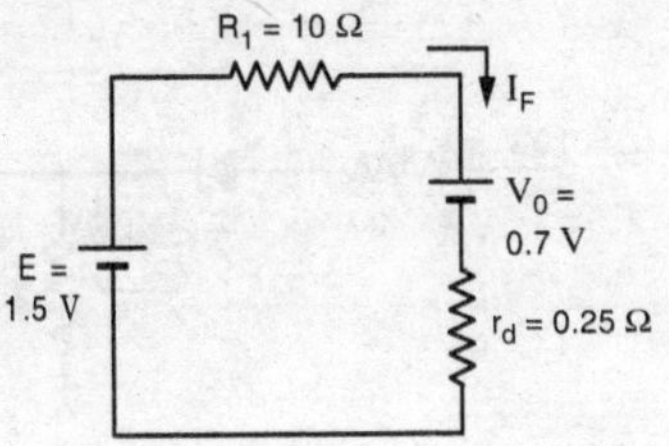

Fig. 1.33

Example 1.13. Determine diode voltage, diode current and voltage across 100 Ω resistor in the series diode configuration shown in Fig. 1.30 with the diode polarity reversed.

Solution : Replacing the diode by open circuit, the equivalent circuit becomes as shown in Fig. 1.34.

Resulting current in the circuit, I = 0 ∵ the circuit is open

Diode current, I_D = Current through resistor, $I_R = I = 0$ A **Ans.**

Voltage across resistor, $V_R = I_R \times R = 0 \times 100 = 0$ V **Ans.**

Voltage drop across diode, $V_D = V_S = 5$ V **Ans.**

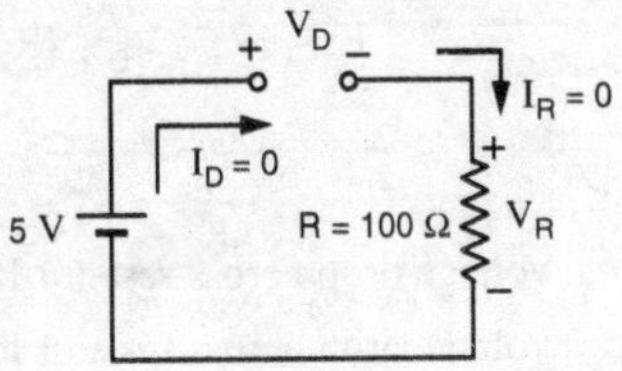

Fig. 1.34

Example 1.14. Determine diode current, diode voltage and voltage across 500 Ω resistor, for the series diode configuration shown in Fig. 1.35.

Solution : Although the applied voltage establishes a current in the same direction as that of diode arrow symbol, but the magnitude of applied voltage is insufficient, being less than potential barrier of silicon diode, to turn the silicon diode "on". The equivalent circuit is shown in Fig. 1.36.

Diode current, $I_D = 0$ A

Resistor current, $I_R = 0$ A

Voltage drop across resistor, $V_R = I_R \times R = 0 \times 500 = 0$ V **Ans.**

Diode voltage, $V_D = V_S = 0.4$ V **Ans.**

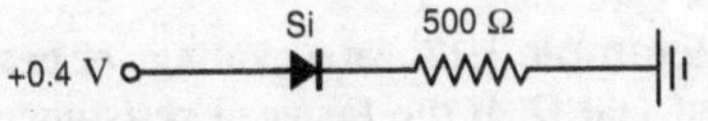

Fig. 1.35

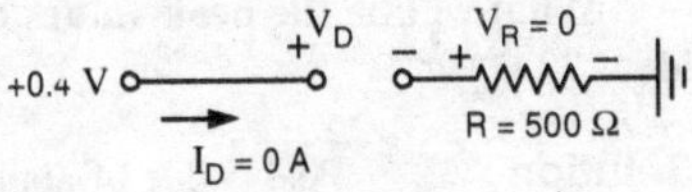

Fig. 1.36

Example 1.15. Find the voltage V_A of the circuit given in Fig. 1.37.

Solution : Assuming diodes of negligible forward resistance, the given circuit is redrawn, as given in Fig. 1.38.

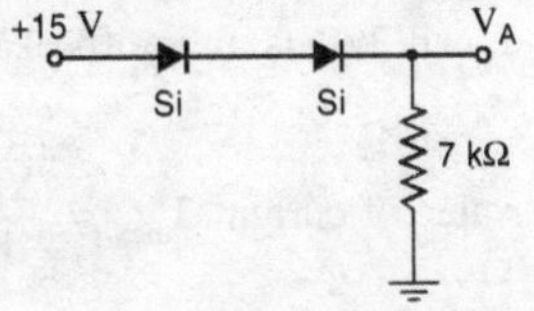

Fig. 1.37

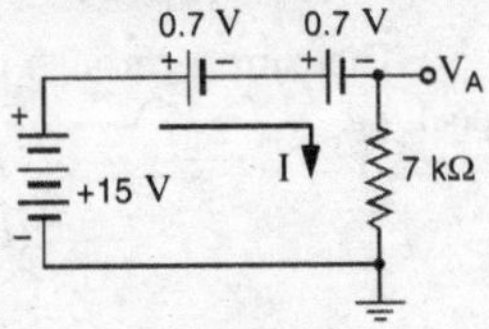

Fig. 1.38

Current through the circuit,

$$I = \frac{V - 2V_0}{R}$$

$$= \frac{15 - 2 \times 0.7}{7\ k\Omega} = 1.943 \text{ mA}$$

∵ barrier potential of silicon diode, $V_0 = 0.7$ V

Potential V_A = Voltage drop across 7 kΩ = 1.943 mA × 7 kΩ = 13.6 V **Ans.**

Example 1.16. Determine I, V_1, V_2 and V_0 for the series dc configuration of the following figure (Fig. 1.39).

[*Electronics Engineering, U.P. Technical Univ. Second Semester 2006-07; Mahamaya Technical Univ. Second Semester 2010-11*]

Solution : Mentally replacing the diode with a resistive element, it is revealed that the resulting current direction matches the arrow of diode symbol and the diode is in conducting state. The equivalent circuit is shown in Fig. 1.40.

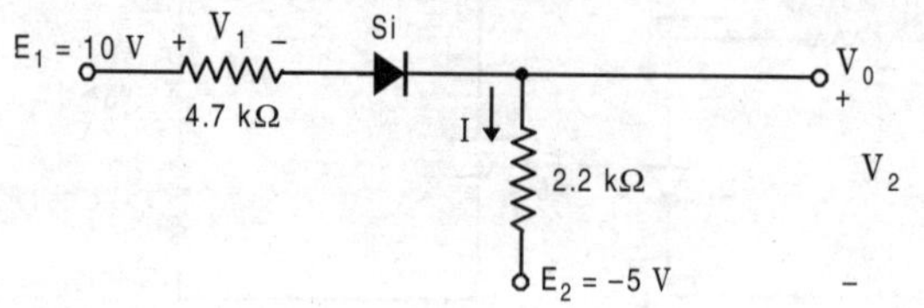

Fig. 1.39

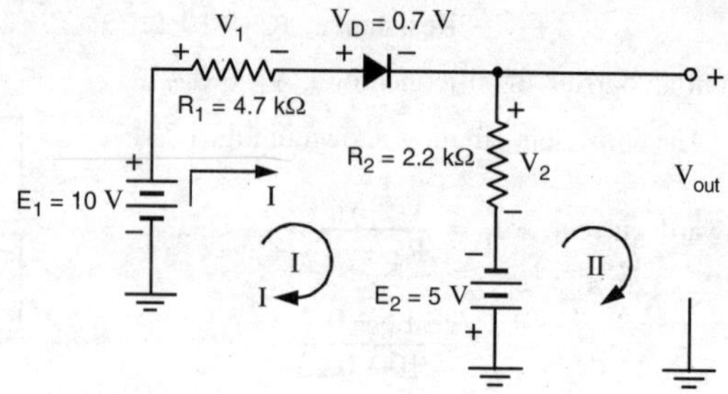

Fig. 1.40

Applying KVL to loop I, we have

$$E_1 - IR_1 - V_D - IR_2 + E_2 = 0$$

$$\text{or } I = \frac{E_1 + E_2 - V_D}{R_1 + R_2} = \frac{10 + 5 - 0.7}{(4.7 + 2.2) \times 10^3} = 2.072 \text{ mA } \textbf{Ans.}$$

Voltage drop across resistor R_1, $V_1 = IR_1 = 2.072 \times 10^{-3} \times 4.7 \times 10^3 = 9.74$ V **Ans.**

Voltage drop across resistor R_2, $V_2 = IR_2 = 2.072 \times 10^{-3} \times 2.2 \times 10^3 = 4.56$ V **Ans.**

Applying KVL to output loop, we have

$$-E_2 + V_2 - V_{out} = 0$$

or Output voltage, $V_{out} = V_2 - E_2 = 4.56 - 5 = -0.44$ V **Ans.**

Example 1.17. An ac voltage of peak value 25 V is connected in series with a silicon diode and a load resistance of 600 Ω. If the forward resistance of the diode is 15 Ω determine (*i*) the peak value of current through the diode (*ii*) peak value of the output voltage.

What will be the peak values of current through the diode and output voltage if the diode is assumed to be ideal ?

Solution : Peak value of applied voltage, $V_{max} = 25$ V

Forward resistance of diode, $r_f = 15\ \Omega$

Load resistance, $R_L = 600\ \Omega$

Potential barrier voltage, $V_0 = 0.7$ V

The current through the diode will be maximum (peak) at the instant the input or applied voltage attains positive peak *i.e.*, $V_{in} = 25$ V

$$(i) \text{ Peak value of current, } I_{max} = \frac{V_{max} - V_0}{R_L + r_f} = \frac{25 - 0.7}{600 + 15} = 0.0395 \text{ A or } 39.5 \text{ mA } \textbf{Ans.}$$

$$(ii) \text{ Peak value of output voltage, } V_{out\ max} = I_{max} \times R_L = 0.0395 \times 600 = 23.7 \text{ V } \textbf{Ans.}$$

When the diode is ideal *i.e.*, $V_0 = 0$ and $r_f = 0$

$$\text{Peak value of current, } I'_{max} = \frac{V_{max}}{R_L} = \frac{25}{600} = 0.04167 \text{ A or } 41.67 \text{ mA } \textbf{Ans.}$$

$$\text{Peak value of output voltage, } V_{out\ max} = I'_{max} \times R_L = 0.04167 \times 600 = 25 \text{ V } \textbf{Ans.}$$

Example 1.18. Calculate V_0 and I_D for the following (Fig. 1.41).

[*U.P. Technical Univ. Electronics Engineering First Semester 2006-07*]

Solution : Converting 10 mA current source having a 2.2 kΩ resistor connected across it into an equivalent voltage source of 10 mA × 2.2 kΩ *i.e.*, 22 V in series with a resistance of 2.2 kΩ as shown in Fig. 1.42, we have

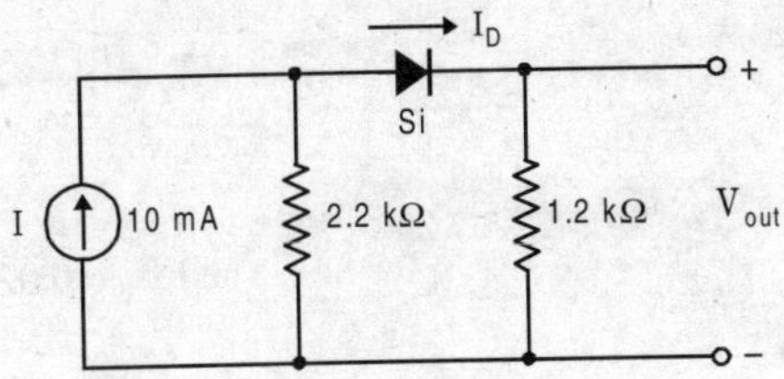

Fig. 1.41

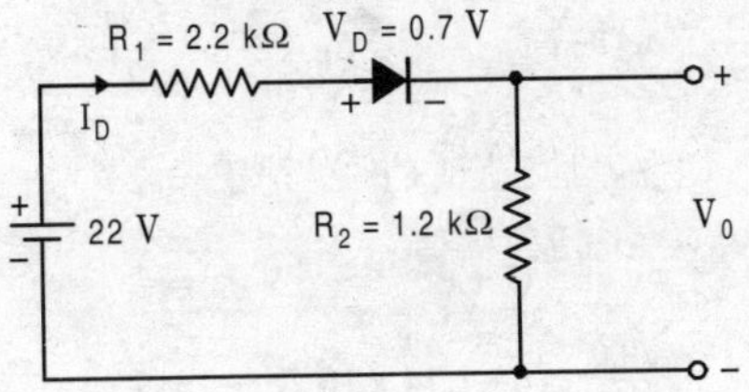

Fig. 1.42

$$\text{Diode current, } I_D = \frac{V_S - V_D}{R_1 + R_2} = \frac{22 - 0.7}{(2.2 + 1.2) \times 10^3} = 6.265 \times 10^{-3} \text{ A or } 6.265 \text{ mA } \textbf{Ans.}$$

Output voltage, V_{out} = Voltage drop across 1.2 kΩ resistor = $I_D \times 1.2 \times 10^3 = 6.265 \times 10^{-3} \times 1.2 \times 10^3 = 7.52$ V **Ans.**

Example 1.19. Determine diode current, diode voltage and output voltage for the circuit shown in Fig. 1.43.

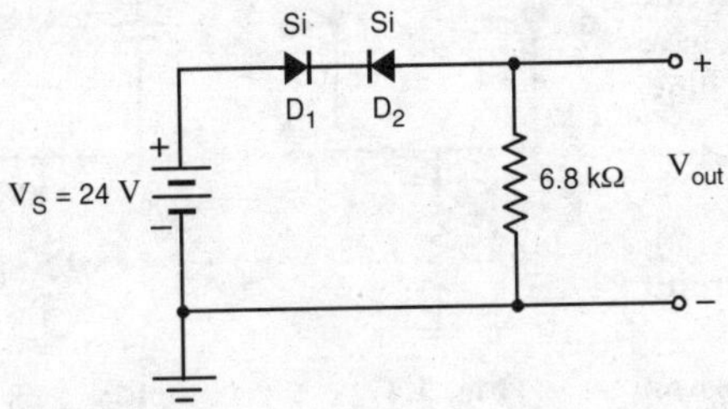

Fig. 1.43

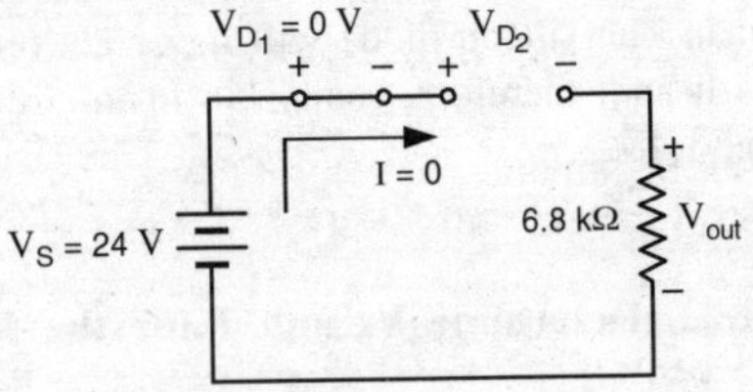

Fig. 1.44

Solution : Although the current established by the applied voltages matches the direction of symbol arrow of diode D_1 but not that of diode D_2. Thus there is an open circuit and the equivalent circuit is shown in Fig. 1.44.

Since the equivalent circuit shown in Fig. 1.44 is open one,

Diode current, I_0 = Output current, I_R = Resulting circuit current, I = 0 A **Ans.**

Diode voltage, $V_{D1} = 0$ because there is no bias

Output voltage, $V_{out} = IR = 0 \times 6.8\ k\Omega = 0$ V **Ans.**

Diode voltage, $V_{D2} = V_S - V_{D1} - V_{out} = 24 - 0 - 0 = 24$ V **Ans.**

Example 1.20. Find the voltage drop across each of the silicon junction diodes shown in Fig. 1.45 at room temperature. Assume that no reverse saturation current flows in the circuit and the magnitude of the reverse breakdown voltage is greater than 5 V. *[U.P.S.C. I.E.S. Electronics & Telecommunication Engineering-I, 2001]*

Solution : Diode D_1 is forward biased and diode D_2 is reversed biased. Let the diode currents be I_1 and I_2 respectively, as shown in Fig. 1.46. Diode currents are given by the expressions

$$I_1 = I_0 (e^{V_{D1}/\eta V_T} - 1) \quad ...(i)$$

and

$$I_2 = I_0 (e^{-V_{D2}/\eta V_T} - 1) \quad ...(ii)$$

where I_0 is the reverse saturation current.

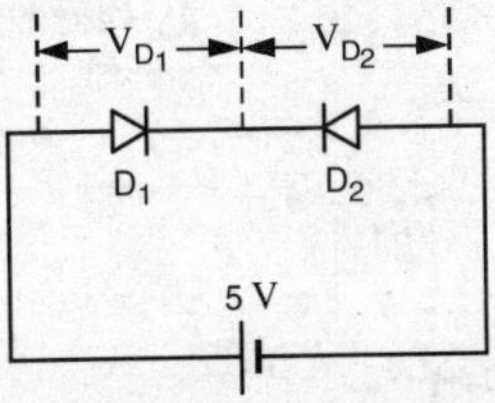

Fig. 1.45

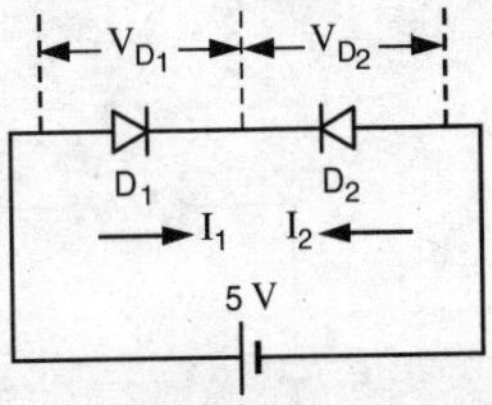

Fig. 1.46

Now from circuit shown in Fig. 1.46

$$I_1 + I_2 = 0$$

or $$I_0 (e^{V_{D1}/\eta V_T} - 1) + I_0 (e^{-V_{D2}/\eta V_T} - 1) = 0$$

or $$e^{V_{D1}/\eta V_T} + e^{-V_{D2}/\eta V_T} = 2$$

$$\text{or} \quad e^{V_{D_1}/\eta V_T} + e^{-(5 - V_{D_1})/\eta V_T} = 2 \qquad \because V_{D_1} + V_{D_2} = V_S = 5 \text{ V}$$

$$\text{or} \quad e^{V_{D_1}/\eta V_T}(1 + e^{-5/\eta V_T}) = 2$$

$$e^{V_{D_1}/2 \times 0.026}(1 + e^{-5/2 \times 0.026}) = 2 \qquad \because \eta \text{ for Si} = 2 \text{ and } V_T = 0.026 \text{ V}$$

$$\text{or} \quad e^{V_{D_1}/0.052}(1 + 1.74 \times 10^{-42}) = 2$$

$$\text{or} \quad V_{D_1} = 0.052 \log_e 2 = 0.036 \text{ V} \quad \textbf{Ans.}$$

$$V_{D_2} = V_S - V_{D_1} = 5 - 0.036 = 4.964 \text{ V} \quad \textbf{Ans.}$$

Example 1.21. Determine the voltage V_A in the circuit shown in Fig. 1.47.

Solution : It appears that on switching on the supply both the diodes turn 'on', but this is not true. When the supply voltage is switched on, germanium diode whose barrier potential is 0.3 V will turn on first and a level of 0.3 V will be maintained across the parallel branch. The silicon diode will never get opportunity to have 0.7 V across it and, therefore, remain in open-circuit state, as illustrated in Fig. 1.48.

So, $V_A = 15 - 0.3 = 14.7$ V **Ans.**

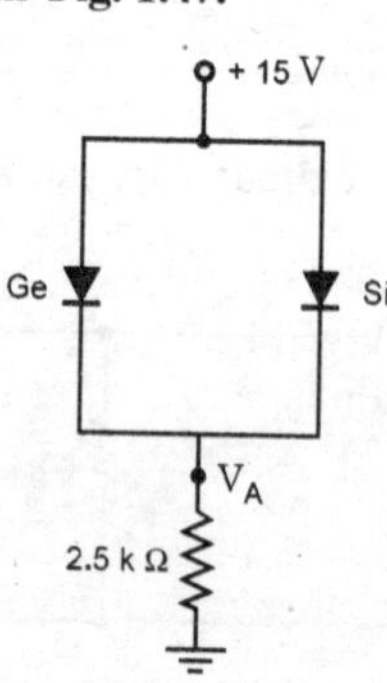

Fig. 1.47

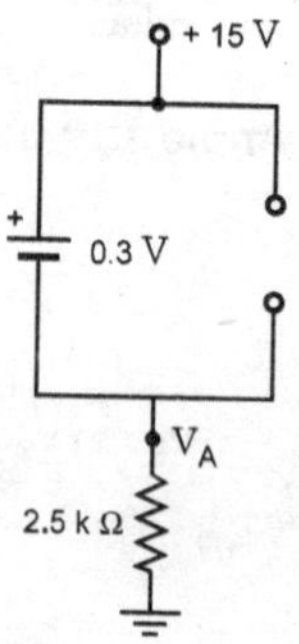

Fig. 1.48

Example 1.22. Determine V_0 and I for the following circuit (Fig. 1.49)

[U.P. Technical Univ. Electronics Engineering Odd Semester 2009-10]

Solution : As discussed in Example 1.21, the silicon diode will remain open circuited and the equivalent circuit is shown in Fig. 1.50.

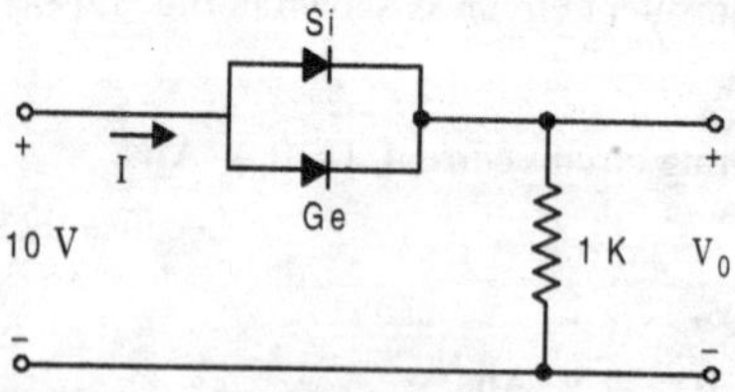

Fig. 1.49

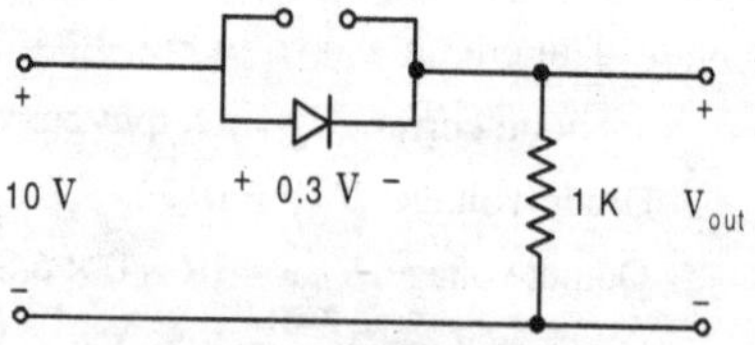

Fig. 1.50

Output voltage, $V_{out} = V_S -$ barrier potential of Ge diode $= 10 - 0.3 = 9.7$ V **Ans.**

$$\text{Current, } I = \frac{V_{out}}{R_L} = \frac{9.7}{1 \times 10^3} = 9.7 \text{ mA} \quad \textbf{Ans.}$$

Example 1.23. Determine V_0, I_1, I_{D1} and I_{D2} for the parallel diode configuration of Fig. 1.51.

[U.P. Technical Univ. Electronics Engineering First Semester 2008-09]

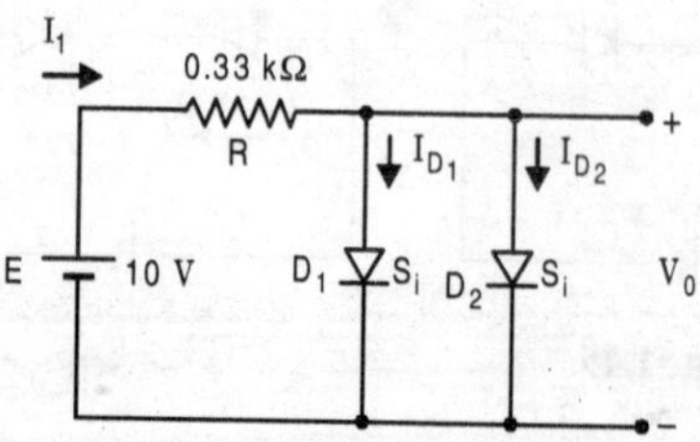

Fig. 1.51

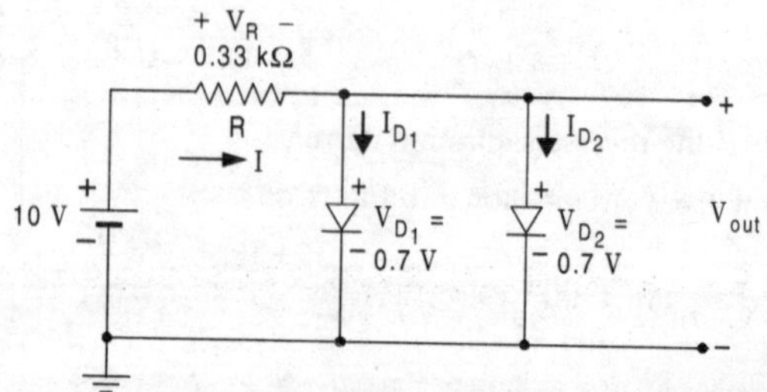

Fig. 1.52

Solution : The current established by the supply source through each diode acts in the same direction and matches that of arrow in each diode symbol and the supply voltage exceeds the potential barrier of each diode, so both diodes are in conducting state. The equivalent circuit is shown in Fig. 1.52.

Output voltage, V_{out} = Junction barrier potential of silicon = 0.7 V **Ans.**

Supply current, $I = \frac{E - V_D}{R} = \frac{10 - 0.7}{0.33 \times 10^3} = 28.2$ mA **Ans.** neglecting diode forward resistance

Assuming diodes D_1 and D_2 of similar characteristics, we have

Diode currents, $I_{D1} = I_{D2} = \frac{I_1}{2} = \frac{28.2}{2} = 14.1$ mA **Ans.**

Example 1.24. Determine the output voltage V_0 in the circuit shown below (Fig. 1.53) for the following values of input voltage :

(*i*) $V_1 = V_2 = 5$ V; (*ii*) $V_1 = V_2 = 0$ V

A silicon diode is to be used with $R_F = 35\ \Omega$, $V_\gamma = 0.6$ V and $R_r = \infty$. *[Rajasthan Univ. 2003]*

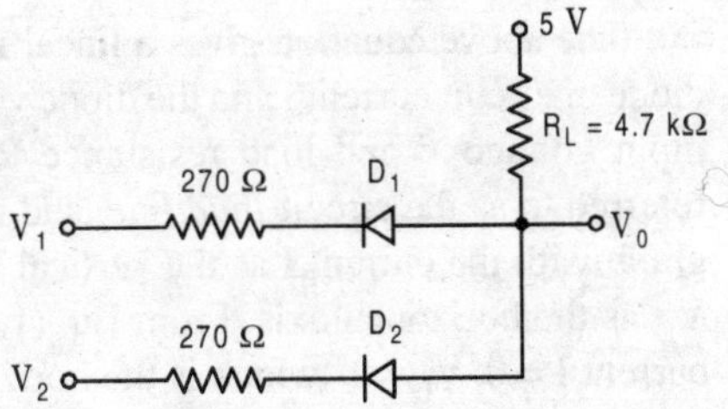

Fig. 1.53

Solution : (*i*) Diodes D_1 and D_2 will be reverse biased till potential of anode is less than the potential of cathode.

Potential of cathodes of diodes D_1 and D_2 = 5 + 0.6 = 5.6 V

Potential of anodes of diodes D_1 and D_2 = V_{CC} = 5 V

i.e., potential of anodes of diodes D_1 and D_2 is less than that of cathodes of diodes D_1 and D_2 and, therefore, both the diodes will be reverse biased and offer infinite resistance. Thus no current will flow through R_L *i.e.*, current through R_L, I = 0.

Output voltage, $V_{out} = V_{CC} - IR_L = 5 - 0 \times 4.7 \times 10^3 = 5$ V **Ans.**

(*ii*) When $V_1 = V_2 = 0$, both the diodes will be forward biased and the equivalent circuit will be as shown in Fig. 1.54.

From the circuit shown in Fig. 1.54

$$V_{CC} = IR_L + V_\gamma + \frac{I}{2}(R_F + R_1)$$

Substituting the values of V_{CC}, R_L, V_γ, R_F and R_1, we have

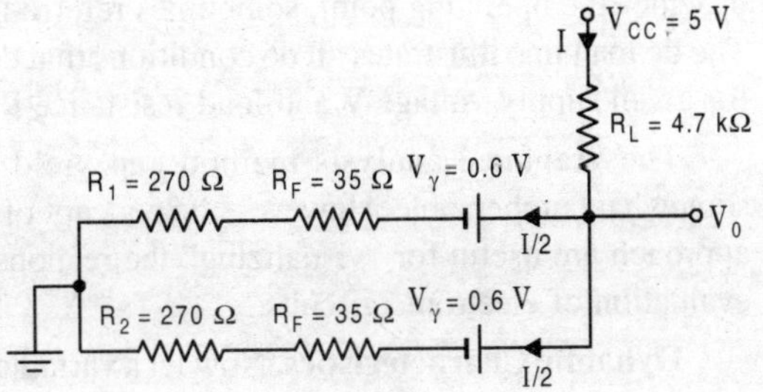

Fig. 1.54

$$5 = I \times 4.7 \times 10^3 + 0.6 + \frac{I}{2} \times (35 + 270)$$

$$\text{or} \quad I = \frac{5 - 0.6}{4{,}700 + 152.5} = 9.067 \times 10^{-4} \text{ A}$$

Output voltage, $V_{out} = V_{CC} - IR_L = 5 - 9.067 \times 10^{-4} \times 4.7 \times 10^3 = 5 - 4.262 = 0.738$ V **Ans.**

1.16. DIODE AS CIRCUIT ELEMENT

The P-N junction diode is considered as a circuit element. One of the most important application of diode is in rectification (*i.e.*, in conversion of sinusoidal ac supply into dc supply).

The basic diode circuit, shown in Fig. 1.55, consists of a diode in series with a load resistance R_L across an input voltage of V volts.

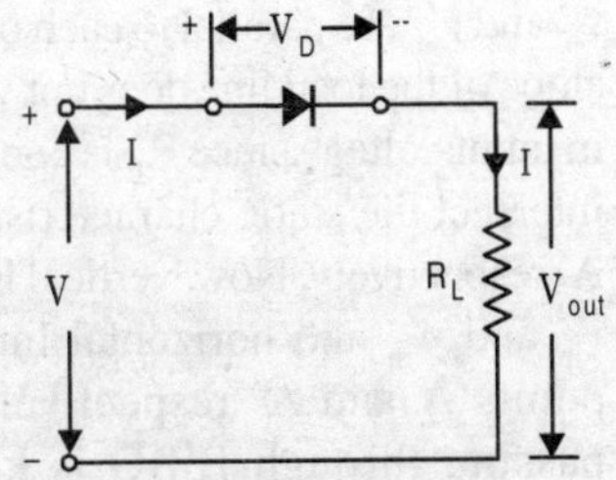

Fig. 1.55. ***Basic Diode Circuit***

1.16.1. Load Line Concept

Kirchhoff's voltage law is applicable both to linear and nonlinear circuits, so we may write

$$V = V_D + V_{out}$$

$$= V_D + IR_L \quad ...(1.54)$$

where I is the circuit current.

$$\text{or } I = \frac{V}{R_L} - \frac{V_D}{R_L} \quad ...(1.55)$$

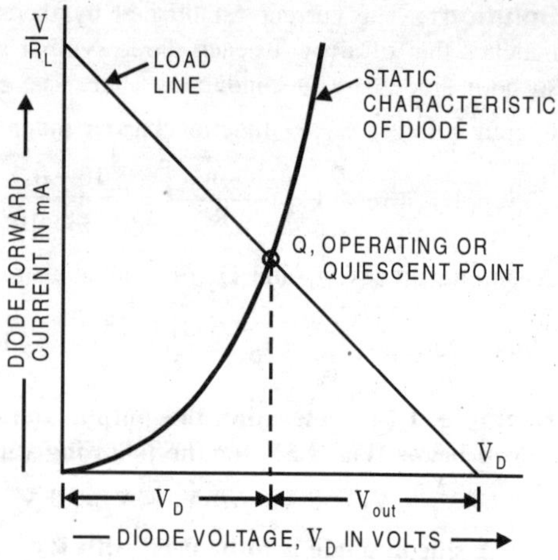

Fig. 1.56. ***The Diode and Load Line Characteristics For The Circuit Given in Fig: 1.55***

The above equation gives a linear relation between the diode or circuit current I and the diode voltage V_D for a given input voltage V and load resistance R_L. This equation is referred to as the circuit *load line,* and is usually plotted on graph with the current I as the vertical axis and the voltage V_D as the horizontal axis. From Eq. (1.55), we see that for current I = 0, V_D = V which is the horizontal axis intercept.

Also from this equation, if $V_D = 0$, then current $I = \frac{V}{R_L}$ which is the vertical axis intercept. The load line can be drawn between these two points. The slope of the load line is determined by R_L, the negative value of the slope is equal to $\frac{1}{R_L}$.

The intersection of load line with the static characteristic of diode provides the operating point, sometimes referred to as quiescent or Q point. The dc load line illustrates all dc conditions that could exist within the circuit for given supply voltage V and load resistance R_L.

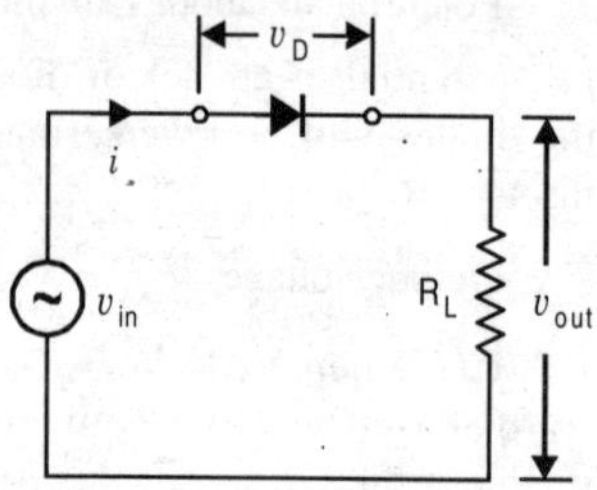

Fig. 1.57. ***Basic Diode Circuit***

The graphical analysis method can yield accurate results, but it is somewhat cumbersome. However, the concept of load line and the graphical approach are useful for "visualizing" the response of the circuit and the load line is used extensively in the evaluation of electronic circuits.

Dynamic Characteristics. Now let a variable voltage v_{in} be applied to a diode circuit, depicted in Fig. 1.57. A plot of current i vs input voltage v_{in} is called the *dynamic characteristic.* The curve is drawn for different values of input voltage v_{in}.

Figure 1.58 depicts the load lines for two voltages v_{in} and v'_{in}. It is observed from Fig. 1.58 that the load lines drawn for the two input voltages v_{in} and v'_{in} are parallel to each other. This is because slope of the load line does not vary with the change in input voltage since R_L is constant. The load lines intersect the static characteristics at points A and A′ respectively. Now, vertical lines are drawn from v_{in} and v'_{in} and horizontal lines are drawn from points A and A′ respectively. A curve drawn passing through B′BO is known as dynamic characteristic.

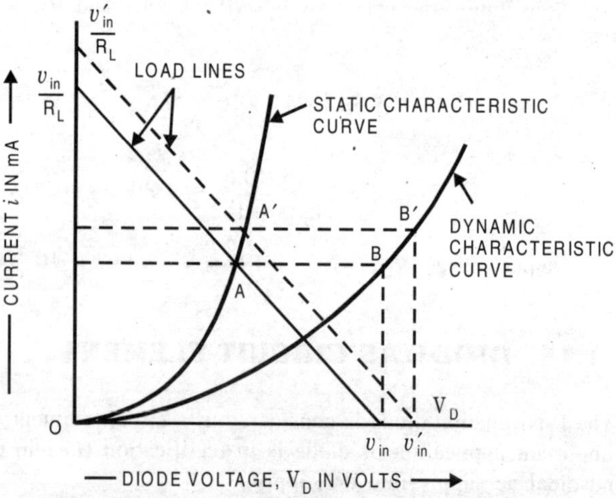

Fig. 1.58. ***Dynamic Characteristic***

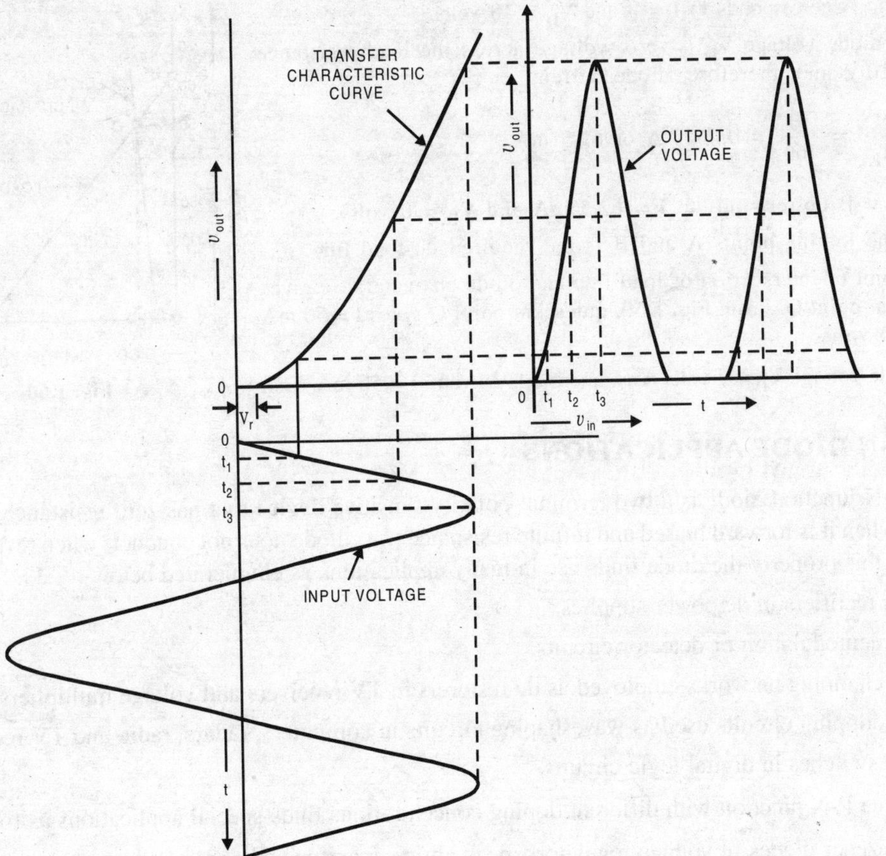

Fig. 1.59. ***Graphical Method For Obtaining Output Voltage Waveform Using Transfer Curve***

Transfer Characteristic. The curve that relates the output voltage v_{out} to the input voltage v_{in} of any circuit is known as the *transfer* or *transmission characteristic*. Figure 1.59 illustrates the graphical method for determination of output voltage waveform for a given input voltage waveform (say, sinusoidal waveform) using transfer characteristic. The input signal waveform is drawn with its time axis vertically downward, so that the voltage axis is horizontal. The different input voltages are at different time instants t_1, t_2, t_3 etc. and corresponding vertical lines are drawn to meet the transfer curve. Now horizontal lines are drawn from the intersection points on transfer curve to provide the output voltage corresponding to time instants t_1, t_2, t_3 etc. Thus output voltage waveform can be obtained by joining these points. From Fig. 1.59, it can be visualized that (*i*) the output voltage is distorted at the start due to nonlinear transfer characteristic and (*ii*) for $v_{in} < V_r$, the diode acts as a clipper and a portion of the input signal does not appear at the output.

Example 1.25. A diode is connected in series with a load resistance of 160 Ω across a dc supply voltage of 10 V. Draw a dc load line and determine the voltage drop across the diode and diode current. Diode characteristic is given.

Solution : From circuit shown in Fig. 1.55.

$$\text{Supply voltage, } V = V_D + IR_L$$

When diode current I is zero, the voltage across the diode, $V_D = V = 10$ V because voltage drop across the load resistance $R_L = 0$ volts.

So point A corresponds to $I = 0$ and $V_D = 10$ volts.

When diode voltage V_D is zero, voltage across the load resistance $R_L = V = 10$ V and, therefore, diode current,

$$I = \frac{V}{R_L} = \frac{10}{160} = 0.0625 \text{ A or } 62.5 \text{ mA}$$

So point B corresponds to $I = 62.5$ mA and $V_D = 0$ volts.

The line joining points A and B is the required dc load line.

The point of intersection of load line and diode characteristic gives the quiescent point Q. From Fig. 1.60, quiescent point Q gives $I = 50$ mA and $V_D = 2$ volts.

So diode voltage $V_D = 2$ volts **Ans.** and diode current, $I = 50$ mA **Ans.**

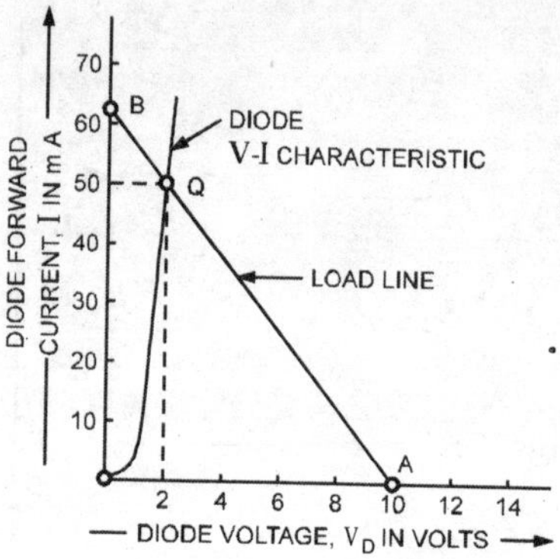

Fig. 1.60

1.17. P-N DIODE APPLICATIONS

An ideal P-N junction diode is a two terminal polarity sensitive device that has zero resistance (*i.e.,* diode conducts) when it is forward biased and infinite resistance (*i.e.,* diode does not conduct) when reverse biased. Because of this property the diode finds use in many applications as enumerated below :

(*i*) As rectifiers in dc power supplies.

(*ii*) In demodulation or detector circuits.

(*iii*) In clamping networks employed as dc restorers in TV receivers and voltage multipliers.

(*iv*) In clipping circuits used as waveshaping circuits in computers, radars, radio and TV receivers.

(*v*) As switches in digital logic circuits.

The same P-N junction with different doping concentrations finds special applications as follows :

(*i*) As zener diodes in voltage regulators, peak clippers, and in switching operation.

(*ii*) As tunnel diodes as a relaxation oscillator at microwave frequencies.

(*iii*) As light-emitting diodes (LEDs) in digital displays.

(*iv*) As LASER diodes in optical communications.

(*v*) As varactor diodes in tuning sections of radio and TV receivers.

(*vi*) As detectors (PIN photodiode) in optical communication circuits.

1.18. DIODE DATA SHEETS

For selection of a diode for a particular application, the data sheets provided by the device manufacturers must be consulted.

Most data sheets start off with the device type number at the top of the sheet, a short descriptive title and a list of major applications of the device. The information is usually followed by mechanical data in the form of illustration showing the package shape and dimensions. The absolute maximum ratings of the diode at 25°C are listed next. These are maximum voltages, currents, temperatures etc., that the device can bear/carry without breaking down. It is very important, for the safe operation of the device, that the maximum ratings given in the data sheets are never exceeded. For reliability the maximum ratings should not even be approached. Also, the maximum ratings must be adjusted downward for operation at temperatures exceeding 25°C. Following the absolute maximum ratings, there is normally a complete list of electrical characteristics for the device. Again, these are specified at 25°C, and allowances are necessary for variation of temperature.

Some of the important parameters are considered below :

1. *Peak reverse voltage or peak inverse voltage, V_R or V_{RM}*. This is the absolute peak of voltage that must be applied in reverse across the diode.

2. *Steady-state forward current, I_0 or I_F*. This is the maximum current that may be passed continuously through the diode.

3. *Peak forward surge current, $I_{FM\ (surge)}$*. The surge current is very much higher than the normal maximum forward current. This rating defines the maximum value and the time interval for such surges in current level. It is a current that may flow briefly when a circuit is first switched on.

4. *Peak repetitive forward current* is the maximum instantaneous value of repetitive forward current. Note that since it is at this level for a brief period of time, its level can be higher than the continuous level.

5. *Average rectified current.* A half-wave-rectified signal has an average value defined by $I_{av} = 0.318\ I_{peak}$. The *average current rating* is lower than the *continuous* or *peak repetitive* forward currents because a half-wave current waveform will have instantaneous values much higher than the average value.

6. *Static maximum voltage drop, $V_{F\ (max)}$*. This is the maximum forward voltage drop for a given forward current and device temperature.

7. *Continuous power dissipation at 25°C, P.* This is the maximum power that the device can safely dissipate on a continuous basis in free air. This rating must be downgraded for operation at higher temperatures, and may be upgraded when the device is mounted on a heat sink.

8. *Reverse recovery time, t_{rr}*. This is the maximum time for the device to switch from on to off.

Depending on the type of diode considered, additional data may also be provided, such as frequency range, noise level, switching time, thermal resistance levels, and peak repetitive values.

Appendix 1.1 shows a data sheet for rectifier diodes, which indicates that these devices can pass an average forward current of 1 A and have maximum reverse voltages ranging from 50 V to 1,000 V. The noteworthy point is that the data sheet does not bear the reverse recovery time t_{rr}. The reason is that rectifier diodes are generally intended for low frequency (50 Hz to perhaps 400 Hz) applications in which switching time is not important. Obviously, these devices can be described as relatively high-current diodes, and some are also high-voltage diodes.

Highlights

1. A diode is a unidirectional device permitting the easy flow of current in one direction but restraining the flow in opposite direction. A semiconductor diode is simply a P-N junction with connecting leads or terminals on its two sides. P-N junction is also called the crystal diode because it is grown out of crystal.

2. A space-charge region or depletion region is formed on either of the metallurgical junction separating the P and N junctions. This region is essentially depleted of any mobile electrons or holes. A net positive charge density, due to the positively charged donor impurity ions, exists in the N region and a negative charge density, due to the negatively charged acceptor impurity ions, exists in the P region. An electric field exists in the depletion region due to the net space charge density. The direction of the electric field is from the N region to the P region.

   A pd exists across the space-charge region. Under zero applied bias, the pd known as built-in potential barrier, maintains thermal equilibrium and holds back majority carrier electrons in the N-region and majority carrier holes in the P region.

   The pd existing across the space charge region is known as the contact potential. The built-in potential barrier or the contact potential for the step junction is given by

$$V_{bi} \text{ or } V_0 = V_T \log_e \frac{N_A N_D}{n_i^2}$$

3. A forward-biased PN junction has the positive terminal applied to the P side and the negative to the N side. The reverse is true for a reverse-biased junction.

   When a forward bias voltage is applied across a PN junction, the potential barrier is lowered so that holes from the P-region and electrons from the N region can flow across the junction.
4. An applied reverse bias voltage (N region positive w.r.t. the P region) increases the potential barrier, increases the space-charge width, and increases the magnitude of the electric field.
5. Metal-semiconductor junction can also form ohmic contacts, which are low-resistance junctions providing conduction in both directions with very little voltage drop across the junction.
6. There are two mechanisms by which breakdown can occur at a reverse biased P-N junction : avalanche breakdown and zener breakdown. The breakdown phenomenon is reversible and harmless so long as the safe operating temperature is maintained.
7. An ideal diode is a two terminal polarity sensitive device that has zero resistance when it is forward biased and infinite resistance when reverse biased.
8. The current in a P-N diode is bipolar in character (because it is made up of both positive and negative charge carriers) and the total current is constant throughout the device, but the proportion of current due to holes and that due to electrons varies with the distance.
9. In the forward-bias condition the diode current increases exponentially with the increase in voltage across the diode. Diode current, when the diode is forward biased, is given by

$$I = I_0(e^{V/V_T} - 1) \text{ where } V_T = \frac{T}{11{,}600}$$

and $I_0 = Ae\left[\frac{D_h}{L_h N_D} + \frac{D_e}{L_e N_A}\right] n_i^2$

Reverse saturation current may be also given by

$$I_0 = A_0 T^3 e^{-V_{GO}/V_T}$$

10. The reverse saturation current I_0 will just double in magnitude for every 10-fold increase in temperature. The temperature coefficient for the forward voltage of a P-N junction is approximately –1.8 mV/°C for silicon diode and –2.02 mV/°C for germanium diode.
11. The static and dynamic resistances of a diode may be determined from the characteristic curves and represent the dc and ac opposition to the flow of current offered by the diode.

    The dc resistance of a diode is determined by the ratio of the diode voltage and current at the point of interest and is not sensitive to the shape of the curve. The dc resistance decreases with the increase in diode current or voltage.

    The ac resistance of a diode is sensitive to the shape of V-I characteristic curve in the region of interest and decreases for higher level of diode current or voltage.
12. An equivalent circuit of a diode is a linearization of the diode's characteristic curve and typically consists of an ideal diode in series with a battery (representing the cut-in, offset or threshold voltage) and a resistance equal to the dynamic resistance of the diode over its linear portion.

    The cut-in or threshold voltage is about 0.7 V for silicon diodes and 0.3 V for germanium diodes.

Exercises

1. What is a crystal diode ? Why is silicon preferred over germanium in the manufacture of semiconductor devices ?
2. What is P-N junction diode ? How potential barrier is formed in a P-N junction diode ?

   [M.D.U. Rohtak Electrical Engg. Mat. and Semicon. Devi. 2006-07]

3. Explain the formation of potential barrier in a P-N junction diode. Derive an expression for contact potential. Describe qualitatively how the width of this layer changes with the applied voltage.

[M.D. Univ. Rohtak Analog Electronics May 2006]

4. Discuss physics of P-N junction and analyse it when it is reverse and forward biased.

[Pb. Technical Univ. Electronic Devices & Circuits May 2007]

5. Draw the energy band diagram for a P-N junction under open-circuited condition. Clearly indicate various energy levels in P-region, space region and N-region. How will it be modified if P-N Junction is forward biased ?

6. Discuss the formation and volt-amp characteristics of P-N junction. What happens when a P-N junction is biased in (*i*) forward direction, (*ii*) backward direction ?

7. What is a knee voltage and what is its significance ?

[M.D. Univ. Rohtak Basic Electronics December 2009]

8. Discuss behaviour of P-N junction diode under forward and reverse biasing.

[R.G. Technical Univ. Basic Electrical & Electronics Engineering Dec. 2010]

9. Explain P-N junction diode in forward and reverse bias. Explain V-I characteristics of a P-N junction diode. **[M.D. Univ. Rohtak Analog Electronics December 2005]**

10. Explain the breakdown mechanism in a diode. **[G.G.S.I.P. Univ. Analog Electronics December 2009]**

11. Explain how the process of avalanche breakdown occurs in a P-N junction diode. How is it different from zener breakdown ? **[Pb. Technical Univ. Basic Electrical and Electronics Engg. Dec. 2008]**

12. Differentiate between zener breakdown and Avalanche breakdown.

[R.G. Technical Univ. Basic Electronics Dec. 2010]

13. What is a diode ? Explain its forward and reverse bias characteristics. Discuss the breakdown mechanism of semiconductor diode. **[R.G. Technical Univ. Basic Electronics April 2009]**

14. Draw and explain V-I characteristics of diode with the help of diode equation of current.

[M.D. Univ. Rohtak, Basic Electronics December 2009]

15. Explain how diode works as on and off switch. Also derive diode current equation.

[M.D. Univ. Analog Electronics December 2006]

16. What is an ideal diode ? How does an actual diode differ from an ideal diode ?

[Pb. Technical Univ. Pulse and Digital Switching Circuits December 2004]

17. Plot the hole current, the electron current, and the total current as a function of distance on the both sides of a P-N junction. Indicate the transition region.

18. Sketch typical forward and reverse characteristics for a germanium diode and for a silicon diode. Compare the characteristics and explain why the reverse saturation current in a silicon diode is much smaller than that in a comparable germanium diode.

[U.P. Technical Univ. Electronics Engg. First Semester 2008-09]

19. Discuss working of an P-N junction diode circuit element. Name the processes involved in causing junction breakdown if it is reverse biased and explain V-I characteristics for Ge and Si-diodes for all regions. **[Pb. Technical Univ. Electronic Devices & Circuits May 2007]**

20. Explain the effect of temperature on V-I characteristics of a P-N junction diode.

[G.G.S.I.P. Univ. Analog Electronics December 2009]

21. Using diode equation show that the dynamic resistance of an ideal P-N junction diode under forward bias condition is inversely proportional to the forward current.

[G.G.S.I.P. Univ. Analog Electronics-I, December 2004]

22. Draw the piecewise linear model and V-I characteristics of a diode.

[G.G.S.I.P. Univ. Analog Electronics, December 2009]

23. What is meant by dc and ac resistance of a crystal diode ? How the diode resistances can be determined from its V-I characteristics ? **[M.D. Univ. Rohtak Analog Electronics December 2008]**

24. Explain the following terms :

(*i*) ideal diode (*ii*) knee voltage (*iii*) dynamic resistance (*iv*) reverse saturation current (*v*) zener breakdown (*vi*) avalanche breakdown. **[M.D. Univ. Rohtak Analog Electronics May 2007]**

25. How does dynamic resistance of a diode vary with current ? Draw the piecewise linear characteristic of a diode. **[M.D. Univ. Rohtak Analog Electronics May 2006]**

26. Sketch the piecewise linear characteristics of a diode. What are the approximate cut-in voltages for silicon and germanium ? **[U.P. Technical Univ. Electronics Engg. First Semester 2007-08]**

27. Consider a circuit consisting of diode D, a resistance R and a signal source v_i in series. Define

(1) Static characteristics, (2) Dynamic characteristics, (3) Transfer characteristics.

[Rajasthan Technical Univ. 2007]

Short Answer Type Questions With Answers

Q. 1. What is a P-N junction ?

Ans. The contact surface between the layers of P-type and N-type semiconductor pieces placed together so as to form a P-N junction is called the P-N junction.

Q. 2. Why is silicon preferred over germanium in the manufacture of semiconductor devices ?

Ans. The silicon semiconductor devices have, in general, higher PIV and current ratings and wider temperature range than germanium semiconductor devices and, therefore, silicon is preferred over germanium in the manufacture of semiconductor devices.

Q. 3. What does the arrowhead represent in a schematic symbol of a P-N junction ?

Ans. The arrowhead in the schematic symbol of a P-N junction indicates the direction of conventional current flow when the diode is forward biased.

Q. 4. How diode acts as a rectifier ? **[Pb. Technical Univ. May 2006]**

Ans. Rectifier is a device which converts the sinusoidal ac voltage into unidirectional pulsating voltage. Diode is an electronic device consisting of two elements known as cathode and anode. Since in a diode electrons can flow in only one direction *i.e.,* from cathode to anode, the diode provides the unilateral conduction necessary for rectification. This is true for diodes of all types—vacuum, gas-filled, crystal or semiconductor diodes.

Q. 5. In a semiconductor diode P-side is grounded and N-side is applied a potential of –3 V. Will the diode conduct ? Explain. **[P.T.U. May 2008]**

Ans. The *p-n* diode connected with *p*-side grounded and with *n*-side to –ve potential of 3 V, the diode will conduct, because *n*-side is at negative potential w.r.t. positive side.

Q. 6. What is potential barrier in a diode ? How it gets established ? **[Pb. Technical Univ. December 2006]**

Ans. The free electrons crossing the P-N junction create negative ions on the P-side by giving some atoms one more electron than their total number of protons. The electrons also leave positive ions behind them on the N-side. As negative ions are created on the P-side of the junction, the P-side acquires a negative potential. Similarly, the positive ions are created on N-side and the N-side acquires a positive potential. The negative potential on the P-side prevents the migration of any more electrons from the N-type material to the P-type material. Similarly, the positive potential on the N-side prevents any further migration of holes across the boundary. Thus, the initial diffusion of charge carriers results in a barrier potential at the junction.

Q. 7. How do the transition region width and contact potential across a P-N junction vary with the applied bias voltage ?

Ans. When the P-N junction is forward biased, the transition region width is reduced and the contact potential is also reduced with the increase in applied bias voltage.

When the P-N junction is reverse biased, the transition region is widened, and the contact potential is increased with the increase in applied bias voltage.

Q. 8. Which type of charges are present on the two opposite faces of the junction ?

Ans. Positive charge on N-side and negative charge on P-side of the junction.

Q. 9. What types of carriers are present in space charge region ?

Ans. No mobile carrier is present in the space charge region.

Q. 10. Why is space region called the depletion region ?

Ans. The region around the junction is completely ionised on formation of P-N junction. As a result, there are neither free electrons on the N-side nor the holes on the P-side. Since the region around the junction is depleted of mobile charges, it is called the depletion region.

Q. 11. Why an electric field is produced in a depletion region of a P-N junction ?

Ans. The separation of positive and negative space charge densities in a P-N junction results in an electric field.

Q. 12. What is space-charge width ?

Ans. The space-charge region extends into the N and P-regions from the metallurgical junction. The distance is known as the space-charge width.

Q. 13. The electric field in the space-charge region decreases with forward bias and increases with reverse bias. Why ?

Ans. Because applied electric field opposes built-in field.

Q. 14. Define cutin voltage of a P-N junction diode ? **[U.P. Technical Univ. First/Second Semester 2001-02]**

Ans. The forward voltage, at which the current through the P-N junction starts increasing rapidly, is called the cutin voltage.

Q. 15. What do you understand by reverse saturation current of a diode ?

Ans. Reverse saturation current of a diode is due to minority carriers and is caused when the diode is reverse biased. Only a very small voltage is required to direct all available minority carriers across the junction, and when all minority carriers are flowing across, further increase in bias voltage will not cause increase in current. This current is referred to as a reverse saturation current.

Q. 16. Define peak inverse voltage ?

Ans. Peak inverse voltage is the maximum voltage that can be applied to the P-N junction without damage to the junction. If the reverse voltage across the junction exceeds its peak inverse voltage (PIV), the junction may get destroyed owing to excessive heat.

Q. 17. Define the limitations in the operating conditions of P-N junction.

Ans. Every P-N junction has limiting values of (*i*) maximum forward current (*ii*) peak inverse voltage (PIV) and (*iii*) maximum power rating. The P-N junction provides satisfactory performance when operated within these limiting values. The P-N junction diode may get destroyed due to excessive heat if any of these values is exceeded.

Q. 18. What is a junction diode ? Draw ideal diode characteristics. **[Pb. Technical Univ. May 2007]**

Ans. Diode is a two terminal device which conducts exponentially in one direction and has a poor conduction in reverse direction. It will be an ideal case if its conduction in one direction is like a perfect conductor and in other direction as perfect insulator. Ideal diode characteristic is shown in Fig. 1.14 (*b*).

Q. 19. What is the main difference between the characteristic of a simple switch and those of an ideal diode ? **[Pb. Technical Univ. December 2005]**

Ans. The important difference between the characteristic of a simple switch and those of an ideal diode is that an ideal diode acts like an automatic switch—the switch is closed when the diode is forward biased and is opened when reverse biased.

Q. 20. Name the breakdown mechanism in a lightly doped P-N junction under reverse biased condition.

Ans. Avalanche breakdown.

Q. 21. Name the breakdown mechanism in a heavily doped P-N junction under reverse biased condition.

Ans. Zener breakdown.

Q. 22. What is reverse saturation current ?

Ans. Reverse current of a diode is due to minority carriers and is caused when the diode is reversed biased. The reverse current is constant, independent of the applied reverse bias, consequently it is called the reverse saturation current.

Q. 23. Is reverse saturation current of a junction diode independent of reverse bias voltage ?

Ans. Yes.

Q. 24. Germanium is more temperature dependent than silicon. Why ?

Ans. Because the reverse saturation current in case of a germanium diode is approximately 1,000 times larger.

Q. 25. What is the effect of temperature on the reverse saturation current of a diode ?

Ans. Reverse saturation current, theoretically, increases by 8 per cent per °C for silicon and 11 per cent per °C for germanium. But from experimental data it is found that the reverse saturation current increases 7 per cent per °C for both silicon and germanium. This is because a surface leakage current component of reverse saturation current is independent of temperature. Since $(1.07)^{10} \simeq 2.0$, the reverse saturation current approximately doubles for every 10°C rise in temperature.

Q. 26. Explain the effect of temperature on the V-I characteristics of the diode.

Ans. Reverse saturation current I_0 and volt equivalent of temperature V_T increase with the increase in temperature. I_0 doubles for every 10°C rise in temperature. This causes lowering of threshold or cutin voltages. With the increase in temperature, forward characteristic becomes more ideal. In reverse bias, the breakdown voltage increases with the increase in temperature. Refer to Fig. 1.20.

Q. 27. What is static resistance of a diode ?

Ans. The static or dc resistance of a diode is the resistance offered by it to the direct current. It is defined as the ratio of the diode voltage and current at the point of interest and is not sensitive to the shape of V-I characteristic curve. It decreases with the increase in diode current or voltage.

Q. 28. Define dynamic resistance of a P-N junction diode in forward biased condition.

Ans. The resistance offered by a P-N junction diode to the changing forward current is defined as the dynamic resistance.

It may be given as

$$\text{AC or dynamic resistance, } r = \frac{\text{Small change in forward voltage}}{\text{Small change is forward current}}$$

Thus ac resistance of a diode is sensitive to the shape of *v-i* characteristic curve in the region of interest. It decreases for higher level of diode current or voltage.

Q. 29. Explain what do you understand by the expression "linearization of a nonlinear device".

Ans. A nonlinear device, such as a diode, has a nonlinear voltage-current characteristic. In a diode V and I are related by equation

$$I = I_0 (e^{V/\eta V_T} - 1)$$

Diode shows nonlinearity until the threshold or cutin voltage V_T is reached. Voltage-current characteristics of the diode follow a linear relationship after it has attained the cutin voltage, just as for a resistance of very low magnitude. It means that diode exhibits nonlinearity for lower voltages ($V < V_T$) but linearity for voltages exceeding V_T.

The equivalent circuit for the linear approximation is a constant-voltage source in series with a resistor [Fig. 1.23 (*b*)].

Problems

1. Determine the built-in potential of a silicon P-N junction at 300 K with doping density of $N_A = 1 \times 10^{16}$ cm^{-3} and $N_D = 1 \times 10^{15}$ cm^{-3} for intrinsic concentration of 1.5×10^{10} cm^{-3}. **[Ans.** 0.635 V]

2. A germanium diode draws 50 mA with a forward bias of 0.27 V. The junction is at room temperature of 27ºC. Determine the reverse saturation current of the diode. **[Ans.** 1.545 μA]

3. Determine the reverse saturation current density in an abrupt silicon diode with following data. $N_D = 1.25 \times 10^{21}$ m^{-3}; $N_A = 1.05 \times 10^{22}$ m^{-3}; $D_e = 3.4 \times 10^{-3}$ m^2 s^{-1}; $D_h = 1.2 \times 10^{-3}$ m^2 s^{-1}; $L_e = 6.8 \times 10^{-4}$ m ; $L_h = 2.8 \times 10^{-4}$ m; $n_i = 1.6 \times 10^{16}$ m^{-3}; $e = 1.6 \times 10^{-19}$ C. **[Ans.** 0.16 μA]

4. Determine the voltage at which the reverse current in a germanium P-N junction diode attains a value of 95% of its saturation value at room temperature of 300 K. Assume V_T = 2.6 mV. **[Ans.** –0.078 V]

5. Determine the forward resistance of a silicon diode, when the forward current is 4 mA at T = 300 K. **[Ans.** 12.93 Ω]

6. Determine ac resistance for a semiconductor diode with a forward bias of 0.25 V. Reverse saturation current at room temperature is of 1.2 μA. **[Ans.** 1.445 Ω]

7. Determine the current flowing in the circuit shown in Fig. 1. **[Ans.** 1.94 mA]

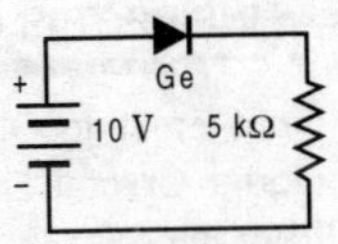

Fig. 1

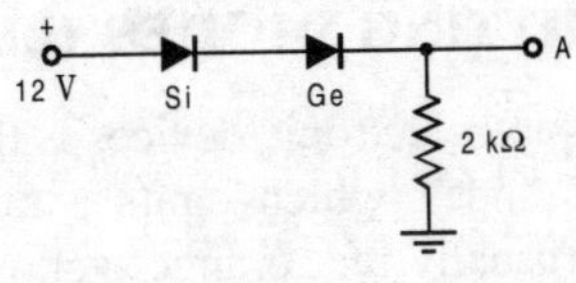

Fig. 2

8. Determine the current flowing in the circuit shown in Fig. 2. Also determine the potential of point A. **[Ans.** 5.5 mA, 11 V]

9. A silicon diode has a saturation current of 5 μA at room temperature of 300 K. Determine its value at 400 K. **[Ans.** 5.12 mA]

□□□

CHAPTER

2

SPECIAL PURPOSE DIODE

2.1. LIGHT-EMITTING DIODES (LEDs)

INSIDE THIS CHAPTER

The best known of all optoelectronic devices is the so called LED (lights-emitting diode), which emits a fairly narrow bandwidth of visible (usually red, orange, yellow or green) or invisible (infrared) light when its internal diode junction is stimulated by a forward electric current/voltage (power). LEDs have typical power-to-light energy conversion efficiencies some 10 to 50 times greater than of a simple tungsten lamp and have very fast response times (approximately 0.1 μs compared with tens or hundreds of milliseconds for a tungsten lamp), and are thus widely employed as visual indicators and as moving light displays.

The operation of light-emitting diode (LED) is based on the phenomenon of electroluminance, which is the emission of light from a semiconductor under the influence of an electrical field. The recombination of charge carriers takes place in a forward P-N junction as the electrons cross from the N-region and recombine with holes existing in P-region. Free electrons are in the conduction band of energy levels, while holes are in the valence energy band. Therefore, the electrons are at high energy levels than the holes. For the electrons to recombine with the holes, they must give some of their energy. Typically, these electrons give up energy in the form of heat and light.

In silicon and germanium diodes, most of the electrons give up their energy in the form of heat. However, with GaAsP, and GaP semiconductors, the electrons give up their energy by emitting photons. If the semiconductor is transluscent, the light will be emitted and the junction becomes a source of light, *i.e.,* a *light-emitting diode* (LED). LEDs emit no light when *reverse biased*. In fact, operating LEDs in reverse bias mode will quickly destroy them.

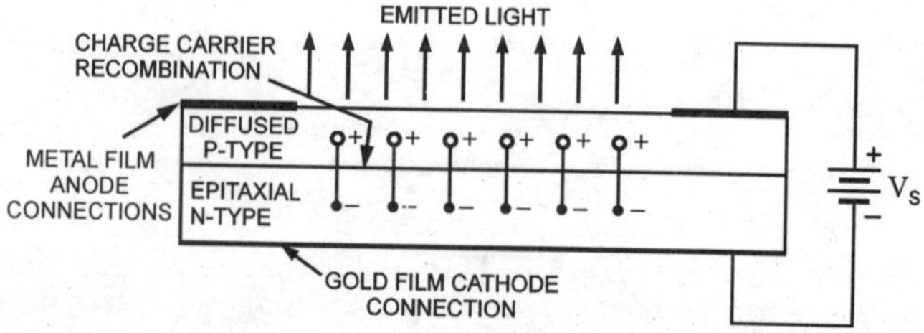

Fig. 2.1. *LED*

A cross-sectional view of a typically diffused LED is given in Fig. 2.1. The semiconductor material used is gallium arsenide (GaAs), gallium arsenide phosphide (GaAsP) or gallium phosphide (GaP). An N-type epitaxial layer is grown upon a substrate, and the P-region is produced by diffusion. Recombination of charge carriers occur in the P-region so this region is required to be kept at the top. Thus, the P-region becomes the device surface. The metal anode connections are made at the outer edges of P-layer so as to allow more surface area for the light to come out. A gold film is applied to the substrate bottom to reflect as much as possible of the light toward the surface of the device and to provide a cathode connection. LEDs are manufactured with domed lenses so as to reduce the reabsorption problem. LEDs are always encased in order to protect their delicate wires. LEDs made from GaAs emit invisible infrared light, LEDs constructed of GaAsP tend to emit either red or yellow light. The GaP LEDs give either red or green light. LEDs can be manufactured to produce light of virtually any colour. Table 2.1 provides a list of common compound semiconductors and the light they produce. In addition, the typical range of forward bias voltage for each is listed.

TABLE 2.1. Light-Emitting Diodes

Colour	Construction	Typical Forward Voltage in Volts
Amber	AllnGaP	2.1
Blue	GaN	5.0
Green or red	GaP	2.2
Yellow	GaAsP	2.2
Red	GaAsP	1.8
White	GaN	4.1
Yellow	AllnGaP	2.1

Figure 2.2 (*a*) shows a circuit symbol for a LED. The arrows indicate radiation emitted by the diode. Figs. 2.2 (*b*) and 2.2 (*c*) give two curves used to determine LED operating characteristics. Figure 2.2 (*b*) is forward bias V-I curve for a typical LED employed in burglar alarms. Forward bias of approximately one volt is required to give significant forward current. Figure 2.2 (*c*) provides radiant power-forward current curve. The radiant output power is rather small and indicates a very low efficiency of electrical to radiant energy conversion.

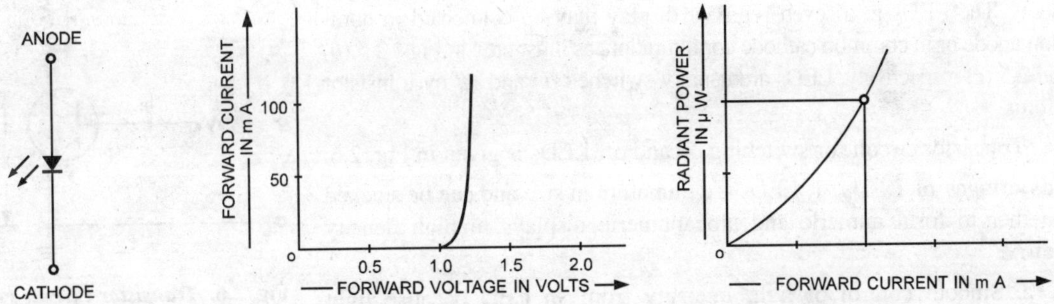

(*a*) (*b*) *Forward Bias V-I Curve* (*c*) *Output Characteristic Curve*

Fig. 2.2. ***LED***

Figure 2.3 shows a source connected to a series resistor R_S and a LED. The outward arrows symbolise the radiated light. The forward resistance of LED is very low which means that once the forward bias exceeds, the current through it will increase rapidly for only a very small increase in voltage. Thus it becomes necessary to use an external series current limiting resistor. The value of series resistor R_S can be determined from the following equation

$$R_S = \frac{V_S - V_D}{I} \quad ...(2.1)$$

where V_S is the supply voltage, V_D is LED forward bias voltage and I is the desired current.

For most of the commercially available LEDs, the typical voltage drop (diode voltage) is from 1.5–2.5 V for currents between 10 and 50 mA. The exact voltage drop depends on the LED current, colour, tolerance, etc.

A LED can be used as an indicator in ac circuit by wiring it in inverse parallel with a normal diode, as shown in Fig. 2.3, to prevent the LED being reverse biased; for a given brightness, the value of R_s should be halved relative to that of the dc circuit.

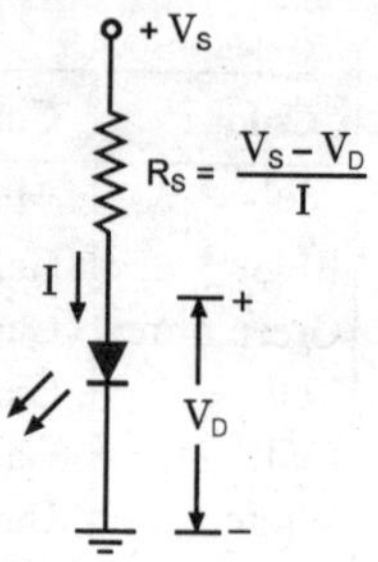

Fig. 2.3. *LED Circuit*

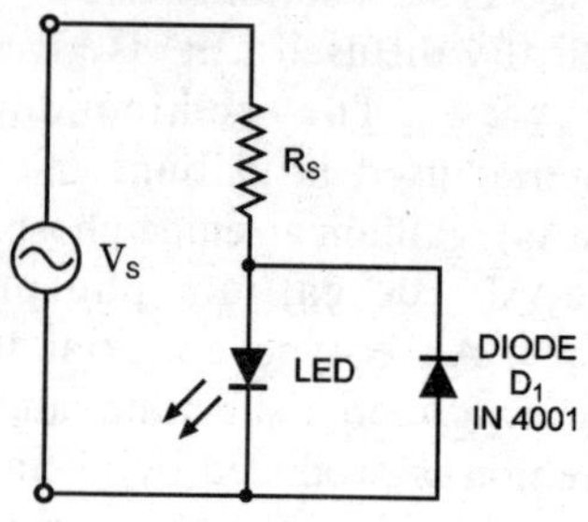

Fig. 2.4. *LED As An Indicator in AC Circuit*

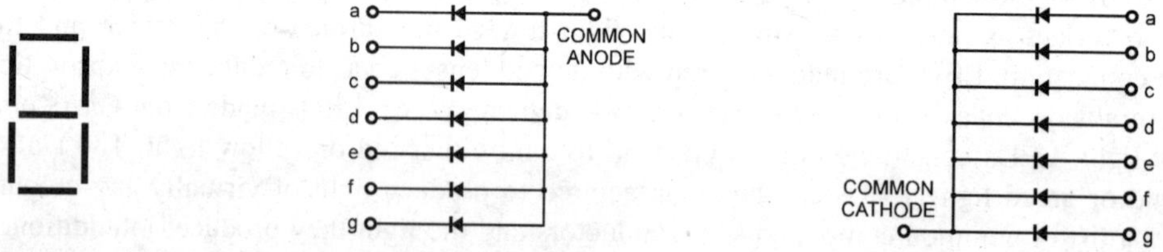

(*a*) *Display Format* (*b*) *Common Anode Configuration* (*c*) *Common Cathode Configuration*

Fig. 2.5

LED displays are available in many different sizes and shapes. The light-emitting region is available in lengths from 2.5 to 25 mm. Numbers can be created by such segments. Typical arrangement of a seven segment LED numerical display is shown in Fig. 2.5 (*a*). Any desired numeral from 0 to 9 can be displayed by passing current through the appropriate segments. The LEDs in a seven segment display may be connected in common anode or in common cathode configuration, as illustrated in Figs. 2.5 (*b*) and 2.5 (*c*) respectively. LEDs are usually switched *on* and *off* by transistor circuit.

Transistor circuit for switching on and off LEDs is given in Fig. 2.6.

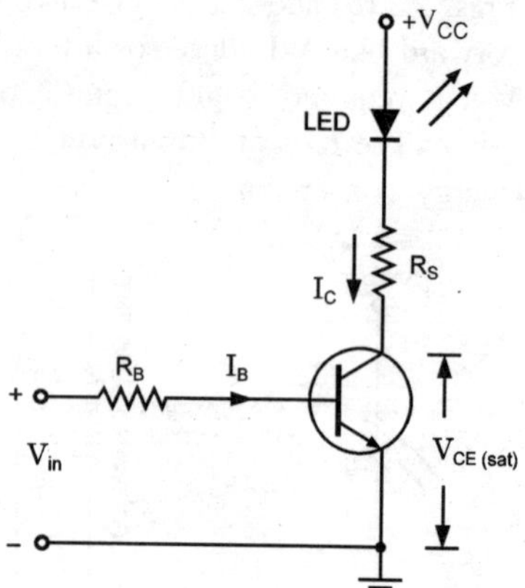

Fig. 2.6. *Transistor Circuit For Switching On and OFF a LED*

Advantages of LEDs. 1. LEDs are miniature in size and can be stacked together to form numeric and alphanumeric displays in high density matrix.

2. Smooth control of light intensity from an LED because light output from an LED is a function of current flowing through it.
3. LEDs are economical and have a high degree of reliability because they are manufactured with the same type of technology as is used for transistors and ICs.
4. LEDs are rugged and can, therefore, withstand shocks and vibrations.
5. LEDs can be operated over a wide temperature range (0–70ºC).
6. The switching time (both on and off) is less than 1 ns. So they are very suitable where dynamic operation of large number of arrays is involved.
7. LEDs are available in different colours like red, yellow, green and amber.
8. They need moderate power. They are used where low dc power is available.
9. They occupy small area.

10. LED devices can be driven by transistor-transistor logic (TTL) whereas gas discharge devices need intermediate transistor stages to allow them to be driven by 5 V TTL.

The *drawbacks of LEDs* are that they may get damaged by overvoltage or overcurrent. They have wide optical bandwidth compared to the LASER (≃ 10 nm). Their temperature depends on the radiant output power and wavelength. LEDs are not suited for large area displays, primarily because of their high cost. For large displays, devices using gas filled plasma are used.

LEDs that emit visible light are widely used in instrument display panel indicator, digital watches, calculators, multimeters, intercom, telephone switchboards, etc. LEDs that emit invisible infrared light find application in remote control schemes, object detectors, burglar alarm systems; and other areas requiring invisible radiation.

The LEDs used for fiber optics are usually of the gallium arsenide (GaAs) type, with various dopants added to shift the center wavelength of the radiation spectrum. Dopants used are phosphorus (P), indium (In), and aluminium (Al). Gallium arsenide phosphide (GaAsP) diodes can be made with bandgaps in the range of 1.5 to 2.0 eV (0.62 – 0.83 μm wavelength). Gallium indium arsenide (GaInAs), indium arsenide phosphide (InAsP), or aluminium indium arsenide (AlInAs) diodes cover the range from 0.5 to 1.5 eV (0.83 to 2.5 μm). Some of these structures may also be used to form laser diodes.

Example 2.1. An LED is connected across a voltage source of +10 V through a series resistance of 820 Ω. Calculate the LED current. Assume the voltage drop across an LED of 1.5 V.

[*U.P. Technical Univ. Analog & Digital Electronics 2009-10*]

Solution : Supply voltage, $V_S = 10$ V

Series resistance, $R_S = 820\ \Omega$

Voltage drop across LED, $V_D = 1.5$ V

$$\text{LED current, } I = \frac{V_S - V_D}{R_S} = \frac{10 - 1.5}{820} = 10.4 \text{ mA} \quad \textbf{Ans.}$$

2.2. LIQUID CRYSTAL DISPLAYS

Another type of displays is liquid crystal displays (LCDs) that are used in similar applications where LEDs are used. The applications of LCDs are display of numeric and alphanumeric characters in segmental and dot matrix displays.

Liquid crystals have been called "the fourth state of matter" (after solids, liquids and gases) because they have certain crystal properties normally found in solids, yet flow like liquids. Unlike LEDs and other electroluminescent devices, LCDs do not generate light energy, but simply alter or control existing light to make selected areas appear bright or dark.

The liquid crystal display has the distinct advantage of having a low power consumption than the LED. It is typically of the order of microwatts for the display in comparison to the some order of milliwatts for LEDs. Low power consumption requirement has made it compatible with MOS integrated logic circuit. Its other advantages are its low cost and good contrast. The main drawbacks of LCDs are additional requirement of light source, a limited temperature range of operation (between 0 and 60ºC), low reliability, short operating life, poor visibility in low ambient lighting, slow speed and the need for an ac drive.

There are two basic ways in which liquid crystals are used for controlling properties of light and thereby change its appearance. Accordingly, there are two types of LCD displays—*dynamic-scattering display* and *field-effect display.*

When *dynamic scattering display* is energized, the molecules of energized area of the display become turbulent and scatter light in all directions. Consequently, the activated areas take on a frosted glass appearance resulting in a silver display. Of course, the unenergized areas remain translucent.

Field effect LCD contains front and back polarizers at right angles to each other. Without electrical excitation, the light coming through the front polarizer is rotated 90° in the fluid, passes through the rear polarizer, and is reflected back by the mirror. When an electrostatic field is applied, the LC fluid molecules rotate 90° so that the light is not rotated 90° and is, therefore, absorbed by the rear polarizer. This causes the appearance of dark digits on a light background.

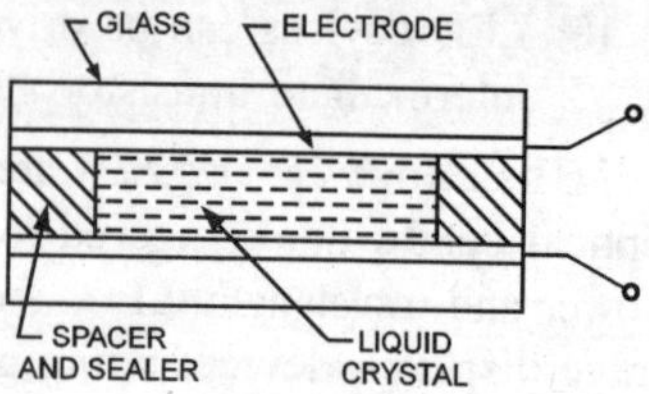

Fig. 2.7. ***Basic Construction of an LCD***

The basic construction of a liquid crystal display is depicted in Fig. 2.7. A thin layer of liquid crystal, 10-20 μm thick, is sandwiched between two glass sheets. To enable application of an electric field across the crystal, a transparent conducting material is deposited on the inside surface of each glass sheet. The conducting material forms the electrodes to which an external voltage is connected when it is desired to change the molecular structure of the liquid crystal. The electrodes are etched in patterns or in individually accessible segments that can be selectively energized to develop a desired display, similar to a seven segment LED display. Materials used to form the electrodes include *stannic-oxide* (SnO_2) and *indium oxide* (In_2O_3).

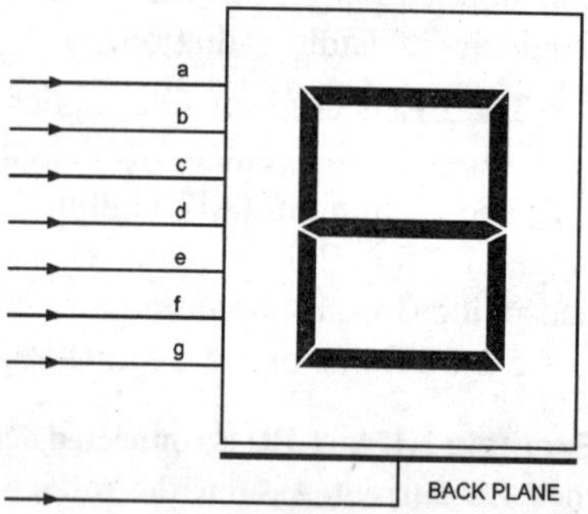

Fig. 2.8

Some LCDs are designed so that light passes completely through them, while others have a mirrored surface that reflects light back to the viewer. In the former case, known as the *transmissive* mode of operation, light originates from a light source on one side of the crystal and is changed in the desired pattern as it passes through to emerge on the other side. In the second case, referred to as *reflective mode,* light enters one side, is changed within the material, and is reflected by a mirror to emerge from the same side it entered. This mode is particularly useful for small displays like watches, where ambient light is present on one side only.

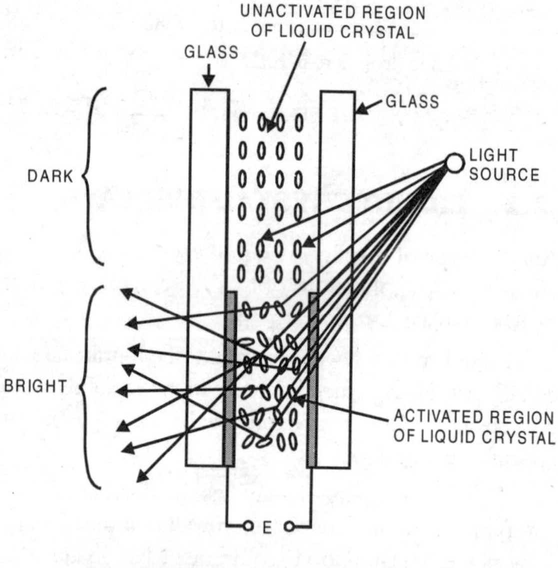

Fig. 2.9. ***Dynamic-Scattering LCD in Transmissive Mode Operation***

A dynamic scattering LCD operated in the transmissive mode is depicted in Fig. 2.9. In the region where no electric field is applied, the rod-shaped molecules have the common alignment shown, and in the region that has been activated by an externally applied field, the molecules have random orientations. The light source is so displaced that the unactivated region appears dark. In the energized region, the molecules make the light to be reflected in many directions (scattered), and it escapes with a bright diffuse appearance. In this way, the pattern appears bright on a dark back ground. A special light-control film can be placed on the illuminated glass to permit the use of a diffuse light source.

A dynamic-scattering LCD operated in the reflective mode has essentially the same construction as depicted in Fig. 2.9, except that a mirrored surface replaces or is added behind one of the glass sheets. However, undesired reflections limit the readability of such displays, which must, therefore, be fitted with special *retroreflectors* in practical applications.

LCDs that depends on light absorption rather than light scattering is constructed using so-called *twisted-nematic* crystals. The electrodes in such displays are coated so that the molecules of the liquid crystal undergo a 90° rotation from one side of the cell to the other. Consequently, light passing through the material undergoes a 90° change in polarization. However, if the crystal is energized by an external electric field, the molecules are reoriented in such a way that there is no alteration in polarization. The absorption type LCD operates on the principle that horizontally polarized light is not visible through a vertically polarizing filter, *i.e.,* it is absorbed. As depicted in Fig. 2.10, light enters the LCD (in transmission mode) through a vertical polarizer and can exit through another vertical polarizer provided the liquid crystal is energized. In the unactivated region, the vertically polarized light becomes horizontally polarised and is absorbed in the vertical filter. Thus, the unactivated region appears dark and the activated region appears bright. Since it is the electric field resulting from an externally applied voltage that reorients the molecules, the twisted-nematic displays are usually known as field-effect type LCD displays.

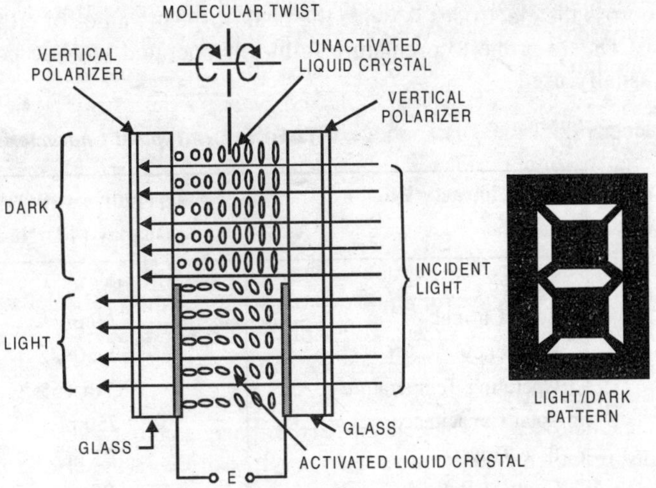

Fig. 2.10. ***Twisted-Nematic Field-Effect LCD Based on Light Absorption : Transmissive Mode Operation***

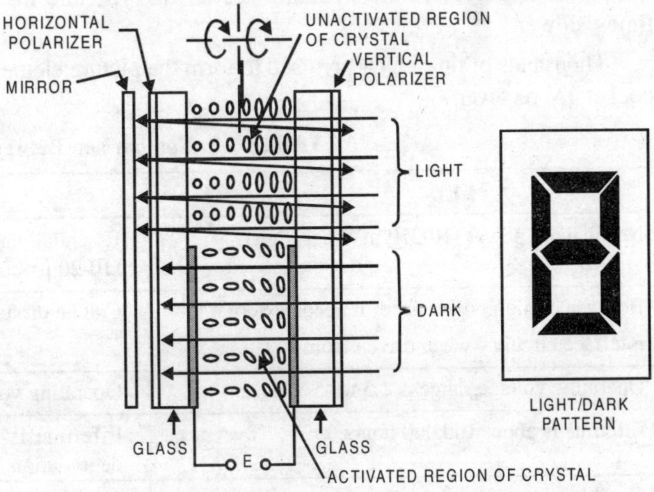

Fig. 2.11. ***Twisted-Nematic Field-Effect LCD Based on Light Absorption : Reflective Mode Operation***

A light absorbing, field-effect LCD operated in reflective mode is shown in Fig. 2.11. Like transmissive LCD, incident light is vertically polarized, but in this mode a horizontal polarizer is placed between one of the glass sheets and a reflector. The vertically polarized light becomes horizontally polarized while passing through an unactivated region of the crystal. This light can pass through the horizontal polarizer and is reflected back into the crystal. The unactivated crystal then shifts the polarization back to vertical, so the light emerges through the vertical polarizer where it entered. The energized region of the crystal does not change the vertical polarization, so light passing through it cannot pass through the horizontal polarizer and is not reflected. As a result, the pattern produced in the energised region appears dark on a light background.

Practical LCDs are energized by ac waveforms because dc energization results in a plating effect shortening the lifetime. The dynamic scattering type needs a relatively high voltage, 25-30 V at 50-60 Hz. One method used to energize the crystal is to drive each of the opposing electrodes with a square wave. If the square waves are in phase, then the net voltage across the crystal is zero, and it remains de-energized. When the driver circuit causes the square waves to be 180° out of phase, the net voltage across the electrodes is twice the peak-to-peak values of either wave, and the crystal is energized. Most LCDs are prone to premature failure if operated with dc potentials, hence an ac square-wave drive is usually used.

TABLE 2.2. ***Typical Characteristics of LCDs***

Characteristic	Dynamic-Scattering Display (4 Digits)	Twisted-Nematic Display (4½ Digits)
Voltage	15–30 V	3 V
Total Current	60 μA	6 μA
Frequency	50–60 Hz	30–1,500 Hz
Operating Temperature	–5 °C to +65 °C	–15 °C to +60 °C
Total Capacitance	750 pF	2 nF
Rise Time	25 ms	100 ms
Contrast Ratio	20 : 1	50 : 1

The noteworthy point is that the twisted-nematic type LCD display needs about one-tenth, the voltage and current of the dynamic-scattering type and has better contrast, but is approximately four times slower.

Thousands of tiny LCDs are used to form the picture elements (pixels) of the screen in one type of B&W pocket TV receiver.

TABLE 2.3. Comparison Between LED and LCD

LED	LCD
Needs more power (10-250 mW per digit).	Essentially acts as a capacitor and needs very less power (10-20 μW per digit).
Because of high power need, it needs external interface circuitry when driven from ICs.	Can be driven directly from IC chips.
Operating voltage range is 1.5 to 5 V dc.	Operating voltage range is 3–20 V ac.
Lifetime is about 100,000 hours.	Lifetime is limited to 50,000 hours due to its chemical degradation.
Good brightness level.	Moderate brightness level.
Temperature range for operation is –40 to 85°C	Temperature range limited to –20 to 60°C.
Emits light in red, orange, yellow, green, blue and white.	Invisible in darkness–needs external or internal light source.
Response time is 50-500 ns	Response time is 50-200 ms. Because of slow decay time, one can even observe the fading out of segments when switched off.
Viewing angle is 150°.	Viewing angle is 100°.
Font style : 7-16 segments; dot matrix.	Font style : 7-16 segments; dot matrix.

Important Features of LCDs. LCDs are characterised by low power consumption, low cost, large area and low operating speed. They have the following important features.

1. LCDs are very slow devices and have a turn-on time of a few milliseconds, and a turn-off time of tens of milliseconds.

2. As the light source for a reflective LCD is the ambient light itself, the only power required is that needed to cause turbulence in the cell, which is very small, typically 1 μW/cm.
3. The materials of nematic liquid crystals (NLC) which is the most popular liquid crystal structure, has high resistivity (exceeding 10^{10} Ω). So the current requirement for scattering light in an NLC is very marginal (typically 0.1 μA/cm^2).
4. The electric field needed to activate LCDs is typically of the order of 1000 V/mm. This is equivalent to an LCD terminal voltage of 10 V when the NLC layer is 10 μm thick.
5. Unlike LED displays, which are usually quite small, liquid crystal displays (LCDs) can be fabricated in almost any convenient size. The maximum power consumed for a typical LCD used in electronics equipment is about 20 μW per segment, or 140 μW per numeral when all seven segments are energized. Comparing this value to about 400 mW per numeral for a LED display (including series resistor).

2.3. PHOTODIODES

Photodiode is a two-terminal semiconductor P-N junction device and is designed to operate with reverse bias. The operating principle of a semiconductor photodiode is that when a reverse-biased P-N junction is illuminated, the current varies almost linearly with the light flux. The device consists of a P-N junction embedded in a clear plastic, as shown in Fig. 2.12. Radiation is allowed to fall upon one surface across the junction. The remaining sides of the plastic are either painted black or enclosed in a metallic case. The entire unit is extremely small and has dimensions of the order of 25 mm. The basic biasing arrangement and symbols for the device are given in Fig. 2.13.

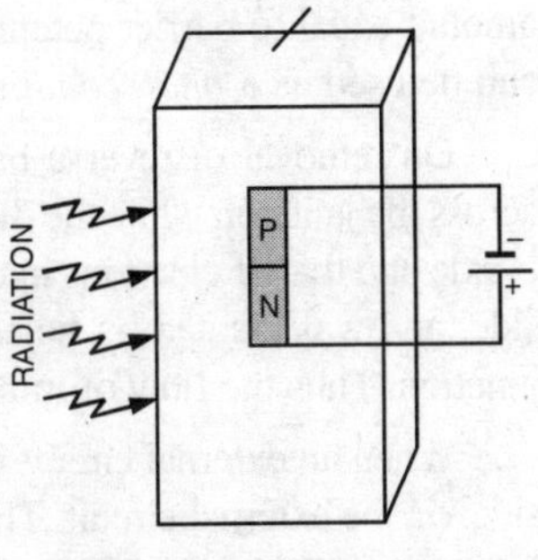

Fig. 2.12. ***Semiconductor Diode Construction***

The output voltage is taken from across a series-connected load resistor R. This resistance may be connected between the diode and ground or between the diode and the positive terminal of the supply, as illustrated in Fig. 2.13.

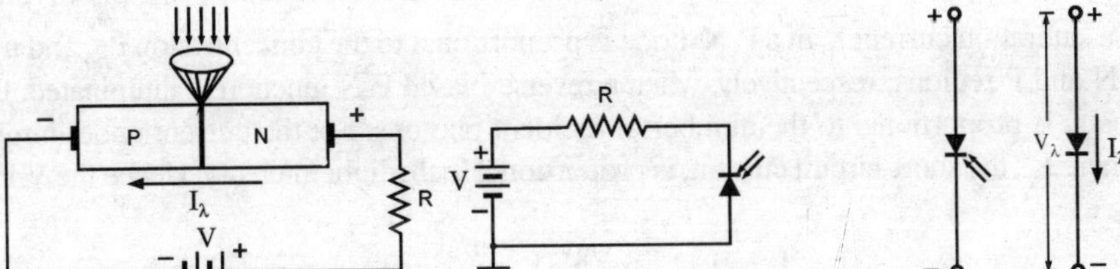

(a) Basic Biasing Arrangement *(b) Symbols*

Fig. 2.13

When the P-N junction is reverse biased, a reverse saturation current flows due to thermally generated holes and electrons being swept across the junction as the minority carriers. With the increase in temperature of the junction more and more hole-electron pairs are created and so the reverse saturation current I_0 increases. The same effect can be had by illuminating the junction. When light energy bombards a P-N junction, it dislodges valence electrons. The more light striking the junction the larger the reverse current in a diode. It is

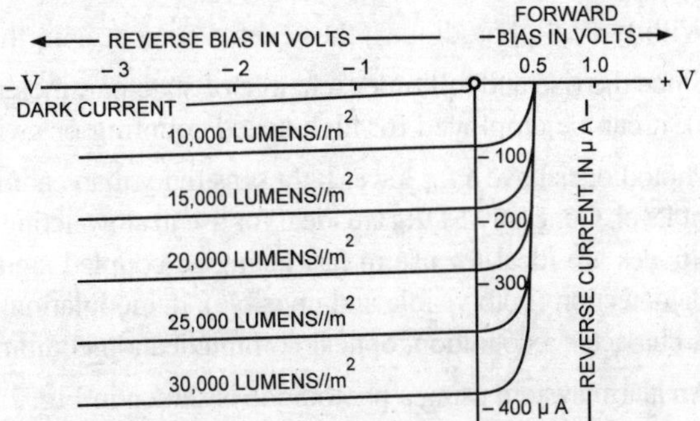

Fig. 2.14. ***Photodiode Characteristics***

due to generation of more and more charge carriers with the increase in level of illumination. This is clearly shown in Fig. 2.14 for different intensity levels. The *dark current* is the current that exists when no light is incident. It is to be noted here that current becomes zero only with a positive applied bias equals to V_0. The almost equal spacing between the curves for the same increment in luminous flux reveals that the reverse saturation current I_0 increases linearly with the luminous flux as shown in Fig. 2.15. Increase in reverse voltage does not increase the reverse current significantly, because all available charge carriers are already being swept across the junction. For reducing the reverse saturation current I_0 to zero, it is necessary to forward bias the junction by an amount equal to barrier potential. Thus the photodiode can be used as a *photoconductive device*.

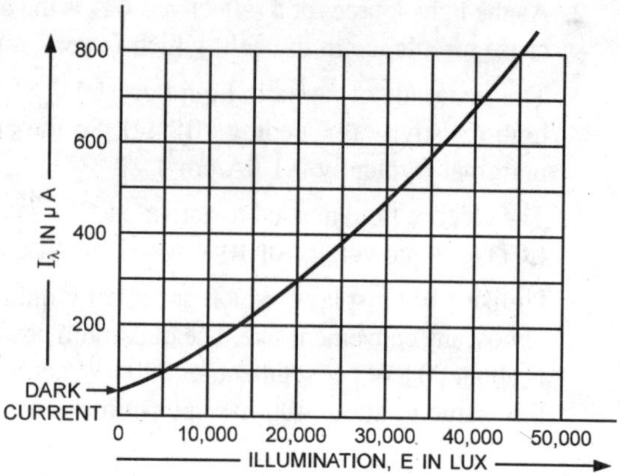

Fig. 2.15

On removal of reverse bias applied across the photodiode, minority charge carriers continue to be swept across the junction while the diode is illuminated. This has the effect of increasing the concentration of holes in the P-side and that of electrons in the N-side. But the barrier potential is negative on the P-side and positive on the N-side, and was created by holes flowing from P to N-side and electrons from N to P-side during fabrication of junction. Thus the flow of minority carriers tends to reduce the barrier potential.

When an external circuit is connected across the diode terminals, the minority carriers return to the original side via the external circuit. The electrons which crossed the junction from P to N-side now flow out through the N-terminal and into the P-terminal. This means that the device is behaving as a voltage cell with the N-side being the negative terminal and the P-side the positive terminal. Thus, the photodiode is a *photovoltaic* device as well as *photoconductive* device. When the device is operated with a reverse bias, it operates as a photoconductive device and when operated without the reverse bias, it operates as a photovoltaic device.

The reverse saturation current I_0 in a P-N diode is proportional to the concentration p_{no} and n_{po} of minority carriers in the N and P regions, respectively, when a reverse-biased P-N junction is illuminated, the number of hole-electron pairs is proportional to the number of incident photons. So, the current under large reverse bias is $I = I_0 + I_{sc}$, where I_{sc}, the short-circuit current, is proportional to the light intensity. Hence the V-I characteristic is given by

$$I = I_{sc} + I_0(1 - e^{V/\eta V_T}) \qquad ...(2.2)$$

where I, I_{sc} and I_0 represent the magnitude of the reverse current, and V is positive for a forward voltage and negative for a reverse bias. The parameter η is unity for germanium and 2 for silicon, and V_T is the volt equivalent of temperature.

With typical photodiodes, the reverse current is in the tens of microamperes.

Since the rise and fall times (change of state parameters) are very small for this device (in the nanosecond range), it can be employed for high speed counting or switching applications.

Photodiodes have a far lower light sensitivity than cadmium sulphide LDRs, but give a fair quicker response in light level. Generally LDRs are ideal for use in slow acting direct coupled light-level sensing applications, while photodiodes are ideal for use in fast acting ac coupled signalling applications. Typical photodiode applications include detection (both visible and invisible), demodulation, switching, logic circuits that need stability and high speed, character recognition, optical communication equipment, IR remote control circuits, encoders etc.

An alarm system using a photodiode is shown in Fig. 2.16 (*a*). The reverse current I_λ will continue to flow as long as the light beam is not interrupted. When the light beam gets interrupted, I_λ falls to the dark current level and sounds the alarm.

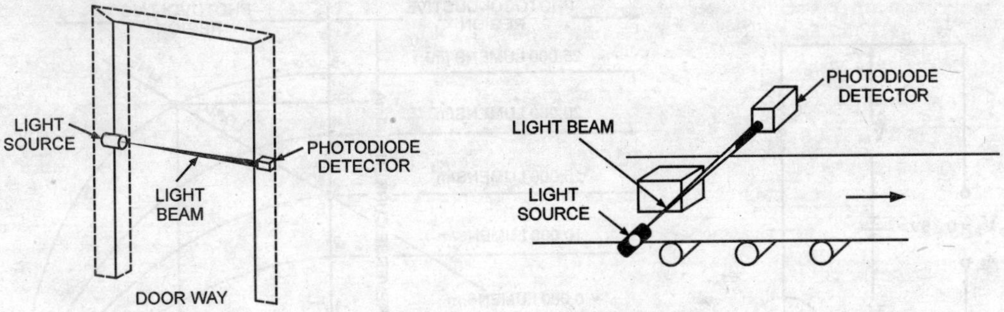

(a) Alarm System Using a Photodiode *(b) Counter Operation Using a Photodiode*

Fig. 2.16

In Fig. 2.16 (*b*) a photodiode is used to count items on a conveyor belt. As each item passes, the light beam is interrupted, I_λ falls to the dark current level and the counter is increased by one count.

Example 2.2. A photodiode is connected as shown in Fig. 2.17. When the device is illuminated with light intensity of 100 lumens/cm², a current of 1.8 mA flows in the circuit. Calculate the resistance offered by the photodiode. If the illumination is doubled, what happens to the resistance value ?

[*U.P.S.C.I.E.S. Electronics & Telecommunication Engineering-I, 2003*]

Solution : Photodiode current, $I_d = 1.8 \text{ mA} = 1.8 \times 10^{-3}$ A

Voltage applied, $V_s = 0.4$ V

Series resistor, $R_1 = 100\ \Omega$

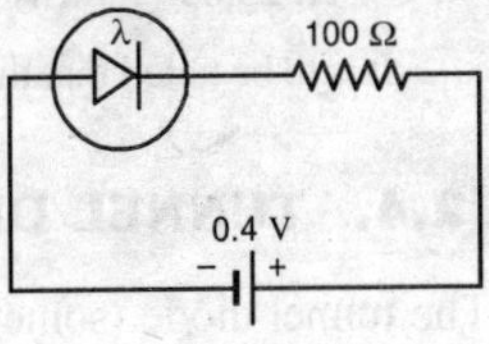

Fig. 2.17

If R is the resistance offered by the photodiode in ohms then

$$I_d = \frac{V_s}{R + R_1}$$

$$\text{or } R = \frac{V_s}{I_d} - R_1 = \frac{0.4}{1.8 \times 10^{-3}} - 100 = 122.22\ \Omega \ \textbf{Ans.}$$

With double illumination, photodiode current I_d will be doubled *i.e.*, $2 \times 1.8 \times 10^{-3} = 3.6 \times 10^{-3}$ A and now resistance offered by photodiode is given as

$$R' = \frac{V_s}{2I_d} - R_1 = \frac{0.4}{3.6 \times 10^{-3}} - 100 = 11.11\ \Omega \ \textbf{Ans.}$$

Example 2.3. A photodiode is connected in a circuit shown in Fig. 2.18 (*a*). The supply polarity reverse biases the device. The characteristics of the photodiode are shown in Fig. 2.18 (*b*). Draw the 250 Ω dc load line for the circuit and determine the diode currents and voltages at 2,000 and 25,000 lumens/m² illumination.

Solution : When diode current, $I_d = 0$

Voltage across resistance R_1 (250 Ω) $= V_{R1} = I_d R_1 = 0$

and diode voltage, $V_d = V_S$,

Supply voltage $= -0.75$ V

When diode voltage, $V_d = 0$

Voltage across 250 Ω resistance, $V_{R1} = V_S = -0.75$ V

$$\text{and diode current, } I_d = \frac{V_{R1}}{R_1} = \frac{-0.75}{250} = -3.0 \text{ mA}$$

The two points ($V_d = -0.75$ V, $I_d = 0$) and ($V_d = 0$, $I_d = -3.0$ mA) are plotted as points A and B respectively. The straight line joining points A and B is 250 Ω load line. **Ans.**

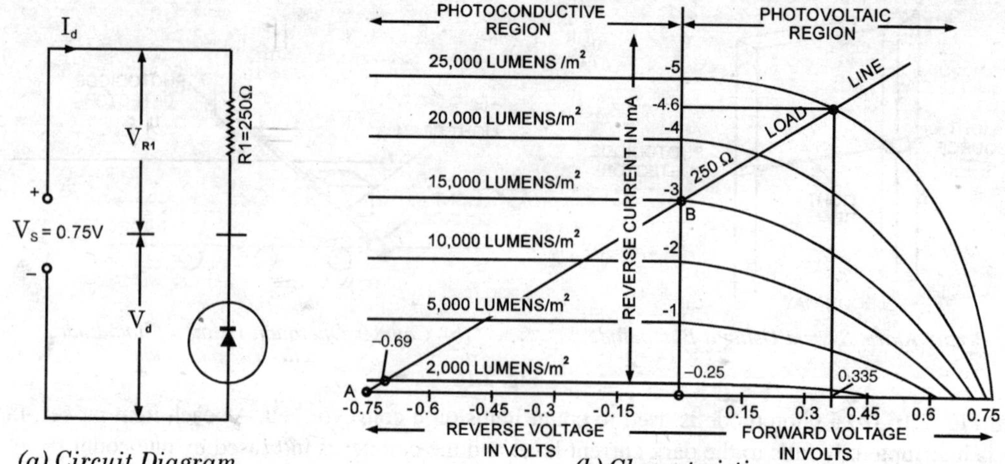

(a) Circuit Diagram *(b) Characteristics*

Fig. 2.18

From the load line we get,

At 2,000 lumens/m^2 : I_d = – 0.25 mA; V_d = – 0.69 V **Ans.**

At 25,000 lumens/m^2 : I_d = – 4.6 mA; V_d = 0.35 V **Ans.**

It is to be noted that the polarity of the diode, voltage V_d, changes from – ve to + ve at the highest level of illumination.

2.4. TUNNEL DIODES

The tunnel diode (sometimes called the *Esaki diode* after its inventor, Dr. Leo Esaki) is a high conductivity, heavily doped two terminal P-N junction. This diode differs from the normal P-N junction diode in the amount of impurity concentration. The impurity concentration in a normal P-N junction diode is about 1 part in 10^8, whereas in the tunnel diode the impurity concentration is about 1 part in 10^3. Because of heavy doping depletion layer width is reduced* to an extremely small value of the order of 10^{-5} mm : reverse breakdown voltage is also reduced to a very small value (approaching zero) resulting in appearance of the diode to be broken for any reverse voltage, and a –ve resistance section is produced in volt-ampere characteristic of the diode. It is found that the reduced depletion layer can result in carriers '*punching through*' the junction with the velocity of light even when they do not possess enough energy to overcome the potential barrier. The result is that large forward current is produced at relatively low forward voltage (less than 100 mV). Such a mechanism of conduction in which charge carriers (possessing very little energy) punch through a barrier directly instead of climbing over it is called *tunneling*. That is why such diodes are called the *tunnel diodes*. Because of heavy doping the tunnel diode can conduct in reverse as well as in forward direction but it is usually used in forward-biased mode.

Construction. Tunnel diodes are usually fabricated from germanium, gallium arsenide or gallium antimonide. Silicon is not used in the construction of tunnel diodes because the ratio of peak value of forward current to the value of valley current (*i.e.*, I_P/I_V) is maximum in case of gallium arsenide (approximately 20) and comparatively smaller for germanium (roughly 10) but very small in case of silicon (about 3).

Tunnel diodes are ordinarily manufactured by alloying from gallium arsenide. The source materials are highly doped semiconductor crystals with an impurity concentration of the order of 10^{25} per cubic metre or more. Arrangement is shown in Fig. 2.19.

* The width of junction barrier varies inversely as the square root of impurity concentration.

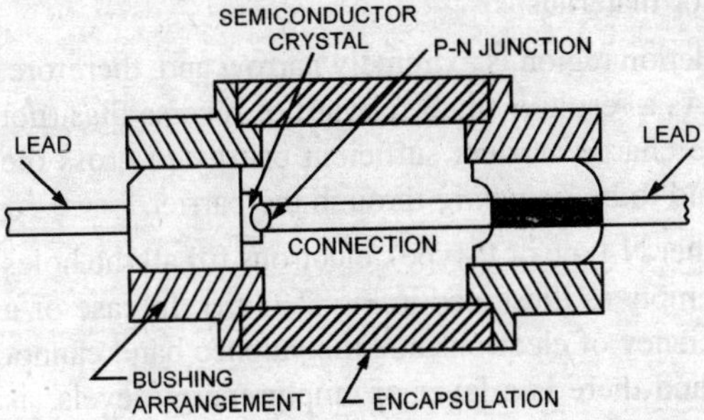

Fig. 2.19

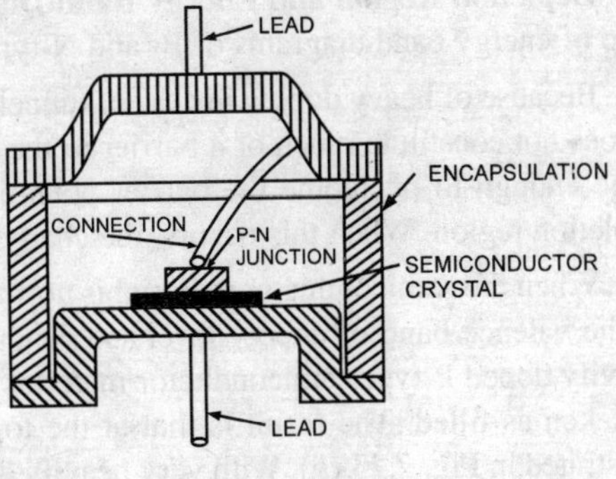

Fig. 2.20. ***Pellet Tunnel Diode***

An alternative arrangement is shown in Fig. 2.20. Such a tunnel diode is called the *pellet tunnel diode.*

It is to be noted that the tunnel diode, being a low power device, can be easily damaged by heat and static electricity.

The commonly used schematic symbols for the tunnel diodes are shown in Fig. 2.21.

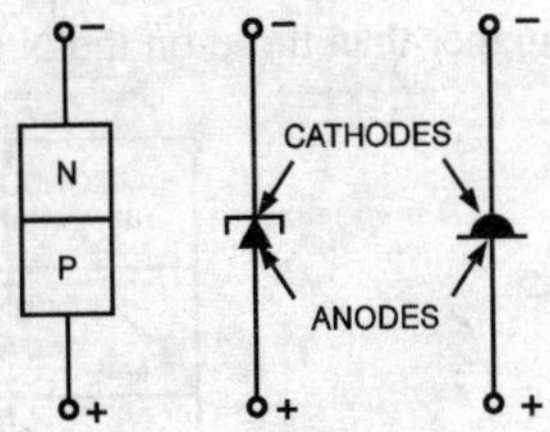

Fig. 2.21. ***Tunnel Diode Symbols***

Volt-Ampere Characteristic. Forward V-I characteristic of a tunnel diode is illustrated in Fig. 2.22. Forward bias produces immediate conduction—as soon as forward bias is applied, significant current is produced. The current attains quickly its peak value, I_P when the applied forward bias attains a value of V_P volts. The current variation in the vicinity of origin is due to quantum mechanical tunneling of electrons through narrow space charge region of the junction. With the further increase in forward voltage, the diode current starts decreasing till it attains its minimum value, called the *valley current* I_V corresponding to *valley voltage* V_V. Thus from peak point A to valley point B the diode current falls with the increase in voltage resulting in *negative resistance* in this region. In fact this portion A B of the characteristic constitutes the most useful property of the diode. In this region the diode, instead of absorbing power, produces power. So the tunnel diode can be used as a very high-frequency oscillator. For voltages higher than valley voltage V_V current starts increasing as in any normal diode.

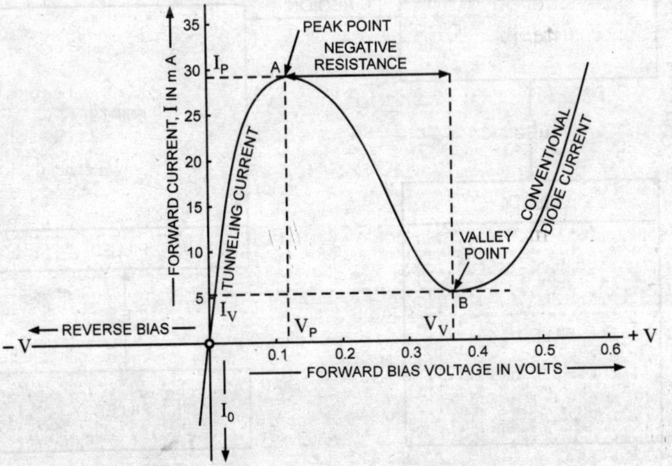

Fig. 2.22. ***V-I Characteristic of Tunnel Diode***

Depletion Region and Energy Band Diagrams. The tunneling phenomenon can be explained with the help of energy band diagrams of P- and N-type semiconductor materials.

Because of heavy doping in case of tunnel diode, the depletion region is extremely narrow and, therefore, it does not constitute much of a barrier to the electron flow. As a result, a small forward or reverse bias (not large enough to overcome the barrier potential) can provide charge carriers sufficient energy to cross the depletion region. When this occurs, the charge carriers are said to be *tunneling* through the barrier.

When the semiconductor material is normally doped (either N-type or P-type), electrons fill all the holes in the valence band of energy level and conduction band is empty as illustrated in Fig. 2.23 (*a*). In case of a heavily doped P-type semiconductor material there is a deficiency of electrons and the valence band cannot be taken as filled. The result is that at the top of valence band there is a layer of empty energy levels, as illustrated in Fig. 2.23 (*b*). With very heavily doped N-type semiconductor material, there is an abundance of electrons. The result is that electrons fill the valence band and create a layer of filled energy level at the bottom of the conduction band, as illustrated in Fig. 2.23 (*c*).

The energy band diagram for a heavily doped unbiased P-N junction is shown in Fig. 2.24. Note that the depletion region is very narrow and the filled levels on the N-side are exactly opposite those on the P-side. In this condition, no tunneling occurs as there are no empty lower energy levels to which electrons from either side might cross the depletion region. It is also to be noted that the conduction and valence bands on P-side are (negatively) higher than those on the N-side. This is due to formation of depletion region and barrier potential as a result of

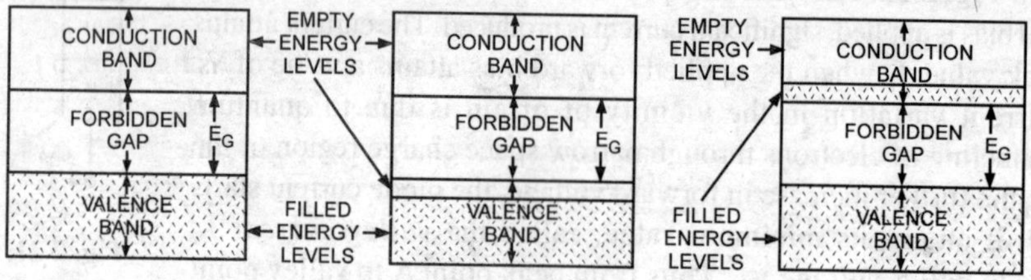

(a) Normally Doped P and N-Type Materials *(b) Very Heavily Doped P-Type Material* *(c) Very Heavily Doped N-Type Material*

Fig. 2.23. ***Energy Band Diagrams For Normally Doped and For Very Heavily Doped Semiconductor Materials***

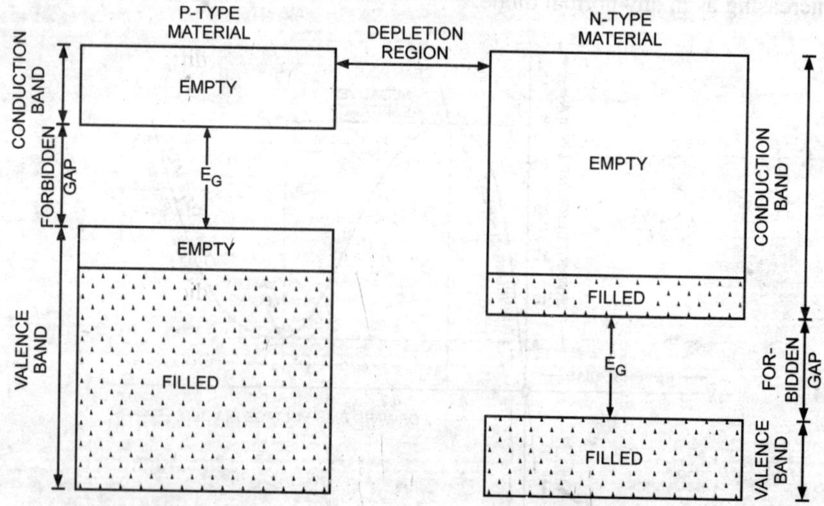

Fig. 2.24. ***Energy Band Diagrams For a Tunnel Diode***

electrons crossing from the N-region to the P-region. The N-region has lost negative charges and P-region has gained them.

When the junction is *reverse biased* the P-region moves up with respect to the N-region. As a result, filled energy levels on the P-side become opposite empty energy levels on the N-side. At this stage electrons tunnel through the narrow space charge region from the higher energy levels on the P-side to the lower energy levels on the N-side. Despite the fact that the junction is reverse biased, significant current flows. With the increase in reverse bias, more and more electrons tunnel from the P-side to N-side and a larger current flows. Thus the reverse characteristic of a tunnel diode is linear, just like that of a resistor (Fig. 2.22).

When the tunnel diode is forward biased, its initial behaviour is similar to that when it is reverse biased. Now, some of the filled energy levels on the N-side shift to a high energy level than empty level on the P-side (Fig. 2.24). Electron tunneling now occurs from the N-side to the P-side. With the increase in forward bias, more and more electrons tunnel from the N-side to the P-side. Eventually, however, a maximum level of tunneling is reached when a band of filled energy levels at the bottom of conduction band on the N-side is directly opposite the band of empty energy levels at the top of the valence band on the P-side. At this stage diode current attains its peak value I_P and the forward bias is V_P ($\simeq$ 100 mV). With the further increase in forward bias, part of the band of filled energy levels on the N-side is raised to an energy level corresponding to forbidden energy gap on the P-side. Electrons cannot tunnel to a forbidden energy level; thus the current flow due to tunneling is reduced. With continued increase in forward bias, tunneling goes on reducing. When all of the band of filled levels at the bottom of the conduction band on the N-side is raised to a level corresponding to the forbidden energy gap on the P-side, the current flow due to tunneling is reduced to the minimum level, called the *valley current* I_V. Now, however, the normal process of current flow across the forward biased junction begins to take over, as the bias becomes large enough to overcome the barrier potential. For voltages exceeding valley voltage V_V, the current starts increasing as in a normal diode.

Diode Parameters. We have seen that between I_P and I_V (points A and B on the V-I characteristic of tunnel diode illustrated in Fig. 2.22) the diode current decreases with the increase in voltage applied across the diode. This region of the characteristic is called the *negative resistance region*, and the negative resistance R_N of the tunnel diode is its most important property. Its value may be determined by computing the slope of the characteristic in the negative resistance region. The value of negative resistance slightly differs from point to point on the negative resistance portion of the characteristic. So R_N is usually specified at the centre of the negative resistance region.

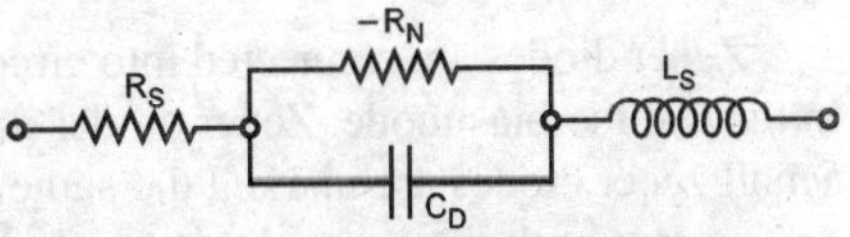

Fig. 2.25. ***Equivalent Circuit For a Tunnel Diode***

Parameters for a typical tunnel diode are given below :

Peak current I_P : 1 – 100 mA; Peak voltage V_P : 50 – 200 mV; Valley current I_V : 0.1 – 5 mA; Valley voltage V_V : 350 – 500 mV; Forward bias V : 0.5 – 1 V; Negative resistance R_N : –10 to –200 Ω.

Tunnel Diode Equivalent Circuit. The equivalent circuit for a tunnel diode biased in the negative resistance region is given in Fig. 2.25. R_S represents the resistance of the connecting leads to the tunnel diode and the semiconductor material, and is of the order of 5 Ω; L_S represents the inductance of the connecting leads and is of the order of 0.5 nH; and C_D is junction diffusion capacitance and is of the order of 5 to 100 pF. These factors limit the frequency at which the diode can be operated. Frequency may range from 100 MHz to 10,000 MHz.

Advantages and Disadvantages of Tunnel Diodes. The advantages of tunnel diodes are low cost, low noise, simplicity, high speed of operation, ease of operation, low power and environmental immunity.

The disadvantages of the tunnel diodes are its low output-voltage swing and being a two-terminal device, it cannot provide isolation between the input and output circuits.

Applications. The tunnel diode can be used as an amplifier, an oscillator, or a switch. Because of very fast response to inputs, it is almost exclusively a high frequency component. The tunnel diode is, however, not widely used because it is a low current device and also that other better devices are available for the operations.

2.5. ZENER DIODES

Zener diode, also sometimes called the *breakdown diode*, is a P-N junction diode specially designed for operation in the breakdown region in reverse bias condition. The diode may use either zener breakdown mechanism or avalanche breakdown mechanism, already discussed in detail in Art. 1.7. Despite the fact that avalanche and zener constitute two types of breakdown diodes, the name zener is commonly applied to both types of diodes. The breakdown diode may be silicon or germanium one but silicon is preferred over germanium because of higher operating temperature and current capability. The knee point is also more sharp in case of a silicon diode. These diodes are analogous to gas discharge tubes in which large current appears on reaching the breakdown potential.

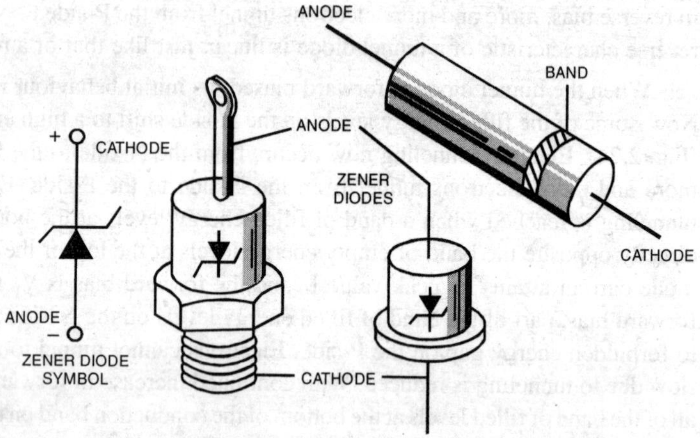

Fig. 2.26. ***Zener Diode***

Zener diodes are connected into circuits with their *cathodes* as the *positive* terminal because they operate in the reverse-bias mode. Zener diodes are packaged in the same way as conventional semiconductor diodes. Small zener diodes are marked the same way as conventional diodes with a cathode band. The larger diodes are marked with the zener diode symbol shown in the illustration (Fig. 2.26). In addition to have a voltage rating (reverse breakdown), zener diodes have a power rating that must not be exceeded.

The symbol of a zener diode is given in Fig. 2.26. This is similar to that of a normal diode except that the line representing the cathode is bent at both ends *i.e.,* the bar is turned into Z-shape; stands for zener.

2.5.1. Volt-Ampere Characteristics

It has already been discussed in Art. 1.7 that when the reverse bias on a crystal diode is gradually increased, a point is reached when the junction breaks down and a reverse current increases abruptly, as illustrated in Fig. 2.27. The breakdown region is the knee of the reverse characteristic, as illustrated in Fig. 2.27. This breakdown of junction was first explained by the American scientist, C. Zener. So, the breakdown voltage is sometimes called, *zener voltage* V_Z and the sharply increased current is called the *zener current* I_Z.

Volt-ampere characteristics of typical silicon and germanium diodes are illustrated in Fig. 2.27. Breakdown diode is operated over the breakdown region and the current is limited by an external resistance. The breakdown or zener voltage V_Z depends upon the amount of doping. A heavily doped diode will have very thin depletion layer, so zener voltage will be very low. By controlling the thickness of depletion layer and doping concentration it is possible to make the zener diode to break down at a specified voltage, called the zener voltage (ranging from 2.4 V to 200 V). The forward characteristics of a zener diode are similar to those of an ordinary P-N junction diode, as illustrated.

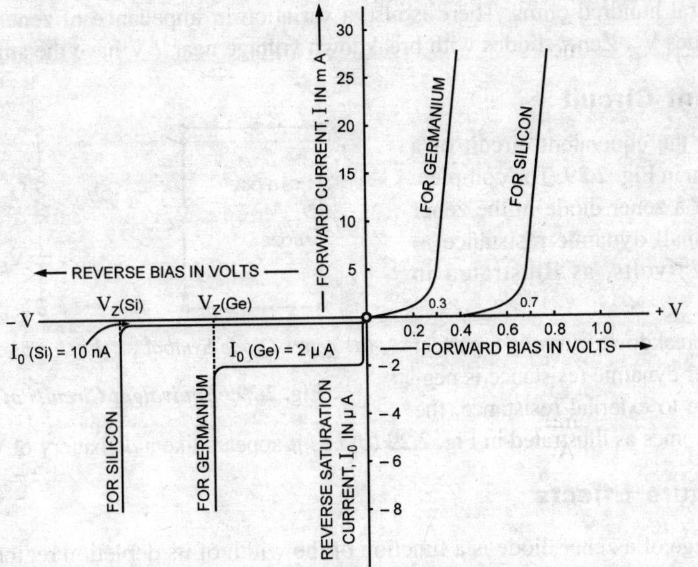

Fig. 2.27. ***Volt-Ampere Characteristics of Typical Silicon and Germanium Zener Diodes***

During the operation in the breakdown region, it does not burn out immediately. As long as the current through the diode is limited by the external circuit within permissible values, it does not burn out.

2.5.2. Zener Impedance

This is very important parameter of a zener diode. Zener impedance Z_Z is essentially the dynamic resistance of a zener diode (similar to dynamic resistance of an ordinary P-N diode). It is defined as the reciprocal of the slope of the zener curve

$$i.e., \; r_Z = \frac{\Delta V_Z}{\Delta I_Z} \qquad ...(2.3)$$

where ΔV_Z and ΔI_Z are the small variations in voltage and current respectively.

It is usually measured just above the knee in zener region and is named as Z_{ZK}. When measured at test point, it is designated Z_{ZT}. It decreases with increase in zener current. For this reason, the diode should be operated with as much reverse current as possible, consistent with rating limitations.

Small value of zener impedance is usually desirable as it indicates steep curve where relatively large change in current, ΔI_Z occurs with a small change in voltage ΔV_Z.

Ideally the breakdown curve should be a perfect vertical line and r_Z should be zero. In practice, however, r_Z has a small value ranging from

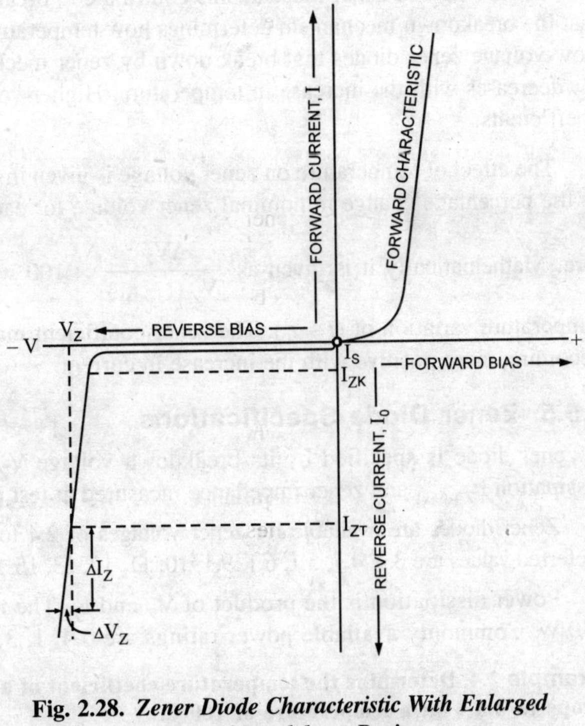

Fig. 2.28. ***Zener Diode Characteristic With Enlarged View of Breakdown Region***

a few ohms to several hundred ohms. There is also a variation in impedance of zener diodes among those having different values V_Z. Zener diodes with breakdown voltage near 7 V have the smallest impedances.

2.5.3. Equivalent Circuit

Circuit symbol and the equivalent circuit of a zener diode are given in Fig. 2.29. The complete equivalent circuit of a zener diode in the zener region includes a small dynamic resistance r_Z and a battery of V_Z volts, as illustrated in Fig. 2.29 (*b*).

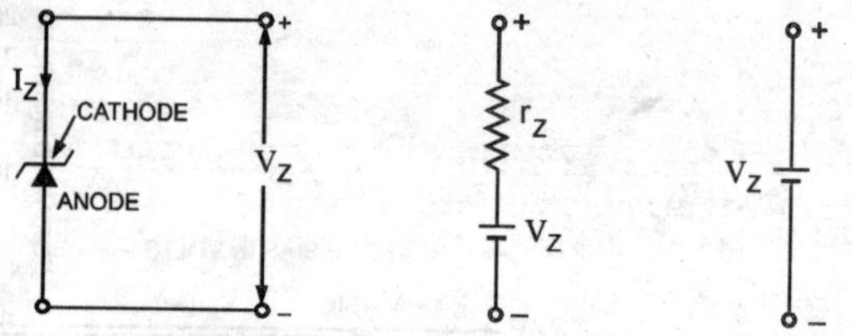

(*a*) *Zener Diode Symbol* (*b*) *Complete* (*c*) *Approximate*

Fig. 2.29. ***Equivalent Circuits of Zener Diode***

Assuming that breakdown curve is a perfect vertical line *i.e.*, zener dynamic resistance is negligible in comparison to external resistance, the equivalent circuit becomes as illustrated in Fig. 2.29 (*c*) *i.e.*, it appears like a *dc* battery of V_Z volts.

2.5.4. Temperature Effects

The breakdown voltage of a zener diode is a function of the width of its depletion region*, which is controlled during manufacturing by the degree of impurity doping. As discussed in chapter 1, the mechanism by which breakdown occurs depends on the breakdown voltage itself. For V_Z less than 5 V, the high electric field intensity across narrow depletion region (about 3×10^7 V/m) strips carriers directly from their bonds, a phenomenon usually known as zener breakdown. For V_Z exceeding 8 V, breakdown occurs as a result of collision between high-energy carriers, the mechanism known as *avalanching*. Between 5 V and 8 V, both the avalanching and zener mechanisms contribute to breakdown. The practical significance of these facts is that the breakdown mechanism determines how temperature variations affect the value of zener voltage V_Z. Low-voltage zener diodes that break down by zener mechanism have negative temperature coefficient *i.e.*, V_Z decreases with the increase in temperature. Higher-voltage avalanche zeners have positive temperature coefficients.

The effect of temperature on zener voltage is given in terms of temperature coefficient, which is defined as the percentage change in nominal zener voltage for each degree Celsius of change in junction temperature. Mathematically it is given as $\dfrac{\Delta V_Z}{V_Z(t - t_0)} \times 100$ where ΔV_Z is the change in zener voltage due to temperature variation of $(t - t_0)$. The zener coefficient may be positive or negative, depending on current, becoming more positive with the increase in current.

2.5.5. Zener Diode Specifications

A zener diode is specified by its breakdown voltage V_Z, breakdown current, I_{ZK}, the maximum power dissipation $P_{Z\,(max)}$ and zener impedance measured at test point, Z_{ZT}.

Zener diodes are available for zener voltages of 2.4 to 200 V with accuracies between 5 and 20%. The preferred values are 3.3, 4.7, 5.1, 6.1, 9.1, 10, 11, 12, 13, 15, to 200 V.

Power dissipation is the product of V_Z and I_Z. The maximum power rating varies from 150 mW to 250 W. Commonly available power ratings are 0.4, 1, 3, 10, and 50 watts.

Example 2.4. Determine the temperature coefficient of a 5 V zener diode at 25°C, if the nominal voltage drops to 4.8 V at a temperature of 100°C.

---

* Heavy doping increases conductivity, which narrows the depletion region and, therefore, reduces the voltage at which breakdown occurs.

Solution : Change in zener voltage, $\Delta V_Z = 4.8 - 5 = -0.2$ V

Change in temperature, $(t - t_0) = 100 - 25 = 75°C$.

$$\text{Temperature coefficient of zener diode} = \frac{\Delta V_Z}{V_Z\,(t - t_0)} \times 100 = \frac{-0.2}{5 \times 75} \times 100 = -0.0533\% \textbf{ Ans.}$$

Example 2.5. A zener diode has an impedance of 40 Ω in the range for I_Z = 1 mA to 10 mA. The voltage corresponding to I_Z of 1 mA is 9 V. For constant impedance over the given range, what minimum and maximum zener voltages can be expected if the diode is used in an application where the zener current varies from 2 mA to 8 mA ?

Solution : The voltage change between I_Z = 1 mA and I_Z = 2 mA,

$$\Delta V_Z = \Delta I_Z r_Z = (2 \text{ mA} - 1 \text{ mA}) \times 40\ \Omega = 0.04 \text{ V}$$

So minimum zener voltage, $V_{Z\,min} = V + \Delta V_Z = 9 + 0.04 = 9.04$ V **Ans.**

The voltage change between I_Z = 2 mA and 8 mA

$$\Delta V_Z = \Delta I_Z r_Z = (8 \text{ mA} - 2 \text{ mA}) \times 40\ \Omega = 0.24 \text{ V}$$

So maximum zener voltage, $V_{Z\,max} = V_{Z\,min} + \Delta V_Z = 9.04 + 0.24 = 9.28$ V **Ans.**

2.6. ZENER DIODE AS A VOLTAGE REGULATOR

Voltage regulation is a measure of a circuit's ability to maintain a constant output voltage even when either input voltage or load current varies.

Figure 2.30 shows how a zener diode can be used as a voltage regulator to provide a constant voltage from a source whose voltage may vary appreciably. A resistor R_S is necessary to limit the reverse current through the diode to safer value. The voltage source V_S and the resistor R_S are so selected that the diode operates in the breakdown region. The diode voltage in this region, which is also the voltage across the load R_L, is called *zener voltage* V_Z and the diode current is called the *zener current* I_Z. The series resistor R_S absorbs the output voltage fluctuations so as to maintain voltage across the load constant. Zener diode is reverse connected across the input voltage whose variations are to be regulated.

As long as voltage across the load resistor R_L is less than the breakdown voltage V_Z, the zener diode does not conduct, the resistors R_S and R_L constitute a potential divider across V_S. At an increased supply voltage V_S, the voltage drop across load resistor (or zener diode) becomes greater than the zener breakdown voltage. It then operates in its breakdown region. The series resistor R_S limits the zener current I_Z from exceeding its rated maximum $I_{Z\,max}$. Current through resistor R_S is given as

$$I_S = \frac{V_S - V_Z}{R_S} \qquad \text{...(2.4)}$$

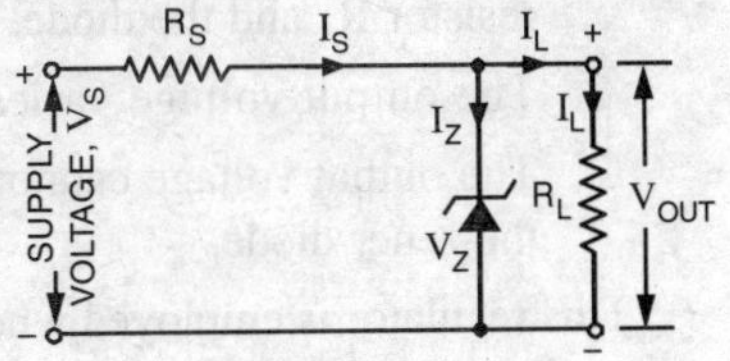

Fig. 2.30

The current from the power supply splits at the junction of zener diode and the load resistor R_L.

$$\text{So } I_S = I_Z + I_L \qquad \text{...(2.5)}$$

When the zener diode operates in its breakdown region, the voltage across it, V_Z remains fairly constant even though the current I_Z flowing through it may vary considerably.

When the supply voltage V_S increases, the current through both the zener diode and load resistance R_L increases. At the same time, however, the zener diode resistance decreases and the current through the diode increases more than proportionately. As a result, a greater voltage drop will occur across the series resistor R_S and the output voltage V_{OUT} (voltage across the diode or load resistance R_L) will become very close to the original value. The reverse is also true. Thus a zener diode can maintain the output voltage V_{OUT} within a fraction of a volt when the supply or input voltage V_S may vary over a range of several volts. The zener diode will maintain a constant voltage across the load as long as the supply voltage is more than the zener voltage.

Let us examine the other cause of the output voltage variation. When the load resistance R_L decreases for constant input voltage V_S, load current I_L increases. This additional current is not supplied from the source of supply but the demand of additional load current is met by decrease in zener current I_Z. This keeps the voltage drop across series resistance R_S constant and so the output voltage V_{OUT}. The worst case occurs for minimum source voltage and maximum load current because the zener current reduces to a minimum. In such a case,

$$I_{S\,min} = \frac{V_{S\,min} - V_Z}{R_{S\,max}}$$

$$\text{or} \quad R_{S\,max} = \frac{V_{S\,min} - V_Z}{I_{S\,min}} \quad \text{...(2.6)}$$

From Eq. (2.5) $\quad I_Z = I_S - I_L$

In worst case, this may be written as

$$I_{Z\,min} = I_{S\,min} - I_{L\,max}$$

The critical point occurs when $I_{L\,max} = I_{S\,min}$. At this point, the zener current, I_Z reduces to zero, and regulation is lost.

By substituting $I_{L\,max}$ for $I_{S\,min}$ in Eq. (2.6), we have

$$R_{S\,max} = \frac{V_{S\,min} - V_Z}{I_{L\,max}} \quad \text{...(2.7)}$$

where $R_{S\,max}$ is the critical value of series resistance, $V_{S\,min}$ is the minimum source voltage, V_Z is zener voltage and $I_{L\,max}$ is maximum load current.

The *critical resistance,* $R_{S\,max}$ is the maximum allowable series resistance. The series resistance R_S must always be less than the critical value; otherwise; breakdown operation is lost, and the regulator stops its operation.

The advantages of zener diode voltage regulators over other voltage regulators are that they are smaller, lighter, more rugged and have a longer life. They are also simpler and inherently cheaper.

Such regulators have the following drawbacks :

1. Their efficiency is low for heavy load currents because there is a considerable power loss in series resistor R_S and the diode.
2. The output voltage varies slightly due to zener impedance r_Z.
3. The output voltage cannot be choosen independently but depends upon the breakdown voltage of the zener diode.

This regulator is employed when there are small variations in load current and supply voltage.

Example 2.6. For the circuit shown in Fig. 2.31, find

(*a*) the output voltage,

(*b*) the voltage drop across R_S,

(*c*) the current through zener.

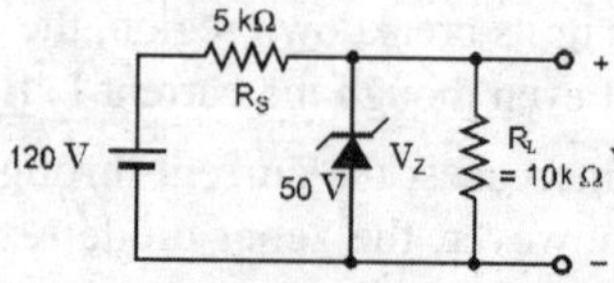

Fig. 2.31

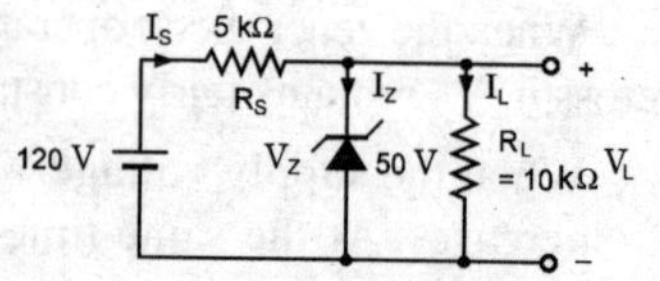

Fig. 2.32

Solution : From the figure shown in Fig. 2.32 we have

(*a*) Output voltage, $V_L = V_Z = 50$ V **Ans.**

(*b*) Voltage drop across $R_S = I_S \times R_S = V_S - V_Z = 120 - 50 = 70$ V **Ans.**

(*c*) Load current, $I_L = \frac{V_L}{R_L} = \frac{50\text{ V}}{10\text{ k}\Omega} = 5$ mA

Current through R_S, $I_S = \frac{70\text{ V}}{5\text{ k}\Omega} = 14$ mA

Hence, current through zener diode is given by

$$I_Z = I_S - I_L = 14\text{ mA} - 5\text{ mA} = 9\text{ mA}$$ **Ans.**

Example 2.7. For the zener regulator shown in Fig. 2.33, find the maximum and minimum current flowing through the zener diode.

[*G.G.S.I.P. Univ. Analog Electronics, December 2009*]

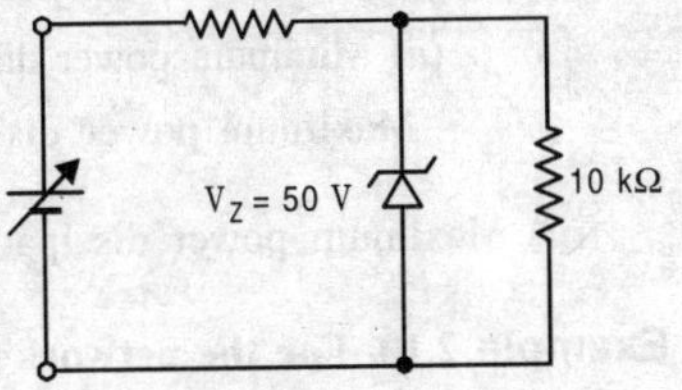

Fig. 2.33

Solution : Load current, $I_L = \frac{V_{out}}{R_L} = \frac{V_Z}{R_L} = \frac{50}{10\text{ k}\Omega} = 5$ mA

Zener current will be maximum when input voltage is maximum *i.e.,* 125 V.

Corresponding current through series resistance of 5 kΩ

$$I_{S\,max} = \frac{V_{S\,max} - V_{out}}{R_S} = \frac{125 - 50}{5\text{ k}\Omega} = 15\text{ mA}$$

Corresponding zener current, $I_{Z\,max} = I_{S\,max} - I_L = 15 - 5 = 10$ mA **Ans.**

The zener current will be minimum when input voltage is minimum *i.e.,* 85 V. Corresponding current through series resistance (5 kΩ).

$$I_{S\,min} = \frac{85 - 50}{5\text{ k}\Omega} = 7\text{ mA}$$

Corresponding zener current, $I_{Z\,min} = I_{S\,min} - I_L = 7 - 5 = 2$ mA **Ans.**

Example 2.8. For the zener diode network of Fig. 2.34, determine V_L, V_R, I_Z and P_Z.

[*Mahamaya Technical Univ. Electronics Engg. Second Semester 2010-11*]

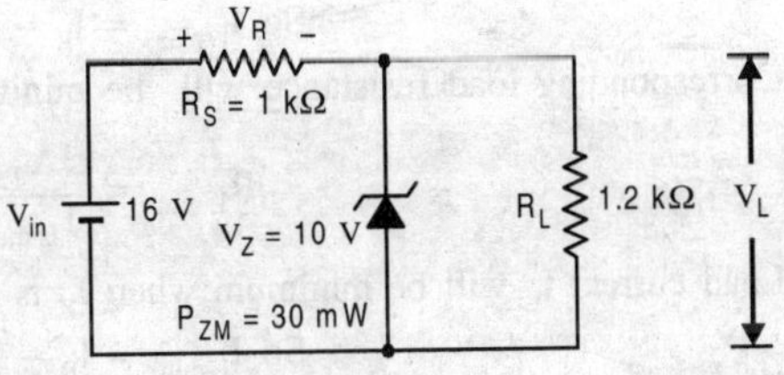

Fig. 2.34

Solution : Voltage across load if the zener diode is open circuited

$$V_L = \frac{V_{in} \times R_L}{R + R_L} = \frac{16 \times 1.2}{1 + 1.2} = 8.727\text{ V}$$

Since voltage across load is less than the breakdown voltage of zener diode *i.e.,* $V_L < V_Z$, the zener diode will not conduct and voltage across load, $V_L = 8.727$ V **Ans.**

Voltage across series resistor, $V_R = 16 - 8.727 = 7.273$ V **Ans.**

Zener diode current, $I_Z = 0$ **Ans.** $\because$ Zener diode is open circuited

Zener diode power, $P_Z = V_Z \times I_Z = V_Z \times 0 = 0$ watt **Ans.**

Example 2.9. For the circuit shown in Fig. 2.35, $R_S = 20\ \Omega$, $V_Z = 18$ V and $R_L = 200\ \Omega$. If V_{in} can vary from 20 V to 30 V, find

(*i*) The minimum and maximum currents in zener diode.

(*ii*) The minimum and maximum power dissipated in the diode.

(*iii*) The maximum rated power dissipation that R_S should have.

[*G.G.S.I.P. Univ. Analog Electronics December 2009*]

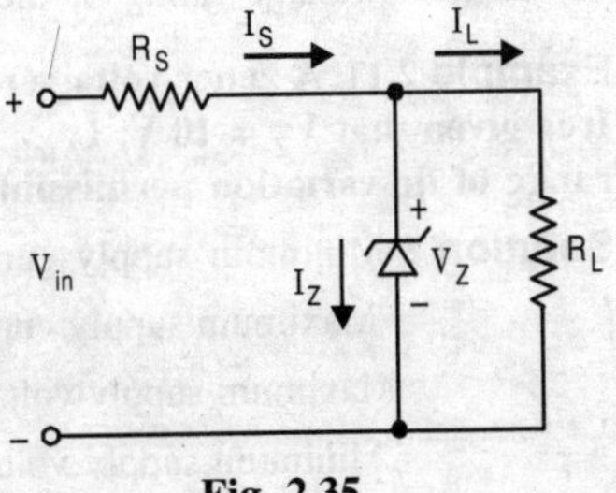

Fig. 2.35

Solution : Load current, $I_L = \frac{V_{out}}{R_L} = \frac{V_Z}{R_L} = \frac{18}{200} = 0.09$ A

Maximum supply current, $I_{S\,max} = \frac{V_{in\,max} - V_Z}{R_S} = \frac{30-18}{20} = 0.6$ A

(*i*) Corresponding zener current, $I_{Z\,max} = I_{S\,max} - I_L = 0.6 - 0.09 = 0.51$ A **Ans.**

Minimum supply current, $I_{S\,min} = \frac{V_{in\,min} - V_Z}{R_S} = \frac{20-18}{20} = 0.1$ A

Corresponding zener current, $I_{Z\,min} = I_{S\,min} - I_L = 0.1 - 0.09 = 0.01$ A or 10 mA **Ans.**

(*ii*) Minimum power dissipation in diode = $V_Z \times I_{Z\,min} = 18 \times 0.01 = 0.18$ W **Ans.**

Maximum power dissipation in diode = $V_Z \times I_{Z\,max} = 18 \times 0.51 = 9.18$ W **Ans.**

(*iii*) Maximum power dissipation rating of $R_S = I_{S\,max}^2 R_S = 0.6^2 \times 20 = 7.2$ W **Ans.**

Example 2.10. For the network of Fig. 2.36, determine the range of R_L and I_L that will result in V_{RL} being maintained at 10 V. Also determine the maximum wattage rating of diode.

[G.B. Technical Univ. Electronics Engineering First Semester 2011-12]

Solution : Input voltage, $V_{in} = 50$ V

Series resistance, R = 1 kΩ

$V_Z = 10$ V

$I_{Zm} = 32$ mA

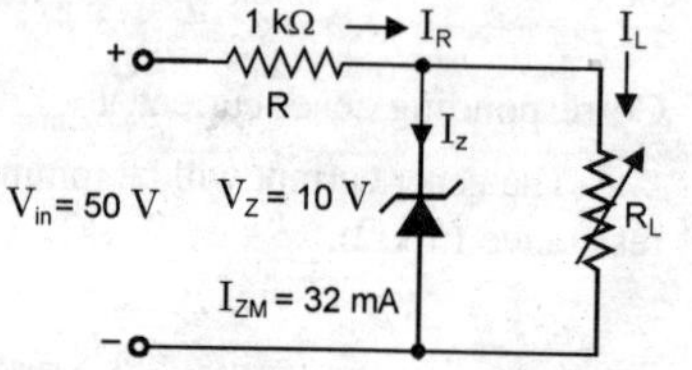

Fig. 2.36

Supply current, $I_R = \frac{V_{in} - V_L}{R} = \frac{50-10}{1{,}000} = 40$ mA

Load current I_L will be maximum when zener current $I_Z = 0$

So $I_{L\,max} = I_R - I_{Z\,min} = 40 - 0 = 40$ mA

Corresponding load resistance will be minimum and

$$R_{L\,min} = \frac{V_L}{I_{L\,max}} = \frac{V_Z}{I_{L\,max}} = \frac{10}{40 \times 10^{-3}} = 250\ \Omega$$

Load current I_L will be minimum when I_Z is maximum *i.e.,* 32 mA.

So $I_{L\,min} = I_R - I_{Z\,max} = 40 - 32 = 8$ mA

Corresponding load resistance will be maximum and

$$R_{L\,max} = \frac{V_L}{I_{L\,min}} = \frac{10}{8 \times 10^{-3}} = 1.25\ k\Omega$$

Hence range of R_L : From 250 Ω to 1.25 kΩ **Ans.** and range of I_L : From 8 mA to 40 mA **Ans.**

Maximum wattage rating of diode = $V_Z \times I_{Z\,max} = 10 \times 32 \times 10^{-3} = 0.32$ W **Ans.**

Example 2.11. A zener voltage regulator has variable load R_L requiring load current to vary 10 mA to 85 mA. It is given that $V_Z = 10$ V; $I_{Z\,min} = 15$ mA; $I_{Zmax} = 100$ mA and the series resistance $R_S = 40\ \Omega$. Calculate the range of dc variation permissible and zener power dissipation. *[G.G.S.I.P. Univ. Analog Electronics December 2010]*

Solution : Minimum supply current, $I_{S\,min} = I_{L\,max} + I_{Zmin} = 85 + 15 = 100$ mA

Maximum supply current, $I_{S\,max} = I_{L\,min} + I_{Zmax} = 10 + 100 = 110$ mA

Maximum supply voltage, $V_{S\,max} = I_{S\,max} R_S + V_Z = 110 \times 10^{-3} \times 40 + 10 = 14.4$ V

Minimum supply voltage, $V_{S\,min} = I_{S\,min} R_S + V_Z = 100 \times 10^{-3} \times 40 + 10 = 14$ V

i.e., supply voltage variation is from 14 V to 14.4 V **Ans.**

Example 2.12. Determine V_L, I_L, I_Z and I_R for the following circuit.

[U.P. Technical Univ. Electronics Engg. First Semester 2009-10]

Solution : Voltage across load if zener diode is open circuited.

$$V_L = V_{in} \times \frac{R_L}{R + R_L} = 20 \times \frac{180}{220 + 180} = 9 \text{ V}$$

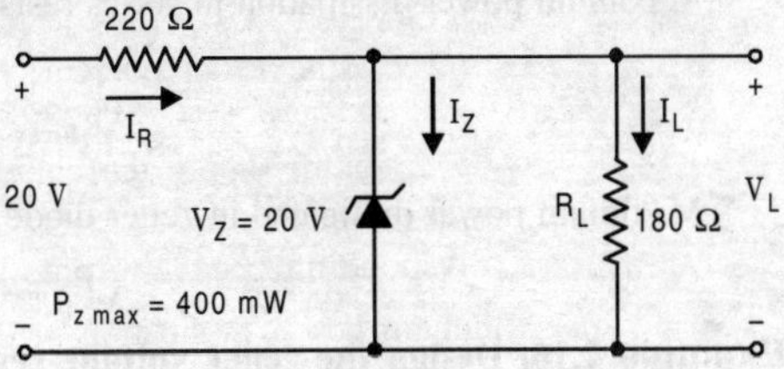

Fig. 2.37

Since voltage across load is less than the breakdown voltage of zener diode *i.e.,* $V_L < V_Z$, the zener diode will not conduct and voltage drop across load, V_L = 9 V **Ans.**

Zener diode current, I_Z = 0 **Ans.** ∵ Zener diode is open circuited

Load current, $I_L = \frac{V_{in}}{R + R_L} = \frac{20}{220 + 180} = 0.05$ A **Ans.**

Current through resistance R,

$I_R = I_L = 0.05$ A **Ans.**

Example 2.13. For the circuit shown in Fig. 2.38, find the voltage drop across the resistance of 5 kΩ.

[U.P. Technical Univ. Electronics Engineering Second Semester 2009-10]

Solution : Removing the zener diode from the circuit, we have

$$V_L = \frac{10}{10 + 5} \times 100 = 66.67 \text{ V} \qquad \therefore \; V_L > V_Z \; (i.e., \; 50 \text{ V})$$

So zener diode is in ON state.

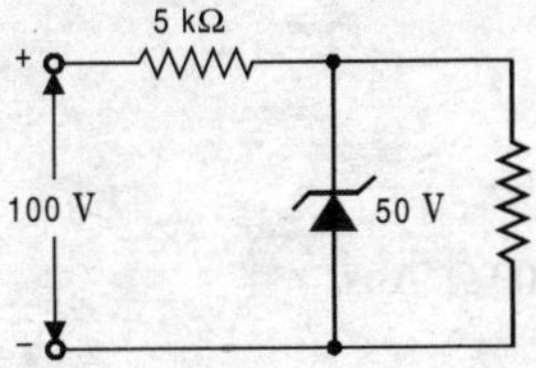

Fig. 2.38

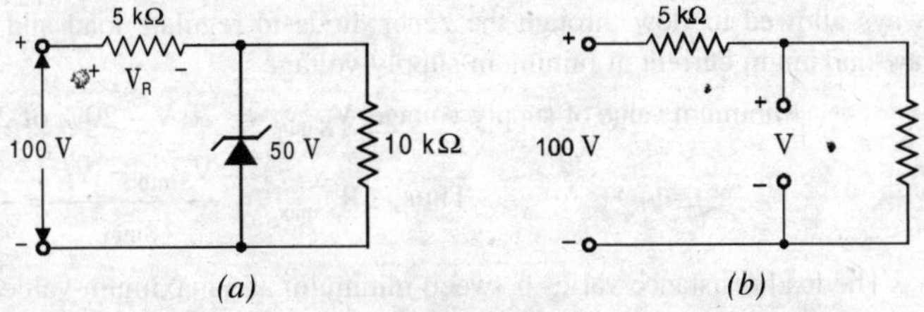

Fig. 2.39

Thus the voltage drop across the 10 kΩ load resistor

$V_L = V_Z = 50$ V

Voltage drop across 5 kΩ resistance

$= V_S - V_L = 100 - 50 = 50$ V **Ans.**

Example 2.14. A zener diode regulator has V_Z = 12 V. The input voltage may vary from 20 to 35 V and load current from 15 to 75 mA. To hold voltage constant under all conditions, what should be the value of series resistance ? *[Pb. Technical Univ. Analog Electronics December 2005]*

Solution : Maximum value of required series resistance,

$$R_{S\,max} = \frac{V_{S\,min} - V_Z}{I_{L\,max}} = \frac{20 - 12}{75 \times 10^{-3}} = 106.7 \; \Omega$$

Select R_S = 100 Ω, standard value **Ans.**

Example 2.15. Draw and design a zener voltage regulator circuit to provide output of 12 V for a maximum load current of 0.5 A when the input voltage variation is 15 to 18 V. *[Pb. Technical Univ. Analog Electronics May 2007]*

Solution : Assume $I_{Z\,min}$ = 5% of $I_{L\,max}$ = 0.05 × 0.5 = 0.025 A

Source resistance (maximum value), $R_{S\,max} = \frac{V_{S\,min} - V_Z}{I_{S\,min}} = \frac{V_{S\,min} - V_Z}{I_{L\,max} + I_{Z\,min}} = \frac{15 - 12}{0.5 + 0.025} = 5.7 \; \Omega$

Select R_S = 5.6 Ω (standard value)

$$I_{Z\,max} = \frac{V_{S\,max} - V_Z}{R_S} - I_L = \frac{18-12}{5.6} - 0.5 = 0.57 \text{ A}$$

Maximum power dissipation in series resistance,

$$P_{R_{S\,max}} = \frac{(V_{S\,max} - V_Z)^2}{R_S} = \frac{(18-12)^2}{5.6} = 6.4 \text{ W}$$

Maximum power dissipated in zener diode

$$P_{Z\,max} = I_{Z\,max} \times V_Z = 0.57 \times 12 = 6.84 \text{ W} \quad \textbf{Ans.}$$

Example 2.16. Design the zener voltage regulator shown in Fig. 2.40, to meet the following specifications : Load voltage 6.8 V, source volts V_S is 20 V ± 20% and load current is 30 mA ± 50%. The zener requires a minimum current of 1 mA to breakdown. The diode D has a forward conducting voltage of 0.6 V.

[GATE 1994]

Solution :

$$I_{Z\,min} = 1 \text{ mA}$$

This current through zener diode should flow even when the load draws the maximum current

i.e., 30 mA + 50% of 30 mA = 45 mA

Thus minimum value of supply current,

$$I_{S\,min} = I_{Z\,min} + I_{L\,max} = 1 + 45 = 46 \text{ mA}$$

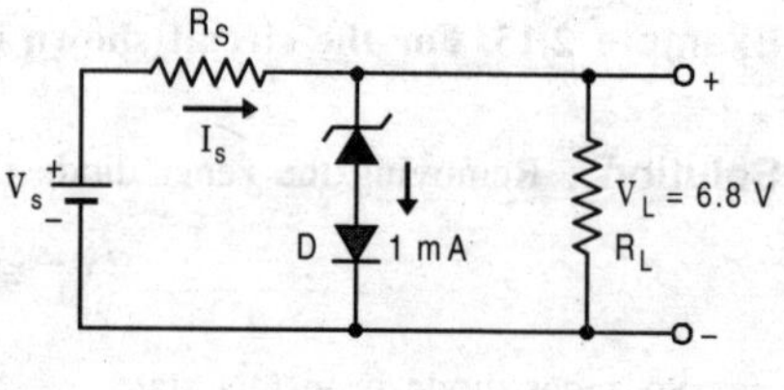

Fig. 2.40

The value of R_S should be of a maximum value such that $I_{S\,max}$ is always allowed to flow through the zener diode to regulate load and to draw maximum current at minimum supply voltage.

Minimum value of supply voltage, $V_{S\,min}$ = 20 V – 20% of 20 V = 16 V

$$\text{Thus,} \quad R_{S\,max} = \frac{V_{S\,min} - V_L}{I_{S\,min}} = \frac{16-6.8}{46 \times 10^{-3}} = 200\ \Omega \quad \textbf{Ans.}$$

The load resistance varies between minimum and maximum values of current

$$R_{L\,max} = \frac{V_L}{I_{L\,min}} = \frac{6.8}{15 \times 10^{-3}} = 453\ \Omega \quad \textbf{Ans.}$$

$$\because \quad I_{L\,min} = 30 \text{ mA} - 50\% \text{ of } 30 \text{ mA} = 15 \text{ mA}$$

$$\text{and} \quad R_{L\,min} = \frac{V_L}{I_{L\,max}} = \frac{6.8}{45 \times 10^{-3}} = 151\ \Omega \quad \textbf{Ans.}$$

Maximum value of supply current, $I_{S\,max} = \dfrac{V_{S\,max} - V_L}{R_{S\,max}} = \dfrac{(20 \text{ V} + 20\% \text{ of } 20 \text{ V}) - 6.8 \text{ V}}{200\ \Omega}$

$$= \frac{24-6.8}{200} = 0.086 \text{ A or } 86 \text{ mA}$$

Maximum value of zener diode current, $I_{Z\,max} = I_{S\,max} - I_{L\,min} = 86 - 15 = 71$ mA

Thus, zener's diode maximum rating, $P_{Z\,max} = V_Z I_{Z\,max} = (V_L - V_D) I_{Z\,max} = (6.8 - 0.6) \times 71 = 440$ mW **Ans.**

Highlights

1. Light-emitting diodes (LEDs) emit light under forward bias conditions but needs 2 V to 4 V for good emission.

   LEDs may emit visible radiation when the PN junction is diffused in GaAsP, infrared if constructed from GaAs, and a laser beam if sufficiently high pulsed current is applied to a GaAs junction.

2. LCDs have a much lower power consumption level than LEDs, but their lifetime is much shorter, and they require an internal or external light source. The applications of LCDs are display of numeric and alphanumeric characters in segmental and dot matrix displays.
3. Photodiode is a two-terminal semiconductor P-N junction and is designed to operate with reverse bias. The photodiode is a photovoltaic device as well as photoconductive device. When the device is operated with a reverse bias, it operates as a photoconductive device and when operated without the reverse bias, it operates as a photovoltaic device.
4. The tunnel diode (sometimes called the *Esaki diode* after its inventor, Dr. Leo Esaki) is a high conductivity, heavily doped two terminal P-N junction. This diode differs from the normal P-N junction diode in the amount of impurity concentration. The impurity concentration in a normal P-N junction diode is about 1 part in 10^8, whereas in the tunnel diode the impurity concentration is about 1 part in 10^3. It can conduct in reverse as well as forward direction but is usually used in forward biased mode.
5. A zener diode is a P-N junction operated in the reverse-biased mode to take advantage of its sharply defined breakdown voltage. The direction of conduction for a zener diode is opposite to that of the arrow in the symbol, and the zener voltage has a polarity opposite to that of a forward-biased diode.
6. The zener voltage V_Z is specified at some test value of current I_{ZT}, at which the diode will exhibit some dynamic impedance Z_Z known as zener impedance and given by $Z_Z = \frac{\Delta V_Z}{\Delta I_Z}$ where ΔV_Z and ΔI_Z are the small variations in voltage and current respectively.
7. Zener diodes have a temperature coefficient, which generally is negative for V_Z below 5 V but positive above 6 V, and is expressed in per cent of V_Z per °C with the change in zener voltage given by equation $\alpha_Z = \frac{\Delta V_Z}{V_Z(t-t_0)} \times 100$. Reference zener diodes are available with α_Z as low as 0.0005% per °C.
8. A zener diode may be used to regulate the load voltage at the value V_Z by acting as a "bypass value" to counteract line voltage or load current variations. Zener diodes are also useful in voltage reference circuits. The percent regulation, a figure of merit for the circuit, is a function of the range of input voltage and load resistance values, and of the individual device parameters.

Exercises

1. Explain the performance characteristics of LED. Give its important features and typical applications.
[G.G.S.I.P. Univ. Delhi Analog Electronics 2011-12]
2. What is LED ? Discuss its advantages, disadvantages and applications.
[Pb. Technical Univ. Electronic Devices and Circuits Dec. 2008; May 2009]
3. Explain the theory and working of an LED. Give the advantages of LEDs.
[M.D. Univ. Electronic Measurements and Instrumentation May 2007, December 2007]
4. Draw the structure of an LED and explain its operation. What are the conditions to be satisfied by the device for emission of visible light ?
[M.D. Univ. Rohtak Electronic Measurement and Instrumentation May 2009]
5. Discuss the principle of operation of LCD and photodiode.
[Pb. Technical Univ. Electronic Devices & Circuits May 2010]
6. What is LCD ? Discuss its advantages, disadvantages and applications.
[Pb. Technical Univ. Electronic Devices and Circuits May 2009]
7. Briefly discuss the use of LED and LCD as display devices in instrumentation. Comment on their relative merits and demerits. **[U.P.S.C. I.E.S. Elec. Engineering I 1992]**
8. What is a photodiode ? With proper characteristic curves and relevant diagrams explain the operation of the devices. What are its applications. **[U.P. Technical Univ. Solid State Devices & Circuits 2009-10]**

9. Why photodiode is used in reverse bias conditions ? Give any two applications of it. **[U.P. Technical Univ. Analog & Digital Electronics 2009-10]**
10. What is a photodiode ? Draw typical I-V characteristic curves at two illumination levels and explain how it works as a photoresistor and also as a photovoltaic converter. **[U.P.S.C. I.E.S. Electrical and Telecommunication Engg.-I, 2003]**
11. Explain with a neat sketch the operation of a photodiode.
12. Draw and explain the characteristics of a tunnel diode and its symbol. What is tunneling ? Explain. **[U.P. Technical Univ. Solid State Devices and Circuits 2007-08]**
13. Explain the operation of a tunnel diode. Draw the I-V characteristics of the diode and list the possible applications of the device. **[U.P. Technical Univ. Solid State Devices & Circuits 2008-09]**
14. Explain the working of tunnel diode. Give its two applications. **[U.P. Technical Univ. Analog & Digital Electronics 2009-10]**
15. Draw the V-I characteristics of tunnel diode and indicate the useful region in the curve. **[U.P. Technical Univ. Analog & Digital Electronics 2009-10]**
16. Sketch the terminal current-voltage characteristics of the following diodes under both forward and reverse biased conditions.
    (*i*) P-N junction diode.
    (*ii*) Zener diode.
    (*iii*) Tunnel diode. **[U.P.S.C. I.E.S. Electronics & Telecomm. Engineering-I, 2001]**
17. Explain with reference to the zener diode characteristic, zener voltage, breakdown current and dynamic impedance. How does the dynamic impedance vary with zener current ? Draw the equivalent circuit of an ideal zener diode in the breakdown region.
18. Draw the V-I characteristics of zener diode and explain the phenomenon of zener breakdown and avalanche breakdown. **[G.G.S.I.P. Univ. Analog Electronics, December 2009]**
19. Explain in detail the effect of temperature on zener diode. **[Pb. Technical Univ. Analog Electronics May 2007]**
20. Draw the equivalent circuit of zener diode as voltage regulator and explain. **[Pb. Technical Univ. Analog Electronics May 2006]**
21. What do you mean by zener diode ? Explain how zener diode maintains constant voltage across the load. **[U.P. Technical Univ. Fundamentals of Electronic Devices 2009-10]**
22. Explain the following diodes with their working and applications.
    (*i*) Tunnel diode (*ii*) Zener diode **[R.G. Technical Univ. Bhopal Basic Electronics April 2009]**

Short Answer Type Questions With Answers

Q. 1. What is radiative recombination in LED ? **[Pb. Technical Univ. December 2008]**

Ans. Whenever LED is forward biased, a forward current starts flowing through it. This current flows because of electrons supplied by the battery. The electrons supplied by the battery are free electrons and they reside in the conduction band. The energy level of conduction band is higher than that of valence band. These electrons recombine with holes to become valence electrons and come to a lower energy level. While coming back to the valence band, they give up the excess heat in the form of light (refer to Fig. 2.1). This process is called recombination in LED because of which light emits in LED.

Q. 2. What precautions are required to be observed in the use of LEDs ?

Ans. LEDs should neither be reverse biased nor operated near their maximum current rating. The leads of a LED should never be bent closer than about 2 mm from the encapsulation.

Q. 3. Why LCDs are not operated from ac supply of frequency lower than 25 Hz and higher than 50 Hz ?

Ans. If the frequency of supply to LCDs is lower than 25 Hz, a visible flicker would be produced and if it exceeds 50 Hz, the current drawn will be more.

Q. 4. How are LCD displays advantageous over LED displays ? **[Pb. Technical Univ. May 2009]**

Ans. The LCDs have the distinct advantage of having low power consumption than the LEDs. It is typically of the order of microwatts for the display in comparison to the some order of mW for LEDs. Other advantages are lower cost and good contrast.

Q. 5. What are the drawbacks of LCDs ?

Ans. The main drawbacks of LCDs are additional requirement of light source, limited temperature range of operation (0–60°C), poor reliability, short operating life, poor visibility in low ambient lighting, slow speed and the need for an ac drive.

Q. 6. What is "dark current" of a photodiode ? **[Pb. Technical Univ. December 2005]**

Ans. The dark current is the current that exists when no light is incident.

Q. 7. State the working principle of photodiode. **[Pb. Technical Univ. May 2007]**

Ans. Photodiode is a two-terminal semiconductor P-N junction device and is designed to operate with reverse bias. When light energy bombards a P-N junction, it dislodges valence electrons. The more light striking the junction, the larger the reverse current in a diode. It is due to generation of more and more charge carriers with the increase in level of illumination. This is the working principle of photodiode.

Q. 8. How the current is reduced to zero in a photodiode ?

Ans. The reverse saturation current I_0 flowing through a photodiode is reduced to zero by applying a forward bias voltage of magnitude equal to barrier potential.

Q. 9. State differences between LED and photodiodes. **[Pb. Technical Univ. May 2008]**

Ans. Photodiode is a two-terminal semiconductor P-N device and is designed to operate with reverse bias. It works when light falls on depletion layer of P-N junction of photodiode. Photodiode is a photovoltaic device as well as photoconductive device. When it is operated with a reverse bias, it is photoconductive device and when operated without the reverse bias, it is a photovoltaic device.

LED stands for light-emitting diode. It emits a fairly narrow bandwidth of visible (usually red, orange, yellow or green) or invisible (infrared) light when its internal diode junction is stimulated by a forward electric current/voltage (power). The operation of LED is based on the phenomenon of electroluminence, which is the emission of light from a semiconductor under the influence of electric field.

Q. 10. What is tunnel diode ?

Ans. The tunnel diode (sometimes called the *Esaki diode* after its inventor, Dr. Leo Esaki) is a high conductivity, heavily doped two terminal P-N junction. This diode differs from the normal P-N junction diode in the amount of impurity concentration. The impurity concentration in a normal P-N junction diode is about 1 part in 10^8, whereas in the tunnel diode the impurity concentration is about 1 part in 10^3. It can conduct in reverse as well as forward direction but is usually used in forward biased mode.

A tunnel diode's characteristics exhibit a negative resistance region and it may be used as an ultra-fast switching device or in ultrahigh frequency oscillators.

Q. 11. What is tunneling ? **[Pb. Technical Univ. December 2005]**

Ans. The mechanism of conduction in a semiconductor diode in which charge carriers (possessing very little energy) punch through a barrier directly instead of climbing over it is called tunneling.

Q. 12. What are the applications of tunnel diodes ?

Ans. Tunnel diodes are used as amplifiers, oscillators or switching devices, being an exclusive high-frequency component because of its very fast response of inputs.

Q. 13. What is a zener diode ?

Ans. Zener diode is a P-N junction diode specially designed for operation in its breakdown region.

Q. 14. What is zener voltage ?

Ans. The voltage at which the zener diode breaks down is called the zener voltage.

Q. 15. What is meant by the temperature coefficient ?

Ans. The effect of temperature on zener voltage is given in terms of temperature coefficient which is defined as the percentage change in nominal zener voltage for each degree centigrade of change in junction temperature.

Mathematically it is given as $\frac{\Delta V_Z}{V_Z(t-t_0)} \times 100$ where ΔV_Z is the change in zener voltage due to temperature variation of $(t - t_0)$.

Q. 16. What happens to the series current, load current and zener current when the dc input voltage of a zener regulator increases ?

Ans. Zener current and series current increase while the load current remains unchanged.

Q. 17. Why is zener diode used as voltage regulator ?

Ans. Zener diode has the property of behaving like a dc battery in 'on' state (*i.e.* when the voltage across zener diode exceeds its zener voltage rating V_Z). In 'on state', the voltage across zener diode remains constant until the voltage across it drops less than V_Z. This property of zener diode makes its use as voltage regulator.

Q. 18. Explain how zener diode maintains constant voltage across the load ?

Ans. Zener diode has the property of behaving like a dc battery in 'on' state. If the zener diode is shunted across the load R_L and the voltage across zener diode is more than zener voltage V_Z then zener diode is in 'on' state, and any variation in voltage across the zener diode due to variations either in supply voltage or in load resistance is not able to change the output voltage. Thus zener diode maintains voltage constant across the load.

Problems

1. A zener diode shunt regulator is shown in Fig. 1. Determine
   (*i*) load voltage,
   (*ii*) voltage drop across series resistance,
   (*iii*) current through the diode. **[Agra Univ. 1999]**
   [Ans. (*i*) 12 V (*ii*) 18 V (*iii*) 51 mA**]**

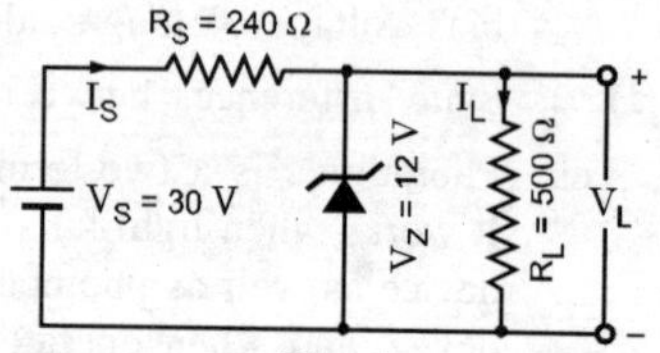

Fig. 1

2. Determine the maximum and minimum values of zener current if in the circuit shown in Fig. 2.30 the load resistance R_L = 5 kΩ, series resistance is 2 kΩ, input voltage varies between 80 V and 100 V and the output voltage is 50 V. **[Ans.** 15 mA and 5 mA**]**
3. In the circuit shown in Fig. 2.30 the series resistance is 2 kΩ, load resistance is 5 kΩ, input voltage is 60 V and output voltage is 35 V. Determine (*i*) voltage drop across series resistance (*ii*) load current (*iii*) current through series resistance and (*iv*) zener diode current. **[Ans.** (*i*) 25 V (*ii*) 7.0 mA (*iii*) 12.5 mA (*iv*) 5.5 mA**]**
4. Determine the temperature coefficient of a 7.5 V zener diode at 25°C if the nominal voltage increases to 7.8 V at a temperature of 105°C. **[Ans.** + 0.05%**]**
5. For the circuit of the following Fig. 2, find (*a*) the output voltage (*b*) voltage across R_S and (*c*) the current through zener diode. **[Ans.** (*a*) 8 V (*b*) 4 V (*c*) 0 A**]**
6. Explain 'Avalanche' breakdown. Draw the *v-i* characteristics of zener diode and explain how does a zener regulate a voltage ? The input voltage for the following Fig. 3 varies from 35 V to 45 V, V_Z = 20 V, r_Z = 5 Ω, I_L (min) = 0 mA, I_L (max) = 100 mA, I_Z (min) = 10 mA, I_Z (max) = 400 mA. Find the values of R and P_z (max). **[U.P. Technical Univ. Electronics Engineering First Semester 2005-06]**
   [Ans. 136.4 Ω, 1.67 W**]**

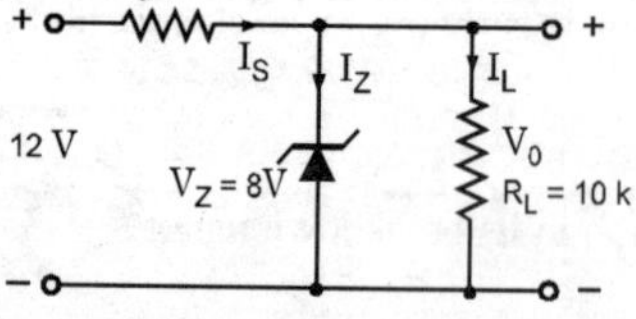

Fig. 2

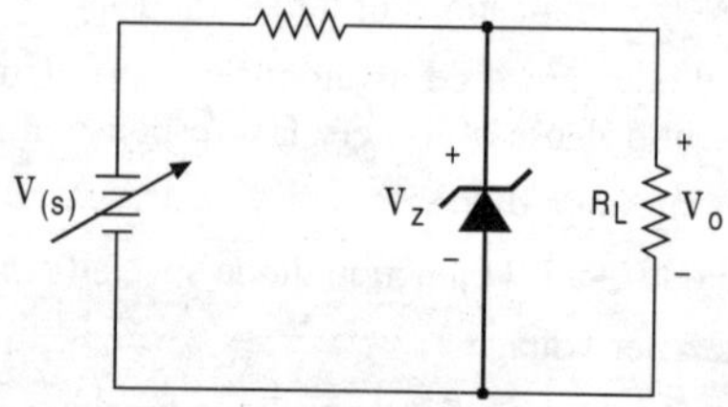

Fig. 3

Unit-II

Transistor Characteristics And Biasing

CONTENTS OF THIS UNIT

UNIT-II : TRANSISTORS, CHARACTERISTICS AND BIASING : Transistor, Types of Transistor; Transistor Current Components; Transistor as an Amplifier; Transistor Characteristics in CB, CE and CC Modes, Operating Point, Bias Stability, Various Biasing Circuits, Stabilization Against I_{CO}, V_{BE} and β. Construction and Applications of Junction Field-Effect Transistors (JFETs), UJT and MOSFETs.

CHAPTER

3

BIPOLAR JUNCTION TRANSISTORS (BJTs)

3.1. INTRODUCTION

INSIDE THIS CHAPTER

The transistor is a solid state device, whose operation depends upon the flow of electric charge carriers within the solid. Transistor is capable of amplification and in most respect it is analogous to a vacuum triode. The main difference between the two is that the transistor is a current controlled device whereas vacuum triode is a voltage controlled device. The transistor is only about 6 decade old, yet it has replaced vacuum tubes in almost all applications. The reasons are obviously its advantages over vacuum tubes such as compact size, light weight, rugged construction, more resistive to shocks and vibrations, instantaneous operation (no heating required), low operating voltage, high operating efficiency (no heat loss) and long life with essentially no ageing effect if operated within permissible limits of temperature and frequency. However, transistors, in comparison to vacuum triodes, have some drawbacks also such as loud hum noise, restricted operating temperature (up to 75ºC) and operating frequency (up to a few MHz only).

It consists of a silicon or germanium (preferably silicon because of its smaller cutoff current I_{CBO}, smaller variations in I_{CBO} due to variations in temperature and higher operating temperature) crystal in which a layer of N-type material is sandwiched between two layers of P-type material, as shown in Fig. 3.1 (*a*). Alternatively a transistor may consist of a layer of P-type material sandwiched between two layers of N-type material, as shown in Fig. 3.1 (*b*). In the former case the transistor is referred to as a P-N-P *transistor* and in latter case, as an N-P-N *transistor*. Each type of transistor has two P-N junctions—one junction between the emitter and base, called the *emitter-base junction* or simply the *emitter junction* and the other junction between the base and collector, called the *collector-base junction* or simply the *collector junction.* Thus a transistor is like two P-N junction diodes connected back to back. Each of these P-N junction diodes have all the characteristics discussed in earlier chapters. The two junctions give rise to three regions provided with three terminals called the *emitter*, *base* and *collector*, as shown in Fig. 3.1. The emitter, base and collector correspond in a general way to the cathode, grid and plate or anode of a vacuum triode.

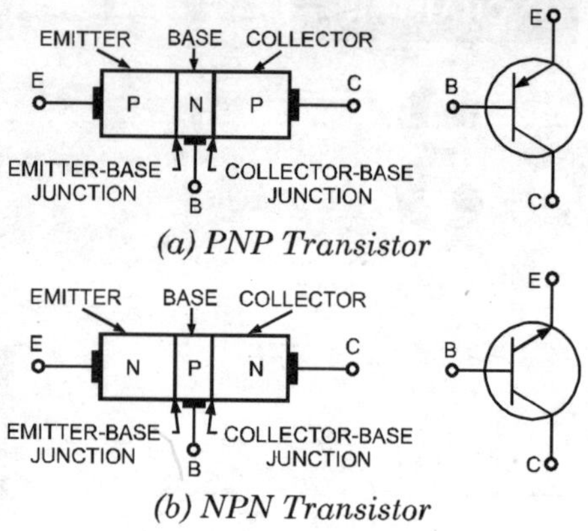

Fig. 3.1

3.2. TRANSISTOR TERMINALS

As mentioned above, transistor is a single crystal in which there are two P-N junctions, as shown in Fig. 3.1. The idea behind is to have first section to supply the charges (either holes or electrons) to be collected by the third section through the middle section. One side section supplying free charges is called the *emitter*, other side section collecting these charges is called the *collector* and the middle section which is formed between the emitter and collector is called the *base*.

1. *Emitter.* It is left hand section (or region) of the transistor and its main function is to supply majority charge carriers (electrons in case of NPN transistors and holes in case of P-N-P transistors) to the base. The emitter is always forward biased w.r.t. base so that it is able to supply majority charge carriers to the base. The emitter is heavily doped so that it may be able to inject a large number of charge carriers. It is of moderate size in order to maintain heavy doping without diluting it or mesh formation in it.

2. *Collector.* It is the right hand section of the transistor and its main function is to collect majority charge carriers. Collector is always reverse biased so as to remove the charge carriers away from its junction with the base. It is moderately doped to avoid the chances of mesh formation even after taking the carriers from the emitter. Large in size to withstand the temperature generated at the collector.

3. *Base.* It is the middle section of the transistor and is very lightly doped to reduce the recombination within the base so as to increase collector current and is very thin (of the order of μm) in comparison to either emitter or collector so that it may pass most of the injected charge carriers to the collector.

Base forms two junctions *i.e.,* emitter junction and collector junction, each having its own barrier voltage. The forward-biased emitter-base junction offers low resistance to the emitter current whereas the collector-base junction, which is reverse biased, offers high resistance to the collector current. As the resistance of emitter-base junction is very small as compared to that of collector-base junction, therefore, the forward bias applied to the emitter-base junction is usually very small whereas the reverse bias on the collector-base junction is much large.

In most of the transistors, the collector region is made physically larger than emitter region because collector is to dissipate much power.* Due to this difference, collector and emitter are not interchangeable. However, for sake of convenience, it is customary to show emitter and collector to be of equal size.

As regards the symbols, arrowhead is always at the emitter. The direction indicates the conventional direction of current flow (from emitter to base in case of P-N-P transistor and from base to emitter in case of NPN transistor).** Generally, no arrowhead is marked for collector since its reverse leakage current is always opposite to the direction of emitter current.

* During operation of a transistor, much heat is generated at the collector junction.

** In a P-N-P transistor emitter is positive w.r.t. base while in case of an N-P-N transistor base is positive w.r.t. emitter.

3.3. TRANSISTOR ACTION

Transistors of both types (P-N-P and N-P-N) behave exactly the same way except change in biasing and majority carriers. In P-N-P transistors the conduction is by holes whereas in N-P-N transistors conduction is by electrons. However, N-P-N transistors are preferred due to their better high frequency response. Basic connections and biasing for P-N-P and N-P-N type transistors are shown in Fig. 3.2 (*a*) and (*b*) respectively. Let us consider an N-P-N transistor, which is in common use, for discussion.

The emitter is forward biased and as a result a large forward current flows across the emitter junction due to flow of majority carriers *i.e.*, electrons from the emitter region enter the base region and holes from base region into the emitter region. However, since the conductivity of emitter region is larger than that of base region, the electrons from the emitter region out the number of holes from the base region. It may be assumed with good accuracy that forward current in the emitter junction flows due to the movement of electrons from the emitter region to the base region. We can, therefore, say that electrons from the emitter region due to forward bias in the emitter junction cross into the base region where electrons are in minority. Injection of electrons makes the electrons concentration on the emitter junction very large and on the collector junction the concentration of the electrons is extremely small and as such the gradient of electrons concentration is very large in base. Injected electrons diffuse into collector region due to extremely small thickness of base which is much less than the diffusion length. Most of the electrons cross into the collector region. Collector is reverse biased and creates a strong electrostatic field between base and collector. The field immediately collects the diffused electrons which enter the collector junction. Flow of electrons into the base region when confronted with the holes, a few electrons (say 1 to 10%) combined and neutralize and rest of the electrons say 90 to 99% of the injected electrons diffuse into the collector region and collected by the collector electrode. To maintain the base neutrality base electrode provides equal number of electrons which have combined with holes and results in a base current. Thus the emitter current I_E is equal to the sum of collector current I_C and base current I_B. The ratio of collector current I_C to emitter current I_E is called the α and it is measure of the possible current amplification. In a transistor α cannot be higher than unity, but practical values of 0.95 to 0.99 are attained in commercial transistors.

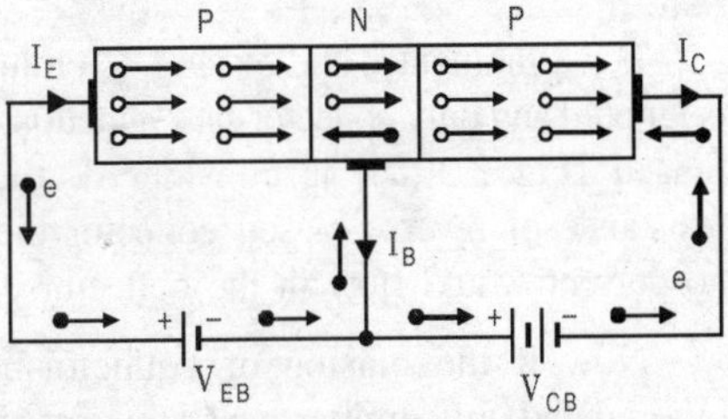

(a) Basic Connections of P-N-P Transistor

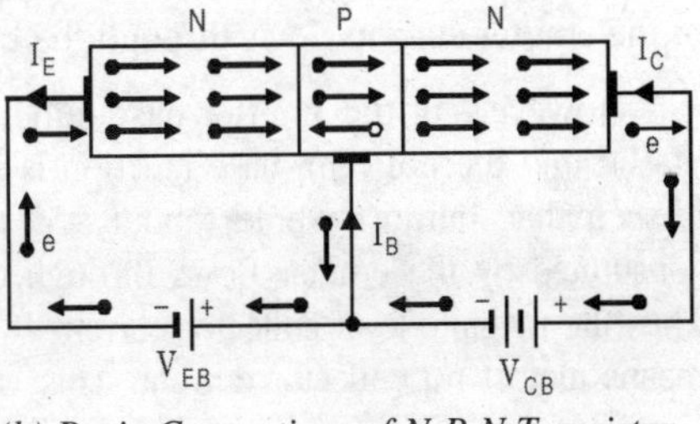

(b) Basic Connections of N-P-N Transistor

For connecting batteries properly in the transistor circuits the following scheme may be remembered.

The battery polarities to the emitter and base are the same as the letter of the emitter and base are impurity designations. Thus, in the P-N-P type, the emitter receives the P (positive) terminal, and the base receives the N (negative) terminal. Similarly, in N-P-N type, the emitter receives the N (negative) terminal, the base receives the P (positive)

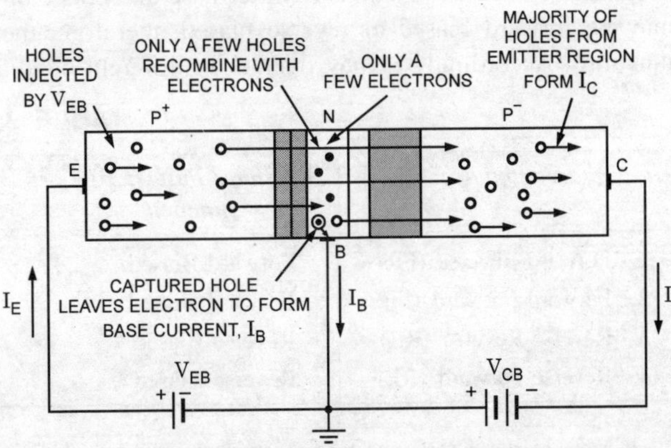

Fig. 3.2. (*c*) ***Charge Carrier Movement in a P-N-P Transistor***

terminal. The collector receives a battery polarity opposite to its impurity type designation. Thus P-type collector receives the N (negative) terminal of the battery and the N-type collector receives the P (positive) terminal.

Operation of P-N-P Transistor. The P-N-P transistor behaves exactly the same way as an N-P-N device, with the exception that the majority charge carriers are holes. As illustrated in Fig. 3.2 (*c*), holes are emitted from the P-type emitter across the forward-biased emitter-base junction into the base. In the lightly doped N-type base, the holes find few electrons to recombine with. Some of the holes flow out via the base terminal, but most are drawn across the collector by the positive-negative electric field at the reverse-biased collector-base junction. As in the case of NPN device, the forward bias at the emitter base junction controls the collector and emitter currents.

3.4. TRANSISTOR BIASING

As already mentioned above, for normal operation of a transistor emitter-base junction is always forward biased and collector-base junction is always reverse biased.

For a moment if the biasing of emitter-base junction is ignored and only collector-base junction, which is reverse biased [Fig. 3.3 (*a*)] is considered, then due to high resistance of reverse biased collector diode, practically* no current would flow in the collector circuit.

Now if the biasing of collector-base junction is ignored and only emitter-base junction, which is forward biased [Fig. 3.3 (*b*)] is considered then practically whole of the emitter current flows through the base *i.e.,* $I_B = I_E$.

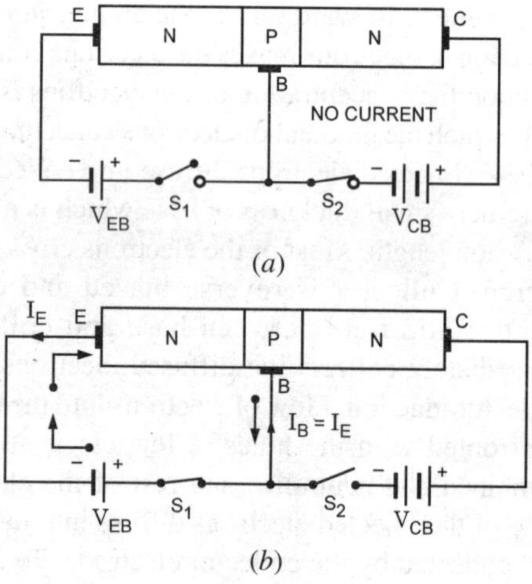

Fig. 3.3

However, if the emitter-base junction is forward biased and the collector-base junction is reverse biased it is seen that almost whole (about 95%) of the current constituted by the emitter flows through collector circuit. Thus the magnitude of collector current I_C depends upon the magnitude of emitter current I_E. No emitter current means almost no collector current. This is precisely what happens in a transistor.

3.4.1. Different Operating Conditions For a Transistor

A transistor has two junctions (emitter-base and collector-base junctions), and each of these two junctions, may be forward biased or reverse biased, therefore, there are four possible ways of biasing these two junctions. Accordingly it may operate in different conditions as listed in Table 3.1.

TABLE 3.1

Condition	*Emitter-Base (EB) Junction*	*Collector-Base (CB) Junction*	*Region of Operation*
1. Forward-Reverse (FR)	Forward-Biased	Reverse-Biased	Active
2. Forward-Forward (FF)	Forward-Biased	Forward-Biased	Saturation
3. Reverse-Reverse (RR)	Reverse-Biased	Reverse-Biased	Cutoff
4. Reverse-Forward (RF)	Reverse-Biased	Forward-Biased	Inverted

* In actual practice a very small reverse saturation current would flow in the collector circuit. This is called collector cutoff current and is because of minority carriers.

1. Forward-Reverse Biasing. As already mentioned above, for normal operation of a transistor emitter-base junction is always forward biased and collector-base junction is always reverse biased. This type of biasing is known as *FR biasing*. For this purpose a battery V_{EB} is connected between emitter and base while a battery V_{CB} is connected between the collector and base, as shown in Fig. 3.4. In Fig. 3.4 (*a*), the emitter-base junction of N-P-N transistor is forward biased by connecting the negative terminal of battery V_{EB} to the emitter and positive terminal to the base while collector-base junction is reverse biased by connecting the positive terminal of the battery V_{CB} to the collector and negative terminal to the base. Similarly in Fig. 3.4 (*b*) the emitter-base junction of P-N-P transistor is forward biased by connecting the positive terminal of battery V_{EB} to the emitter and negative terminal to the base while collector-base junction is reverse biased by connecting the negative terminal of battery V_{CB} to the collector and positive terminal to the base.

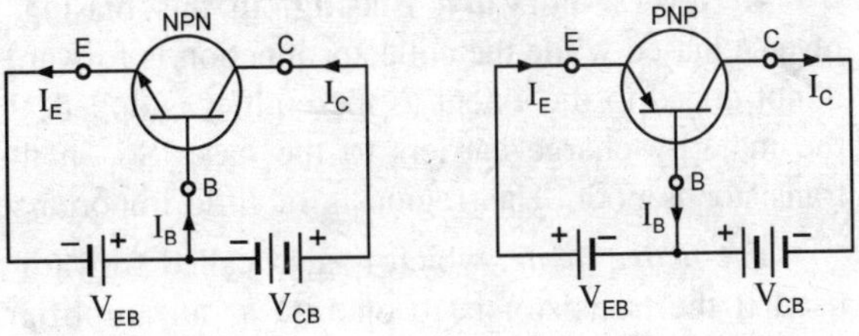

(*a*) *NPN Transistor* (*b*) *PNP Transistor*

Fig. 3.4. ***Active Region***

Typical base-to-emitter voltages for both N-P-N and P-N-P transistors are 0.7 V for silicon and 0.3 V for germanium. Typical collector-to-base voltages might be anything between 3 V and 20 V for most types of transistors, although in many cases the collector-to-base voltage may exceed 20 volts even.

The voltages across the terminals of a transistor follow the usual double subscript voltage convention. For example V_{BE} represents the voltage of base with respect to the emitter. If V_{BE} is equal to 0.7 V it means the base is 0.7 V more positive.

In FR biasing the transistor is in active region and the collector current I_C, depends upon the emitter current I_E. Generally transistor is operated in this region for amplification.

For connecting batteries properly in the transistor circuits the following scheme may be remembered.

The battery polarities to the emitter and base are the same as the letter of the emitter and base impurity designations. Thus, in the P-N-P type, the emitter receives the P (positive) terminal, and the base receives the N (negative) terminal. Similarly, in N-P-N type, the emitter receives the N (negative) terminal, the base receives the P (positive) terminal. The collector receives a battery polarity opposite to its impurity type designation. Thus P-type collector receives the N (negative) terminal of the battery and the N-type collector receives the P (positive) terminal.

2. Forward-Forward Biasing. In this biasing, both collector and emitter are forward biased, as illustrated in Fig. 3.5. The transistor operates in saturation region. In this region, the collector current becomes independent of the base current. Thus the transistor acts like a close switch.

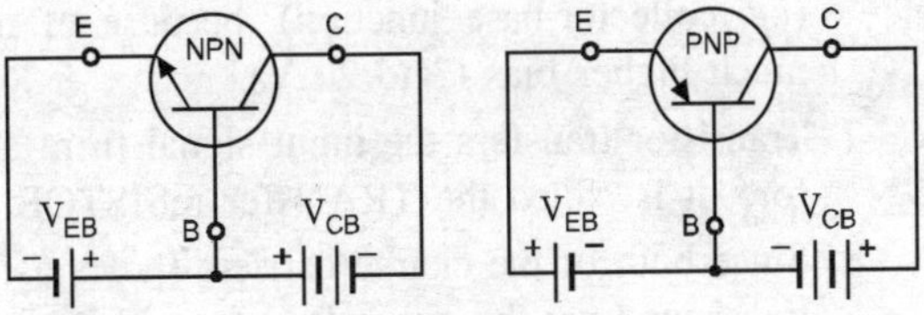

Fig. 3.5. ***Saturation Region***

3. Reverse-Reverse Biasing. In this biasing, both the junctions are reverse biased. In this biasing, the transistor has practically zero current because the emitter does not emit charge carriers into the base and no charge carriers are collected by the collector except a few thermally generated minority carriers. Thus, the transistor acts like an open switch.

4. Reverse-Forward Biasing. In this biasing, the emitter junction is reverse biased while the collector junction is forward biased. As the collector is not doped to the extent as the emitter is doped, therefore, it cannot inject the majority charge carriers to the base. So, in this region, the action of transistor is poor. This region is of little importance.

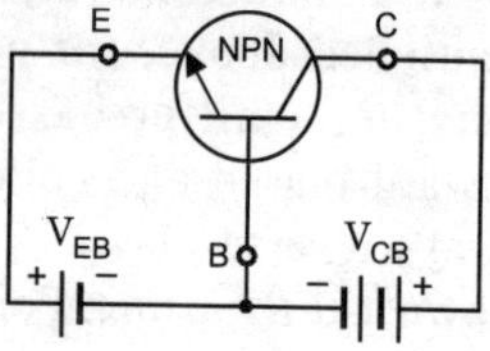

Fig. 3.6. *Cutoff Region*

The *active mode*, which is also called forward active mode, is the one used if the transistor is to operate as an amplifier. Switching operations (*e.g.*, logic circuits) utilize both the *cutoff mode* and the *saturation mode*. The reverse active (or inverse active) mode has very limited application but is conceptually significant.

3.5. IMPORTANT POINTS REGARDING WORKING OF TRANSISTOR

The noteworthy points regarding working of transistors are summarized below.

1. Current conduction in N-P-N transistor is by electrons and the conventional current flow will be in the opposite direction. Current conduction in P-N-P transistor is by holes but in external leads, the current conduction will be by flow of electrons.
2. The collector current I_C is always less than emitter current I_E, being the difference of emitter current I_E and base current I_B.
3. The base current is only a small fraction (usually 5%) of emitter current.
4. As a standard convention, all the currents entering into the transistor are considered to be positive. Conversely currents that flow out of the transistor are taken to be negative. It means that if the actual conventional current flows in the outward direction, a negative sign is included along with its magnitude. Thus in an N-P-N transistor, the emitter current I_E is to be taken negative because it flows out of the transistor [Figs. 3.3 (*b*) and 3.4 (*a*)] while both the base current I_B and collector current I_C are to be taken positive because they flow into the transistor. Similarly in a P-N-P transistor I_E is to be taken positive and both base current I_B and collector current I_C are to be taken negative [Figs. 3.2 (*c*) and 3.4 (*b*)]. However, to avoid confusion, the actual direction of flow of current is indicated in the diagrams.
5. Emitter-base junction is always forward biased and collector-base junction is always reverse biased for active region of operation.
6. The input circuit (*i.e.,* emitter-base junction), because of forward bias, offers low resistance and so needs usually very small bias (approximately 0.7 V for Si and 0.3 V for Ge). The output circuit (*i.e.,* collector-base junction), because of reverse bias, offers high resistance and, therefore, needs much higher bias (3 to 20 V).
7. Transistor transfers the input signal from a low resistance circuit to a high resistance circuit, therefore, it is called the TRANsfer reSISTOR (TRANSISTOR).
8. Since both of the charge carriers (holes as well as electrons) are involved in current flow through a transistor (may be either P-N-P or N-P-N), these devices are sometimes called the *bipolar junction transistors* (BJTs).
9. Although, it is said that a transistor is a device with two PN diodes connected back to back, but it does not mean that two discrete diodes connected back to back, as shown in Fig. 3.7, can work as a transistor. In such a case, each diode has two equally doped regions, so that the overall circuit has four equally doped regions. This would not work as base region is not the same as in a transistor.

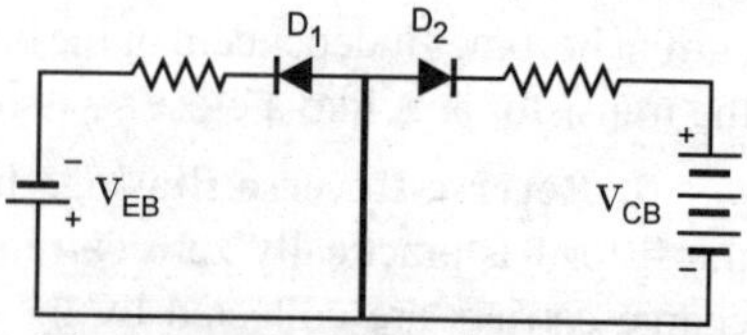

Fig. 3.7. *Two Diodes Connected Back to Back*

The key to the transistor action is the lightly doped thin base between the heavily doped emitter and moderately doped collector.

In an N-P-N transistor, the free electrons passing through the base to the collector region have a short lifetime. As long as the base is thin, the free electrons can reach the collector. But in case of two discrete back to back connected diodes there are four doped regions instead of three and there is nothing that resembles a thin base region between an emitter and a collector.

Hence two discrete diodes connected back to back can never work as a transistor.

10. The choice of N-P-N transistor is made more often because majority charge carriers are electrons whose mobility is much more (almost double) than that of holes.

3.6. TRANSISTOR CURRENT COMPONENTS

The various current components which flow across the forward-biased emitter junction and reverse biased collector junction are indicated in Fig. 3.8. The current flowing into the emitter is referred to as the *emitter current* and designated as I_E. The emitter current I_E consists of two constituents (hole current I_{hE} due to crossing of holes from emitter into the base and electron current I_{eE} due to crossing of electrons from base into the emitter).

i.e., $I_E = I_{hE} + I_{eE}$ ...(3.1)

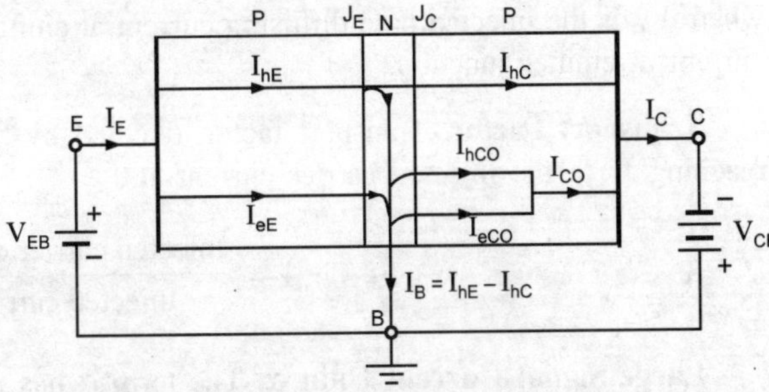

Fig. 3.8. ***Current Components in a PNP Transistor For a Forward Biased Emitter Junction and a Reverse Biased Collector Junction***

In a commercial transistor the emitter is heavily doped and the base is very lightly doped and, therefore, electron current I_{eE} is negligibly small in comparison to hole current I_{hE}. Thus the entire emitter current I_E in a P-N-P transistor is due to flow of holes from emitter to the base. It is to be noted that the indicated direction of I_E external to the transistor is the conventional current direction.

All the holes crossing the emitter junction J_E do not reach the collector junction J_C because some of them combine with the electrons in the N-type base. If I_{hC} is the hole current at collector junction J_C, there must be a bulk recombination current I_{hE} less I_{hC} leaving the base, as shown in Fig. 3.8. In fact electrons enter the base region through the base lead to replenish those electrons which have been lost by recombination with the holes injected into the base across the emitter junction J_E. The holes on reaching the collector junction J_C cross this junction and enter the collector P-region.

If the emitter were open circuited, I_E would have been zero and then I_{hC} also would have been zero. Under these conditions, the base and collector would act as a reverse-biased diode and the collector current I_C would be equal to that reverse saturation current I_{CBO} (or simply I_{CO}).

When the emitter circuit is closed *i.e.,* when $I_E \neq 0$, then from Fig. 3.8

Collector current, $I_C = I_{hC} + I_{CO}$ ...(3.2)

For a P-N-P transistor, I_{CO} consists of holes moving across junction J_C from base region to collector region and electrons crossing junction J_C in the opposite direction.

Thus $I_{CO} = I_{hCO} + I_{eCO}$ ...(3.3)

In the active region operation with emitter junction J_E forward biased, the collector current is given as

$$I_C = \alpha I_E + I_{CO} \quad ...(3.4)$$

where α is the fraction of the total current I_E which constitutes I_{hC}.

It is to be noted that the indicated direction of emitter current I_E external to the transistor is the conventional current direction. Base current I_B and collector current I_C are also shown external to the transistor conventional current direction. According to the Kirchhoff's first law total current flowing into the transistor must be equal to the total current flowing out of it. So emitter current I_E is equal to the sum of collector and base currents I_C and I_B.

$$i.e., \quad I_E = I_C + I_B \qquad ...(3.5)^*$$

Let us now define different parameters which relate the current components discussed above.

Emitter Efficiency. The emitter or injection efficiency denoted by γ is defined as the ratio of current of injected carriers at J_E to the total emitter current

$$i.e., \quad \gamma = \frac{\text{Current of injected carriers at } J_E}{\text{Total emitter current}} = \frac{I_{hE}}{I_{hE} + I_{eE}} = \frac{I_{hE}}{I_E} \qquad ...(3.6)$$

where I_{hE} is the injected hole diffusion current at emitter junction and I_{eE} is the injected electron diffusion current at emitter junction.

Transport Factor. Transport factor, denoted by β, is defined as the ratio of injected carrier current reaching J_C to the injected carrier current at J_E

$$i.e., \quad \beta = \frac{\text{Injected carrier current reaching } J_C}{\text{Injected carrier current at } J_E} = \frac{I_{hC}}{I_{hE}} \qquad ...(3.7)$$

Large Signal Current Gain α. The term α has already been defined above. However, α can alternatively be defined from Eq. (3.4). From Eq. (3.4) α may be defined as the ratio of collector current increment from cutoff value ($I_C = I_{CO}$) to I_C and the emitter current increment from cutoff value ($I_E = 0$) to I_E. Thus, we may write

$$\alpha = \frac{I_C - I_{CO}}{I_E - 0} = \frac{I_C - I_{CO}}{I_E} \qquad ...(3.8)$$

Term α is called the *large signal current gain* of a common base transistor. The parameter α is extremely important in transistor theory. It is to be pointed here that α is not constant, but varies with emitter current I_E, collector-base voltage V_{CB}, and the temperature. Typically α lies in the range 0.95 to 0.99.

From Eqs. (3.2) and (3.8) we have

$$\alpha = \frac{I_{hC}}{I_E} = \frac{I_{hC}}{I_{hE}} \times \frac{I_{hE}}{I_E} = \beta\gamma \qquad ...(3.9)$$

$$\because \text{From Eq. (3.7) } \frac{I_{hC}}{I_{hE}} = \beta \text{ and from Eq. (3.6) } \frac{I_{hE}}{I_E} = \gamma$$

For an efficient transistor we would like β and γ to be very near unity *i.e.,* the emitter current should be due mostly to holes ($\gamma = 1$) and most of the injected holes should eventually participate in the collector current ($\beta \simeq 1$).

The product $\beta\gamma$ is defined as the factor α, called the *current gain ratio,* which represents the emitter-to-collector current amplification. There is no real amplification between these currents, since α is lesser than unity. On the other hand, the relation between collector current and base current is more promising for amplification.

DC Current Gain. If I_{CO} is negligibly small in comparison to I_C, α approximately equals I_C/I_E. This is referred to as the *dc current gain* of the common base (CB) transistor and is denoted by α_{dc}.

* Some authors taking standard transistor convention, write this equation as $I_E = -(I_C + I_B)$

Thus $$\alpha_{dc} = \frac{I_C}{I_E} \quad ...(3.10)$$

α_{dc} is always positive and less than unity.

Small Signal Current Gain α_{ac}. Small signal current gain α_{ac} is defined as the ratio of change in collector current to change in emitter current

i.e., $$\alpha_{ac} = \frac{\Delta I_C}{\Delta I_E} \text{ for constant collector-base voltage } V_{CB} \quad ...(3.11)$$

α_{ac} is always positive but less than unity (very close to unity). It also varies with I_E, V_{CB} and temperature.

Generalized Expression For Collector Current. We know that a small reverse current flows across a P-N junction diode due to *thermally generated minority carriers* that are propelled by the barrier potential. When the junction is reverse biased, this reverse current increases slightly. For moderate reverse-bias voltages, the reverse current attains its saturation value, I_S (or I_0). Because the collector-base junction of a transistor is reverse biased, there is likewise a reverse current due to thermally generated carriers. Of course this "reverse" current, in the context of a transistor, is in the same direction as the main (collector) current flowing through the device due to the injection of minority carriers into the base. The total collector current is, therefore, the sum of these two components: the injected minority carriers and the thermally generated minority carriers

i.e., $$I_C = \alpha I_E + I_{CO}$$

The reverse current I_{CBO} is given by the volt-ampere relationship of Eq. (1.35) with I_0 replaced by $-I_{CO}$ and V by V_C where symbol V_C represents the voltage drop across junction J_C from the P-side to the N-side and the complete expression for collector current I_C may be written as

$$I_C = \alpha I_E + I_{CO}(1 - e^{V_C/\eta V_T}) \quad ...(3.12)$$

If V_C is negative and is of large magnitude in comparison with that of V_T, Eq. (3.12) reduces to Eq. (3.4).

The physical interpretation of Eq. (3.12) is that the P-N junction diode current crossing the collector junction J_C is supplemented by a fraction α of the current I_E flowing in the emitter.

Example 3.1. For a transistor if I_E = 1 mA and α = 0.98, determine the value of I_B.

[Pb. Technical Univ. Basic Electrical & Electronics Engg. May 2006]

Solution :

Emitter current, I_E = 1 mA

Current amplification factor, α = 0.98

Collector current, $I_C = \alpha I_E = 0.98 \times 1$ mA = 0.98 mA

Base current, $I_B = I_E - I_C = 1.0 - 0.98 = 0.02$ mA **Ans.**

3.7. CURRENT AMPLIFICATION FACTORS

There are two current amplification factors, the alpha factor (α) and the beta factor (β) defined below :

The alpha factor (α) is also called the *current amplification factor* and is the ratio of collector current (I_C) to emitter current (I_E).

In general, $$\alpha = \frac{\Delta I_C}{\Delta I_E} \; V_{CB} \text{ remaining constant} \quad ...(3.13)$$

The beta factor (β) is the current gain factor (*also called the transport factor*) of a common emitter circuit and is defined as the ratio of collector current (I_C) and base current (I_B).

In general, $$\beta = \frac{\Delta I_C}{\Delta I_B} \quad ...(3.14)$$

3.8. RELATIONSHIP BETWEEN α AND β

As defined above $\beta = \dfrac{\Delta I_C}{\Delta I_B}$

and $\alpha = \dfrac{\Delta I_C}{\Delta I_E} = \dfrac{\Delta I_C}{\Delta I_C + \Delta I_B}$ $\quad \because \Delta I_E = \Delta I_B + \Delta I_C$

or $\dfrac{1}{\alpha} = \dfrac{\Delta I_C + \Delta I_B}{\Delta I_C} = 1 + \dfrac{\Delta I_B}{\Delta I_C} = 1 + \dfrac{1}{\beta} = \dfrac{1+\beta}{\beta}$

$$\text{Hence } \alpha = \frac{\beta}{1+\beta} \quad \text{and} \quad \beta = \frac{\alpha}{1-\alpha} \qquad ...(3.15)$$

Thus, if $\alpha = 0.98$ $\quad \beta = \dfrac{0.98}{0.02} = 49$

and if $\alpha = 0.99$ $\quad \beta = \dfrac{0.99}{0.01} = 99$

i.e., a small variation in α corresponds to a large variation in β. It is, therefore, better to determine β experimentally and calculate therefrom the corresponding value of α by means of expression

$$\alpha = \frac{\beta}{1+\beta} \qquad ...(3.16)$$

Example 3.2. If the base current in transistor is 30 μA when the emitter current is 7.2 mA, what are the values of α and β ? [*G.B. Technical Univ. Electronics Engineering First Semester 2011-12*]

Solution : Base current, $I_B = 30\ \mu A = 0.03$ mA

Emitter current, $I_E = 7.2$ mA

Collector current, $I_C = I_E - I_B = 7.2 - 0.03 = 7.17$ mA

$\alpha = \dfrac{I_C}{I_E} = \dfrac{7.17}{7.20} = 0.9958$ **Ans.**

$\beta = \dfrac{I_C}{I_B} = \dfrac{7.17}{0.03} = 239$ **Ans.**

Example 3.3. Determine I_E, α and β of the common base transistor given $I_C = 7$ mA, $I_B = 0.1$ mA. [*Mahamaya Technical Univ. Electronics Engineering Second Semester 2011-12*]

Solution : Base current, $I_B = 0.1$ mA

Collector current, $I_C = 7$ mA

Emitter current, $I_E = I_C + I_B = 7 + 0.1 = 7.1$ mA

$\alpha = \dfrac{I_C}{I_B} = \dfrac{7}{7.1} = 0.9859$ **Ans.**

$\beta = \dfrac{I_C}{I_B} = \dfrac{7}{0.1} = 70$ **Ans.**

Example 3.4. In a common base configuration of a PNP transistor the current amplification factor is 0.988. Determine the base current if emitter current is 1.2 mA. Neglect collector leakage current. [*Rajasthan Univ. 2001*]

Solution : Current amplification factor, $\alpha = 0.988$

Emitter current, $I_E = 1.2$ mA

Collector leakage current, $I_{CO} \simeq 0$

From Eq. (3.4)

$$\text{Collector current, } I_C = \alpha I_E + I_{CO} = 0.988 \times 1.2 + 0 = 1.1856 \text{ mA}$$

$$\text{Base current, } I_B = I_E - I_C = 1.2 - 1.1856 = 0.0144 \text{ mA} = 14.4 \ \mu\text{A} \quad \textbf{Ans.}$$

Example 3.5. For a given transistor $\alpha_{dc} = 0.98$, $I_E = 2$ mA. Calculate β_{dc} and I_B. *[Rajasthan Univ. 2004]*

Solution : Emitter current, $I_E = 2$ mA

$$\alpha_{dc} = 0.98$$

$$\beta_{dc} = \frac{\alpha_{dc}}{1-\alpha_{dc}} = \frac{0.98}{1-0.98} = 49 \quad \textbf{Ans.}$$

$$\text{and base current, } I_B = \frac{I_C}{\beta_{dc}} = \frac{\alpha_{dc} I_E}{\beta_{dc}} = \frac{0.98 \times 2}{49} = 0.04 \text{ mA or } 40 \ \mu\text{A} \quad \textbf{Ans.}$$

Example 3.6. For a PNP transistor the base current and collector current are respectively 45 μA and 5.45 mA. Determine (*i*) values of α, β and I_E; (*ii*) Base current required to make collector current of 10 mA.

[Rajasthan Univ. 1994, 2003]

Solution : Base current, $I_B = 45 \ \mu\text{A} = 45 \times 10^{-6} \text{ A} = 0.045$ mA

Collector current, $I_C = 5.45$ mA

(*i*) Emitter current, $I_E = I_B + I_C = 0.045 + 5.45 = 5.495$ mA **Ans.**

$$\text{Current amplification factor, } \alpha = \frac{I_C}{I_E} = \frac{5.45}{5.495} = 0.9918 \quad \textbf{Ans.}$$

$$\text{Current gain factor, } \beta = \frac{I_C}{I_B} = \frac{5.45}{0.045} = 121 \quad \textbf{Ans.}$$

(*ii*) Base current required to make collector current I_C of 10 mA,

$$I_B = \frac{I_C}{\beta} = \frac{10}{121} = 0.0826 \text{ mA or } 82.6 \ \mu\text{A} \quad \textbf{Ans.}$$

3.9. BASE SPREADING RESISTANCE

With two depletion layers penetrating the base, the base holes are confined to the thin region of P-type semiconductor illustrated in Fig. 3.9. The resistance of this thin region (or section) is called the *base spreading resistance* $r_{bb'}$. With the increase in reverse bias V_{CB} on the collector diode, the width of the P-section is reduced, which is equivalent to an increase in $r_{bb'}$.

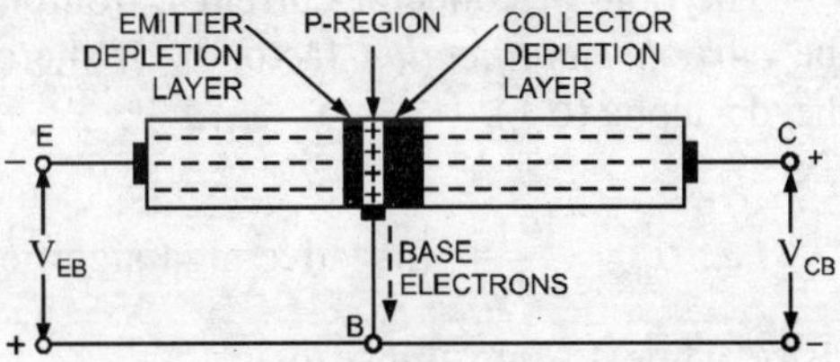

Fig. 3.9. ***Base Spreading Resistance***

The recombination current in the base I_B must flow through $r_{bb'}$. When it does so it causes a voltage drop. The value of base spreading resistance $r_{bb'}$ depends upon the width of the P-section as well as doping of the base. In rare case $r_{bb'}$ may be as high as 1,000 Ω. Typically, it is in the range of 50 – 150 Ω. The effects of $r_{bb'}$ are important in high-frequency circuits. At low frequencies $r_{bb'}$ usually has little effect and so it is ignored.

3.10. TRANSISTOR CIRCUIT CONFIGURATION

As already mentioned, a transistor is a three-terminal device (having three terminals namely emitter, base and collector) but we require four terminals—two for the input and two for the output for connecting it in a circuit. Hence one of the terminals of the transistor is made common to the input and output circuits.

Thus there are three types of configurations for operation of a transistor. These configurations are *common base* (CB), *common emitter* (CE) and *common collector* (CC) configurations. The term 'common' is used to denote the electrode that is common to input and output circuits. Since the common terminal is usually grounded, these configurations are also referred to as *grounded base*, *grounded emitter* and *grounded collector* configurations, as shown in Figs. 3.10 (*a*), 3.10 (*b*) and 3.10 (*c*) respectively. Each configuration has its own merits and demerits. It is to be remembered that regardless of the circuit configurations, the emitter is always forward biased and the collector is always reverse biased.

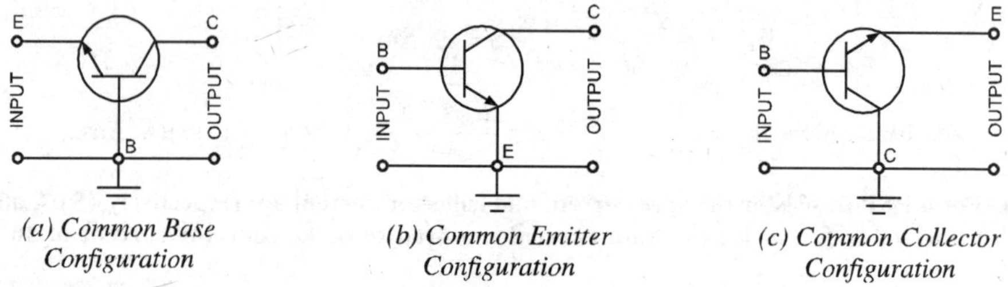

Fig. 3.10. ***Different Circuit Configurations For NPN Transistor***

3.11. COMMON BASE (CB) CONFIGURATION

In common base (CB) configuration, input is connected between emitter and base and output is taken across collector and base, as illustrated in Fig. 3.11 for an N-P-N transistor. Thus base is common between input and output circuits.

In this arrangement (Fig. 3.11) the emitter-base junction is forward biased and collector-base junction reverse biased. The emitter current I_E flows in the input circuit and the collector current I_C in the output circuit. A change in emitter current produces a similar change in the collector current due to change in diffusion of electrons or holes across the junctions. The impedances of these two circuits are different and thus some voltage and power gains can be achieved since a current variation in the low resistance emitter-base circuit has similar current variations in the high resistance collector-base circuit.

The ratio of collector current I_C (output current) to the emitter current I_E (input current) is called the *current amplification* factor α. If there is no input ac signal then the ratio of I_C to I_E is called the dc alpha (α_{dc}).

i.e., $\alpha_{dc} = \dfrac{-I_C}{I_E}$ (negative sign signifies that I_E flows out of transistor while I_C into transistor).*

If we write α_{dc} simply by α and ignore –ve sign,

$$\alpha = \frac{I_C}{I_E} \qquad ...(3.17)$$

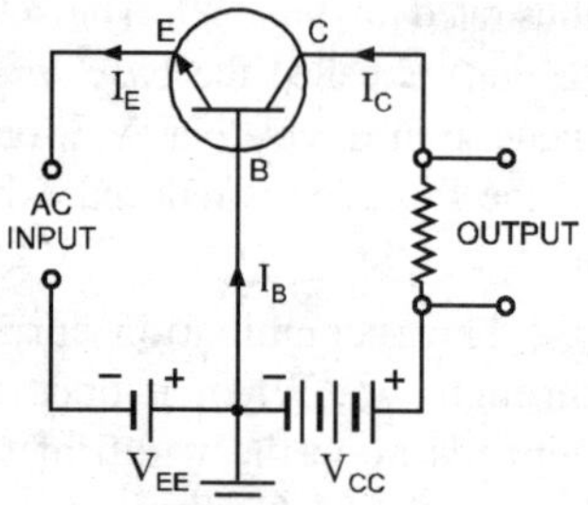

Fig. 3.11. ***Common Base NPN Transistor***

Incidently, there is also an ac alpha (α_{ac}) for a transistor and it refers to the ratio of change in collector current to change in emitter current

$$i.e., \; \alpha_{ac} = \frac{-\Delta I_C}{\Delta I_E} \text{ for constant collector-base voltage } V_{CB}. \qquad ...(3.18)$$

* In case of a P-N-P transistor emitter current I_E flows into the transistor and collector current I_C flows out of it.

α* of a transistor is a measure of its quality–higher the value of α better is the transistor in the sense that I_C approaches I_E, its value ranges from 0.95 to 0.99. The value of α can be increased by making base thin and lightly doped.

The collector current consists of two parts (*i*) the current produced by normal transistor action *i.e.*, component depending upon emitter current (αI_E) which is produced by majority carriers and (*ii*) the leakage current due to movement of minority carriers across base-collector junction on account of it being reverse biased. This leakage current flows in the same direction as that of I_C and is abbreviated as I_{CBO}** (collector-base current with emitter open).

$\therefore$ Total collector current, $I_C = \alpha I_E + I_{CBO}$ ...(3.19)

The current I_{CBO}, sometimes referred to as I_{CO}, is usually very small and may be neglected in transistor calculations.

From Eq. (3.19)

$$\alpha = \frac{I_C - I_{CO}}{I_E} \simeq \frac{I_C}{I_E} \quad ...(3.20)$$

neglecting I_{CO} in comparison with I_C

$$I_B = I_E - I_C = I_E - (\alpha I_E + I_{CO})$$

$$= (1 - \alpha) I_E - I_{CO} \quad ...(3.21)$$

$\because$ from Eq. (3.19) $I_C = \alpha I_E + I_{CO}$

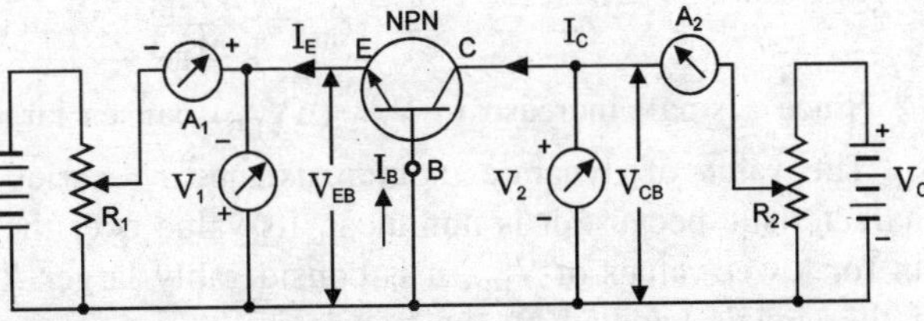

Fig. 3.12. ***Circuit Arrangement For Determination of Input and Output Characteristic Curves of a Transistor in CB Configuration***

3.11.1. Characteristics of Common Base Configuration

The performance of transistors, when connected in a circuit, may be determined from their characteristic curves that relate different dc currents and voltages of a transistor. Such curves are known as *static characteristic curves*. There are two important characteristics of a transistor viz. input characteristics and output characteristics.

Both of the input and output characteristics can be determined with the same circuit arrangement shown in Fig. 3.12.

Input Characteristics. The curve drawn between emitter current I_E and emitter-base voltage V_{EB} for a given value of collector-base voltage V_{CB} is known as *input characteristic*. For determination of the input characteristics, the output voltage V_{CB} is maintained constant, and the input voltage V_{EB} is set at several convenient levels. For each level of input voltage, the input current I_E is noted. Then the curve is plotted between emitter current I_E and emitter base voltage V_{EB}, as shown in Fig. 3.13.

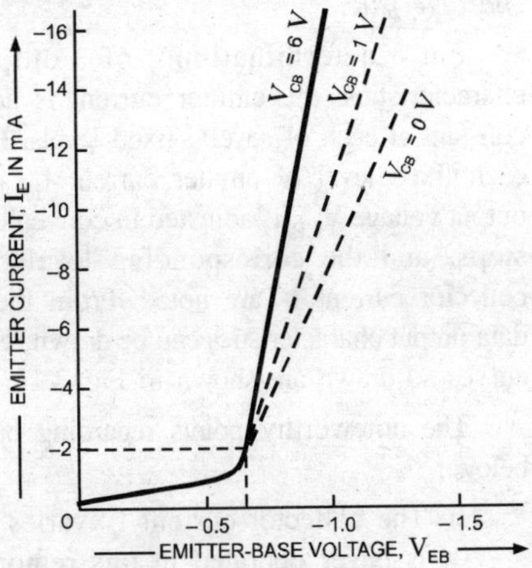

Fig. 3.13. ***Input Characteristics For Common Base NPN Transistor***

For a given value of V_{CB} the curve is just like that of a forward biased P-N junction because emitter-base junction in forward bias mode is similar to a P-N junction. This characteristic may be used to determine input resistance of the transistor.

---

* For all practical purposes α_{dc}, α_{ac} and α have the same values.

** I_{CBO} is exactly like the reverse saturation current I_S or I_0 of a reverse biased diode discussed earlier and is extremely temperature dependent as it is made up of thermally generated minority carriers.

From Fig. 3.13 it is seen that with the increase in V_{CB}, it conducts better, although the effect is not very significant. This is because large collector-base (reverse bias) voltages cause the depletion layer at the collector-base junction to penetrate deeper into the base of the transistor, thus reducing the distance and the resistance between the emitter-base and collector-base regions.

There exists a *cutin*, *offset* or *threshold* voltage V_γ below which the emitter current is very small. Substantial emitter current flows only when emitter-base junction is forward biased beyond the knee of the characteristic (approximately 0.7 V for Si and 0.3 V for Ge).

The emitter current I_E increases rapidly with the small increase in V_{EB} which means low dynamic input resistance because input resistance of a transistor is given as the reciprocal of the slope of its input characteristic *i.e.*, mathematically

$$r_{in} = \frac{\Delta V_{EB}}{\Delta I_E} \quad \ldots V_{CB} \text{ constant} \qquad \ldots(3.22)$$

Since a small increase in V_{EB} (ΔV_{EB}) causes large change in I_E (Δ I_E), r_{in} is quite low.

The value of dynamic input resistance r_{in} varies from point to point in the initial part of the characteristic because it is nonlinear. Its value over the linear part of the characteristic is about 100 Ω but for lower values of V_{EB}, it is considerably larger. This dependence of r_{in} on V_{EB} causes distortion of the signals handled by the transistors.

Output Characteristics. The curve drawn between collector current I_C and collector-base voltage V_{CB} for a given value of emitter current I_E is known as *output characteristic*.

For determination of output characteristics, the emitter current is held constant at each of several fixed levels. For each fixed level of emitter current I_E, the output voltage V_{CB} is adjusted in convenient steps, and the corresponding levels of collector current I_C are noted. From these data output characteristics can be drawn. The curves so drawn are shown in Fig. 3.14.

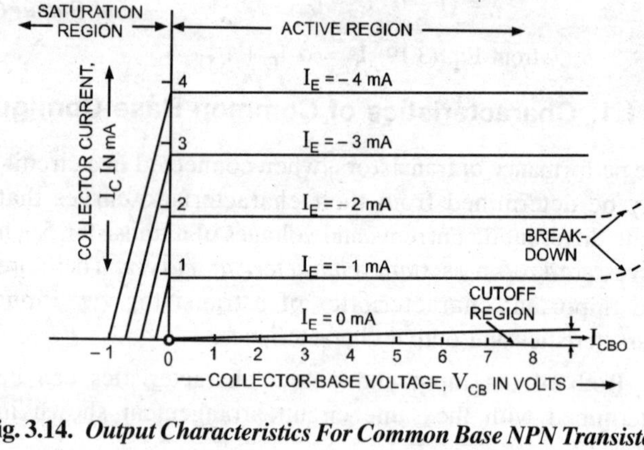

Fig. 3.14. ***Output Characteristics For Common Base NPN Transistor***

The noteworthy points regarding output characteristics of common base (CB) configuration are given below :

(*i*) The collector current I_C varies with V_{CB} only for very low voltage (below 1 V) but transistor is never operated in this region.

(*ii*) In active region (emitter forward biased and collector reverse biased) collector current I_C is almost equal to I_E and appears to remain constant when V_{CB} is increased. In fact, there is very small increase in I_C with increase in V_{CB}. This is because the increase in V_{CB} expands the collector-base depletion region and thus shortens the distance between the two depletion regions. With emitter current I_E held constant, however, the increase in I_C is so small that it is usually noticeable only for large variations in V_{CB}. I_C is slightly lesser than I_E in magnitude.

Transistor is normally operated in active region.

(*iii*) Although the collector current I_C is practically independent of V_{CB} over the transistor operating range, however, if V_{CB} is increased beyond a certain value, I_C eventually increases rapidly because of avalanche or zener (or both) effects. This condition is known as *punch-through* or *reach-through*. When it occurs large currents can flow, possibly destroying the device. The extension of the depletion region is, of course, the direct consequence of the increase in V_{CB}. Thus it is very essential to maintain V_{CB} below the maximum safe limit specified by the manufacturer of the device.

(*iv*) A very large change in collector voltage causes a very small change in collector current *i.e.*, output resistance of CB configuration is very high (of the order of few hundred kΩ)—the dynamic output resistance r_{out} being given as the ratio of ΔV_{CB} and ΔI_C for a given value of I_E.

(*v*) In *cutoff* region (emitter and collector junctions both reverse biased) small collector current I_C flows even when emitter current $I_E = 0$. This is the collector leakage current I_{CBO} or I_{CO}.

(*vi*) In saturation region (both emitter and collector junctions forward biased) collector current I_C flows even when $V_{CB} \simeq 0$. Even when the externally applied bias voltage is reduced to zero, there is still a barrier potential existing at the collector-base junction, and this assists in the flow of I_C. To stop it the collector-base junction has to be forward biased. Consequently, as illustrated in Fig. 3.14, collector current I_C is reduced to zero when V_{CB} is increased negatively.

We have already determined that a small signal voltage impressed in the low-resistance input (emitter) circuit of a transistor causes a relatively large emitter current. Almost the same amount of current will flow in the high-resistance output (collector) circuit of transistor, where the voltage may be very high. Evidently, then, both the output voltage and power can be quite large, as compared to the tiny input voltage and power present at the emitter.

CB configuration is rarely used in audio-frequency (AF) circuits because its current gain is less than unity and its input and output resistances are quite different.

Current Gain Characteristics. The current gain characteristics, also referred to as *forward transfer characteristics,* can be determined experimentally (by keeping V_{CB} constant at a convenient value, recording I_C value measured for different settings of I_E and then plotting I_C versus I_E) or from the common-base output characteristics as shown in Fig. 3.15. A vertical line is drawn through a selected value of V_{CB} and corresponding values of I_E and I_C are read off along this line. Then a graph is plotted between I_C versus I_E, and the characteristic is labelled with the V_{CB} used.

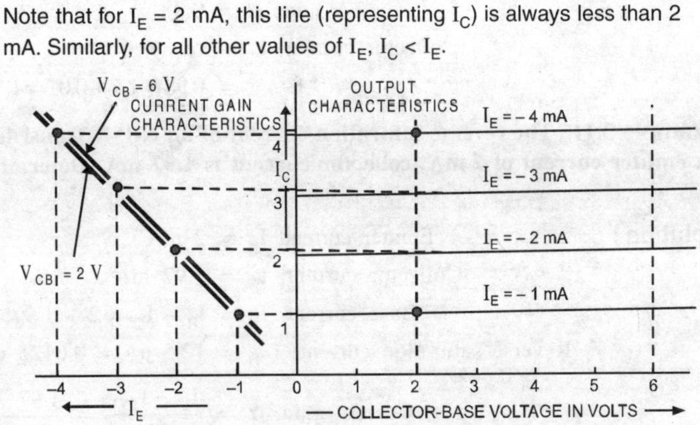

Fig. 3.15. ***Current Gain Characteristics For Common Base NPN Transistor***

Example 3.7. A transistor is connected in CB configuration. When the emitter voltage is changed by 200 mV, the emitter current changes by 5 mA. During this variation, the collector-to-base voltage is kept fixed. Calculate the dynamic input resistance of transistor. [*Rajasthan Univ. 2003*]

Solution : The dynamic input resistance for CB configuration is given as

$$r_{in} = \left.\frac{\Delta V_{EB}}{\Delta I_E}\right|_{V_{CB} = \text{constant}} = \frac{200 \text{ mV}}{5 \text{ mA}} = 40\ \Omega \textbf{ Ans.}$$

Example 3.8. In a grounded base configuration (Fig. 3.11) the voltage drop across load resistance of 4 kΩ is 3 V. Determine base current. α = 0.96. *[Rajasthan Univ. 2003]*

Solution :

$$\text{Load resistance } R_L = 4\ k\Omega$$

$$\text{Voltage drop across load resistance, } I_C R_L = 3\ V$$

$$\text{Collector current, } I_C = \frac{I_C R_L}{R_L} = \frac{3\ V}{4 k\Omega} = 0.75\ mA$$

$$\text{Emitter current, } I_E = \frac{I_C}{\alpha} = \frac{0.75}{0.96} = 0.78\ mA$$

$$\text{Base current, } I_B = I_E - I_C = 0.78 - 0.75 = 0.03\ mA \textbf{ Ans.}$$

Example 3.9. An N-P-N transistor with α = 0.98 is operated in the CB configuration. If the emitter current is 3 mA and reverse saturation current is 10 μA. What are the base current and collector current ?

[G.B. Technical Univ. Electronics Engineering Second Semester 2010-11]

Solution :

$$\text{Emitter current, } I_E = 3\ mA$$

$$\text{Reverse saturation current, } I_{CO} = 10\ \mu A = 0.01\ mA$$

$$\alpha = 0.98$$

$$\text{Total collector current, } I_C = \alpha I_E + I_{CBO} = 0.98 \times 3 + 0.01 = 2.95\ mA \textbf{ Ans.}$$

$$\text{Base current, } I_B = I_E - I_C = 3 - 2.95 = 0.05\ mA \textbf{ Ans.}$$

Example 3.10. In a common base configuration, the emitter current is 1.6 mA. The collector current with emitter circuit open is 10 μA. Determine the total collector current. *[MREC 2002]*

Solution :

$$\text{Emitter current, } I_E = 1.6\ mA$$

$$\text{Collector current with emitter circuit open, } I_{CO} = 10\ \mu A = 0.01\ mA$$

$$\text{Assume } \alpha = 0.98$$

$$\text{Collector current, } I_C = \alpha I_E + I_{CO}$$ Refer to Eq. (3.4)

$$= 0.98 \times 1.6 + 0.01 = 1.578\ mA \textbf{ Ans.}$$

Example 3.11. The reverse saturation current of an N-P-N transistor in common base circuit is 12.5 μA. For an emitter current of 2 mA, collector current is 1.97 mA. Determine the current gain and base current.

[Rajasthan Univ. 2003]

Solution :

$$\text{Emitter current, } I_E = 2\ mA$$

$$\text{Collector current, } I_C = 1.97\ mA$$

$$\text{Base current, } I_B = I_E - I_C = 2 - 1.97 = 0.03\ mA \textbf{ Ans.}$$

$$\text{Reverse saturation current, } I_{CO} = 12.5\ \mu A = 0.0125\ mA$$

$$\text{Current gain, } \alpha = \frac{I_C - I_{CO}}{I_E} = \frac{1.97 - 0.0125}{2} = 0.979 \textbf{ Ans.}$$

Example 3.12. A silicon transistor has I_{CBO} = 1 μA at 27°C. The leakage current doubles every 6°C rise in temperature. If the I_E = 1 mA, find the I_B at 57°C. Given that α = 0.95. *[Rajasthan Technical Univ. 2007]*

Solution :

$$\text{Emitter current, } I_E = 1\ mA$$

$$\text{Current amplification factor, } \alpha = 0.95$$

$$\text{Leakage current at 57°C, } I_{CBO_{57}} = I_{CBO_{27}} \times 2^{\frac{57-27}{6}}$$ ∵ I_{CBO} doubles every 6°C rise in temperature

$$= 1 \times 10^{-6} \times 2^5 = 32 \times 10^{-6}\ A = 0.032\ mA$$

$$\text{Collector current, } I_C = \alpha I_E + I_{CBO} = 0.95 \times 1 + 0.032 = 0.982\ mA$$

$$\text{Base current at 57°C, } I_B = I_E - I_C = 1 - 0.982 = 0.018\ mA = 18\ \mu A \textbf{ Ans.}$$

3.12. COMMON EMITTER CONFIGURATION

In common emitter configuration input is connected between base and emitter while the output is taken between collector and emitter, as shown in Fig. 3.16. Thus emitter is common to input and output circuits.

In this configuration, the bias voltages are applied between collector and emitter and base and emitter. Emitter-base junction is forward biased and the base is made more positive than the emitter by V_{BB}, collector-emitter junction is reverse biased and the collector is made more positive than emitter by V_{CC}. The value of V_{CC} must be greater than that of V_{BB}.

In this circuit arrangement (Fig. 3.16) the base current I_B flows in the input circuit and collector current I_C flows in the output circuit. In this arrangement current gain between the input and output sides is obtained; and since the input resistance is again less than the output resistance (though difference is not as much as in case of common base arrangement) there will be high voltage and power gains like those in equivalent vacuum tube circuits. The common emitter produces a reversal between input and output signals. Common emitter (CE) configuration is commonly used because its current, voltage and power gains are quite high and output to input impedance ratio is moderate.

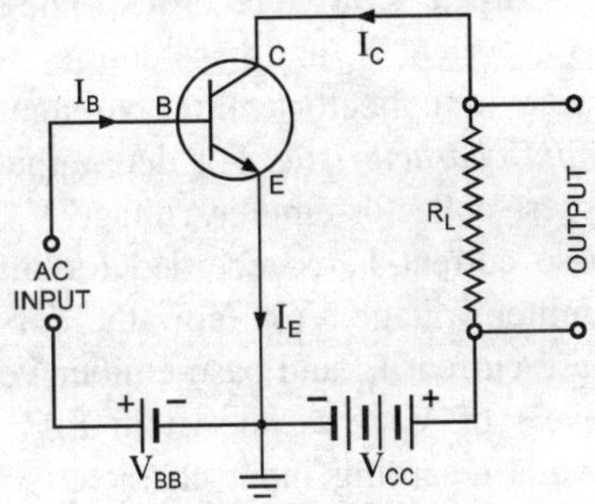

Fig. 3.16. ***Common Emitter NPN Transistor***

The ratio of change in collector current (output current) and change in base current (input current) is called the *base current amplification factor* β

i.e., $$\beta = \frac{\Delta I_C}{\Delta I_B} \qquad \text{...(3.23)}$$

Almost in all transistors, the base current is less than 5% of emitter current I_E so β is usually higher than 20. Usually β ranges from 20 to 100.

Also $$\beta = \frac{I_C}{I_B} = \frac{I_C}{I_E - I_C} = \frac{I_C/I_E}{1-\frac{I_C}{I_E}} = \frac{\alpha}{1-\alpha} \qquad \text{...(3.24)}$$

In CE configuration, a small collector current flows even when base current is zero (*i.e.*, base lead is open). This is the collector *cutoff* current and denoted by I_{CEO}. I_{CEO} must be larger than I_{CBO}.

The current equations are $$I_E = I_B + I_C \qquad \text{...(3.25)}$$

and $$I_C = \alpha\, I_E + I_{CBO} = \alpha\,(I_B + I_C) + I_{CBO}$$

or $$I_C\,(1 - \alpha) = \alpha\, I_B + I_{CBO}$$

or $$I_C = \frac{\alpha}{1-\alpha} I_B + \frac{I_{CBO}}{1-\alpha} \qquad \text{...(3.26)}$$

and total collector current, $$I_C = \beta\, I_B + I_{CEO} \qquad \text{...(3.27)}$$

Comparing Eqs. (3.26) and (3.27) we have

$$\beta = \frac{\alpha}{1-\alpha}$$ as already derived in Eq. (3.24)

and $$I_{CEO} = \frac{1}{1-\alpha} I_{CBO} \qquad \text{...(3.28)}$$

Substituting the value of $I_{CEO} = \frac{1}{1-\alpha} I_{CBO}$ in Eq. (3.27) we have

$$I_C = \beta I_B + \frac{1}{1-\alpha} I_{CBO} = \beta I_B + (\beta + 1)\, I_{CBO} \qquad \text{...(3.29)}$$

$\because$ from Eq. (3.24) $\frac{1}{1-\alpha} = \beta + 1$

3.12.1. Characteristics of Common Emitter Configuration

The circuit used for determination of transistor common emitter characteristics is shown in Fig. 3.17. The input voltage is applied between base and emitter terminals, and the output is taken from collector and emitter terminals. Input and output voltages and currents are measured by voltmeters and ammeters as shown.

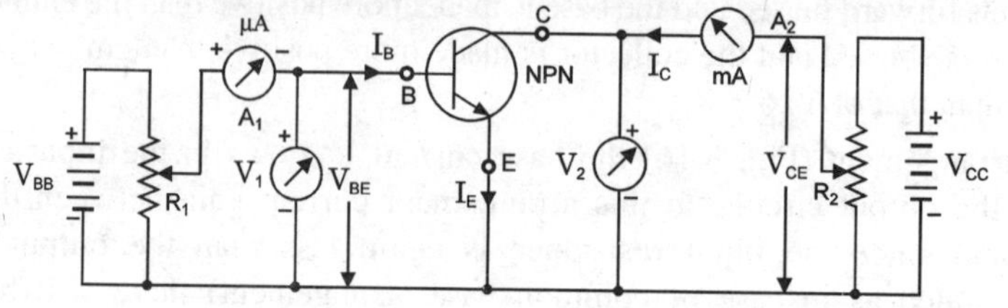

Fig. 3.17. ***Circuit Arrangement For Determination of Input and Output Characteristics of a Transistor in CE Configuration***

Input Characteristics. The curve drawn between base current I_B and base-emitter voltage V_{BE} for a given value of collector-emitter voltage V_{CE} is known as the *input characteristic*. For determination of input characteristics, collector-emitter voltage V_{CE} is held constant and base current I_B is recorded for different values of base-emitter voltage V_{BE}. Now the curves are drawn between base current I_B and base-emitter voltage V_{BE} for different levels of V_{CE}, as shown in Fig. 3.18. The noteworthy points regarding input characteristics are given below :

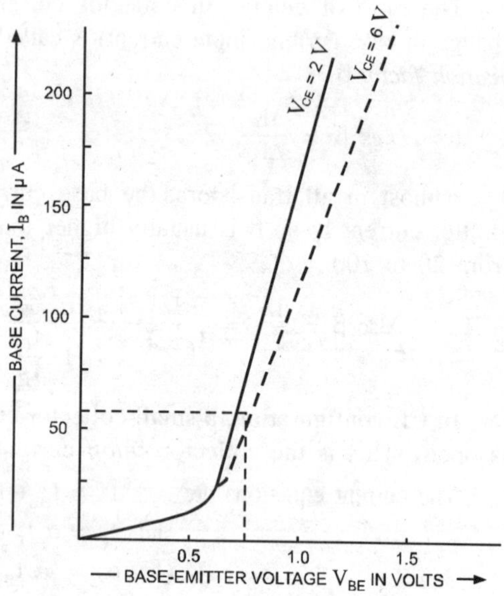

Fig. 3.18. ***Input Characteristics For Common Emitter NPN Transistor***

(*i*) The input characteristics of CE transistors are quite similar to those of a forward biased diode because the base-emitter region of the transistor is a diode and it is forward biased.

(*ii*) In comparison to common base (CB) arrangement base (or input) current increases less rapidly with the increase in base-emitter voltage, V_{BE}. This indicates that input resistance is larger in common emitter configuration than that in common base configuration. Due to initial nonlinearity of the curve, input resistance varies from point to point in the initial part of the characteristic. Its value over the linear part of the curve is of the order of few hundred ohms.

(*iii*) An increment in value of V_{CE} causes the input current I_B to be lower for a given level of V_{BE}. This is because the higher levels of V_{CE} provide greater collector-base junction reverse bias, causing greater depletion region penetration into the base and thus reducing the distance between the collector-base and emitter-base regions. As a result more of the charge carriers from the emitter flows across the collector-base junction, and few flow out through the base lead. Usually effect of change in V_{CE} on the input characteristic is ignored.

The ratio of change in base-emitter voltage (ΔV_{BE}) to the resulting change in base current (ΔI_B) at constant collector-emitter voltage (V_{CE}) is known as *dynamic input resistance* (r_{in}). It is calculated as the reciprocal of the slope of the input characteristic at a given V_{BE}.

$$\text{i.e., } r_{in} = \left.\frac{\Delta V_{BE}}{\Delta I_B}\right|_{V_{CE} = \text{constant}} \quad ...(3.30)$$

Output Characteristics. Output characteristic for a common emitter transistor is the curve drawn between collector current I_C and collector-emitter voltage V_{CE} for a given value of base current I_B. For determination of common emitter output characteristics, base current I_B is maintained constant at several convenient levels. At each fixed level of I_B, collector-emitter voltage V_{CE} is adjusted in steps, and the corresponding values of collector current I_C are noted. For each level of I_B, I_C is plotted versus V_{CE}. Thus a family of characteristics is obtained which are typically as illustrated in Fig. 3.19. The noteworthy points regarding output characteristics are given below :

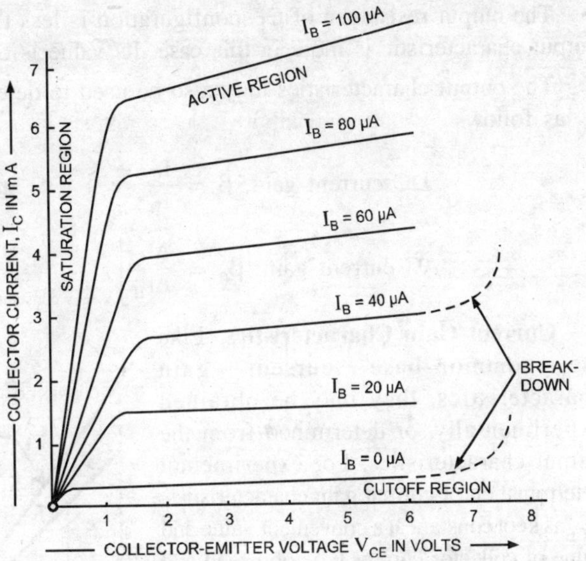

Fig. 3.19. ***Output Characteristics For Common Emitter NPN Transistor***

(*i*) The collector current I_C varies with V_{CE} for V_{CE} between 0 and 1 V and then becomes almost constant and independent of V_{CE}. The transistors are always operated above 1 V.

(*ii*) Output characteristic in CE configuration has some slope while CB configuration has almost horizontal characteristics. This indicates that output resistance in case of CE configuration is less than that in CB configuration.

(*iii*) In active region (collector junction reverse biased and emitter junction forward biased), for small values of base current I_B the effect of collector voltage V_C over I_C is small but for large values of I_B this effect increases.

The shape of the characteristic is same as in CB configuration but with the difference that collector current I_C is larger than input current (base current I_B). Thus the current gain for this configuration is larger than unity. The transistor must always be operated in the active region, when employed as an amplifying device.

(*iv*) With low values (ideally zero) of V_{CE} the transistor is said to be operated in saturation region and in this region base current I_B does not cause a corresponding change in collector current I_C.

(*v*) With much higher V_{CE}, the collector-base junction completely breaks down and because of this avalanche breakdown collector current I_C increases rapidly and the transistor gets damaged.

(*vi*) In cutoff region, small amount of collector current I_C flows even when base current $I_B = 0$. This is called I_{CEO}. Since main current I_C is zero, the transistor is said to be cutoff.

(*vii*) Moderate output to input impedance ratio makes this configuration an ideal one for coupling between various transistor stages.

The ratio of change in collector-emitter voltage (ΔV_{CE}) to the change in collector current (ΔI_C) at constant base current is known as *dynamic* output resistance.

It is calculated as the reciprocal of the slope of output characteristic at a given V_{CE}.

$$\text{i.e., } r_{out} = \left.\frac{\Delta V_{CE}}{\Delta I_C}\right|_{I_B = \text{constant}} \quad ...(3.31)$$

The output resistance of CE configuration is less than that of CB configuration as the slope of the output characteristic is more in this case. Its value is of the order of 50 kΩ.

The output characteristics may also be used to determine the dc current gain β and ac current gain β_0 as follows

$$\text{DC current gain, } \beta = \frac{I_C}{I_B} \quad \text{...(3.32)}$$

$$\text{AC current gain, } \beta_0 = \left.\frac{\Delta I_C}{\Delta I_B}\right|_{V_{CE} = \text{constant}} \quad \text{...(3.33)}$$

Current Gain Characteristics. Like the common-base current gain characteristics, they may be obtained experimentally, or determined from the output characteristics. For experimental determination of current gain characteristics, V_{CE} is kept constant at a convenient value and value of collector current I_C is observed and recorded for different values of base current I_B and then a graph is plotted between I_C and I_B.

The current gain characteristics can be obtained from the output characteristics as shown in Fig. 3.20, and detailed in Art. 3.11 except that I_C is plotted against I_B rather than I_C against I_E.

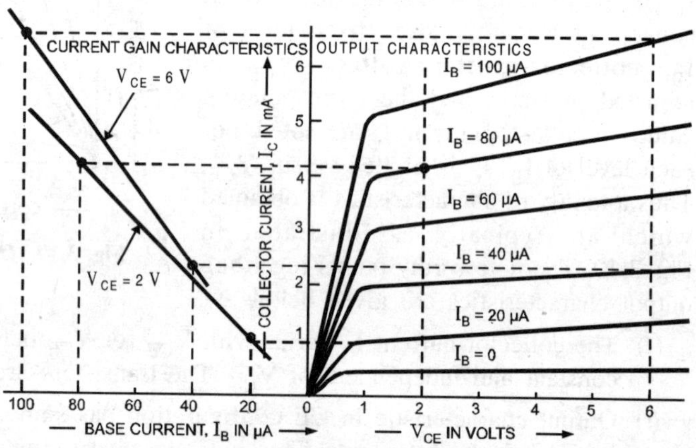

Fig. 3.20. ***Current Gain Characteristics For Common Emitter NPN Transistor***

Example 3.13. A change of 250 mV in base-emitter voltage causes a change of 1 mA in the base current. Determine the dynamic input resistance.

Solution : Change in base-emitter voltage, $\Delta V_{BE} = 250$ mV

Change in base current, $\Delta I_B = 1$ mA

$$\text{Dynamic input resistance, } r_{in} = \frac{\Delta V_{BE}}{\Delta I_B} = \frac{250 \times 10^{-3}}{1 \times 10^{-3}} = 250\ \Omega \textbf{ Ans.}$$

Example 3.14. Increase in collector-emitter voltage from 5 V to 10 V causes increase in collector current from 5 mA to 5.8 mA. Determine the dynamic output resistance.

Solution : Change in collector-emitter voltage, $\Delta V_{CE} = 10 - 5 = 5$ V

Change in collector current, $\Delta I_C = 5.8 - 5.0 = 0.8$ mA

$$\text{Dynamic output resistance, } r_{out} = \frac{\Delta V_{CE}}{\Delta I_C} = \frac{5}{0.8 \times 10^{-3}} = 6.25\ k\Omega \textbf{ Ans.}$$

Example 3.15. A transistor is connected in common emitter (CE) configuration in which collector supply is 8 volts and the voltage drop across resistance R_C connected in the collector circuit is 0.5 volt.

The value of R_C = 800 ohms. If α = 0.96, determine :
(*i*) Collector-emitter voltage (*ii*) Base current.

[*Rajasthan Technical Univ. Electrical and Electronics Engineering March 2009*]

Solution : Collector supply voltage, $V_{CC} = 8$ V

Voltage drop across resistance $= I_C R_C = 0.5$ V

(*i*) Collector-emitter voltage, $V_{CE} = V_{CC} - I_C R_C = 8 - 0.5 = 7.5$ V **Ans.**

$$\text{Collector current, } I_C = \frac{I_C R_C}{R_C} = \frac{0.5}{800} \text{ A} = 0.625 \text{ mA}$$

$$\beta = \frac{\alpha}{1-\alpha} = \frac{0.96}{1-0.96} = 24$$

$$\text{(ii) Base current, } I_B = \frac{I_C}{\beta} = \frac{0.625}{24} = 0.026 \text{ mA or } 26 \ \mu\text{A} \quad \textbf{Ans.}$$

Example 3.16. I_E = 5 mA, I_C = 4.95 mA, I_{CEO} = 200 μA. Calculate β_{dc} and leakage current I_{CBO}.

[*Rajasthan Univ. 2005*]

Solution : Base current, $I_B = I_E - I_C = 5 - 4.95 = 0.05$ mA = 50 μA **Ans.**

$$\text{Current gain factor, } \beta_{dc} = \frac{I_C}{I_B} = \frac{4.95}{0.05} = 99 \ \textbf{Ans.}$$

$$\text{Current amplification factor, } \alpha_{dc} = \frac{\beta_{dc}}{1+\beta_{dc}} = \frac{99}{1+99} = 0.99$$

Collector-to-base leakage current, $I_{CBO} = I_{CEO}(1 - \alpha_{dc}) = 200(1 - 0.99) = 2$ μA **Ans.**

Example 3.17. In an NPN transistor, α = 0.995, I_E = 10 mA and I_{CO} = 0.5 μA. Determine the values of I_C, I_B, β and I_{CEO}.

[*Rajasthan Univ. 2003*]

Solution : Emitter current, I_E = 10 mA

Collector-to-base leakage current, I_{CO} = 0.5 μA = 0.0005 mA

Collector current, $I_C = \alpha I_E + I_{CO} = 0.995 \times 10 + 0.0005 = 9.9505$ mA **Ans.**

Base current, $I_B = I_E - I_C = 10 - 9.9505 = 0.0495$ mA = 49.5 μA **Ans.**

$$\text{Current gain factor, } \beta = \frac{\alpha}{1-\alpha} = \frac{0.995}{1-0.995} = 199 \ \textbf{Ans.}$$

$$\text{Collector current with base lead open, } I_{CEO} = \frac{I_{CBO}}{1-\alpha} = \frac{0.5 \times 10^{-6}}{1-0.995} = 100 \ \mu\text{A} \ \textbf{Ans.}$$ Refer to Eq. (3.28)

Example 3.18. Following observations are obtained for a silicon NPN transistor in common base configuration, as shown in Fig. 3.21. (*a*) With emitter open on application of a voltage of 10 V between collector and base results in a collector current of 0.25 μA; (*b*) With base open, on application of a voltage of 10 V between collector and emitter results in a current of 25 μA. Determine emitter current, base current and α of the transistor when the collector current is 1.2 mA.

[*Rajasthan Univ. 1994, 2001*]

Solution : I_{CBO} = Collector current with emitter open = 0.25 μA

I_{CEO} = Collector current with base open = 25 μA

Current amplification factor,

$$\alpha = \frac{I_{CEO} - I_{CBO}}{I_{CEO}} = \frac{25 - 0.25}{25} = 0.99 \ \textbf{Ans.}$$

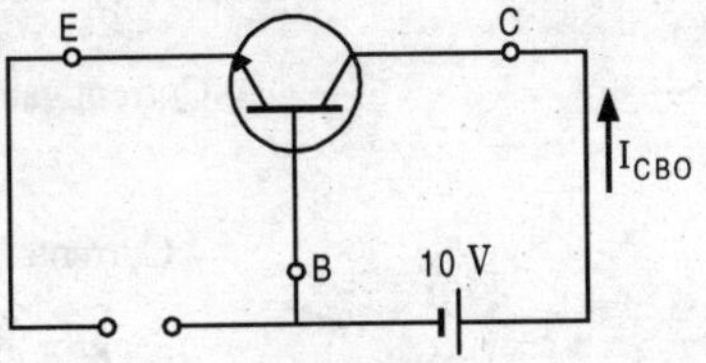

Fig. 3.21

$$\because I_{CEO} = \frac{I_{CBO}}{1-\alpha}$$

$$\text{Emitter current, } I_E = \frac{I_C - I_{CBO}}{\alpha} = \frac{1.2 - 0.25 \times 10^{-3}}{0.99} = 1.212 \text{ mA } \textbf{Ans.} \qquad \because I_C = \alpha I_E + I_{CBO}$$

Base current, $I_B = I_E - I_C = 1.212 - 1.2 = 0.012$ mA = 12 μA **Ans.**

Example 3.19. The collector leakage current in a transistor is 300 μA in CE arrangement. If the transistor is now connected in CB arrangement, what will be the leakage current. Given that β = 100. *[MREC 2000]*

Solution :

$$I_{CEO} = \text{Collector leakage current in CE arrangement} = 300\ \mu A$$

Current gain factor, $\beta = 100$

$$\text{Current amplification factor, } \alpha = \frac{\beta}{1+\beta} = \frac{100}{101} = 0.99$$

Leakage current in CB arrangement $= I_{CBO} = I_{CEO}(1 - \alpha) = 300(1 - 0.99) = 3\ \mu A$ **Ans.**

Example 3.20. The collector and base current of an N-P-N transistor are measured as $I_C = 5$ mA, $I_B = 50$ μA and $I_{CBO} = 1$ μA.

(*i*) Determine α, β and I_E

(*ii*) Determine the new level of I_B required to produce $I_C = 10$ mA.

[U.P. Technical Univ. Electronics Engineering First Semester 2005-06]

Solution : (*i*) Given that

$$I_C = 5\ mA$$

$$I_B = 50\ \mu A$$

$$I_{CBO} = 1\ \mu A$$

$$\text{Since } I_C = \beta I_B + (\beta + 1)\, I_{CBO}$$

$$\text{or } \beta = \frac{I_C - I_{CBO}}{I_B + I_{CBO}} = \frac{5\times10^{-3} - 1\times10^{-6}}{50\times10^{-6} + 1\times10^{-6}} = \mathbf{98} \quad \textbf{Ans.}$$

$$\alpha = \frac{\beta}{1+\beta} = \frac{98}{99} = \mathbf{0.99\ Ans.}$$

$$I_E = I_B + I_C = 50 \times 10^{-6} + 5 \times 10^{-3} = \mathbf{5.05\ mA} \quad \textbf{Ans.}$$

$$(ii)\ I_B = \frac{I_C - (\beta+1)\, I_{CBO}}{\beta} = \frac{10\times10^{-3} - (98+1)\times1\times10^{-6}}{98} = \mathbf{101\ \mu A} \quad \textbf{Ans.}$$

Example 3.21. A transistor is connected in common emitter configuration as an amplifier. The parameters of the transistor specified are :

$I_B = 25$ μA, $I_{CBO} = 100$ nA, β = 100. Find its I_C, I_E, I_{CEO} and α, where symbols have their usual meaning.

[U.P.S.C. I.E.S. Electronics & Telecommunication Engineering-II, 2001]

Solution :

$$\text{Collector current, } I_C = \beta I_B + (\beta + 1) I_{CBO}$$

$$= 100 \times 25 \times 10^{-6} + (100 + 1) \times 100 \times 10^{-9}$$

$$= (2{,}500 + 10.1) \times 10^{-6} = 2{,}510.1\ \mu A \text{ or } 2.51\ mA \ \textbf{Ans.}$$

$$\text{Emitter current, } I_E = I_C + I_B = 2{,}510.1 + 25 = 2{,}535\ \mu A \text{ or } 2.535\ mA \ \textbf{Ans.}$$

$$\text{Current gain, } \alpha = \frac{\beta}{\beta+1} = \frac{100}{101} = 0.99$$

$$\text{Current, } I_{CEO} = \frac{I_{CBO}}{1-\alpha} = \frac{100\times10^{-9}}{1-0.99} = 10\ \mu A \ \textbf{Ans.}$$

3.13. COMMON COLLECTOR CONFIGURATION

In this circuit arrangement, input is applied between base and collector while the output is taken across emitter and collector. Thus the collector forms the terminal common to both the input and output circuits. The circuit is essentially the same as that of CE configuration except that load resistor R_L is now placed in the emitter circuit instead of collector circuit. The operation of the circuit in terms of the currents that flow is basically the same as that for the CE configuration. In this arrangement base current I_B flows in the input circuit and emitter current I_E flows in the output circuit. So the change in emitter current ΔI_E to the change in base current ΔI_B gives the current amplification factor γ. The circuit provides nearly the same current gain as in case of CE arrangement because $\Delta I_E \simeq \Delta I_C$.

With base current I_B equal to I_{CO}, the emitter current I_E is zero so no current flows in the load resistor R_L. With the increase in input current I_B, the transistor passes through the active region and finally reaches saturation. Almost the complete supply voltage, except for a small voltage drop across the transistor appears across the load resistor R_L.

The current equations are

$$I_C = \alpha\, I_E + I_{CBO}$$

$$\text{and} \quad I_E = I_B + I_C = I_B + \alpha\, I_E + I_{CBO}$$

$$\text{or} \quad I_E\,(1 - \alpha) = I_B + I_{CBO}$$

$$\text{or} \quad I_E = \frac{I_B}{1-\alpha} + \frac{I_{CBO}}{1-\alpha} = (\beta + 1)\, I_B + (\beta + 1)\, I_{CBO} \qquad \ldots(3.34) \quad \because \frac{1}{1-\alpha} = \beta + 1$$

$$\text{Current gain,} \quad \gamma = \frac{\Delta I_E}{\Delta I_B}$$

$$= \frac{\Delta I_E}{\Delta I_E - \Delta I_C}$$

$$= \frac{\Delta I_E / \Delta I_C}{\frac{\Delta I_E}{\Delta I_C} - 1} = \frac{1/\alpha}{1/\alpha - 1} = \frac{1}{1-\alpha} = \beta + 1 \qquad \ldots(3.35)$$

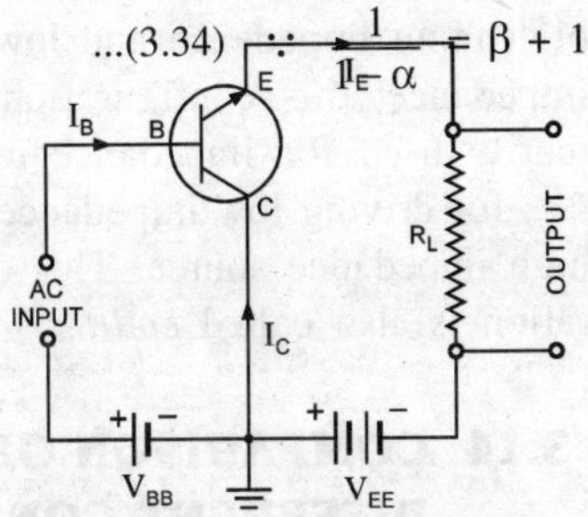

Fig. 3.22. ***Common Collector NPN Transistor***

3.13.1. Characteristics of Common Collector Configuration

The circuit used for determination of transistor common collector configuration is shown in Fig. 3.23. For determination of input characteristics, V_{CE} is kept at a suitable fixed level and base current I_B is recorded for different values of collector-base voltage V_{CB}. Now the curves are drawn between base current and collector-base voltage, as shown in Fig. 3.24.

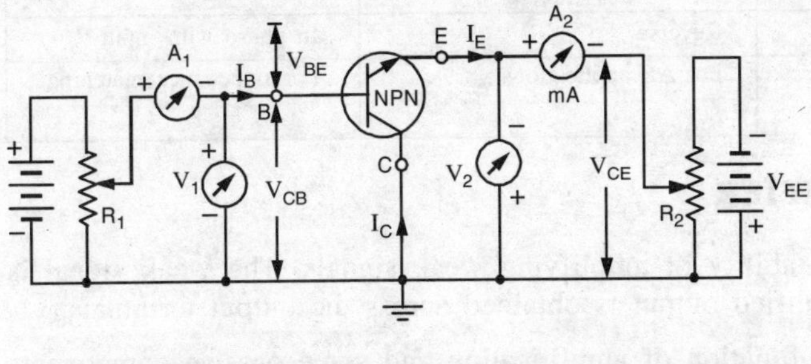

Fig. 3.23. ***Circuit Arrangement For Determination of Input and Output Characteristics of a Transistor in CC Configuration***

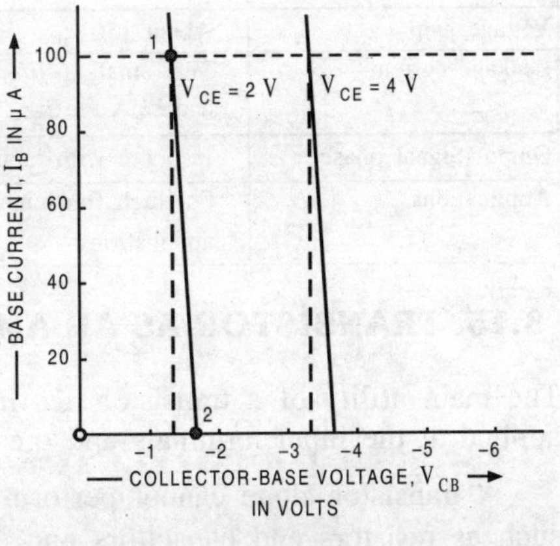

Fig. 3.24. ***Input Characteristics For Common Collector NPN Transistor***

The output characteristics are plotted between emitter current I_E and collector-emitter voltage V_{CE}. The common collector gain characteristics are plotted between emitter-current and base current. Figure 3.25 depicts the output and current gain characteristics.

The common collector (CC) arrangement gives very high input impedance and very low output impedance and, therefore, its voltage gain is always less than unity. Hence this configuration is seldom used for amplification. However, owing to relatively high input impedance and low output impedance, this configuration is primarily used for impedance matching *i.e.,* for driving low impedance from a high impedance source. This configuration is also called *emitter follower.*

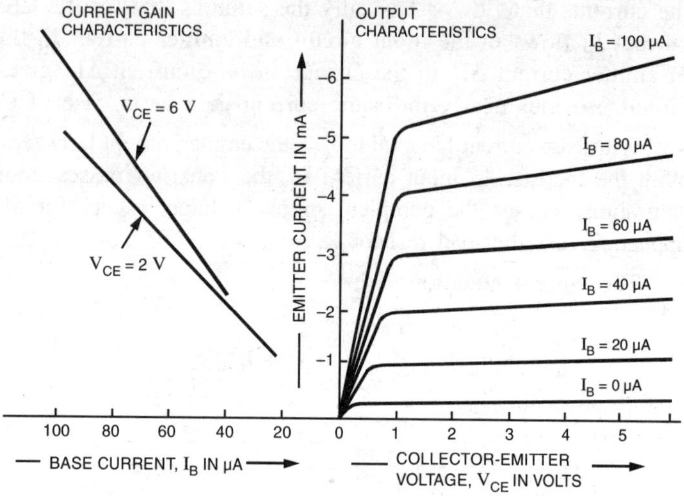

Fig. 3.25. ***Output and Current Gain Characteristics of Transistor in CC Configuration***

3.14. COMPARISON OF CHARACTERISTICS OF TRANSISTORS IN DIFFERENT CONFIGURATIONS

The essential characteristics of each of basic configurations are tabulated below :

Configuration / *Characteristics*	*Common Base*	*Common Emitter*	*Common Collector*
Input impedance	Low (about 100 Ω)	Medium (about 800 Ω)	Very high (about 750 kΩ)
Output impedance	Very high (about 500 kΩ)	High (about 50 kΩ)	Low (about 50 Ω)
Current gain	Less than unity but usually more than 0.9 (about 0.98)	High (about 80)	High (about 100)
Voltage gain	About 150	About 500	Less than unity
Leakage current	Very small (5 μA for Ge and 1 μA for Si)	Very large (500 μA for Ge and 20 μA for Si)	Very large
Output Signal phase	In phase with input	Reverse	In phase with input
Applications	For high frequency applications	For AF applications	For impedance matching

3.15. TRANSISTOR AS AN AMPLIFIER

The main utility of a transistor lies in its ability of amplifying weak signals. The weak signal is applied at the input terminals and the amplified output is obtained across the output terminals.

A transistor alone cannot perform the function of amplification and some passive components such as resistors and capacitors and a biasing battery is to be connected. Common emitter (CE) configuration, because of its high current, voltage and power gains, is much suited for most of the amplifier circuits. A common emitter N-P-N transistor amplifier circuit is given in Fig. 3.26.

The weak signal is applied between emitter-base junction and output is taken across a load resistor R_L connected in series with collector supply voltage V_{CC} (Fig. 3.26). In order to obtain faithful amplification, it is necessary that the input circuit remains always forward biased regardless the polarity of the ac input signal. So a battery V_{BB} is inserted in the circuit with the polarity indicated in addition to the signal voltage. This dc voltage is called the *bias voltage* and its magnitude is such as to keep the input circuit always forward biased regardless the polarity of the input signal.

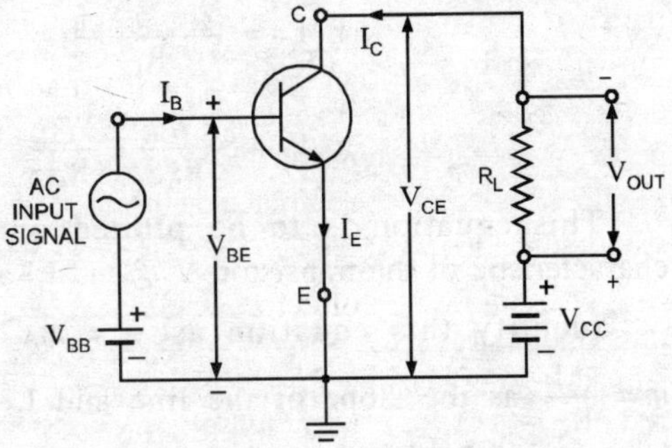

Fig. 3.26. ***Common Emitter NPN Transistor Amplifier Circuit***

The input circuit being forward biased has low resistance and a small change ΔV_{in} in input signal voltage causes a relatively large change ΔI_E in emitter current. This causes almost the same change in collector current because of transistor action. The collector current flowing through a high load resistance R_L develops a large voltage across it. The change in output voltage across load resistance R_L may be many times the change in input signal voltage. Thus voltage amplification $A = \frac{\Delta V_{out}}{\Delta V_{in}}$ will be greater than unity, and the transistor acts as an amplifier. This is further illustrated below by considering typical circuit values.

Let the load resistance R_L be of 10 kΩ and a change of 0.1 V in input signal voltage cause a change of 0.5 mA in emitter current. This change of 0.5 mA in emitter current will also change collector current I_C by approximately 0.5 mA. This change of 0.5 mA in collector current will produce a change of $0.5 \times 10^{-3} \times 10 \times 10^3$ *i.e.,* 5 V in output voltage appearing across the load resistor R_L of 10 kΩ. Thus the change of 0.1 V in input signal voltage causes a change of 5 V in the output voltage giving a voltage amplification of $\frac{5}{0.1} = 50$.

Necessity of Biasing. For most of the applications, transistors are required to operate as *linear amplifiers* (*i.e.,* to amplify output voltage as a linear function of the input voltage). To achieve this, it is necessary to operate the transistor over region of its characteristic curves which are linear, parallel and equi-spaced for equal increments of the parameters. Such an operation can be ensured by proper selection of zero signal operating point and limiting the operation of the transistor over the linear portion of the characteristics. For proper selection of zero signal operating point, proper biasing *i.e.,* application of dc voltages at emitter-to-base junction and collector-to-base junction is required.

If the transistor is not biased properly, it would work inefficiently and produce distortion in the output signal.

3.16. TRANSISTOR LOAD LINES

The concept of load line is very important in understanding the working of a transistor. It is defined as the locus of operating point on the output characteristic of the transistor. It is the line on which the operating point moves when ac signal is applied to the transistor.

3.16.1. DC Load Line

In the circuit shown in Fig. 3.26, V_{CC} is the supply voltage to collector, R_C (or R_L) is the collector resistance (or load resistance) and V_{CE} is the collector-to-emitter voltage. Applying Kirchhoff's second law to the output or collector circuit we have,

$$V_{CC} = V_{CE} + I_C R_C \quad ...(3.36)$$

$$\text{or } I_C = \frac{V_{CC} - V_{CE}}{R_C}$$

$$\text{or } I_C = \frac{-V_{CE}}{R_C} + \frac{V_{CC}}{R_C} \qquad ...(3.37)$$

This equation is to be plotted on the output characteristic of the transistor, V_{CE} and I_C are variables.

Identify this equation as $y = mx + c$ where $m = \frac{-1}{R_C}$ as the slope of the line and $I_C = \frac{V_{CC}}{R_C}$ as intercept of the line on vertical current axis (Fig. 3.27).

Consider the following two particular situations :

(*i*) When $V_{CE} = 0$, $I_C = \frac{V_{CC}}{R_C}$ ...saturation point A

(*ii*) When $I_C = 0$, $V_{CE} = V_{CC}$ ...cutoff point B

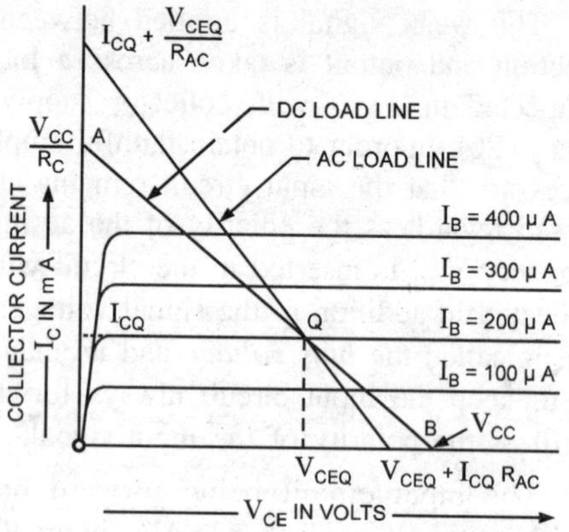

Fig. 3.27. ***Common Emitter Transistor Output Characteristics With DC and AC Load Lines***

By joining these two points A and B, *dc load line* is obtained. The dc load line represents the dynamic characteristic of the device. The dc load line gives the values of collector current I_C and collector-emitter voltage V_{CE} corresponding to zero signal conditions.

3.16.2. Quiescent Point

It is a point on the dc load line which represents dc collector-emitter voltage V_{CE} and collector current I_C in the absence of ac signal. It is also called the *operating point* because the variations in V_{CE} and I_C take place about this point when signal is applied. The best position for this point is midway between cutoff and saturation points where $V_{CE} = \frac{1}{2} V_{CC}$. Point Q is quiescent point marked on the output characteristics shown in Fig. 3.27. Selection of the operating point is done as per application for which the device is to be used. For example in case of a small signal amplifier, in which power is conserved, operating point is selected so as to give lowest quiescent value of I_C, while for an amplifier operated to deliver small amount of power, operating point is selected so that available quiescent current is about one-half of the maximum permissible collector current I_C.

3.16.3. AC Load Line

When an ac signal is applied, the transistor voltage V_{CE} and collector current I_C vary above and below the quiescent point Q. So point Q is common to both dc and ac load lines. The *ac load line* gives the values of V_{CE} and I_C when an ac signal is applied. For drawing ac load line, take a convenient collector current change ΔI_C and compute the corresponding collector-emitter voltage change $\Delta V_{CE} = -\Delta I_C.R_C$ to obtain another point lying on the ac load line. Now the ac load can be drawn by joining this point and point Q. AC load is steeper than dc load line but the two lines intersect at the quiescent point Q determined by the biasing dc voltages and currents. AC load line takes into account the ac load resistance while the dc load line considers only the dc load resistance.

Example 3.22. In CE configuration (Fig. 3.26) collector supply voltage V_{CC} = 10 V, load resistance R_L is 8 kΩ. Draw dc load line. Determine the operating point Q for zero signal if base current is 15 μA and β is 40.

Solution :

Collector supply voltage, V_{CC} = 10 V

Load resistance, R_L = 8 kΩ

Zero signal base current, I_B = 15 μA

Current gain factor, $\beta = 40$

Collector-emitter voltage V_{CE} is given as

$$V_{CE} = V_{CC} - I_L R_C$$

For $V_{CE} = 0$,

$$I_C = \frac{V_{CC}}{R_L} = \frac{10\text{ V}}{8\text{ k}\Omega} = 1.25\text{ mA}$$

This gives point A on the load line.

For $I_C = 0$, $V_{CE} = V_{CC} = 10$ V

This gives point B on the load line. AB is the required dc load line.

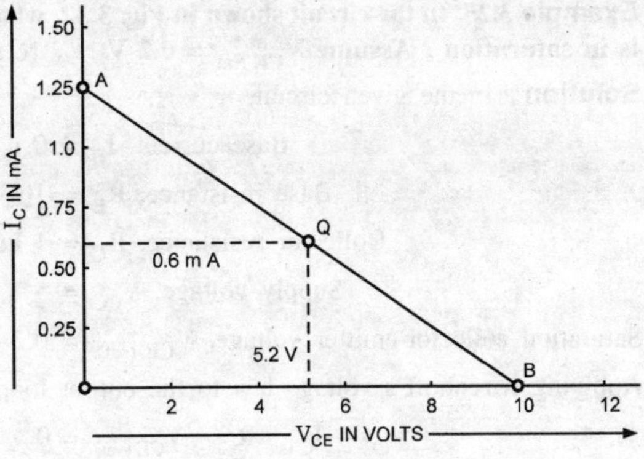

Fig. 3.28

Zero signal collector current,

$$I_C = \beta I_B = 40 \times 15 \times 10^{-6} = 0.6\text{ mA}$$

Zero signal collector-emitter voltage,

$$V_{CE} = V_{CC} - I_C R_L$$
$$= 10 - 0.6\text{ mA} \times 8\text{ k}\Omega = 5.2\text{ V}$$

$\therefore$ Operating point Q is (5.2 V, 0.6 mA) **Ans.**

Example 3.23. Draw the load line for the following figure (Fig. 3.29). What is I_C at saturation point ? Find V_{CE} at cutoff point. *[U.P. Technical Univ. Basic Electronics First/Second Semester 2001-02]*

Solution : (*i*) Applying Kirchhoff's voltage law to the output loop we have

$$V_{CC} = I_C R_C + V_{CE}$$

For cutoff point, $I_C = 0$

and, therefore, $V_{CE} = V_{CC} = 20$ V **Ans.**

$\therefore$ For saturation point, $V_{CE} = 0$

$$\text{So } I_C = \frac{V_{CC}}{R_C}$$

$$= \frac{20}{3.3 \times 10^3} = 6\text{ mA } \textbf{Ans.}$$

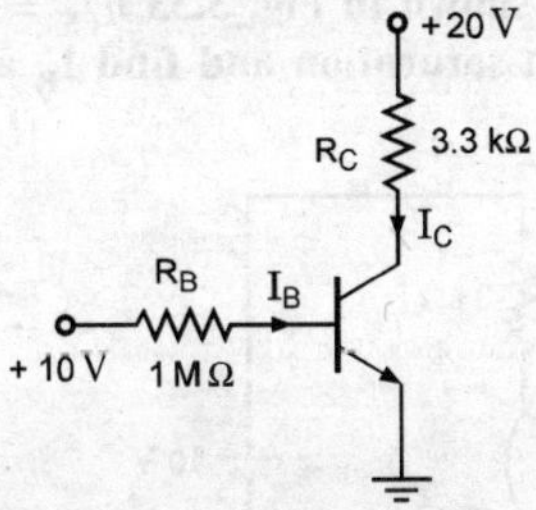

Fig. 3.29

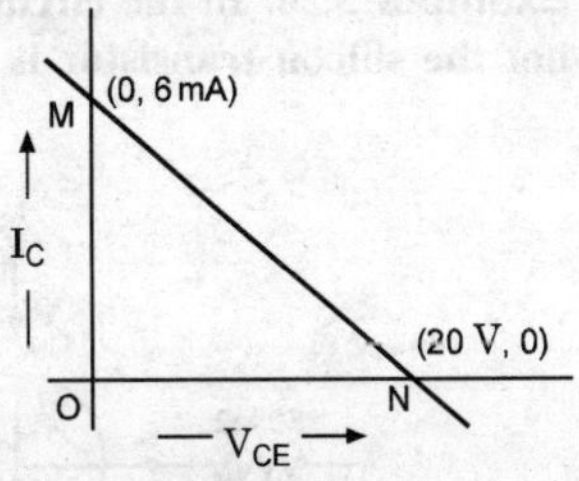

Fig. 3.30

Therefore, load line coordinates are given as (20 V, 0 mA) and (0 V, 6 mA) as shown in Fig. 3.30.

Example 3.24. Determine V_C and V_B for the network given below (Fig. 3.31) with $\beta = 45$ and $V_{BE} = 0.7$ volt. *[U.P. Technical Univ. Basic Electronics First Semester 2001-02]*

Solution : $\beta = 45$ and $V_{BE} = 0.7$ volt

Applying Kirchhoff's voltage law in input loop we have

$$I_B R_B + V_{BE} + V_{EE} = 0$$

Putting values in above equation we have

$$I_B \times 100 \times 10^3 + 0.7 - 9 = 0$$

$$\text{or } I_B = \frac{9 - 0.7}{10^5} = 83\ \mu\text{A}$$

$$I_C = \beta \times I_B = 45 \times 83 \times 10^{-6}\text{ A} = 3.735\text{ mA}$$

Collector voltage, $V_C = V_{CC} - I_C R_C$

$$= 0 - 3.735 \times 10^{-3} \times 1.2 \times 10^3 = -4.482\text{ V } \textbf{Ans.}$$

$$V_B = V_{BE} + V_{EE} = 0.7 - 9 = -8.3\text{ volts } \textbf{Ans.}$$

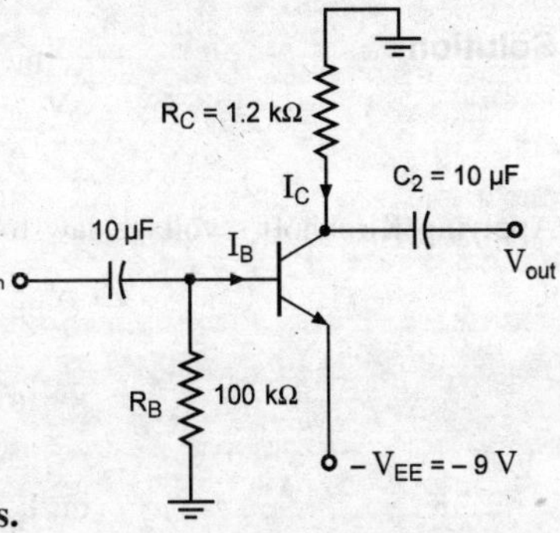

Fig. 3.31

Example 3.25. In the circuit shown in Fig. 3.32, what should be the minimum value of β such that the transistor is in saturation ? Assume $V_{CE\ Sat}$ = 0.2 V. *[U.P.S.C. I.E.S. Electronics & Telecommunication Engineering-I, 2007]*

Solution : In the given circuit,

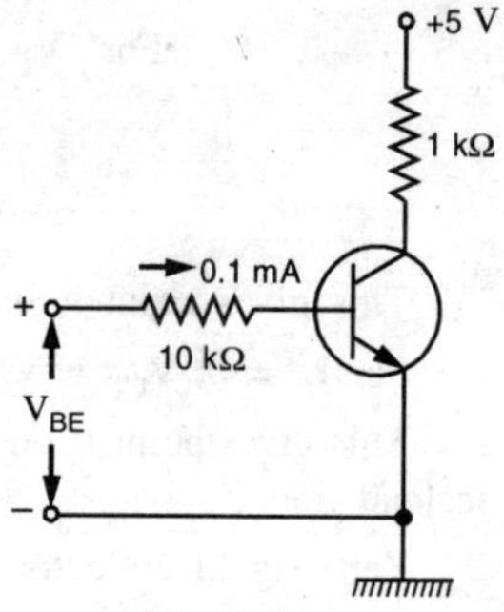

Fig. 3.32

Base current, I_B = 0.1 mA

Base resistance, R_B = 10 kΩ

Collector resistance, R_C = 1 kΩ

Supply voltage, V_{CC} = 5 V

Saturation collector-emitter voltage, $V_{CE\ (sat)}$ = 0.2 V

Applying Kirchhoff's voltage law to the output loop, we have

$$V_{CC} - I_{C\ (sat)}R_C - V_{CE\ (sat)} = 0$$

$$\text{or } I_{C\ (sat)} = \frac{V_{CC} - V_{CE\ (sat)}}{R_C} = \frac{5-0.2}{1\times10^3} = 4.8 \text{ mA}$$

For transistor to be in saturation

$$\text{Base current, } I_B \geq \frac{I_{C\ (sat)}}{\beta}$$

$$\text{or} \quad \beta \geq \frac{I_{C\ (sat)}}{I_B} \geq \frac{4.8}{0.1} \geq 48$$

Hence, minimum value of β = 48 **Ans.**

Example 3.26. In the circuit shown in Fig. 3.33 h_{FE} = 100, V_{BE} = 0.8 V, V_{CE} = 0.2 V, determine whether or not the silicon transistor is in saturation and find I_B and I_C.

[U.P. Technical Univ. Basic Electronics First Semester 2002-03]

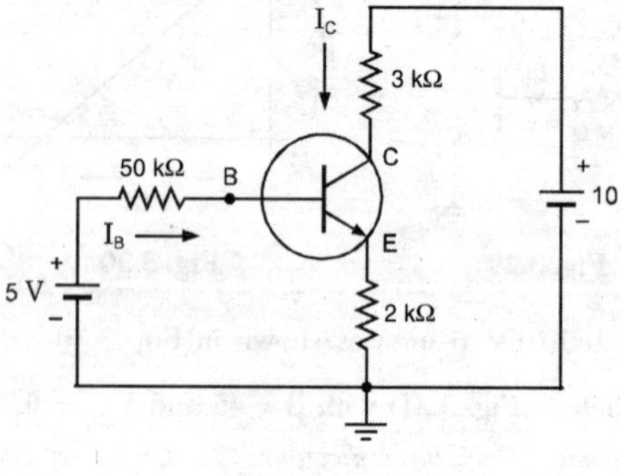

Fig. 3.33

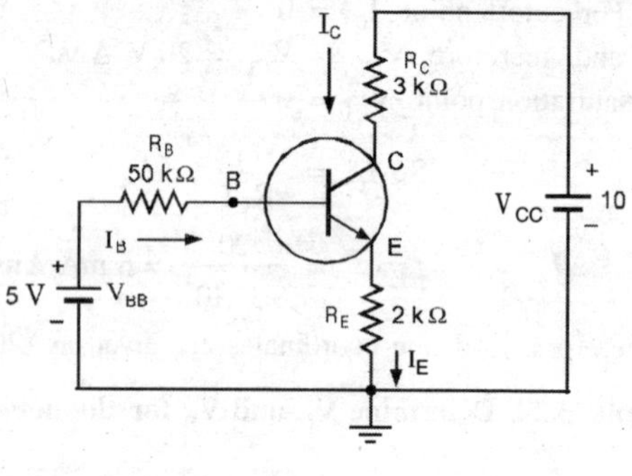

Fig. 3.34

Solution :

V_{BE} = 0.8 V

V_{CE} = 0.2 V

$\beta = h_{FE}$ = 100

Applying Kirchhoff's voltage law to input loop we have

$$V_{BB} = I_B R_B + V_{BE} + I_E R_E = I_B R_B + V_{BE} + (1+\beta)I_B R_E \quad \because I_E = (1+\beta)I_B$$

$$\text{or } I_B = \frac{V_{BB} - V_{BE}}{R_B + (1+\beta)R_E}$$

$$\text{or } I_B = \frac{5-0.8}{[50 + 2(100+1)]\times10^3} = \frac{4.2\times10^{-3}}{252} \text{ A} = 0.0166 \text{ mA} \quad \textbf{Ans.}$$

Applying KVL in the output loop we have

$$V_{CC} - I_C R_C - V_{CE} - I_E R_E = 0$$

$$10 - I_C (3 \times 10^3) - V_{CE} - I_E (2 \times 10^3) = 0$$

$$I_C (3 + 2) \times 10^3 = 10 - V_{CE} \qquad [\because I_E \simeq I_C]$$

$$\text{or } I_C = \frac{10 - 0.2}{5 \times 10^3} = 1.96 \text{ mA}$$

i.e., for the transistor to be in saturation, collector current should be equal to 1.96 mA *i.e.*, $I_{C\,(sat)} = 1.96$ mA

But actual collector current,

$$I_C = \beta I_B = 100 \times 0.0166 \text{ mA} = 1.66 \text{ mA} \quad \textbf{Ans.}$$

Since I_C is lesser than $I_{C(sat)}$, transistor is not in saturation. **Ans.**

3.17. STANDARD NOTATIONS FOR VOLTAGES AND CURRENTS

Standard notations or symbols are usually used for voltages and currents in transistor circuit analysis and applications.

1. Subscripts E (or *e*), B (or *b*) and C (or *c*) usually refer to emitter, base and collector quantities respectively.
2. Upper case (capital) letters with capital subscripts of the proper electrode symbol are used for dc or average values of voltages and currents.
3. Upper case (capital) letters with lower subscripts of the proper electrode symbol are used for rms values of ac voltages and currents.
4. Lower case (small) letters with lower subscripts of proper electrode symbol are used for instantaneous values of ac voltages and currents.

Notations used for various quantities (voltages and currents) in transistor circuit analysis and applications are given below in tabular form.

S. No.	*Particular*	*DC or average*	*RMS ac*	*Instantaneous ac*
1.	Emitter current	I_E	I_e	i_e
2.	Base current	I_B	I_b	i_b
3.	Collector current	I_C	I_c	i_c
4.	Collector-emitter voltage	V_{CE}	V_{ce}	v_{ce}
5.	Emitter-base voltage	V_{EB}	V_{eb}	v_{eb}
6.	Collector-base voltage	V_{CB}	V_{cb}	v_{cb}
7.	Emitter supply voltage	V_{EE}	—	—
8.	Base supply voltage	V_{BB}	—	—
9.	Collector supply voltage	V_{CC}	—	—
10.	Emitter voltage to ground	V_E	V_e	v_e
11.	Base voltage to ground	V_B	V_b	v_b
12.	Collector voltage to ground	V_C	V_c	v_c

3.18. IMPORTANCE OF COLLECTOR-EMITTER VOLTAGE, V_{CE}

The collector-emitter voltage, V_{CE} plays a significant role in indication of region of operation of transistor *i.e.*, whether the transistor is operating in cutoff, saturation or well into saturation regions.

Equality of V_{CE} and V_{CC} (*i.e.*, $V_{CE} = V_{CC}$) indicates that the transistor is operating in *cutoff mode* (*i.e.*, the transistor is turned *off*). Zero value of V_{CE} indicates that the transistor is operating in saturation mode (*i.e.*, the transistor is turned *on* fully). Negative value (less than zero value) of V_{CE} indicates that the transistor is *well in saturation*. In practice, both of these conditions are avoided, except in switching circuits.

In practice, the transistor is operated with $V_{CE} = \frac{1}{2}V_{CC}$ (*i.e.*, at approximately $\frac{1}{2}$ *on*). Thus, variation in base current I_B in either direction controls collector current I_C well in both directions. It means that collector current I_C increases or decreases with the increase or decrease of base current I_B. However, with I_B *off* ($I_B = 0$), I_C is also *off* ($I_C = 0$). On the other hand, if the collector has been turned fully *on* (*i.e.*, transistor is operating in saturation), the collector current is maximum, and the further increase in I_B will not reflect in I_C.

Highlights

1. Semiconductor devices have the following advantages over vacuum tubes : They are (*i*) smaller in size (*ii*) lighter in weight, (*iii*) more rugged and (*iv*) more efficient. In addition, they have (*i*) no warm-up period, (*ii*) no heating requirement and (*iii*) lower operating voltages.
2. A transistor may be compared loosely with a triode in that the emitter may be thought of as a non-heated cathode, the collector as a plate and the base as a grid.
3. There are two complementary bipolar transistors—NPN and PNP. Each transistor has three separately doped regions and two PN junctions. The center region (base) is very narrow, so the two PN junctions are said to be interacting junctions.
4. Since a transistor consists of either PNP or NPN slices, a very simplified model for biasing purposes comprises two diodes "pointing inward or outward" respectively.
5. For normal operation, base-emitter junction is forward biased and the base-collector junction is reverse biased. Under these conditions, majority carriers from the emitter are injected into the base where they become minority carriers. These minority carriers diffuse across the base into the B-C space charge region where they are swept into the collector.

   In this mode of operation, the current at one terminal of the transistor (collector current) is controlled by the voltage applied across the other two terminals of the transistor (base-emitter voltage). This is the basic transistor action.
6. The modes of operation of a BJT are determined by the biases applied to the two junctions. The four modes of operation are : forward active, cutoff, saturation, and inverse active. In the forward-active mode, the B-E junction is forward biased and the B-C junction is reverse biased, and the collector and base currents are related by the common-emitter current gain $\beta(I_C = \beta I_B)$. The relationship is the same for both NPN and PNP transistors, as long as the conventional current directions are maintained. In the cutoff mode of operation, the B-E and B-C junctions are both reverse biased and all currents are zero. In saturation mode, the B-E and B-C junctions are both forward biased and the collector current is no longer a function of base current.
7. The dc emitter current I_E is always the largest one of a transistor, whereas the base current I_B is always the smallest. The collector current I_C is almost as large as emitter current I_E (*i.e.*, $I_C \simeq I_E$). The emitter current is always equal to the sum of the other two (*i.e.*, $I_E = I_B + I_C$).
8. The collector current I_C is made up of two components : the majority component and the minority component (also known as the leakage current).

   i.e., $I_C = I_{C_{majority}} + I_{CO_{minority}}$
9. The arrow in the transistor symbol defines the direction of conventional current flow for the emitter current and thereby defines the directions for the other currents in the device.

   The arrow in the symbol of an NPN transistor points out of the device (not pointing in), whereas the arrow points to the centre of the symbol for a PNP transistor (pointing in).

10. On an average basis, as a second approximation, the base-to-emitter voltage of an operating transistor can be assumed to be 0.7 V in case of silicon device and 0.3 V in germanium device.
11. The quantity α relates the collector and emitter currents $\left(\alpha = \frac{I_C}{I_E}\right)$ and is always approximately equal to one (never greater than 1).
12. The impedance between terminals of a forward-biased junction is always relatively small, whereas the impedance between terminals of a reverse-biased junction is usually quite large.
13. The quantity beta β provides an important relationship between the base and collector currents and is usually between 50 and 400.
14. The dc beta is defined by a simple ratio of dc currents at an operating point $\left(i.e. \beta_{dc} = \frac{I_C}{I_B}\right)$, whereas the ac beta β_{ac} is sensitive to the characteristics in the region of interest. For most applications, however, the two are considered equivalent as a first approximation.
15. A three-terminal device needs two sets of characteristics to completely define its characteristics. The transistor input characteristic is essentially that of a forward biased junction, with some dependence upon the transistor's output voltage.
16. For a transistor used in CB configuration, the collector or output characteristics consist of practically horizontal lines of I_C versus V_{CB} with I_E as the controlling parameter.
17. The CB ac current gain, $\alpha_{ac} = \left.\frac{\Delta I_C}{\Delta I_E}\right|_{V_{CB} = \text{Constant}}$ while $\alpha_{dc} = \frac{I_C}{I_E}$ and is less than unity.
18. The reverse saturation current in a CB configuration = $I_{CBO} = I_{CO}$, while in a CE configuration it is $I_{CEO} = (\beta + 1)I_{CBO}$.
19. The collector or output characteristics for a transistor used in the CE configuration have the collector current I_C versus collector-emitter voltage curves sloping upward slightly with base current as the controlling parameter.
20. The CE ac current gain, $\beta_{ac} = \left.\frac{\Delta I_C}{\Delta I_B}\right|_{V_{CE} = \text{Constant}}$, while $\beta_{dc} = \frac{I_C}{I_B}$.
21. In either CB or CE configuration, majority carriers are swept across the narrow, lightly doped base to the collector. The collector current is controlled by a relatively small base input current that is the result of the forward-biased base-emitter junction, which accounts for the amplifying action in a transistor.

Exercises

1. Draw NPN and PNP transistors. Label all the currents and show the direction of flow. How are all the currents of a transistor related ? **[U.P. Technical Univ. Basic Electro. Second Semester 2000-01]**
2. What is a transistor ? Show the transistor action for the N-P-N and P-N-P transistors. **[Pb. Technical Univ. Electronic Devices & Circuits 2004-05]**
3. Draw the schematic diagram of the NPN transistor and explain its working. **[R.G. Technical Univ. Basic Electronics June 2008]**
4. Draw a schematic diagram of a transistor indicating the different currents. Arrive at the relationship among them. **[R.G. Technical Univ. Basic Electronics Jan.-Feb., 2008]**
5. Explain the working of a PNP transistor with the help of its energy band diagram. Explain 'emitter injection efficiency' and base 'transport factor'. How do they influence the operation of a transistor ?
6. If a transistor has its collector-base junction forward biased and the other junction reverse biased, will it work ? Explain. **[U.P. Technical Univ. Basic Electronics First Semester 2003-04]**

7. Define with respect to BJT the following : (*i*) I_{CO}, (*ii*) α, (*iii*) β, (*iv*) I_{CEO}, (*v*) I_{CBO} **[U.P. Technical Univ. Electronics Engineering First Semester 2005-06]**
8. In how many modes the BJT works ? Also explain the biasing pattern for each of them. **[U.P. Technical Univ. Electronics Engineering First Semester 2006-07]**
9. Explain α and β factors of transistors. Derive expressions for them and state their meaning. **[M.D. Univ. Rohtak Analog Electronics December 2008]**
10. Explain the following regions in a transistor : (*i*) Active, (*ii*) Saturation, (*iii*) Cutoff. **[U.P. Technical Univ. Solid State Devices and Circuits 2005-06]**
11. Explain different components of currents flowing through the structure of a N-P-N transistor. How the emitter injection efficiency and base transport factor influences the amplification factor ? **[U.P. Technical Univ. Semiconductor Materials & Devices 2005-06]**
12. Define base spreading resistance of BJT. **[A.M.I.E. Sec. B. Summer 1997]**
13. Draw the circuits of various transistor configurations. List their important features. **[R.G. Technical Univ. Basic Electrical & Electronics Engineering Dec. 2010]**
14. Draw the input and output characteristic curves of a CB configuration for a transistor. Explain the shape of curves. Clearly indicate cutoff, active and saturation regions. **[Pb. Technical Univ. Electronic Devices & Circuits May 2008]**
15. Explain the operation of NPN transistor in CB configuration. Show that $\alpha = \dfrac{\beta}{1+\beta}$. Also give the reason why the input impedance of CB is less than input impedance of CE configuration. **[U.P. Technical Univ. Basic Electronics 2006-07]**
16. For a transistor in CE configuration, (*i*) Derive the expression between I_C, I_B and I_{CO}, (*ii*) Explain why slope of the output characteristics is more than in CB configuration (*iii*) cutoff condition. **[G.G.S.I.P. Univ. Analog Electronics December 2009]**
17. Draw a simple circuit of a common-collector amplifier and explain its operation. Obtain an expression for its voltage gain.
18. What are the three transistor configurations ? Compare and contrast the characteristics of these configurations. Also state any one application of each of these configurations. **[R.G. Technical Univ. Basic Electronics Dec. 2010]**
19. Discuss the doping levels for the emitter, base and collector regions and their physical size in BJT. Also give comparison of the three configurations of the BJT. **[R.G. Technical Univ. Basic Electronics June 2009]**
20. Compare the important parameters (A_v, A_i, R_i, R_0) of a transistor amplifier in different transistor configurations. **[G.G.S.I.P. Univ. Analog Electronics December 2009]**
21. Explain how transistor acts as an amplifier. **[R.G. Technical Univ. Basic Electronics April 2009]**
22. Draw the load line on the output characteristics of a common-emitter amplifier. What is its significance ? Show how the transistor acts as an amplifier. **[Pb. Univ. Applied Electronics I, December 1995]**
23. Explain the difference in ac and dc load lines, which is more steeper ac load or dc load ? **[G.G.S.I.P. Univ. Analog Electronics December 2009]**

Short Answer Type Questions With Answers

Q. 1. Explain why an ordinary transistor is called bipolar ? **[P.T.U. May 2008]**

Ans. Because the transistor operation is carried out by two types of charge carriers (majority and minority carriers), an ordinary transistor is called bipolar.

Q. 2. Why transistor is called current controlled device ?

Ans. The output voltage, current or power is controlled by the input current in a transistor so it is called the current controlled device.

Q. 3. What is the significance of arrowhead in the transistor symbol ?

Ans. Arrowhead is always marked on the emitter. The direction indicates the conventional direction of current flow (from emitter-to-base in case of P-N-P transistor and from base-to-emitter in case of N-P-N transistor). Generally, no arrowhead is marked for collector since its reverse leakage current is always opposite to the direction of emitter current.

Q. 4. What do you understand by transistor biasing ? Why is it necessary to bias a transistor ? **[P.T.U. May 2008]**

Ans. For normal operation, base-emitter junction should be forward biased and the collector-base junction reverse biased. The proper flow of zero signal collector current and the maintenance of proper collector-emitter voltage during the passage of signal is called the transistor biasing. The amount of bias required is significant for establishment of the operating or Q-point which is dictated by the mode of operation desired. In case the transistor is not biased properly, it would (*i*) work inefficiently, (*ii*) produce distortion in the output signal and (*iii*) with the change in transistor parameters or temperature rise, the operating point may shift and the amplifier output will be unstable.

Q. 5. Which of the transistor currents is always the largest ? Which is always the smallest ? Which two currents are relatively close in magnitude ? **[U.P. Technical Univ. First Semester 2006-07]**

Ans. The emitter current I_E is always the largest one. The base current I_B is always the smallest. The collector current I_C and emitter current I_E are relatively close in magnitude *i.e.,* $I_C \simeq I_E$.

Q. 6. Why silicon type transistors are more often used than germanium type ?

Ans. Because silicon transistor has smaller cutoff current I_{CBO}, small variations in I_{CBO} due to variations in temperature and high operating temperature as compared to those in case of germanium type.

Q. 7. Why is size of collector kept maximum in BJT ? **[Rajasthan Univ. 2004]**

Ans. The potential hill (bend in the Fermi level) between the base and collector region is very steep. Specifically the electrons must lose a considerable energy as they move from the base region to the collector region. Typically the electrons give up their energy in the form of heat, therefore, the collector region must be capable of dissipating this heat energy. It is for this reason that the size of the collector is kept maximum in BJT.

Q. 8. Why the width of the base region of a transistor is kept very small compared to other regions ?

Ans. Base region of a transistor is kept very small and very lightly doped so as to pass most of the injected charge carriers to the collector.

Q. 9. Why emitter is always forward biased ?

Ans. Emitter is always forward biased w.r.t. base so as to supply majority charge carriers to the base.

Q. 10. Why collector is always reverse biased w.r.t. base ?

Ans. Collector is always reverse biased w.r.t. base so as to remove the charge carriers away from the base-collector junction.

Q. 11. What are three regions of operation of a transistor ? **[P.T.U. December 2007]**

Ans. The three regions of operation of a transistor are :

(*i*) Cutoff region, (*ii*) Active region and (*iii*) Saturation region.

Q. 12. Can a transistor be obtained by connecting two semiconductor diodes back to back ?

Ans. No. Because in case of two discrete back to back connected diodes there are four doped regions instead of three and there is nothing that resembles a thin base region between an emitter and a collector.

Q. 13. How α and β are related to each other ?

Ans. α and β are related as below : $\alpha = \dfrac{\beta}{1+\beta}$ or $\beta = \dfrac{\alpha}{1-\alpha}$

Q. 14. Define beta of a transistor.

Ans. The β factor of a transistor is the common-emitter current gain of that transistor and is defined as the ratio of collector current to base current.

$$i.e., \ \beta = \frac{I_C}{I_B}.$$

Q. 15. Indicate whether the β-value of a BJT increases or decreases with the increase in the values of the following parameters : (*i*) base width (*ii*) minority carrier lifetime in the base region, (*iii*) temperature (*iv*) collector current (*v*) collector voltage. **[U.P.S.C. I.E.S. E.T.E.-I, 2007]**

Ans. The β-value of a BJT

(*i*) decreases with the increase in base width.
(*ii*) increases with the increase in minority carrier lifetime in the base region.
(*iii*) increases with the increase in temperature.
(*iv*) increases with the increase in collector current.
(*v*) increases with the increase in collector voltage.

Q. 16. Why there is a maximum limit of collector supply voltage for a transistor ?

Ans. Although collector current is practically independent of collector supply voltage over the transistor operating range, but if V_{CB} is increased beyond a certain value collector current I_C eventually increases rapidly and possibly destroys the device.

Q. 17. Explain why $I_{CEO} >> I_{CBO}$? **[U.P.T.U. Second Semester 2003-04]**

Ans. The collector cutoff current denoted by I_{CEO} is much larger than I_{CBO}. I_{CEO} is given as

$$I_{CEO} = \frac{1}{1-\alpha} I_{CBO}$$

Because α is nearly equal to unity (slightly less than unity), $I_{CEO} >> I_{CBO}$.

Q. 18. Give reasons why common emitter (CE) configuration is widely used in amplifier circuits. **[Pb. Technical Univ. May 2008]**

Ans. CE configuration is mainly used because its current, voltage and power gains are quite high and the ratio of output impedance and input impedance are quite moderate.

Q. 19. Why is CE configuration preferred for cascaded amplifiers ? **[Pb. Technical Univ. May 2007]**

Ans. CE configuration, because of its moderate output to input impedance ratio, is preferred for cascaded amplifiers.

Q. 20. Why CC configuration is called a voltage buffer ? What is its other name ? **[U.P. Technical Univ. Third Semester 2006-07]**

Ans. Because of its high input impedance and low output impedance, the common collector circuit finds wide application as a buffer amplifier between a high impedance source and low impedance load, that is why it is called a voltage buffer. Its other name is emitter follower.

Q. 21. What are the main purposes for which a CC amplifier may be used. **[U.P.S.C. I.E.S. Electrical Engineering II 1997]**

Ans. Because of its high input impedance and low output impedance, the common collector circuit finds wide application as a buffer amplifier between a high impedance source and low impedance load.

Q. 22. Compare three configurations CB, CC and CE in terms of their resistances. **[Pb. Technical Univ. May 2008]**

Ans.

Configuration / Resistances	Common Base	Common Emitter	Common Collector
Input Resistance	Low (about 100 Ω)	Medium (about 800 Ω)	Very high (about 750 kΩ)
Output Resistance	Very high (about 500 kΩ)	High (about 50 kΩ)	Low (about 50 Ω)

Q. 23. What do you understand by collector reverse saturation current ? In which configuration does it have a greater value ? **[G.G.S.I.P. Univ. December 2009]**

Ans. When input current (I_E in case of CB configuration and I_B in case of CE configuration) is zero, collector current I_C is not zero although it is very small. In fact this is the reverse leakage current or collector reverse saturation current (I_{CBO} or simply I_{CO} in CB configuration and I_{CEO} in CE configuration). In case of CE configuration it is much more than that in case of CB configuration.

Q. 24. What do you mean by operating point ? **[A.M.I.E. Sec. B. Electronic Circuits Summer 2006]**

Ans. Quiescent point is a point on the *dc* load line which represents V_{CE} and I_C in the absence of *ac* signal and variations in V_{CE} and I_C take place around this point when ac signal is applied.

Problems

1. Determine emitter current I_E, collector current I_C for a transistor with $\alpha_{dc} = 0.97$ and collector to base leakage current 10 μA. Base current is 50 μA. **[Ans.** 2.00 mA, 1.95 mA]
2. For a certain transistor collector current is 6.25 mA, base current is 100 μA and collector-to-base leakage current 5 μA. Determine emitter current and α_{dc}. **[Ans.** 6.35 m A, 0.983]
3. In a particular transistor the collector current is 5.6 mA and the emitter current is 5.75 mA. Determine α_{dc}. **[Ans.** 0.974]
4. A transistor has $\alpha = 0.98$. If emitter current of the transistor is 1 mA, determine the base current and gain factor β. **[Ans.** 20 μA; 49]
5. Determine β if $\alpha = 0.99$ **[Ans.** 99]
6. A bipolar junction transistor has a collector current of 4 mA and base current of 20 μA. Determine its β. **[Ans.** 200]
7. A BJT has a base current of 200 μA and emiter current of 20 mA. Determine collector current and β. **[Ans.** 19.8 mA; 99]
8. In a transistor $I_C = 0.95$ mA, $I_E = 1$ mA. Determine base current, α and β in CB configuration. **[Ans.** 50 μA; 0.95; 19]
9. In a transistor CE configuration the voltage across 5 kΩ load resistance is 5 V. Determine collector current, base current and emitter current if $\beta = 50$. **[Ans.** 1.0 mA; 20 μA; 1.02 mA]
10. In a grounded base configuration the voltage drop across load resistance of 4 kΩ is 3.2 V. Determine collector current, emitter current and base current if $\alpha = 0.96$. **[Ans.** 0.8 mA; 0.8333 mA; 0.0333 mA]
11. A CE transistor amplifier has input resistance of 100 Ω and output resistance of 6 kΩ. Determine the voltage gain and power gain if current gain factor is 90. **[Ans.** 5,400; 486,000]
12. Increase in collector-emitter voltage from 4 V to 8 V causes increase in collector current from 5 mA to 6 mA. Determine the output resistance. **[Ans.** 4 kΩ]
13. A change of 200 mV in base-emitter voltage causes a change of 60 μA in the base current. Determine its input resistance. **[Ans.** 3.33 kΩ]
14. A transistor is connected in CB configuration. When the emitter voltage is changed by 200 mV, the emitter current changes by 5 mA. During this variation, the collector to base voltage is kept fixed. Calculate the dynamic input resistance of transistor. **[U.P. Tech. Univ. Basic Electronics Second Semester 2001-02]** **[Ans.** 40 Ω]

□□□

CHAPTER

4

TRANSISTOR BIASING AND STABILIZATION

4.1. TRANSISTOR BIASING

INSIDE THIS CHAPTER

The basic function of a transistor is amplification. The process of raising the strength of a weak signal without any change in its general shape is referred to as *faithful amplification*. For faithful amplification it is essential that (*i*) emitter-base junction is forward biased; (*ii*) collector-base junction is reverse biased; and (*iii*) proper zero signal collector current.

For achieving faithful amplification, fulfilment of the following basic conditions is essential.

1. ***Proper Zero Signal Collector Current.*** Zero signal collector current should be at least equal to the maximum collector current due to signal alone *i.e.,* zero signal current ≥ maximum collector current due to signal alone.
2. ***Minimum Proper Base-Emitter Voltage At Any Instant.*** The base-emitter voltage V_{BE} should not fall below 0.3 V for germanium transistors and 0.7 V for silicon transistors at any instant. If the base-emitter voltage V_{BE} falls below these values (potential barrier values) during any part of the signal, that part will be amplified to smaller extent due to small collector current and, therefore, faithful amplification will not be available.
3. ***Minimum Proper Collector-Emitter Voltage At Any Instant.*** The collector-emitter voltage V_{CE} should not fall below knee voltage (0.5 V for germanium and 1.0 V for silicon transistors). With collector-emitter voltage V_{CE} less than knee voltage, the collector-base junction is not properly reverse biased and, therefore, the collector cannot attract the charge carriers emitted by the emitter and hence a larger portion of them goes to the base. Consequently, collector current decreases and base current increases. Thus when V_{CE} falls below knee voltage V_{knee} during any part of the signal, that part will be amplified to lesser extent due to reduced value of β and result in unfaithful amplification. For collector-emitter voltage exceeding knee voltage ($V_{CE} > V_{knee}$), the collector-base junction is properly reverse biased and the value of β remains constant, resulting in faithful amplification.

The conditions 1 and 2 ensure that emitter-base junction remains properly forward biased during all parts of the signal and condition 3 ensures that collector-base junction remains properly reverse biased at all times. By fulfilling these conditions the transistor is made to operate over region of its characteristic curves which are linear, parallel and equi-spaced for equal increments of the parameter.

The proper flow of zero signal collector current and the maintenance of proper collector-emitter voltage during the passage of signal is called the *transistor biasing*.

If the transistor is not biased properly, it would work inefficiently and produce distortion in the output signal. A transistor is biased either with the help of battery or associating a circuit with the transistor. The latter method is more efficient and is frequently used. The circuit used for transistor biasing is called the *biasing circuit*. It may be noted that transistor biasing is very essential for the proper operation of transistor in any circuit.

Example 4.1. An N-P-N silicon transistor has V_{CC} = 10 V and the collector load resistance R_C = 3 kΩ. Determine (*i*) the maximum collector current i_C that can be allowed to flow during the application of signal for faithful amplification and (*ii*) the minimum zero signal collector current required.

What is the maximum input signal if β = 100 and a change of 1 V in base-emitter voltage V_{BE} causes a change of 4 mA in collector current i_C? Determine also the base current (maximum value).

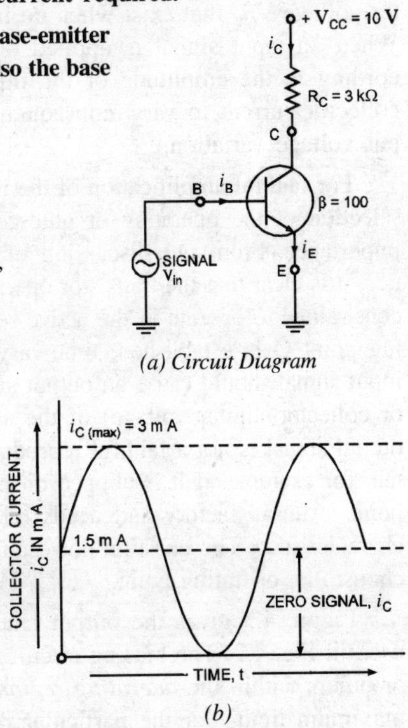

(a) Circuit Diagram

(b)

Fig. 4.1

Solution :

Collector supply voltage, V_{CC} = 10 V

Collector load resistance, R_C = 3 kΩ

Knee voltage for silicon transistor, V_{knee} = 1 V

So maximum voltage drop across resistance R_C that can be allowed,

$$i_{C\,(max)} R_C = V_{CC} - V_{knee} = 10 - 1 = 9 \text{ V}$$

Maximum allowable collector current,

$$i_{C\,(max)} = \frac{9}{R_C} = \frac{9}{3 \text{ k}\Omega} = 3 \text{ mA} \textbf{ Ans.}$$

(*i*) Thus, the maximum collector current that can be allowed during any part of the input signal is 3 mA. If the collector current i_C is allowed to exceed this value, collector-emitter voltage V_{ce} will fall below knee voltage of 1 V and available amplification will not be faithful one.

(*ii*) During negative peak of the input signal, i_C can at the most be allowed to drop to zero value. As the negative and positive half cycles are equal, so the change in collector current i_C due to these will also be equal but in opposite direction.

$$\therefore \text{ Zero signal collector current} = \frac{3 \text{ mA}}{2} = 1.5 \text{ mA} \textbf{ Ans.}$$

$$\text{Maximum base current, } i_{B\,(max)} = \frac{i_{C\,(max)}}{\beta} = \frac{3 \text{ mA}}{100} = 30 \text{ μA}$$

$$\text{Signal voltage (base-voltage)} = \frac{\text{Collector current}}{\text{Change in } i_C \text{ per volt change in } V_{BE}} = \frac{3 \text{ mA}}{4 \text{ mA/V}} = 0.75 \text{ V or } 750 \text{ mV} \textbf{ Ans.}$$

4.2. SELECTION OF OPERATING POINT

To study the effects of biasing conditions on the performance of a transistor, it is necessary to draw a dc load line on the output characteristics of the transistor. Consider the common emitter transistor amplifier circuit shown in Fig. 4.2. The input terminals are base and the emitter and the output is taken from the collector and emitter. Note that the polarity of the transistor terminal voltage in Fig. 4.2 is such that the base-emitter junction is forward biased and collector-base junction is reverse biased. These are the normal bias polarities for the device.

The output characteristics of the transistor are shown in Fig. 4.3. DC load line is drawn on the output characteristics, as mentioned in Art. 3.16.1.

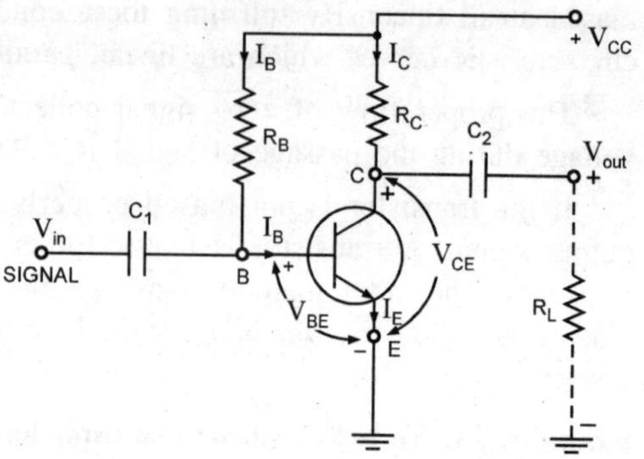

Fig. 4.2. ***Common Emitter Transistor Amplifier***

In designing a circuit a point on the load line is selected as the dc *bias point* or *quiescent point*. This bias point specifies the collector current I_C and collector-emitter voltage V_{CE} that exist when no input signal is applied. When an input signal is applied base current varies according to the amplitude of the input signal and causes collector current to vary, consequently producing an output voltage variation.

For faithful amplification of the input signal a judicious selection of the operating or quiescent point is of utmost importance. From our discussion of transistor characteristics, it is clear that the transistor operates linearly when it is constrained to operate in the active region. Once an operating point Q is established, time-varying excursions of the input signal should cause an output signal (collector current or collector-emitter voltage) of the same waveform. If the output signal is not a faithful reproduction of the input signal, for example, if it is clipped on one side, the operating point is unsatisfactory and needs relocation on the output characteristics. The question now naturally arises as how to choose the operating point.

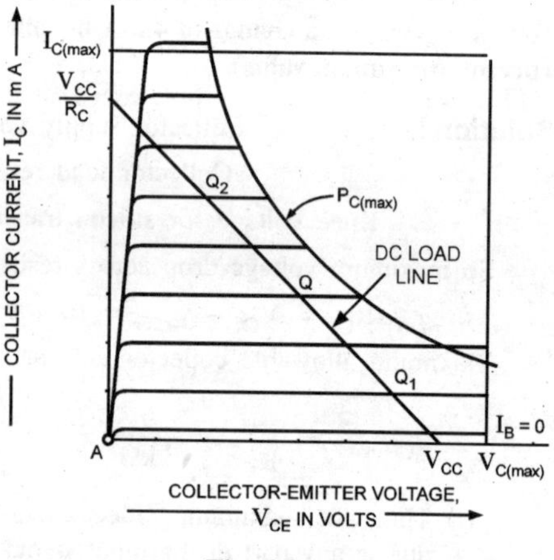

Fig. 4.3

Figure 4.3 gives the output characteristics, with four indicated operating points, of the transistor used in Fig. 4.2. The biasing circuit may be designed to set the device to operate at any of these points or others within the *operating region*. The operating region is the area of current or voltage within the maximum limits for the particular device. The maximum ratings, as mentioned in the manufacturer's specification sheet, are maximum collector current $I_{C\ (max)}$, maximum collector voltage $V_{C\ (max)}$, maximum collector dissipation $P_{C\ (max)}$, and maximum emitter-to-base voltage $V_{BE\ (max)}$. Figure 4.3 shows three of these bounds on typical output characteristics.

The BJT device could be biased to operate outside these maximum limit points but such an operation would result in either a considerable shortening of lifetime of the device or destruction to the device. Confining ourselves to the safe operating region, we may select several operating points, such as Q, Q_1 and Q_2. The exact point often depends on the intended use of the circuit. Let us consider some differences between operation at various points shown in Fig. 4.3 to present some basic ideas about the operating point and, thereby, the bias circuit.

If no bias were used, the device would initially be completely cut off, which would result in zero current through the device and zero voltage across it (*i.e.*, $I_C = 0$ and $V_{CE} = 0$, as represented by point A). It is, therefore, necessary, to bias the device so that it can respond or cause change in current and voltage for the entire range of an input signal.

Now let us consider point Q, located at the centre of the load line. When a signal is applied to the circuit, *in addition to the bias level*, collector voltage and current will vary approximately symmetrically around the quiescent values of V_C and I_C respectively allowing the device to react and amplify both the positive and negative excursions of the input signal. If, as could be the case, the input signal is small, the voltage and current of the device will vary but not enough to drive the device into *cutoff* or *saturation*. Cutoff is the condition in which the device no longer conducts. Saturation is the condition in which voltage across the device is the smallest possible with the current in the device reaching a limiting or saturating value depending on the external circuit. The usual amplifier action desired occurs within the operating region of the device, that is, between saturation and cutoff.

At point Q_1, near the cutoff region, output current and voltage would be allowed to vary but clipped at the negative peaks for a sinusoidally varying input signal. So this is not a suitable operating point.

At point Q_2, near the saturation region, the output signal would be clipped at the positive peaks for a sinusoidally varying input signal. Thus this is also not suitable operating point.

Thus point Q, located at the centre of load line, seems to be the best operating point in terms of linear gain or largest possible voltage and current swing.

4.3. BIAS STABILIZATION

Only the fixing of a suitable operating point is not sufficient but it is also to be ensured that the operating point remains stable *i.e.,* it does not shift due to change in temperature or due to variations in transistor parameters (due to replacement of transistor). Unfortunately it is not possible in practice unless special efforts are made to achieve it.

The maintenance of the operating point stable (independent of temperature variations or variations in transistor parameters) is known as *stabilization.*

4.3.1. Need For Bias Stabilization

The stabilization of operating point is essential because of (*i*) temperature dependence of collector current I_C (*ii*) individual variations and (*iii*) thermal runaway.

(i) Temperature Dependence of Collector Current I_C. The instability of collector current I_C, being equal to $\beta I_B + (1 + \beta) I_{CO}$, due to variations in temperature, is caused because of the following three main factors.

(*a*) Reverse saturation current (leakage current), I_{CO}, which doubles for every 10ºC rise in temperature.

(*b*) Transistor current gain, β, which increases with the increase in temperature.

(*c*) Base-emitter voltage, V_{BE}, which decreases by 2.5 mV per ºC.

Any or all of the above factors can cause the bias point to shift from the values originally fixed by the circuit because of a change in temperature.

Though with the change in temperature collector-emitter voltage V_{CE} also changes but the change is very small and collector current I_C is not affected.

(ii) Individual Variations. The value of β and V_{BE} are not exactly the same for any two transistors even of the same type. For an example BC147 is a silicon transistor with β varying from 100 to 600 (for one transistor β may be 100 and for the other it may be 600). The major reason for these variations is that transistor is a new device and manufacturing techniques have not too much advanced. For an instance, it has not been possible to control the base width and it may vary, although slightly, from one transistor to another one of the same type. Such little variations result in a large change in the transistor parameters such as β, V_{BE} etc.

Thus when a transistor is replaced by another transistor of the same type, the operating point may shift. Such problems do not arise in case of vacuum tube circuits because it is possible to manufacture vacuum tubes with identical characteristics.

(iii) Thermal Runaway. The collector current I_C, being equal to $\beta I_B + (1 + \beta)I_{CO}$, increases with the increase in temperature. This leads to increased power dissipation with further increase in temperature. Being a cumulative process, it can lead to *thermal runaway* resulting in burn-out of the transistor. The self destruction of an unstabilized transistor is called the *thermal runaway*.

However, if by some modification, I_B is made to fall with increase in temperature automatically, then decrease in the term βI_B can be made to neutralize the increase in the term $(1 + \beta)I_{CO}$, thereby keeping I_C almost constant. This will achieve thermal stability resulting in bias stability.

4.3.2. Means of Achieving Stability For Operating Point

Operating point stability may be achieved by adopting either stabilization techniques or compensation techniques.

Stabilization technique makes use of a resistive biasing circuit that permits such a variation of base current (biasing current) I_B as to maintain collector current I_C almost constant inspite of variations in reverse saturation current I_{CO}, base-emitter voltage V_{BE} and β.

Compensation techniques make use of temperature sensitive devices such as diodes, transistors, thermistors etc. Such devices produce compensating voltages and currents so as to make operating point stable.

Additional means are adopted to maintain the ambient and junction temperatures constant. Air-conditioning or thermostat chambers (wherever possible) may be used for maintaining the ambient or case temperature constant. Heat sink and air blast may be used to remove heat from the transistor.

4.4. STABILITY FACTOR

The degree of success achieved in stabilizing I_C in the face of variations in I_{CO} is expressed in terms of stability factor S* and it is defined as the rate of change of collector current w.r.t. I_{CO} keeping β and V_{BE} constant

$$i.e.,\ S = \frac{\partial I_C}{\partial I_{CO}} \approx \frac{\Delta I_C}{\Delta I_{CO}} \simeq \frac{dI_C}{dI_{CO}} \quad \text{at constant } \beta \text{ and } V_{BE} \text{ (or } I_B) \qquad ...(4.1)$$

From above equation it is obvious that smaller is the value of S, higher is the stability. The ideal value of S is unity (because I_C includes I_{CO}) but it is never possible to achieve it in practice.

While defining stability factor S above, V_{BE} and β are assumed to remain constant while they vary with temperature. Hence other two stability factors S_V and S_β are defined as below :

Stability factor S_V is defined as the rate of change of collector current I_C with base-emitter voltage V_{BE}, keeping I_{CO} and β constant

$$i.e.,\ S_V = \frac{\partial I_C}{\partial V_{BE}} = \frac{\Delta I_C}{\Delta V_{BE}} \quad \text{at constant } I_{CO} \text{ and } \beta \qquad ...(4.2)$$

Stability factor S_β is defined as the rate of change of collector current I_C with respect to β keeping I_{CO} and V_{BE} constant *i.e.,*

$$S_\beta = \frac{\partial I_C}{\partial \beta} \approx \frac{\Delta I_C}{\Delta \beta} \quad \text{at constant } I_{CO} \text{ and } V_{BE} \qquad ...(4.3)$$

* The larger the value of S, the more likely the circuit to exhibit thermal instability, so it will be more proper to call S an instability factor.

4.4.1. General Expression For Stability Factor S

In the active region the basic relationship between collector current I_C and base current I_B is given below :

$$I_C = \beta I_B + (1 + \beta)I_{CO} \qquad ...(4.4)$$

Differentiating above Eq. (4.4) with respect to I_C considering β to be constant, we have

$$1 = \beta\frac{dI_B}{dI_C} + (1+\beta)\frac{dI_{CO}}{dI_C} = \beta\frac{dI_B}{dI_C} + \frac{1+\beta}{S}$$

So stability factor,
$$S = \frac{1+\beta}{1-\beta\frac{dI_B}{dI_C}} \qquad ...(4.5)$$

The value of $\frac{dI_B}{dI_C}$ depends upon the biasing arrangement used and for determination of the stability factor S, it is only necessary to find the relationship between I_C and I_B.

4.4.2. General Expression For Stability Factor S_β

Differentiating Eq. (4.4) w.r.t. I_C considering I_{CO} to be constant we have

$$1 = \beta\frac{dI_B}{dI_C} + I_B\frac{d\beta}{dI_C} + I_{CO}\frac{d\beta}{dI_C}$$

or
$$\frac{d\beta}{dI_C}[I_{CO} + I_B] = 1 - \beta\frac{dI_B}{dI_C}$$

or
$$\frac{1}{S_\beta}[I_{CO} + I_B] = 1 - \beta\frac{dI_B}{dI_C} \qquad \because \text{ from Eq. (4.3) } S_\beta = \frac{\delta I_C}{\delta\beta}$$

or
$$S_\beta = \frac{I_{CO} + I_B}{1-\beta\frac{dI_B}{dI_C}} \qquad ...(4.6)$$

4.5. TRANSISTOR BIASING CIRCUITS

There is a large number of circuits for biasing of a transistor. Some of these are quite simple. These circuits differ so as to their ability to keep the quiescent point fixed inspite of variations in transistor characteristics and also effects of temperature variations and ageing. Before discussing different biasing circuits, it becomes imperative to know the basic requirements of a biasing circuit.

Requirements of Biasing Circuit. A biasing network associated with a transistor should fulfill the following requirements :

1. Establish the operating point in the middle of the active region of the characteristics, so that on applying the input signal the instantaneous operating point does not move either to the cutoff region or to the saturation region, even at the extreme values of the input signal.
2. Stabilize the collector current I_C against temperature variations.
3. Make the operating point independent of transistor parameters so that replacement of transistor by another transistor of the same type in the circuit does not shift the operating point.

4.5.1. Simplest Biasing Circuit

The simplest biasing circuit could be as shown in Fig. 4.4. The emitter-base junction is forward biased by the battery V_{BB} and the collector-base junction is reverse biased by the battery V_{CC}. The voltage V_{BE} across the forward biased junction is very low (0.3 V for Ge and 0.7 V for Si), therefore, the battery voltage V_{BB} must also be of the same order. The voltage V_{CC} of battery required for reverse bias junction should be of a much large value than voltage V_{BB} so as to reverse bias the collector-base junction. Since it is extremely difficult to have a battery of voltage as low as 0.3 V or 0.7 V, this is not a practical circuit.

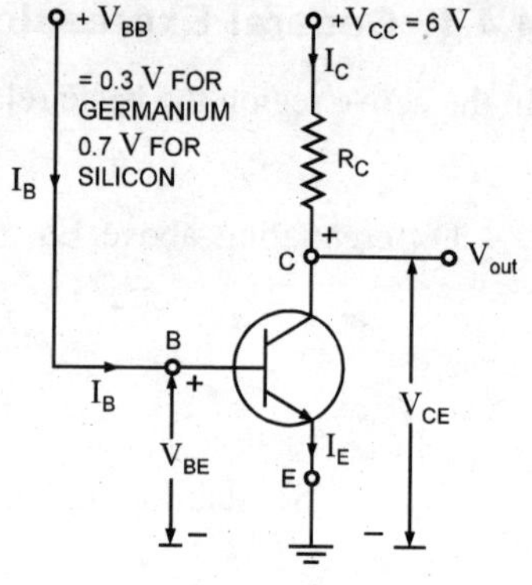

Fig. 4.4

4.5.2. Fixed Bias Circuit

The circuit shown in Fig. 4.5 uses two batteries V_{BB} and V_{CC}. The battery V_{BB} gives potential to the base terminal through resistance R_B (of the order of hundred kΩ) and battery V_{CC} supplies power to the collector through resistance R_C (of the order of kΩ).

The base current can be determined by applying Kirchhoff's second law for the branch circuit, starting from battery V_{BB} to ground.

$$V_{BB} = I_B R_B + V_{BE}$$

$$\text{or} \quad I_B = \frac{V_{BB} - V_{BE}}{R_B} \qquad \text{...(4.7)}$$

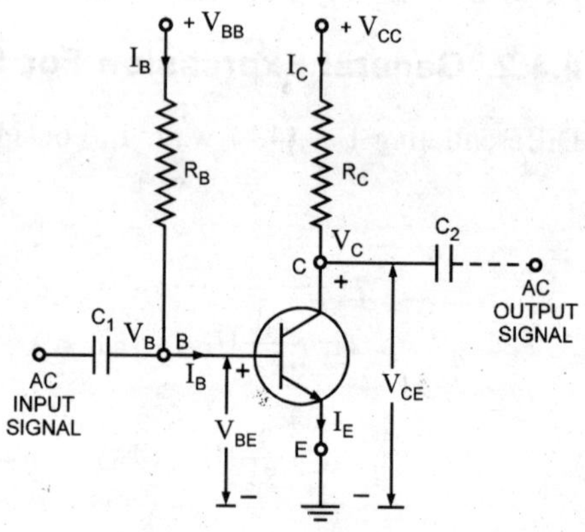

Fig. 4.5. *Fixed Bias Circuit*

Since V_{BE} is very small (0.7 V in case of silicon transistor and 0.3 V in case of Ge transistor) in comparison to V_{BB}, not much error will be committed if it is neglected. The base current is then given by the simple expression

$$I_B \simeq \frac{V_{BB}}{R_B} \qquad \text{...(4.8)}$$

The supply voltage V_{BB} being of fixed value—once the resistance R_B is chosen—the base current I_B is also fixed. Hence the name fixed bias circuit is given.

The collector-emitter section consists of the supply battery V_{CC}, the collector resistor and the transistor collector-emitter junction. The currents through the collector and emitter are about the same since I_B is very small in comparison to either. For linear amplifier operation, the collector current will be β times base current, where β is the transistor current gain *i.e.,*

$$I_C = \beta I_B \qquad \text{...(4.9)}$$

The base current I_B is determined from the operation of the base-emitter section of the circuit. The collector current, as shown above, will be β times greater than the base current and not at all dependent on the resistance in the collector circuit. Thus collector current I_C is controlled by base-emitter section of the circuit.

Applying Kirchhoff's second law to the branch circuit starting from V_{CC} to ground.

$$V_{CC} = I_C R_C + V_{CE}$$

$$\text{or} \quad V_{CE} = V_{CC} - I_C R_C \qquad \text{...(4.10)}$$

The quiescent point is given by I_C, V_{CE}.

The coupling capacitors C_1 and C_2 have reactances quite small at the frequency of ac signal, so that very little voltage drop of ac signal takes place across them. These capacitors are open circuit for direct currents, so that dc biasing potentials of this amplifier stage do not affect the circuits connected at the input as well as output terminals.

The fixed bias circuit is the simplest way to bias a transistor, but from stability point of view it is the worst circuit, because with the rise in temperature of transistor, β changes and also there is increase of leakage current and drop in V_{BE} value.

Since base current I_B controls the collector current I_C and I_C is equal to βI_B, with the change in the value of β, I_C changes (I_B remaining unchanged) and the operating point gets shifted.

The circuit shown in Fig. 4.5 may be modified by using a single battery, as shown in Fig. 4.2. From the practical point of view, it is always preferable to have an electronic circuit operated from a single battery.

The expressions for I_B, I_C and V_{CE} are given below

$$I_B = \frac{V_{CC} - V_{BE}}{R_B} \simeq \frac{V_{CC}}{R_B} \quad \text{...(4.11)}$$

$$I_C = \beta I_B \quad \text{...(4.12)}$$

$$V_{CE} = V_{CC} - I_C R_C \quad \text{...(4.13)}$$

Stability Factor S. From Eq. (4.5) stability factor $S = \dfrac{1+\beta}{1-\beta \dfrac{dI_B}{dI_C}}$ and in fixed biasing method I_B is independent of I_C, as obvious from Eq. (4.8) or Eq. (4.11), so stability factor is given as

$$S = 1 + \beta \quad \text{...(4.14)}$$

If β = 150, then S = 151 which means that collector current I_C increases 151 times as much as I_{CO}. Such a large value of S makes thermal runaway, a definite possibility with this circuit.

Advantages and Disadvantages. Though this biasing circuit is very simple, as very few components (only two resistors and one battery) are required, biasing conditions can be easily set, the calculations are simple and there is no loading of the source by the biasing circuit, as no resistor is used across base-emitter junction but this method of biasing is rarely used. This is due to the following reasons.

1. This method provides poor stability, as there is no means to stop a self increase in collector current I_C owing to temperature rise and individual variations. For example, if β increases either due to temperature rise or due to replacement of transistor, then collector current, I_C also increases by the same factor as base current I_B is constant.
2. There are good chances of thermal runaway. This is due to high stability factor S.

Example 4.2. In a fixed bias circuit, R_B = 150 kΩ, R_C = 2 kΩ, V_{CC} = 12 V. Transistor is silicon with β = 100, find I_B.
[Pb. Technical Univ. Electronic Devices and Circuits December 2006]

Solution : Collector supply voltage, $V_{CC} = 12$ V

Base resistance, $R_B = 150$ kΩ

Collector resistance, $R_C = 2$ kΩ

Base-emitter voltage, $V_{BE} = 0.7$ V ∵ Silicon transistor V_{BE} is assumed to be 0.7 V

$$\text{Base current, } I_B = \frac{V_{CC} - V_{BE}}{R_B} = \frac{12 - 0.7}{150 \times 10^3} = 75.33 \times 10^{-6} \text{ A} = 75.33\ \mu\text{A} \quad \textbf{Ans.}$$

Example 4.3. The transistor in Fig. 4.6 has values of h_{FE} = 100. Determine the Q-point values of I_C and V_{CE}.

[*Mahamaya Technical Univ. Electronics Engineering Second Semester 2011-12*]

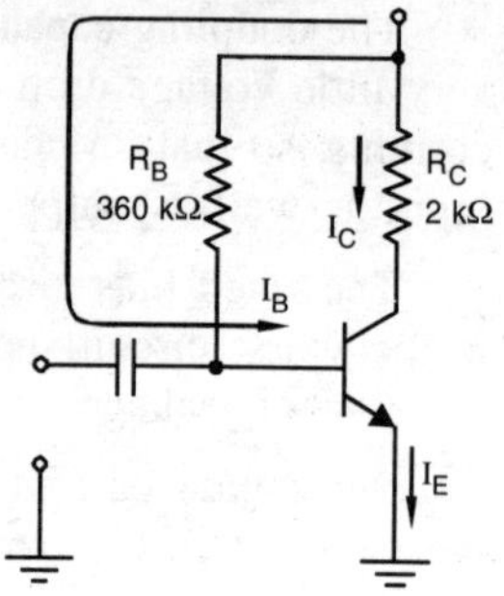

Fig. 4.6

Solution : Base current, $I_B = \dfrac{V_{CC} - V_{BE}}{R_B} = \dfrac{(8 - 0.7)\ V}{360\ k\Omega} = 20.278\ \mu A$

$$\beta = h_{FE} = 100$$

Collector current, $I_C = \beta I_B = 100 \times 20.278 \times 10^{-6} = 2.0278$ mA

Collector-emitter voltage, $V_{CE} = V_{CC} - I_C R_C$

$$= 8 - 2.0278 \times 10^{-3} \times 2 \times 10^3 = 3.944\ V$$

Q-point values of I_C and V_{CE} are 2.0278 mA and 3.944 V respectively **Ans.**

Example 4.4. For the given network in Fig. 4.7 (*a*) determine I_C, V_{CE}, V_B and V_C.

[*Mahamaya Technical Univ. Electronics Engineering First Semester 2011-12*]

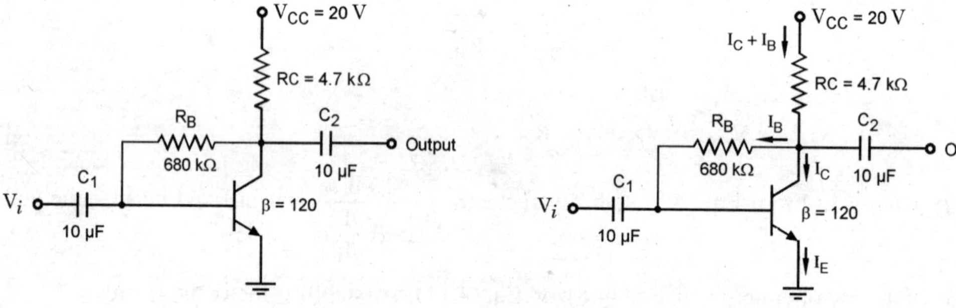

(*a*) (*b*)

Fig. 4.7

Solution : Applying KVL in input side, we have

$$V_{CC} = (I_C + I_B)\,R_C + I_B R_B + V_{BE} = (\beta + 1)\,I_B R_C + I_B R_B + V_{BE}$$

$$\text{or } I_B = \frac{V_{CC} - V_{BE}}{(\beta + 1)\,R_C + R_B} = \frac{20 - 0.7}{(120 + 1) \times 4.7 \times 10^3 + 680 \times 10^3} = 15.5\ \mu A$$

$$\text{or } I_C = \beta I_B = 120 \times 15.5 \times 10^{-6} = \mathbf{1.86\ mA}\ \ \textbf{Ans.}$$

Base voltage, $V_B = V_{BE} = 0.7$ V (∵ emitter is grounded) **Ans.**

Applying KVL to output side, we have

$$V_{CE} = V_{CC} - (I_B + I_C)\,R_C = 20 - (15.5 \times 10^{-6} + 1.86 \times 10^{-3}) \times 4.7 \times 10^3 = \mathbf{11.19\ V}\ \ \textbf{Ans.}$$

Collector voltage, $V_C = V_{CE}$ = **11.19 volts. Ans.** (∵ emitter is directly connected to ground)

Example 4.5. In the fixed bias circuit of transistor, V_{CC} = 15 V, R_B = 300 kΩ, R_L = 2 kΩ. If β = 100, I_{CO} = 20 nA and V_{BE} = 0.7 V. Find the stability factor of Q-point with respect to I_{CO}.

[*Pb. Technical Univ. Electronic Devices and Circuits May 2007*]

Solution : In a fixed bias circuit, the stability factor of operating point w.r.t. I_{CO} is given by equation

$$S = 1 + \beta$$

Refer to Eq. (4.14)

Substituting β = 100 in above equation, we have

$$S = 1 + 100 = 101 \ \ \textbf{Ans.}$$

Example 4.6. In a fixed bias circuit using transistor with α = 0.97, the temperature changes from 30°C to 60°C producing change in I_C from 2.2 mA to 3.8 mA. Assume reverse saturation current changes 1.7 μA/°C. What is stability factor ?

Solution : Change in collector current, $\Delta I_C = 3.8 - 2.2 = 1.6$ mA or 1.6×10^{-3} A

Change in temperature, $\Delta t = 60 - 30 = 30°C$

Change in reverse saturation current for Δt of 30°C

$$\Delta I_{CO} = 30 \times 1.7 \times 10^{-6} = 51 \times 10^{-6} \text{ A}$$

$$\text{Stability factor, } S = \frac{\Delta I_C}{\Delta I_{CO}} = \frac{1.6 \times 10^{-3}}{51 \times 10^{-6}} = 31.37 \textbf{ Ans.}$$

Example 4.7. Draw the dc load line and locate the operating point for fixed biasing transistor circuit shown in the Fig. 4.8 (*a*). What will be its stability factor ? V_{BE} = 0.7 V.

[*U.P. Technical Univ. Electronics Engineering Second Semester 2005-06*]

Solution : Base current, $I_B = \frac{V_{CC} - V_{BE}}{R_B} = \frac{(15 - 0.7) \text{ V}}{820 \text{ k}\Omega} = 17.44 \text{ µA}$.

Collector current,

$$I_C = \beta I_B = 120 \times 17.44 \times 10^{-6} = 2.09 \text{ mA}$$

Collector emitter voltage,

$$V_{CE} = V_{CC} - I_C R_C$$
$$= 15 - 2.09 \times 10^{-3} \times 4.7 \times 10^3 = 5.164 \text{ V}$$

Thus I_{CQ} = 2.09 mA and V_{CEQ} = 5.164 V

For fixed bias stability factor,

$$S = \beta + 1 = 120 + 1 = \textbf{121 Ans.}$$

For dc load line point A $\left(0, \frac{V_{CC}}{R_L}\right)$ and point B (V_{CC}, 0)

$$V_{CE} = V_{CC} = 15 \text{ V} \quad \text{when } I_C = 0$$

$$\text{and } I_C = \frac{V_{CC}}{R_L} = \frac{15\text{V}}{4.7 \text{ k}\Omega} = 3.2 \text{ mA when } V_{CE} = 0$$

Load line is drawn joining points A and B, as shown in Fig. 4.8 (*b*).

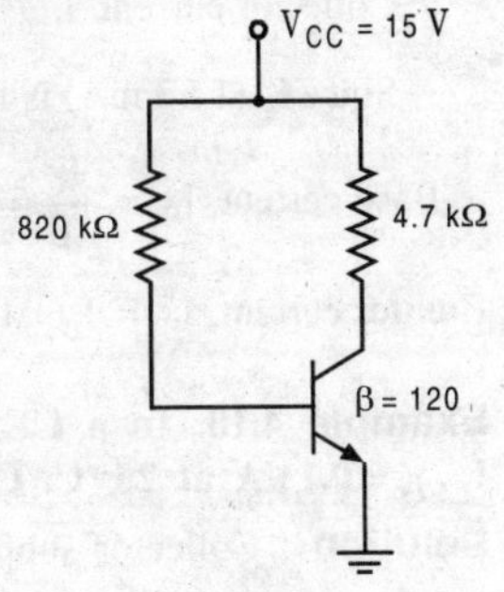

(*a*)

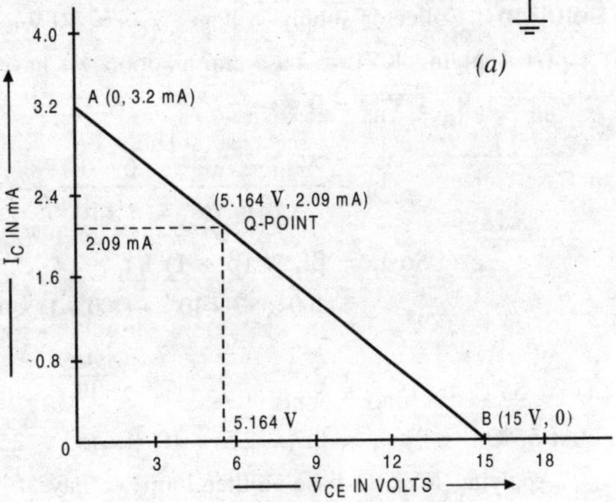

(*b*)

Fig. 4.8

Example 4.8. What is the base current in the following figure (Fig. 4.9) ?

[*U.P. Technical Univ. Basic Electronics First/Second Semester 2001-02*]

Solution : Base supply voltage, V_{BB} = 10 V

Base-emitter voltage, V_{BE} = 0.7 V ∵ transistor is silicon one

$$\text{Base current, } I_B = \frac{V_{BB} - V_{BE}}{R_B} = \frac{10 - 0.7}{330 \times 10^3}$$

$$= 28.2 \text{ µA } \textbf{Ans.}$$

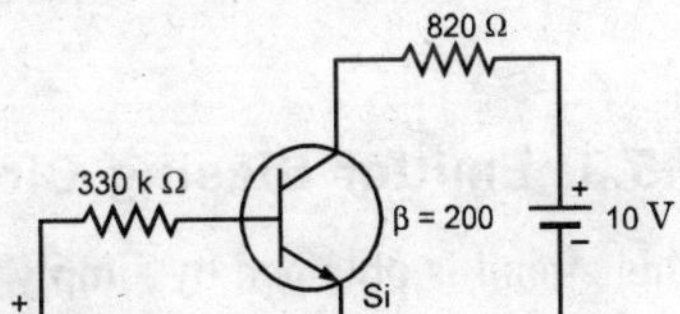

Fig. 4.9

Example 4.9. Consider the transistor circuit of Fig. 4.10 which has a resistance included between emitter and ground. Show that the transistor is operating in active mode. Calculate I_C, I_E and I_B. Given : β = 50.

[*U.P. Technical Univ. Electronics Engineering First Semester 2007-08*]

Solution : Applying KVL to input loop, we have

$$V_{BB} = I_B R_B + V_{BE} + I_E R_E = \frac{I_C}{\beta} R_B + V_{BE} + \frac{\beta + 1}{\beta} I_C R_E \qquad \because I_B = \frac{I_C}{\beta} \text{ and } I_E = \left(1 + \frac{1}{\beta}\right) I_C$$

$$\text{or } I_C = \frac{V_{BB} - V_{BE}}{\frac{R_B}{\beta} + \frac{\beta+1}{\beta} R_E} = \frac{5 - 0.7}{\frac{50 \times 10^3}{50} + \frac{51}{50} \times 1.8 \times 10^3} = 1.52 \text{ mA } \textbf{Ans.}$$

In case the transistor circuit goes into saturation, the maximum

$$\text{Collector current, } I_{C\,max} = \frac{V_{CC}}{R_C + R_E} = \frac{5}{800 + 1{,}800} = 1.92 \text{ mA}$$

Since I_C (1.52 mA) is less than $I_{C\,max}$ (1.92 mA), the circuit is in active mode.

$$\text{Base current, } I_B = \frac{I_C}{\beta} = \frac{1.52 \text{ mA}}{50} = 30.4\ \mu\text{A } \textbf{Ans.}$$

$$\text{Emitter current, } I_E = I_B + I_C = 0.0304 + 1.52 = 1.55 \text{ mA } \textbf{Ans.}$$

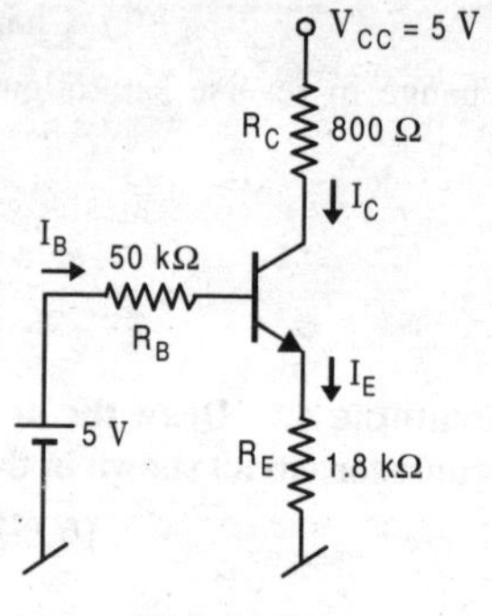

Fig. 4.10

Example 4.10. In a CE transistor circuit V_{CC} = 30 V, V_{BB} = 6V, R_C = 1 kΩ, R_B = 1 MΩ, β_{dc} = 100 and I_{CBO} = 0.1 μA at 25 °C. Compute I_B and I_C at *(i)* 25 °C *(ii)* 55 °C.

Solution : Collector supply voltage, V_{CC} = 30 V

(i) Applying KVL to base-emitter loop we have

$$V_{BB} - I_B R_B - V_{BE} = 0$$

$$\text{or Base current, } I_B = \frac{V_{BB} - V_{BE}}{R_B} = \frac{6 - 0.7}{1 \times 10^6} = 5.3\ \mu\text{A } \textbf{Ans.}$$

$$\text{So } I_C = \beta I_B + (\beta + 1)\, I_{CBO}$$

$$= 100 \times 5.3 \times 10^{-6} + (100 + 1) \times 0.1 \times 10^{-6} = 0.54 \text{ mA } \textbf{Ans.}$$

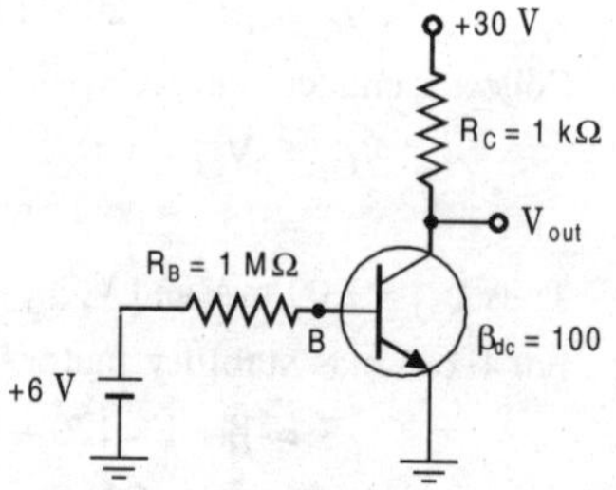

Fig. 4.11

(ii) At 55°C, $I_{CBO} = I_{CBO}$ at 25°C $\times\ 2^{\frac{55-25}{10}} = 0.1 \times 10^{-6} \times 2^3 = 0.8\ \mu\text{A}$

At 55°C, $V_{BE} = 0.7 - 2.5 \times 10^{-3} \times (55 - 25) = 0.625$ V

Applying KVL to base-emitter loop we have

$$\text{Base current, } I_B = \frac{V_{BB} - V_{BE}}{R_B} = \frac{6 - 0.625}{1 \times 10^6} = 5.375\ \mu\text{A } \textbf{Ans.}$$

$$\text{and collector current, } I_C = \beta I_B + (\beta + 1)\, I_{CBO}$$

$$= 100 \times 5.375 \times 10^{-6} + (100 + 1) \times 0.8 \times 10^{-6} = 0.6183 \text{ mA } \textbf{Ans.}$$

4.5.3. Emitter Biasing Circuit

This circuit is obtained by simply introducing an emitter resistor to the fixed bias circuit as shown in Fig. 4.12 (*a*). For analysis, we will first examine the base-emitter loop and then use the results to investigate the collector-emitter loop.

The base-emitter loop of the network given in Fig. 4.12 (*a*) may be redrawn as shown in Fig. 4.12 (*b*). Applying KVL around the indicated loop, we have

$$V_{CC} = I_B R_B + V_{BE} + I_E R_E$$

$$= I_B R_B + V_{BE} + (\beta + 1) I_B R_E$$

$$\because I_E = (\beta + 1) I_B$$

$$\text{or } I_B = \frac{V_{CC} - V_{BE}}{R_B + (\beta + 1) R_E} \quad ...(4.15)$$

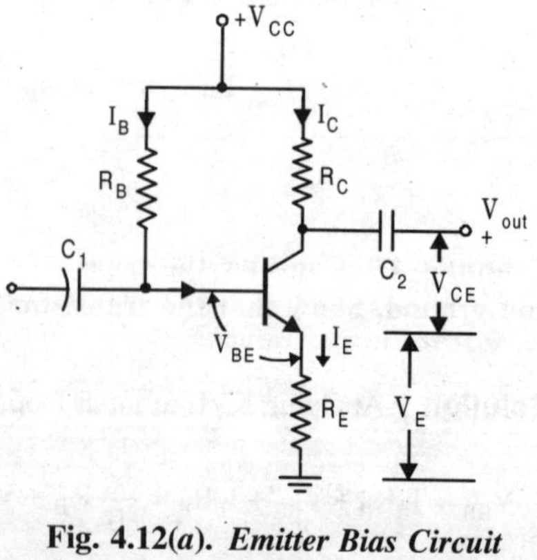

Fig. 4.12(*a*). *Emitter Bias Circuit*

Note that the only difference between the above equation and equation for the fixed bias configuration $\left[I_B = \frac{V_{CC} - V_{BE}}{R_B}\right]$ is the term $(\beta + 1)R_E$.

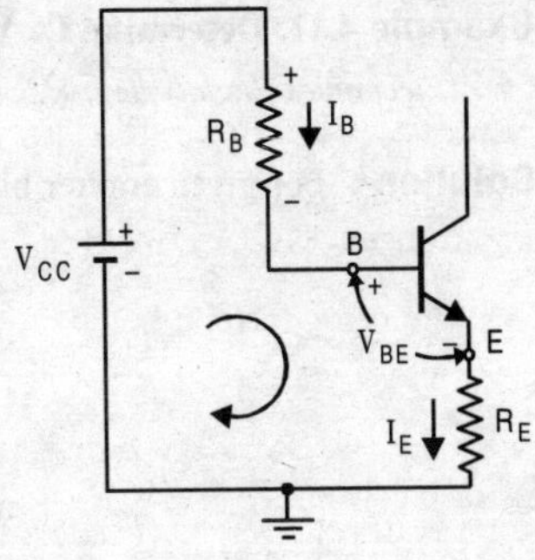

Fig. 4.12(b)

The collector-emitter loop is redrawn in Fig. 4.12 (c).

Applying KVL for the indicated loop, we have

$$I_C R_C + V_{CE} + I_E R_E = V_{CC}$$

$$\text{or } V_{CE} = V_{CC} - I_C R_C - I_C R_E \quad \because I_C \simeq I_E$$

$$\text{or } V_{CE} = V_{CC} - I_C(R_C + R_E) \quad \text{...(4.16)}$$

Voltage from emitter-to-ground is

$$V_E = I_E R_E \quad \text{...(4.17)}$$

while the voltage from collector-to-emitter is

$$V_{CE} = V_C - V_E = V_C - I_E R_E \quad \text{...(4.18)}$$

$$\text{Collector voltage, } V_C = V_{CC} - I_C R_C \quad \text{...(4.19)}$$

$$\text{and Base voltage, } V_B = V_{CC} - I_B R_B \quad \text{...(4.20)}$$

$$= V_{BE} + V_E \quad \text{...(4.21)}$$

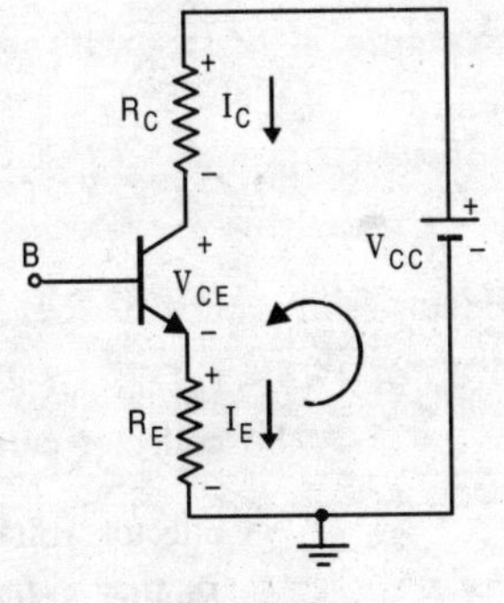

Fig. 4.12(c)

From Eq. (4.15), we have

$$V_{CC} = I_B R_B + V_{BE} + (I_B + I_C)R_E$$

Differentiating above equation w.r.t. I_C (considering V_{BE} to be independent of I_C), we have

$$0 = (R_B + R_E)\frac{dI_B}{dI_C} + R_E$$

$$\text{or } \frac{dI_B}{dI_C} = \frac{-R_E}{R_B + R_E} \quad \text{...(4.22)}$$

From Eq. (4.5) stability factor S is given as

$$S = \frac{1+\beta}{1-\beta\frac{dI_B}{dI_C}}$$

Substituting the value of $\frac{dI_B}{dI_C}$ from Eq. (4.22) in above expression, we have

$$\text{Stability factor, } S = \frac{1+\beta}{1+\beta\frac{R_E}{R_B+R_E}} = (1+\beta)\frac{1+\frac{R_B}{R_E}}{1+\beta+\frac{R_B}{R_E}} \quad \text{...(4.23)}$$

The above equation indicates that stability factor S varies between 1 for smaller values of $\frac{R_B}{R_E}$ and $(1 + \beta)$ for larger values of $\frac{R_B}{R_E}$. For proper operation both V_{CC} and R_E should be larger and R_B smaller.

The addition of the emitter resistor to the dc bias of the BJT provides improved stability, *i.e.*, the dc bias currents and voltages remain closer to where they were fixed by the circuit when the outside conditions, such as temperature, and transistor beta change.

Example 4.11. Determine I_C, V_E, V_B, V_C and I_B for the following circuit (Fig. 4.13).

[*Electronics Engineering; U.P. Technical Univ. First Semester 2009-10; Mahamaya Technical Univ. Second Semester 2010-11*]

Solution : For given emitter bias CE amplifier,

$$R_B = 510\ k\Omega$$
$$R_C = 2.4\ k\Omega$$
$$R_E = 1.5\ k\Omega$$
$$V_{CC} = 20\ V$$
$$\text{and } \beta = 100$$

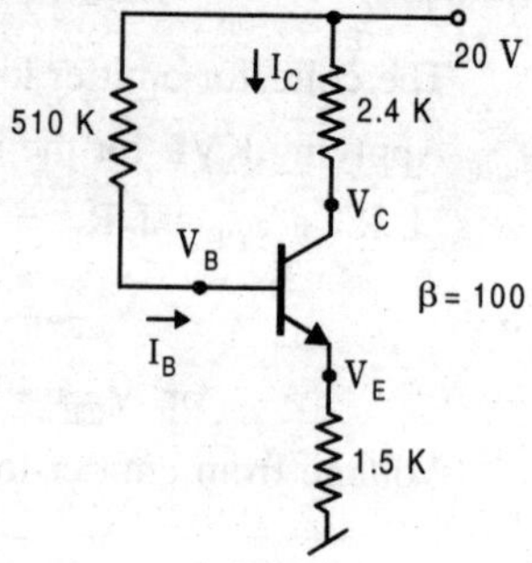

Fig. 4.13

Applying KVL for the input loop

$$V_{CC} = I_BR_B + V_{BE} + I_ER_E$$
$$= I_BR_B + V_{BE} + (\beta + 1)\ I_BR_E$$

$$\text{or } I_B = \frac{V_{CC} - V_{BE}}{R_B + (\beta+1)R_E}$$

$$= \frac{20 - 0.7}{510 \times 10^3 + 101 \times 1.5 \times 10^3} = \frac{19.3}{661.5 \times 10^3} = 0.0292 \text{ mA } \textbf{Ans.}$$

So collector current, $I_C = \beta I_B = 100 \times 0.0292 = 2.92$ mA **Ans.**

Collector voltage, $V_C = V_{CC} - I_CR_C = 20 - 2.92 \times 10^{-3} \times 2.4 \times 10^3 = 13$ V **Ans.**

Emitter voltage, $V_E = I_ER_E = (\beta + 1)I_BR_E = 101 \times 0.0292 \times 10^{-3} \times 1.5 \times 10^3 = 4.42$ V **Ans.**

Base voltage, $V_B = V_{CC} - I_BR_B = 20 - 0.0292 \times 10^{-3} \times 510 \times 10^3 = 5.12$ V **Ans.**

Example 4.12. Find I_C and V_{CE} for the following circuit of $\beta = 80$ for BJT.

[*U.P. Technical Univ. Electronics Engineering First Semester 2005-06*]

Solution : For given emitter bias CE amplifier

$R_B = 180\ k\Omega$; $R_C = 820\ \Omega$; $R_E = 200\ \Omega$; $V_{CC} = 25$ V and $\beta = 80$

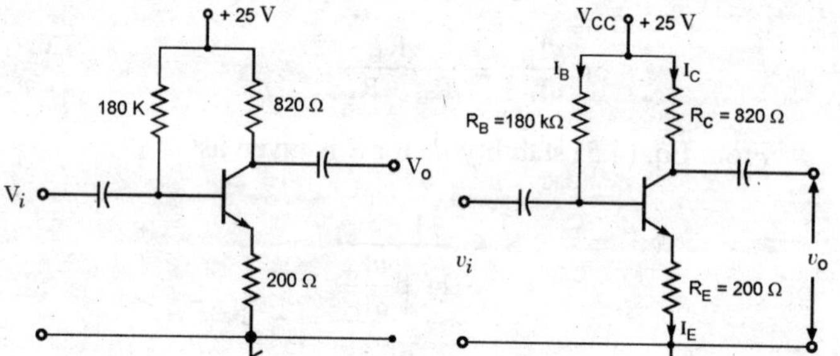

Fig. 4.14

Applying KVL for input loop,

$$V_{CC} = I_BR_B + V_{BE} + I_ER_E$$
$$= I_BR_B + V_{BE} + (\beta + 1)\ I_BR_E$$

Putting values in above equation, we have

$$I_B = \frac{V_{CC} - V_{BE}}{R_B + (\beta+1)R_E} = \frac{25 - 0.7}{180 \times 10^3 + 81 \times 200} = 124\ \mu A$$

So, collector current $I_C = \beta I_B = 80 \times 124 \times 10^{-6} = \mathbf{9.9\ mA}$ **Ans.**

Applying KVL for output loop.

$$V_{CC} = I_CR_C + V_{CE} + I_ER_E$$

$$\text{or, } V_{CE} = V_{CC} - I_C(R_C + R_E) = 25 - 9.9 \times 10^{-3}(820 + 200) = 14.9 \text{ V} \quad (\because I_C \approx I_E)$$

Thus operating point (V_{CE}, I_C) are **(14.9 V, 9.9 mA) Ans.**

Example 4.13. Determine the resistor values for the following network for the indicated operating point and power supply voltage. *[U.P. Technical Univ. Electronics Engineering Second Semester 2006-07]*

Solution : The given circuit is emitter biased CE amplifier circuit and its operating point is

$$V_{CEQ} = 10 \text{ V and } I_{CQ} = 2 \text{ mA}$$

Applying KVL in the outer loop, we have

$$V_{CC} = I_{CQ} R_C + V_{CEQ} + I_{EQ} R_E$$

$$\text{or } V_{CC} = I_{CQ}(R_C + R_E) + V_{CEQ} \quad \because I_{CQ} \simeq I_{EQ}$$

$$\text{or } R_C + R_E = \frac{V_{CC} - V_{CEQ}}{I_{CQ}} = \frac{20 - 10}{2 \times 10^{-3}} = 5 \text{ k}\Omega$$

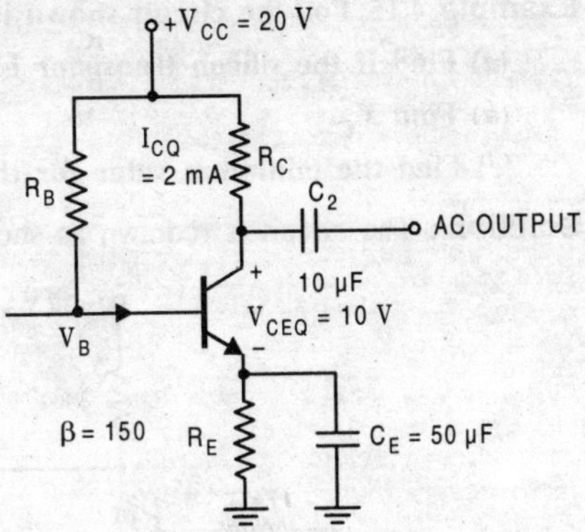

Fig. 4.15

Choosing two practical values for R_C and R_E

$$R_C = 3.9 \text{ k}\Omega \textbf{ Ans.}$$

$$R_E = 1.0 \text{ k}\Omega \textbf{ Ans.}$$

Now applying KVL to the input loop, we have

$$V_{CC} = I_B R_B + V_{BE} + I_E R_E$$

$$\text{or } V_{CC} = \frac{I_C}{\beta} R_B + 0.7 + I_C R_E$$

$$\text{or } R_B = \frac{V_{CC} - 0.7 - I_C R_E}{I_C} \times \beta = \frac{20 - 0.7 - 2 \times 10^{-3} \times 1 \times 10^3}{2 \times 10^{-3}} \times 150$$

$$= 1{,}297 \text{ k}\Omega \simeq 1.2 \text{ M}\Omega \text{ (standard value) } \textbf{Ans.}$$

Example 4.14. For the bias circuit shown in Fig. 4.16(*a*), determine the value of R_B to yield an operating point at $I_C = 5$ mA. Assume $\beta_0 = 60$, $V_{EB} = 0.7$ V, $R_C = 500\ \Omega$, $R_E = 100\ \Omega$ and $V_{CC} = -6$ V.

What is one advantage and one limitation of this type of biasing ? *[U.P.S.C. I.E.S. Electronics & Telecommunication Engineering-II, 2006]*

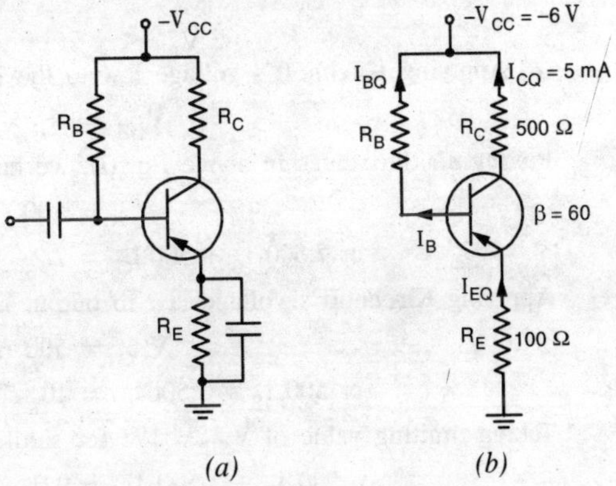

Fig. 4.16

Solution : The circuit is redrawn, with dc biasing as shown in Fig. 4.16 (*b*).

Applying Kirchhoff's voltage law to the collector-emitter loop, we have

$$V_{CC} = I_{CQ} R_C + V_{CEQ} + I_{EQ} R_E$$

$$-6 = -5 \times 10^{-3} \times 500 + V_{CEQ} - \frac{60+1}{60} \times 5 \times 10^{-3} \times 100$$

$$\because I_{EQ} = (\beta + 1) I_{CQ}$$

$$\text{or } V_{CEQ} = -6 + 2.5 + 0.508 = -2.992 \text{ V}$$

Applying Kirchhoff's voltage law to base-emitter loop, we have

$$R_B = \frac{V_{CC} - V_{BE} - I_{EQ} R_E}{I_{BQ}} \qquad \because I_{BQ} = \frac{I_{CQ}}{\beta} \text{ and } I_{EQ} = \frac{\beta+1}{\beta} I_{CQ}$$

$$= \frac{V_{CC} - V_{BE} - \left(\frac{\beta+1}{\beta}\right) I_{CQ} R_E}{\frac{I_{CQ}}{\beta}} = \frac{-6 - (-0.7) - \left(\frac{60+1}{60}\right)(-5 \times 10^{-3}) \times 100}{\frac{-5 \times 10^{-3}}{60}}$$

$$= \frac{4.792 \times 60}{5} = 57.5 \text{ k}\Omega \quad (56 \text{ k}\Omega\text{, standard value}) \textbf{ Ans.}$$

This circuit stabilises the base current and so the operating point can be stabilized but the operating point stability also depends upon other factors like junction temperature and transistor parameters mentioned in Art. 4.3.1. So this circuit is not able to stabilize the operating point.

Example 4.15. For the circuit shown in Fig. 4.17(*a*), assume h_{FE} = 100 and V_{BE} = 0.8 V.

(*a*) Find if the silicon transistor is in cutoff, saturation or active region.

(*b*) Find V_C.

(*c*) Find the minimum value for the base resistor for which the transistor operates in active region.

Solution : The circuit is redrawn as shown in Fig. 4.17 (*b*).

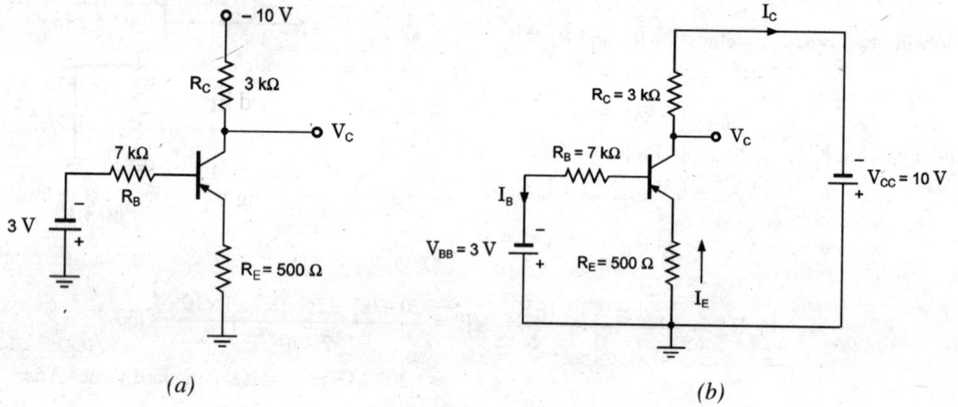

Fig. 4.17

(*a*) Applying Kirchhoff's voltage law to the input loop, we have

$$V_{BB} = R_E \times (I_B + I_C) + R_B \times I_B + V_{BE} \qquad ...(i)$$

Putting all the values in above Eq. (*i*), we have

$$3 = 500\,(I_B + I_C) + 7{,}000\,I_B + 0.8$$

$$\text{or } 7{,}500\,I_B + 500\,I_C = 2.2 \qquad ...(ii)$$

Applying Kirchhoff's voltage law to output loop we have

$$V_{CC} = 500\,(I_B + I_C) + 3{,}000\,I_C + V_{CE}$$

$$\text{or } 500\,I_B + 3{,}500 I_C = 10 - V_{CE}$$

Taking limiting value of V_{CE} = 1 V for saturation we have

$$500\,I_B + 3{,}500\,I_C = 9.0 \qquad ...(iii)$$

Solving Eqs. (*ii*) and (*iii*) we have

$$I_C = 2.55 \text{ mA}$$

$$I_B = 0.123 \text{ mA}$$

$$\text{Gain of amplifier, } \beta = \frac{I_C}{I_B} = \frac{2.55}{0.123} = 20.7$$

The actual gain is very less than h_{FE} = 100,

Therefore, transistor is in saturation region. **Ans.**

(*b*) Since transistor is in saturation region,

$$V_C = -V_{CE} - (I_B + I_C)R_E$$

$$= -1 - (0.123 + 2.55) \times 10^{-3} \times 500 = -2.3365 \text{ V } \textbf{Ans.}$$

(*c*) To bring transistor to active region, base current I_B should be reduced (by increasing resistance R_B). The maximum value of base current, $I_{B\,(max)}$ for which transistor remains in active region

$$I_{B\,(max)} = \frac{I_C}{h_{fE}} = \frac{2.55}{100} = 0.0255 \text{ mA}$$

Applying Kirchhoff's voltage law to input loop we have

$$V_{BB} = I_C R_E + V_{BE} + R_B I_{B\,(max)} \qquad \because I_E \simeq I_C$$

Putting the values we have $3 = 2.55 \times 10^{-3} \times 500 + 0.8 + 0.0255 \times 10^{-3} R_B$

$$\text{or } R_B = \frac{3 - 1.275 - 0.8}{0.0255 \times 10^{-3}} = 36.27 \text{ k}\Omega \quad (39 \text{ k}\Omega \text{ standard value})$$

Therefore, to bring the transistor in active region the minimum value for base resistance should be 39 kΩ **Ans.**

4.5.4. Self Bias or Potential Divider Bias Circuit

This is the most commonly used biasing arrangement. The arrangement is shown in Fig. 4.18 (*a*). The name voltage divider is derived due to the fact that the voltage divider is formed by the resistors R_1 and R_2 across V_{CC}. The emitter resistor R_E provides stabilization. The resistor R_E causes a voltage drop in a direction so as to reverse bias the emitter junction. Since the emitter-base junction is to be forward biased, the base voltage is obtained from supply V_{CC} through R_1–R_2 network. For forward biasing the emitter-base junction R_1 and R_2 are so adjusted that the base terminal becomes more positive than emitter. The net forward bias across the emitter-base junction, V_{BE} is equal to V_B minus dc voltage drop across R_E. The dc bias circuit is independent of transistor current gain factor β. In case of amplifiers, to avoid the loss of ac signal (because of feedback caused by R_E) a capacitor of large capacitance is connected across R_E. The capacitor offers a very small reactance to the ac signal and so it passes through the capacitor.

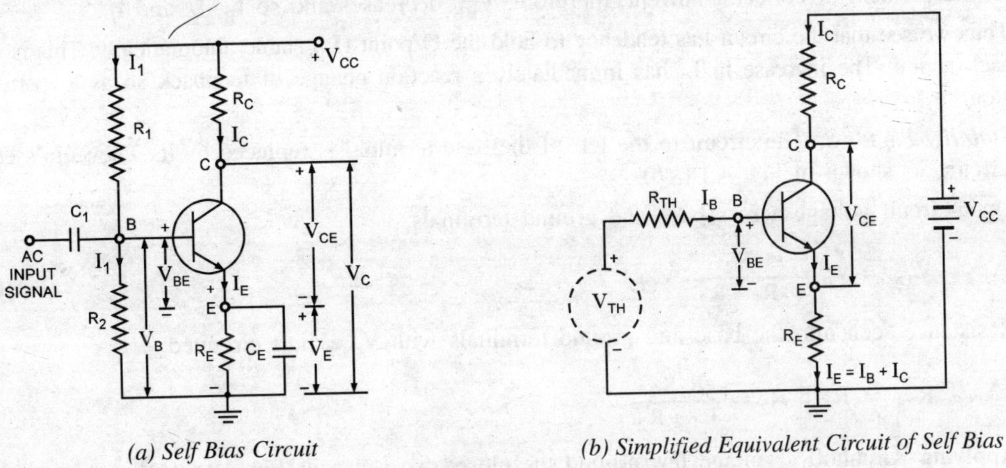

(*a*) *Self Bias Circuit* (*b*) *Simplified Equivalent Circuit of Self Bias*

Fig. 4.18

Circuit Analysis. Assume that the current flowing through resistance R_1 is I_1. As base current I_B is very small so current flowing through resistance R_2 can also be assumed to be equal to I_1

$$\text{and } I_1 = \frac{V_{CC}}{R_1 + R_2}$$

Voltage across resistance R_2, $V_B = \frac{V_{CC}}{R_1 + R_2} \cdot R_2$

Applying Kirchhoff's second (or voltage) law to the base-emitter loop [Fig. 4.18(*a*)] we have

$$V_B = V_{BE} + V_E = V_{BE} + I_E R_E$$

$$\text{or} \quad I_E = \frac{V_B - V_{BE}}{R_E}$$

and collector current, $I_C \simeq I_E \simeq \dfrac{V_B - V_{BE}}{R_E}$ ...(4.24)

Applying Kirchhoff's second (or voltage) law to the collector-emitter loop we have

$$V_{CC} = I_C R_C + V_{CE} + I_E R_E = I_C R_C + V_{CE} + I_C R_E \quad \text{since } I_C \simeq I_E$$

$$= V_{CE} + I_C (R_C + R_E)$$

$$\text{or } V_{CE} = V_{CC} - I_C (R_C + R_E) \quad ...(4.25)$$

From Eqs. (4.24) and (4.25) the values of I_C and V_{CE} can be determined and the quiescent point Q is established.

It is clear from Eq. (4.24) that I_C does not at all depend upon β. Though collector current I_C depends upon V_{BE} but in practice V_{BE} is very small in comparison to V_B and so collector current I_C is practically independent of V_{BE}. Thus collector current I_C in this biasing circuit is almost independent of transistor parameters and hence good stabilisation is ensured.

In this biasing circuit the emitter resistance R_E provides excellent stabilisation. This is explained as below :

Now let the temperature of transistor junction rise when it is loaded. This causes increase in leakage currents and so increase in the value of β. Hence collector current I_C tends to increase. With the increase in the value of I_C, voltage drop across emitter resistance R_E increases. Since voltage drop across R_2 (*i.e.*, V_B) is independent of collector current, therefore, V_{BE} decreases and so I_B, I_E and I_C.

Thus we see that the circuit has tendency to hold the Q point (I_C) stable automatically. This is due to feedback action. The increase in I_C has immediately a reaction change of feedback so as to correct the situation.

Stability Factor S. The circuit to the left of the base terminal is replaced by its Thevenin's equivalent circuit, as shown in Fig. 4.18 (*b*).

Open-circuit voltage across base and ground terminals,

$$V_{Th} = V_B = \frac{R_2}{R_1 + R_2} V_{CC} \quad ...(4.26)$$

Resistance seen into the base and ground terminals with V_{CC} short circuited,

$$R_{Th} = R_1 \parallel R_2 = \frac{R_1 R_2}{R_1 + R_2} \quad ...(4.27)$$

Applying Kirchhoff's voltage law around the closed base circuit [Fig. 4.18 (*b*)] yields

$$V_{Th} = I_B R_{Th} + V_{BE} + I_E R_E = I_B R_{Th} + V_{BE} + (I_B + I_C) R_E \quad ...(4.28)$$

Differentiating above Eq. (4.28) w.r.t. I_C (considering V_{BE} to be independent of I_C) we have

$$0 = R_E + (R_{Th} + R_E) \frac{dI_B}{dI_C}$$

$$\text{or} \quad \frac{dI_B}{dI_C} = \frac{-R_E}{R_{Th} + R_E} \quad ...(4.29)$$

From Eq. (4.5) stability factor S is given as

$$S = \frac{1 + \beta}{1 - \beta \dfrac{dI_B}{dI_C}}$$

Substituting the value of $\frac{dI_B}{dI_C}$ from Eq. (4.29) in above equation we have

$$\text{Stability factor, } S = \frac{1+\beta}{1+\beta\dfrac{R_E}{R_{Th}+R_E}} = (1+\beta)\,\frac{1+R_{Th}/R_E}{1+\beta+\dfrac{R_{Th}}{R_E}} \quad \ldots(4.30)$$

The above equation shows that stability factor S varies between 1 for small values of R_{Th}/R_E and $(1+\beta)$ for larger values of $\frac{R_{Th}}{R_E}$. For proper operation both R_E and V_{CC} should be larger and R_{Th} small. Typical practical value of S for this type of biasing circuit is about 10.

Example 4.16. A silicon transistor with $\beta = 100$ is to be used in self biasing circuit, such that Q point corresponds to $V_{CE} = 12$ V, $I_C = 2$ mA. Find R_E if $V_{CC} = 24$ V and $R_C = 5$ kΩ.

[*Pb. Technical Univ. Electronic Devices and Circuits December 2005*]

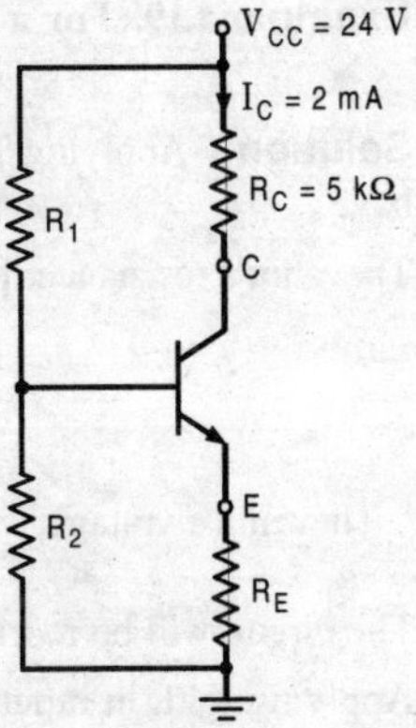

Fig. 4.19

Solution : $I_C = 2$ mA

$$\text{Emitter current, } I_E = \frac{\beta+1}{\beta}I_C = \frac{100+1}{100}\times 2\text{ mA} = 2.02\text{ mA}$$

Applying KVL to outer loop, we have

$$V_{CC} = I_C R_C + V_{CE} + I_E R_E$$

$$\text{or}\quad R_E = \frac{V_{CC} - I_C R_C - V_{CE}}{I_E} = \frac{24 - 2\times 10^{-3}\times 5\times 10^{3} - 12}{2.02} \simeq 1\text{ kΩ } \textbf{Ans.}$$

Example 4.17. In a self biased CE amplifier, $R_C = 5$ kΩ, $R_2 = 9$ kΩ, $R_1 = 81$ kΩ, $\beta = 50$ and $R_E = 810$ Ω. Compute the stability factor.

[*M.D. Univ. Rohtak Analog Electronics May 2007*]

Solution :

$$R_{Th} = \frac{R_1 R_2}{R_1 + R_2} = \frac{81\times 9}{81+9} = 8.1\text{ kΩ}$$

Stability factor for self-biased circuit is given by

$$S = (1+\beta)\frac{1+\dfrac{R_{Th}}{R_E}}{1+\beta+\dfrac{R_{Th}}{R_E}} = (1+50)\times\frac{1+\dfrac{8.1}{0.81}}{1+50+\dfrac{8.1}{0.81}} = 51\times\frac{11}{62} = 9.05\ \textbf{Ans.}$$

Example 4.18. For the voltage-divider bias configuration of Fig. 4.20, determine (*i*) I_C, (*ii*) V_E, (*iii*) V_{CC}, (*iv*) V_{CE}, (*v*) V_B and (*vi*) R_1.

[*Electronics Engineering M.T.U. Second Semester 2010-11; G.B.T.U. First Semester 2011-12*]

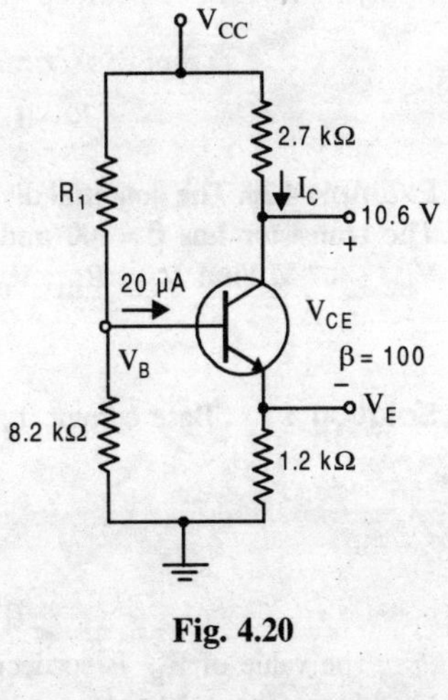

Fig. 4.20

Solution : Base current, $I_B = 20\ \mu A = 20\times 10^{-6}$ A

(*i*) Collector current, $I_C = \beta I_B = 100\times 20\times 10^{-6}$ A

$= 2\times 10^{-3}$ A or 2 mA **Ans.**

Emitter current, $I_E = I_C + I_B = 2 + 0.02 = 2.02$ mA

(*ii*) Emitter voltage, $V_E = I_E R_E = 2.02\times 10^{-3}\times 1.2\times 10^{3}$

$= 2.424$ V **Ans.**

(*iii*) Supply voltage, $V_{CC} = V_C + I_C R_C$

$= 10.6 + 2\times 10^{-3}\times 2.7\times 10^{3} = 16$ V **Ans.**

(*iv*) $V_{CE} = V_C - V_E = 10.6 - 2.424 = 8.176$ V **Ans.**

(*v*) Base voltage, $V_B = V_E + V_{BE} = 2.424 + 0.7 = 3.124$ V

Thevenin's voltage, $V_{Th} = \dfrac{R_2}{R_1 + R_2} V_{CC} = V_B$

or $\dfrac{8.2}{R_1 + 8.2} \times 16 = 3.124$ V

or $R_1 = \dfrac{8.2 \times 16}{3.124} - 8.2 = 33.8\ k\Omega$ **Ans.**

Example 4.19. For a voltage divider biasing circuit shown in Fig. 4.21 (*a*). Find I_C, V_{CE}, I_B, V_E and V_B.

[*U.P. Technical Univ. Electronics Engineering II, 2007-08*]

Solution : Applying Thevenin's theorem to given potential divider network, we have

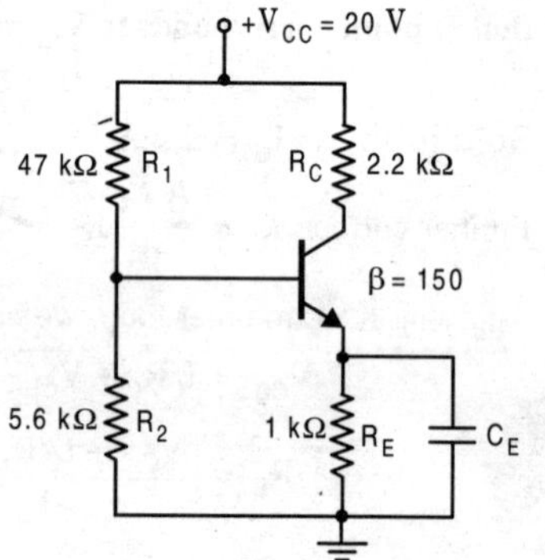

Fig. 4.21 (*a*)

Thevenin's resistance, $R_{Th} = R_1 \parallel R_2$

$$= \frac{R_1 \times R_2}{R_1 + R_2} = \frac{47 \times 5.6}{47 + 5.6}\ k\Omega = 5\ k\Omega$$

Thevenin's voltage, $V_{Th} = \dfrac{R_2}{R_1 + R_2} \times V_{CC} = \dfrac{5.6}{47 + 5.6} \times 20 = 2.13$ V

The circuit will be redrawn as shown in Fig. 4.21 (*b*).

Applying KVL in input loop, we have

$$V_{Th} = I_B R_{Th} + V_{BE} + I_E R_E = I_B R_{Th} + V_{BE} + (1 + \beta) I_B R_E$$

$$\text{or } I_B = \frac{V_{Th} - V_{BE}}{R_{Th} + (\beta + 1) R_E} = \frac{2.13 - 0.7}{5 \times 10^3 + (150 + 1) \times 1 \times 10^3}$$

$$= 9.2\ \mu A \text{ **Ans.**}$$

Collector current, $I_C = \beta I_B = 150 \times 9.2 \times 10^{-6} = 1.375$ mA **Ans.**

Emitter current, $I_E = I_C + I_B = 1.375 + 0.0092 = 1.384$ mA **Ans.**

Emitter voltage, $V_E = I_E R_E = 1.384 \times 10^{-3} \times 1 \times 10^3 = 1.384$ V **Ans.**

Base voltage, $V_B = V_{BE} + V_E = 0.7 + 1.384 = 2.084$ V **Ans.**

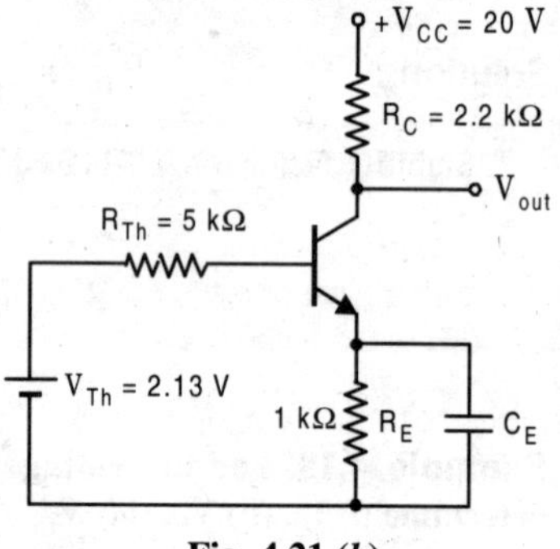

Fig. 4.21 (*b*)

Applying KVL in output loop

$$V_{CE} = V_{CC} - I_C R_C - I_E R_E$$

$$= 20 - 1.375 \times 10^{-3} \times 2.2 \times 10^3 - 1.384 = 15.6 \text{ V **Ans.**}$$

Example 4.20. The potential divider biasing of a transistor is as shown in Fig. 4.22 (*a*). The transistor has $\beta = 100$ and operates at I_C = 3 mA, and V_{CE} = 10 V. Assume V_{BE} = 0.7 V. Find V_{CC}, R_{B1}, V_B and R_E.

[*A.M.I.E. Sec. B. Electronic Circuits Summer 2005*]

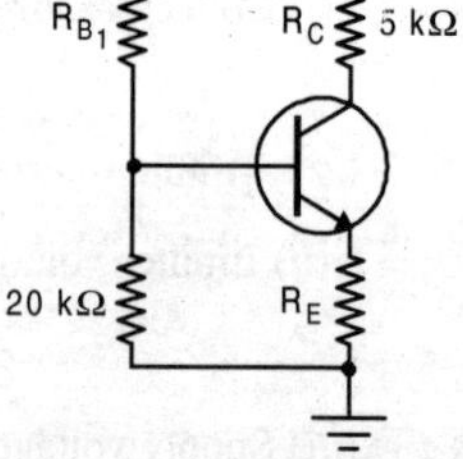

Fig. 4.22(*a*)

Solution : Base current, $I_B = \dfrac{\text{Collector current, } I_C}{\beta}$

$$= \frac{3 \times 10^{-3}}{100} = 0.03 \text{ mA}$$

$$I_1 = 10\ I_B = 10 \times 0.03 = 0.3 \text{ mA}$$

The value of R_{B_1} is so decided that the current I_1 flowing through R_{B_1} and R_{B_2} (20 kΩ) is about 10 times of base current to achieve good stabilization.

For the circuit shown in Fig. 4.22 (*b*), we have

$$V_{CB} + V_{BE} = V_{CE} \quad ...(i)$$

and $$I_1 R_{B_1} = I_C R_C + V_{CB} \quad ...(ii)$$

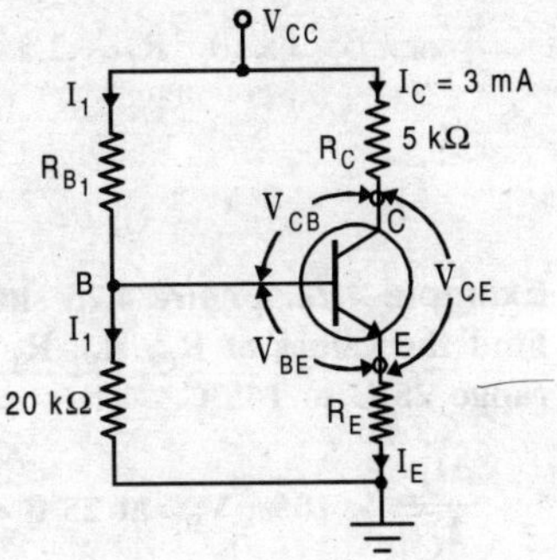

Fig. 4.22(*b*)

So from Eq. (*i*), we have

$$V_{CB} = V_{CE} - V_{BE} = 10 - 0.7 = 9.3 \text{ V}$$

and from Eq. (*ii*), we have

$$I_1 R_{B_1} = I_C R_C + V_{CB} = 3 \times 10^{-3} \times 5 \times 10^3 + 9.3 = 24.3 \text{ V}$$

and $$R_{B_1} = \frac{24.3}{I_1} = \frac{24.3}{0.3 \times 10^{-3}} = 81 \text{ k}\Omega \textbf{ Ans.}$$

$$V_B = I_1 \times R_{B_2} = 0.3 \times 10^{-3} \times 20 \times 10^3 = 6 \text{ V } \textbf{Ans.}$$

$$V_{CC} = I_1 (R_{B_1} + R_{B_2}) = 0.3 \times 10^{-3} (81 + 20) \times 10^3 = 30.3 \text{ V } \textbf{Ans.}$$

Now $I_E R_E = V_B - V_{BE} = 6 - 0.7 = 5.3 = \text{V}$

$$\text{or } R_E = \frac{I_E R_E}{I_E} = \frac{5.3}{3 \times 10^{-3}} = 1.766 \text{ k}\Omega \text{ (1.8 k}\Omega \text{ standard value) } \textbf{Ans.} \qquad \text{Assuming } I_E \simeq I_C$$

Example 4.21. The BJT amplifier of Fig. 4.23 has h_{fe} = 100, V_{BE} = 0.7 V, I_{CO} = 0. Calculate the values of R_1 and R_C such that its I_C = 1 mA and V_{CE} = 2.5 V. *[U.P.S.C. I.E.S. Electrical Engineering II, 2001]*

Solution : Thevenin's resistance,

$$R_{Th} = \frac{R_1 R_2}{R_1 + R_2} = \frac{10 R_1}{10 + R_1}$$

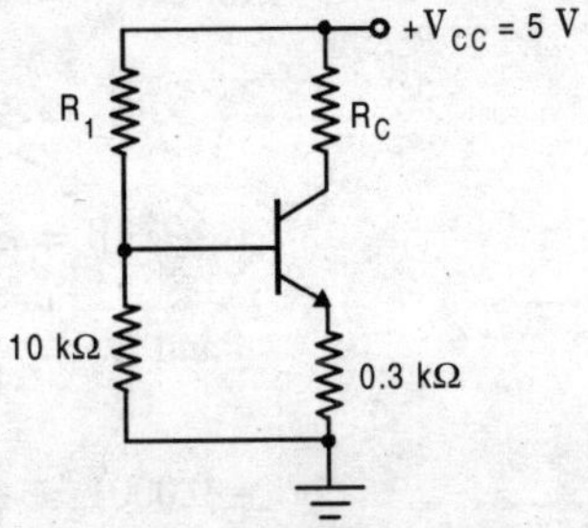

Fig. 4.23

Thevenin's voltage,

$$V_{Th} = V_B = \frac{R_2}{R_1 + R_2} V_{CC} = \frac{10 \times 5}{10 + R_1} = \frac{50}{10 + R_1}$$

Now applying Kirchhoff's voltage law to the base-emitter-ground loop, we have

$$V_B - I_B R_B - V_{BE} - I_E R_E = 0$$

$$\text{or} \quad V_{Th} - \frac{I_C}{\beta} R_{Th} - V_{BE} - \left(1 + \frac{1}{\beta}\right) I_C R_E = 0$$

$$\text{or} \quad \frac{50}{10 + R_1} - \frac{1 \times 10^{-3}}{100} \times \frac{10 R_1}{10 + R_1} \times 10^3 - 0.7$$

$$= \left(1 + \frac{1}{100}\right) \times 1 \times 10^{-3} \times 0.3 \times 10^3$$

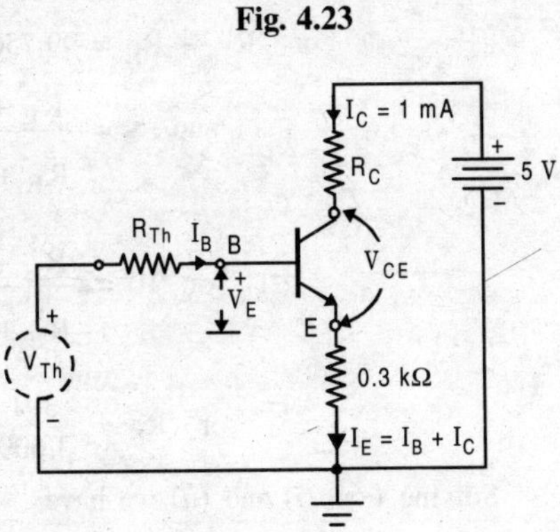

Fig. 4.24

$$\text{or} \quad \frac{50}{10 + R_1} - \frac{R_1}{10(10 + R_1)} = 0.7 + \frac{101}{100} \times 0.3$$

$$\text{or} \quad 500 - R_1 = (0.7 + 0.303)(R_1 + 10) \times 10$$

$$\text{or} \quad 500 - R_1 = 10.03(R_1 + 10) = 10.03 R_1 + 100.3$$

$$\text{or} \quad R_1 = \frac{399.7}{11.03} \text{ k}\Omega = 36.24 \text{ k}\Omega \text{ (39 k}\Omega\text{, standard value) } \textbf{Ans.}$$

Applying Kirchhoff's voltage law to collector-emitter-ground loop, we have

$$V_{CC} - I_C R_C - V_{CE} - I_E R_E = 0$$

or $5-1\times10^{-3}R_C-2.5-\frac{101}{100}\times1\times10^{-3}\times0.3\times10^3 = 0$

or $R_C = \frac{5-2.5-0.303}{1\times10^{-3}} = 2.2\ \text{k}\Omega$ (standard value) **Ans.**

Example 4.22. Figure 4.25 shows a self biased transistor amplifier using a silicon transistor with 0.6 mA. Find the values of R_C, R_E, R_1 and R_2 such that it meets the following specifications over the temperature range 25°C to 145°C.

$\frac{\Delta I_C}{I_C} \le 10\%$, V_{BE} at 25°C = 650 ± 50 mV, I_{CO} at 25°C = 5 nA max, I_{CO} at 145°C = 3.0 μA max. Assume that percentage change in I_C due to V_{BE} and I_{CO} is same (5%). [*GATE 1996*]

Solution : Change in I_C due to V_{BE} and I_{CO} is 5%

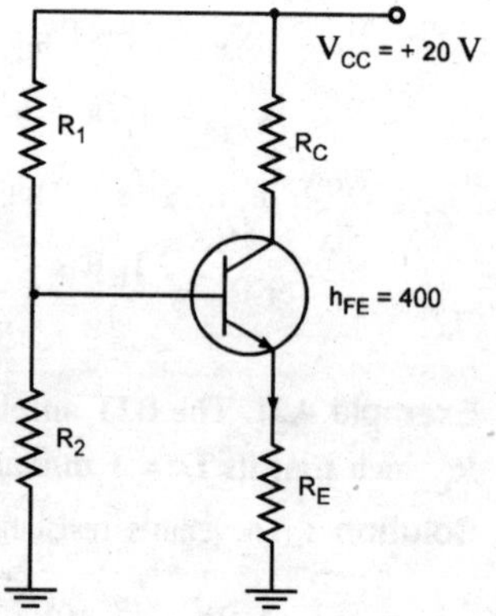

Fig. 4.25

Change in collector current, $\Delta I_C = \frac{5}{100}\times 0.6 = 0.03$ mA

Change in reverse saturation current,

$$\Delta I_{CO} = 3\times10^{-6} - 0.005\times10^{-6} = 2.995\ \mu\text{A}$$

$$\text{Stability factor, } S = \frac{\Delta I_C}{\Delta I_{CO}} = \frac{0.03\times10^{-3}}{2.995\times10^{-6}} = 10$$

For a silicon transistor V_{BE} decreases at the rate of 2.5 mV/°C

$$\text{So } \Delta V_{BE} = -2.5\,(145-25) = -300\ \text{mV}$$

$$\therefore S_V = \frac{\Delta I_C}{\Delta V_{BE}} = \frac{0.03\times10^{-3}}{-300\times10^{-3}} = -0.0001$$

$$\beta = h_{FE} = 400$$

$$\text{and } S_V = \frac{-S}{R_{Th}+R_E}\times\frac{\beta}{\beta+1}$$

$$\text{or } -0.0001 = \frac{-10}{R_{Th}+R_E}\times\frac{400}{401}$$

$$\text{or } R_{Th}+R_E = 99{,}750.6\ \Omega \qquad ...(i)$$

$$\text{and } S = \frac{R_E+R_{Th}}{R_E+\frac{R_{Th}}{\beta+1}}$$

$$\text{or } 10 = \frac{R_E+R_{Th}}{R_E+\frac{R_{Th}}{401}}$$

$$\text{or } R_E = \frac{391}{3{,}609}R_{Th} \qquad ...(ii)$$

Solving Eqs. (*i*) and (*ii*) we have

$R_E = 9.75\ \text{k}\Omega$ **Ans.**

and $R_{Th} = 90\ \text{k}\Omega$

$$\Delta I_C = SI_{CO} + S_V\Delta V_{BE} = 10\times3\times10^{-6} + (-0.0001)\times(-300\times10^{-3})$$

$$= 0.03\times10^{-3} + 0.03\times10^{-3} = 0.06\ \text{mA}$$

$$\text{So } I_C = 0.6 + 0.06 = 0.66\ \text{mA}$$

From Eq. (4.25) we have $V_{CC} = I_C R_C + V_{CE} + I_E R_E$

$$20 = 0.66\ R_C + 10 + \left(0.66 + \frac{0.66}{400}\right) \times 9.75$$

or $R_C = 5.377$ kΩ **Ans.**

$$\because V_{Th} = I_B R_{Th} + V_{BE} + I_E R_E = \frac{0.66}{400} \times 90 + 0.65 + \left(0.66 + \frac{0.66}{400}\right) \times 9.75 = 7.25\ \text{V}$$

$$R_1 = R_{Th}\frac{V_{CC}}{V_{Th}} = 90 \times \frac{20}{7.25} = 248\ \text{k}\Omega\ \textbf{Ans.}$$

$$R_2 = \frac{V_{Th} \times R_1}{V_{CC} - V_{Th}} = \frac{7.25 \times 248}{20 - 7.25} = 141\ \text{k}\Omega\ \textbf{Ans.}$$

As R_1, R_2, R_C and R_E are designed for variation up to 145ºC, the variation of V_{BE} at 25ºC of 650 ± 50 mV will be absorbed in the overall variation.

Example 4.23. A silicon transistor uses potential divider method of biasing. V_{CC} = 12 V, $R_1 = R_2$ = 10 kΩ, R_E = 3 kΩ, R_C = 4 kΩ. Determine operating point using Thevenin's theorem.

[Pb. Technical Univ. Electronic Devices and Circuits December 2006]

Solution : Referring to Figs. 4.18 (*a*) and (*b*)

$$V_{CC} = 12\ \text{V},\ R_1 = R_2 = 10\ \text{k}\Omega,\ R_E = 3\ \text{k}\Omega,\ R_C = 4\ \text{k}\Omega$$

$$R_{Th} = \frac{R_1 R_2}{R_1 + R_2} = \frac{10 \times 10}{10 + 10} = 5\ \text{k}\Omega$$

$$V_{Th} = \frac{R_2}{R_1 + R_2} V_{CC} = \frac{10}{10 + 10} \times 12 = 6\ \text{V}$$

Applying KVL around the closed base circuit [Fig. 4.18 (*b*)], we have

$$V_{Th} = I_B R_{Th} + V_{BE} + I_E R_E = I_B R_{Th} + V_{BE} + (\beta + 1) I_B R_E$$

$$\because I_E = I_C + I_B = \beta I_B + I_B = (\beta + 1) I_B$$

$$I_B = \frac{V_{Th} - V_{BE}}{R_{Th} + (\beta + 1) R_E} = \frac{6 - 0.7}{5 \times 10^3 + (50 + 1) \times 3 \times 10^3}$$

Assuming $\beta = 50$ and $V_{BE} = 0.7$ V

$$= \frac{5.3}{158 \times 10^3} = 0.0335\ \text{mA}$$

Collector current, $I_C = \beta I_B = 50 \times 0.0335 = 1.68$ mA

Emitter current, $I_E = I_B + I_C = 0.0335 + 1.68 = 1.71$ mA

Applying KVL to output loop, we have

$$V_{CE} = V_{CC} - I_C R_C - I_E R_E = 12 - 1.68 \times 10^{-3} \times 4 \times 10^3 - 1.71 \times 10^{-3} \times 3 \times 10^3$$

$$= 12 - 6.72 - 5.13 = 0.15\ \text{V}$$

Thus, operating point (V_{CE}, I_C) is (0.15 V, 1.68 mA) **Ans.**

Example 4.24. If in a CE single stage amplifier, R_L = 1 kΩ, R_E = 200 Ω, β = 100 and V_{CC} = 9 V. Compute the values of biasing resistors R_1 and R_2, so that the quiescent point will be specified by I_{CQ} = 3.75 mA and V_{CEQ} = 4.5 V.

Solution : $I_{CQ} = 3.75$ mA

$$\text{Base current, } I_B = \frac{I_{CQ}}{\beta} = \frac{3.75}{100} = 37.5\ \mu\text{A}$$

From Eq. (4.28) we have $V_{Th} = I_B R_{Th} + V_{BE} + I_E R_E$

or $V_{Th} = 37.5 \times 10^{-6} R_{Th} + 0.7 + (\beta + 1) I_B \times R_E$

$= 37.5 \times 10^{-6} R_{Th} + 0.7 + 37.5 \times 10^{-6} \times 101 \times 200$

or $V_{Th} = 37.5 \times 10^{-6} R_{Th} + 1.4575$

Assuming $R_{Th} = 20\ k\Omega$

$V_{Th} = 37.5 \times 10^{-6} \times 20 \times 10^3 + 1.4575 = 2.2075\ V$

From Eqs. (4.26) and (4.27) $R_1 = \dfrac{R_{Th} V_{CC}}{V_{Th}} = \dfrac{20 \times 9}{2.2075} = 81.54\ k\Omega$ (82 kΩ, standard value) **Ans.**

and $R_2 = \dfrac{R_1 V_{Th}}{V_{CC} - V_{Th}} = \dfrac{82 \times 2.2075}{9 - 2.2075} = 26.65\ k\Omega$ (27 kΩ standard value) **Ans.**

Example 4.25. Determine V_{CEQ} and I_{CQ} for the BJT amplifier of Fig. 4.26. Also find I_B.

Solution : As we have to calculate biasing current I_{CQ}, we are not concerned with signal V_S.

Now for given potential divider network,

$$V_A = \frac{V_{CC}}{R_1 + R_2} \times R_2 = \frac{12}{15 + 2.7} \times 2.7 = 1.83 \text{ volts}$$

Applying Kirchhoff's voltage law to input loop we have

$$V_A = V_{BE} + I_E R_E$$

$$I_E = \frac{V_A - V_{BE}}{R_E} = \frac{1.83 - 0.7}{1 \times 10^3} = 1.13 \text{ mA} \quad ...(i)$$

Applying Kirchhoff's voltage law to output loop we have

$$V_{CC} = I_{CQ} R_C + V_{CEQ} + I_E R_E$$

or $V_{CEQ} = V_{CC} - I_{CQ} (R_C + R_E) \quad \because I_{CQ} \simeq I_E$

$= 12 - 1.13 (1 + 1) = 9.74\ V$ **Ans.**

Thus $V_{CEQ} = 9.74\ V$ and $I_{CQ} = 1.13\ mA$ **Ans.**

and $I_B = \dfrac{I_{CQ}}{\beta} = \dfrac{1.13\ mA}{100} = 11.3\ \mu A$ **Ans.**

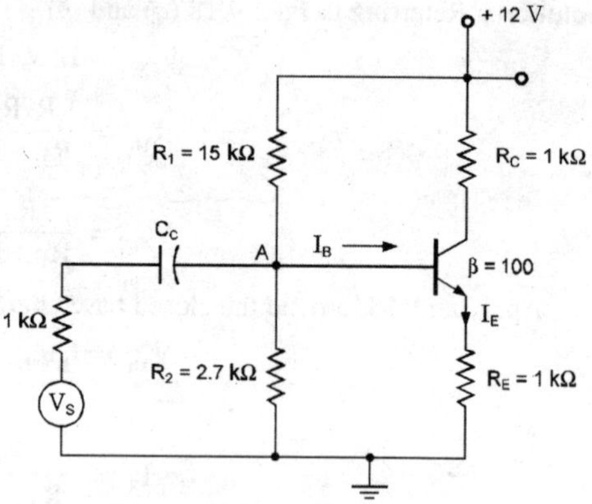

Fig. 4.26

Example 4.26. In a single stage CE amplifier, with single battery supply V_{CC} = 20 V, the device has a maximum h_{FE} value of 20 and I_{CO} = 10 μA. If R_L = 4 kΩ and the quiescent point is required to be at V_{CE} = 6 V and I_C = 2 mA, estimate the values of R_1 and R_2.

Solution : From Eq. (4.4)

Collector current, $I_C = \beta I_B + (1 + \beta) I_{CO}$

So base current, $I_B = \dfrac{I_C - (1+\beta) I_{CO}}{\beta} = \dfrac{2\,mA - (20+1) \times 10\,\mu A}{20} = \dfrac{1.79\,mA}{20} = 89.5\ \mu A$

Emitter current, $I_E = (\beta + 1) I_B = (20 + 1) \times 89.5\ \mu A = 1.88\ mA$

From Eq. (4.25) we have $V_{CC} = I_C R_L + V_{CE} + I_E R_E$

or emitter resistance, $R_E = \dfrac{V_{CC} - I_C R_L - V_{CE}}{I_E} = \dfrac{20 - 2\,mA \times (4\,k\Omega) - 6}{1.88} = 3.2\ k\Omega$

From Eq. (4.28) we have $V_{Th} = I_B R_{Th} + V_{BE} + I_E R_E = I_B R_{Th} + 0.7 + 1.88\ mA \times 3.2\ k\Omega = I_B R_{Th} + 6.7$

Assuming $R_{Th} = 20\ k\Omega$

$$V_{Th} = 89.5 \times 10^{-6} \times 20 \times 10^3 + 6.7 = 8.49 \text{ V}$$

From Eqs. (4.26) and (4.27) we have

$$R_1 = \frac{R_{Th} V_{CC}}{V_{Th}} = \frac{20 k\Omega \times 20 V}{8.49 V} = 47.1 \text{ k}\Omega \quad \text{(47 k}\Omega \text{ standard value) } \textbf{Ans.}$$

$$\text{and } R_2 = \frac{R_1 V_{Th}}{V_{CC} - V_{Th}} = \frac{47.1 \times 8.49}{20 - 8.49} = 34.74 \text{ k}\Omega \quad \text{(33 k}\Omega\text{, standard value) } \textbf{Ans.}$$

Example 4.27. A Ge transistor with $\beta = 49$ has self biasing arrangement as given in Fig. 4.27. Given $V_{CC} = 10$ V, $R_L = 1$ kΩ, $V_{CE} = 5$ V, $I_C = 4.9$ mA as $V_{BE} = 0.2$ V, the stability factor S is derived to be 10. Obtain values of R_1, R_2, R_E. *[G.B. Technical Univ. Electronics Engineering Second Semester 2010-11]*

Solution : Base current, $I_B = \frac{I_C}{\beta} = \frac{4.9}{49} = 0.1$ mA

Emitter current, $I_E = I_C + I_B = 4.9 + 0.1 = 5$ mA

So emitter resistance, $R_E = \frac{V_E}{I_E}$

$$= \frac{V_{CC} - I_C R_L - V_{CE}}{I_E}$$

$$= \frac{10 - 4.9 \times 10^{-3} \times 1 \times 10^3 - 5}{5 \times 10^{-3}} = 20 \ \Omega \ \textbf{Ans.}$$

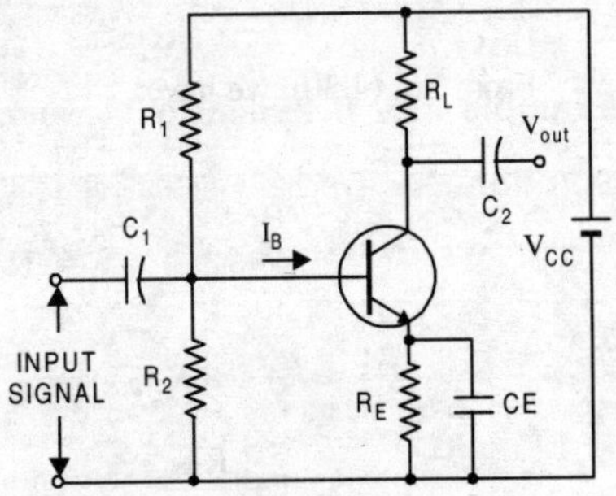

Fig. 4.27

From Eq. (4.30), we have
$$S = (1+\beta)\frac{1 + \frac{R_{Th}}{R_E}}{1 + \beta + \frac{R_{Th}}{R_E}}$$

$$\therefore 10 = (1+49)\frac{1 + \frac{R_{Th}}{20}}{1 + 49 + \frac{R_{Th}}{20}}$$

$$\text{or } R_{Th} = 225 \ \Omega$$

$$\text{Since } V_{Th} = I_B R_{Th} + V_{BE} + I_E R_E = 0.1 \times 10^{-3} \times 225 + 0.2 + 5 \times 10^{-3} \times 20 = 0.3 \text{ V}$$

$$\text{Also } V_{Th} = \frac{R_2}{R_1 + R_2} V_{CC} = \frac{R_{Th}}{R_1} V_{CC}$$

$$\text{or } R_1 = R_{Th} \times \frac{V_{CC}}{V_{Th}} = 225 \times \frac{10}{0.3} = 7{,}500 \ \Omega \text{ or } 7.5 \text{ k}\Omega \ \textbf{Ans.}$$

$$\text{and since } V_{Th} = \frac{R_2}{R_1 + R_2} V_{CC}$$

$$\text{So } R_2 = \frac{V_{Th} \times R_1}{V_{CC} - V_{Th}} = \frac{0.3 \times 7{,}500}{10 - 0.3} = \frac{2{,}250}{9.7} = 232 \ \Omega \ \textbf{Ans.}$$

Example 4.28. The circuit shown uses a silicon transistor with $\beta = 50$, $V_{BE} = 0.6$ V, $V_{CC} = 22.5$ V and $R_C = 5.6$ kΩ. Find the values of the resistors R_E, R_1 and R_2 so that Q point is set at $V_{CE} = 12$ V and $I_C = 1.5$ mA. The stability factor, S must be ≤ 3. *[U.P.S.C. I.E.S. Electrical Engineering-II, 2011]*

Solution : Base current, $I_B = \frac{I_C}{\beta} = \frac{1.5}{50} = 0.03\ mA$

Emitter current, $I_E = I_B + I_C = 0.03 + 1.5 = 1.53\ mA$

From Eq. (4.25), we have

$$R_E = \frac{V_E}{I_E} = \frac{V_{CC} - I_C R_C - V_{CE}}{I_E}$$

$$= \frac{22.5 - 1.5\times10^{-3}\times5.6\times10^3 - 12}{1.53\times10^{-3}}$$

$$= \frac{2.1}{1.53\times10^{-3}} = 1.373\ k\Omega \ \textbf{Ans.}$$

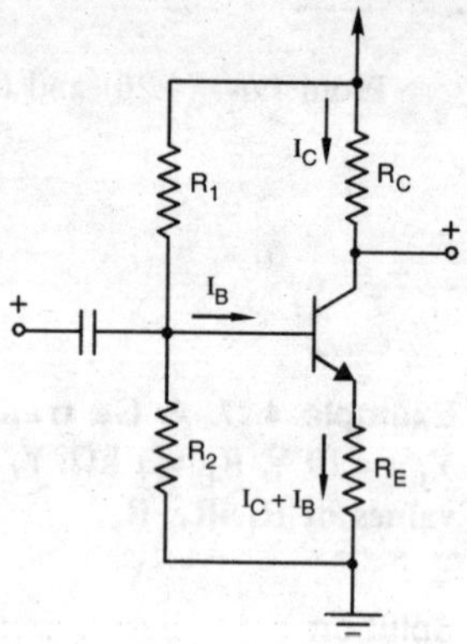

Fig. 4.28

From Eq. (4.30), we have

$$S = (1+\beta)\frac{1+\frac{R_{Th}}{R_E}}{1+\beta+\frac{R_{Th}}{R_E}}$$

$$\text{or} \quad (1+50)\frac{1+\frac{R_{Th}}{1.373}}{1+50+\frac{R_{Th}}{1.373}} \le 3$$

$$\text{or} \quad R_{Th} \le 2.916\ k\Omega$$

$$V_{Th} = I_B R_{Th} + V_{BE} + I_E R_E = 0.03\times10^{-3}\times2.916\times10^3 + 0.6 + 1.53\times10^{-3}\times1.373\times10^3$$

$$= 0.08748 + 0.6 + 2.1 = 2.78748\ V$$

$$R_1 = R_{Th}\times\frac{V_{CC}}{V_{Th}} = \frac{2.916\times22.5}{2.78748} = 23.54\ k\Omega \ \textbf{Ans.}$$

$$R_2 = \frac{V_{Th}R_1}{V_{CC}-V_{Th}} = \frac{2.78748\times23.54}{22.5-2.78748} = 3.3\ k\Omega \ \textbf{Ans.}$$

Example 4.29. A voltage emitter bias circuit has been shown in Fig. 4.29 (*a*). Obtain R_1 and R_2 to provide the following bias conditions to a Ge transistor circuit. Also draw the equivalent circuit of the above arrangement

$I_{CQ} = 1$ mA, $V_{CEQ} = 5$ V, $\beta = 100$, $S = 8$, $R_E = 1\ k\Omega$.

[Pb. Technical Univ. Electronic Devices and Circuits May 2007]

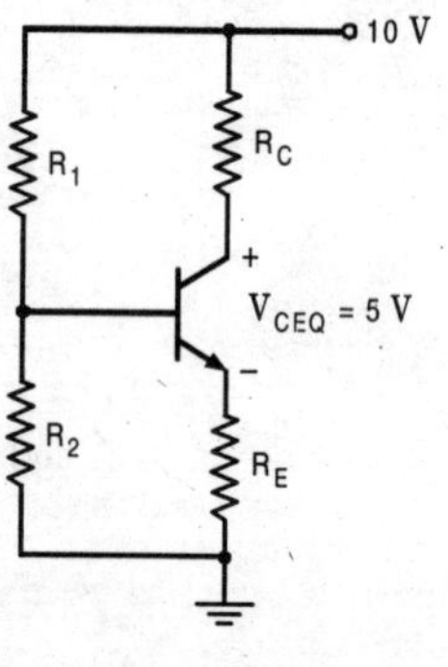

Fig. 4.29(*a*)

Solution : Collector current, $I_C = 1$ mA

Base current, $I_B = \frac{I_C}{\beta} = \frac{1}{100} = 0.01\ mA$

Emitter current, $I_E = I_B + I_C = 0.01 + 1 = 1.01\ mA$

Applying KVL to outer loop, we have

$$V_{CC} = I_C R_C + V_{CE} + I_E R_E$$

$$\text{or} \quad R_C = \frac{V_{CC} - I_E R_E - V_{CE}}{I_C} = \frac{10 - 1.01 \times 10^{-3} \times 1 \times 10^3 - 5}{1 \times 10^{-3}}$$

$$= 3.99 \text{ k}\Omega \text{ (3.9 k}\Omega\text{, standard value)}$$

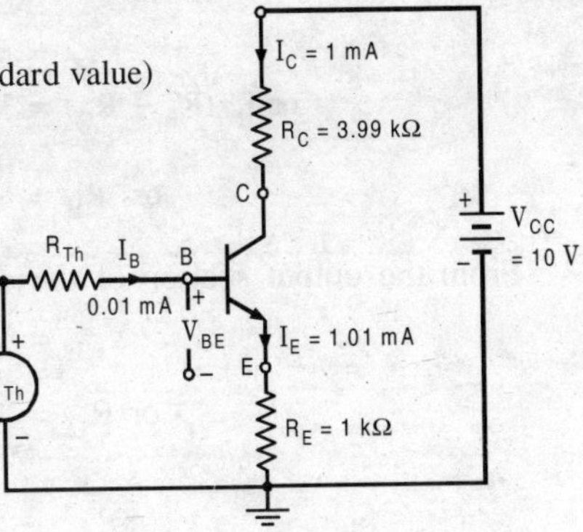

Fig. 4.29(*b*)

Stability factor is given by equation

$$S = (1+\beta)\frac{1 + \frac{R_{Th}}{R_E}}{1 + \beta + \frac{R_{Th}}{R_E}}$$

$$\text{or} \quad 8 = \frac{(1+100)\left(1 + \frac{R_{Th}}{1}\right)}{1 + 100 + \frac{R_{Th}}{1}}$$

$$\text{or } R_{Th} = \frac{8 \times (1+100) - (1+100) \times 1}{(1+100) - 8} = \frac{707}{93} = 7.6 \text{ k}\Omega$$

Applying KVL to the base-emitter circuit, we have

$$V_{Th} = I_B R_{Th} + V_{BE} + I_E R_E$$

$$= 0.01 \times 10^{-3} \times 7.6 \times 10^3 + 0.3 + 1.01 \times 10^{-3} \times 1 \times 10^3 \qquad \because \text{ for Ge, } V_{BE} = 0.3 \text{ V}$$

$$= 0.076 + 0.3 + 1.01 = 1.386 \text{ V}$$

$$\text{Also, } V_{Th} = \frac{R_2}{R_1 + R_2} V_{CC} = \frac{1}{R_1} \times \frac{R_1 R_2}{R_1 + R_2} V_{CC} = \frac{R_{Th}}{R_1} V_{CC}$$

$$\text{or} \quad R_1 = \frac{R_{Th}}{V_{Th}} \times V_{CC} = \frac{7.6}{1.386} \times 10 = 54.83 \text{ k}\Omega \text{ (56 k}\Omega\text{, standard value)} \quad \textbf{Ans.}$$

$$\text{Since} \quad V_{Th} = \frac{R_2}{R_1 + R_2} V_{CC}$$

$$\text{or} \quad R_2 = \frac{V_{Th} \times R_1}{V_{CC} - V_{Th}} = \frac{1.386 \times 56}{10 - 1.386} = 9 \text{ k}\Omega \text{ (10 k}\Omega\text{, standard value)} \quad \textbf{Ans.}$$

Equivalent circuit is shown in Fig. 4.29 (*b*).

4.5.5. Collector-To-Base Bias (or Base Bias With Collector Feedback)

This circuit is like a fixed bias circuit except that the base resistor R_B is returned to the collector terminal instead of supply V_{CC}. It derives its name from the fact that voltage for R_B is derived from the collector. There exists a negative feedback effect which tends to stabilise I_C against changes either as a result of change in temperature or as a result of replacement of the transistor by another one.

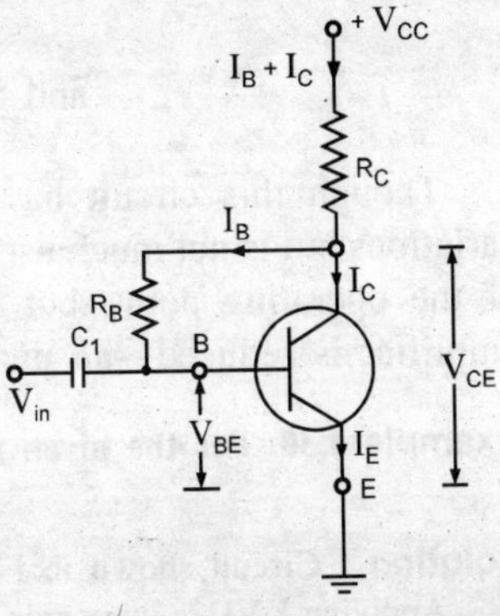

Fig. 4.30. ***Collector-To-Base Bias Circuit***

Let us see how does this circuit operate. If the collector current I_C tends to increase (either due to rise in temperature or due to replacement of transistor), V_{CE} decreases due to larger voltage drop across collector resistor R_C. The result is that base current I_B is reduced. The reduced base current in turn reduces the original increase in collector current I_C. Thus a mechanism exists in the circuit which does not allow collector current I_C to increase rapidly.

Circuit Analysis. The required value of base resistor R_B to give the zero signal current I_C can be determined as follows :

From circuit diagram shown in Fig. 4.30, applying Kirchhoff's voltage law to the input circuit, we have

$$V_{CC} = (I_B + I_C)R_C + I_B R_B + V_{BE}$$

$$\text{or } I_B (R_B + R_C) = V_{CC} - V_{BE} - I_C R_C$$

$$\text{or } R_B = \frac{(V_{CC} - V_{BE} - I_C R_C)\beta}{I_C} - R_C \qquad ...(4.31) \quad \because \beta = \frac{I_C}{I_B}$$

From the output section of the circuit we have

$$V_{CE} = V_{CC} - (I_B + I_C) R_C$$

$$\text{or } R_C = \frac{V_{CC} - V_{CE}}{I_B + I_C} = \frac{V_{CC} - V_{CE}}{I_C\left(\frac{1}{\beta}+1\right)} \qquad ...(4.32)$$

Stability Factor S. From Eq. (4.31)

$$I_B = \frac{I_C}{\beta} = \frac{V_{CC} - V_{BE} - I_C R_C}{R_B + R_C} \qquad ...(4.33)$$

Since V_{BE} is almost independent of collector current, so differentiating above equation w.r.t. I_C we have

$$\frac{dI_B}{dI_C} = \frac{-R_C}{R_B + R_C}$$

Substituting $\frac{dI_B}{dI_C} = \frac{-R_C}{R_B + R_C}$ in Eq. (4.5) we have

$$\text{Stability factor, } S = \frac{1+\beta}{1+\beta\frac{R_C}{R_B + R_C}} \qquad ...(4.34)$$

Value of stability factor so obtained is less than $(1 + \beta)$, obtained from fixed bias circuit. So this method provides improved stability as compared to that of a fixed bias circuit.

Stability Factor $S_{V_{BE}}$. From Eq. (4.33)

$$I_C\left[\frac{R_B + R_C}{\beta} + R_C\right] = V_{CC} - V_{BE}$$

$$\text{and } S_{V_{BE}} = \frac{\partial I_C}{\partial V_{BE}} = \frac{-\beta}{R_B + R_C(1+\beta)}$$

Though this circuit has a tendency to stabilize the operating point against temperature and β variations but is not much used. The base resistor R_B not only provides a dc feedback for the stabilization of the operating point, but it also causes an ac feedback. As a result of this, voltage gain of the amplifier is reduced—an undesirable feature. This is the reason that this circuit is not much used.

Example 4.30. For the given network shown in the Fig. 4.31 determine I_C, V_{CE}, V_B and V_C.

[U.P. Technical Univ. Electronics Engineering Second Semester 2005-06]

Solution : Circuit shown in Fig. 4.31 is collector-to-base bias circuit.

Applying KVL in input side, we have

$$V_{CC} = (I_C + I_B) R_C + I_B R_B + V_{BE} = (\beta + 1) I_B R_C + I_B R_B + V_{BE}$$

$$\text{or } I_B = \frac{V_{CC} - V_{BE}}{(\beta+1)R_C + R_B} = \frac{20-0.7}{(120+1)\times 4.7\times 10^3 + 680\times 10^3} = 15.5\ \mu A$$

or $I_C = \beta\, I_B = 120 \times 15.5 \times 10^{-6} =$ **1.86 mA Ans.**

Base voltage, $V_B = V_{BE} = 0.7$ V (∵ emitter is grounded) **Ans.**

Applying KVL to output side, we have

$$V_{CE} = V_{CC} - (I_B + I_C)\, R_C$$

$$= 20 - (15.5 \times 10^{-6} + 1.86 \times 10^{-3}) \times 4.7 \times 10^3 = \textbf{11.19 V Ans.}$$

Collector voltage

$$V_C = V_{CE} \quad (\because \text{ emitter is directly connected to ground})$$

$$= \textbf{11.19 volts Ans.}$$

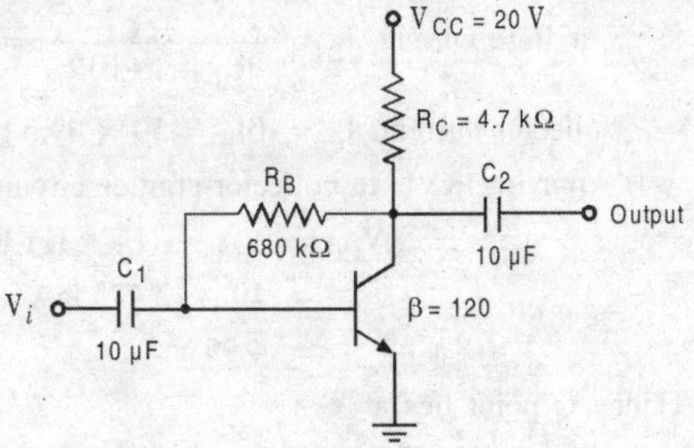

Fig. 4.31

Example 4.31. A transistor with β = 100 is used in CE configuration. The collector circuit resistance is $R_C = 1$ kΩ and $V_{CC} = 20$ V. Assuming $V_{BE} = 0$, find the value of collector to base resistance, such that quiescent collector-emitter voltage is 4 V. Also determine the stability factor in this case.

Solution : Applying KVL to collector-emitter loop we have

$$V_{CC} = I'_C\, R_C + V_{CE}$$

$$\text{or } I'_C = \frac{V_{CC} - V_{CE}}{R_C} = \frac{20-4}{1\text{k}\Omega} = 16 \text{ mA}$$

$$\text{and } I'_C = I_C + \frac{I_C}{\beta}$$

$$\text{or } I_C = \frac{I'_C}{1+\frac{1}{\beta}} = \frac{16\text{mA}}{1+\frac{1}{100}} = \frac{16\text{mA}}{1.01} = 15.84 \text{ mA}$$

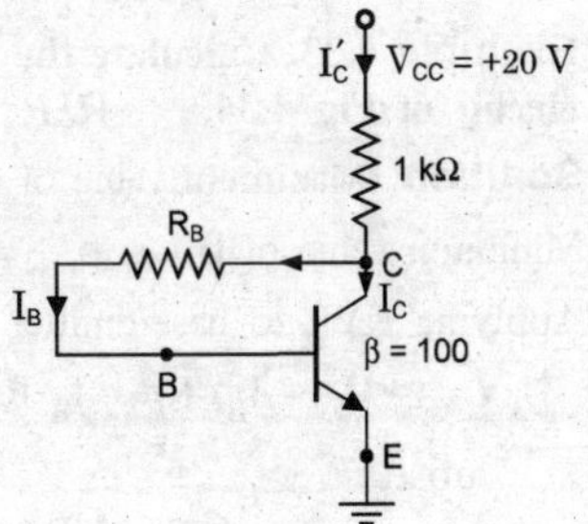

Fig. 4.32

$$I_B = I'_C - I_C = 16 - 15.84 = 0.16 \text{ mA}$$

Applying Kirchhoff's voltage to base-emitter loop we have

$$I_B R_B = V_{CE} - V_{BE}$$

$$= 4 - 0 = 4 \quad \because V_{BE} = 0 \text{ and } V_{CE} = 4 \text{ V}$$

$$\text{or } R_B = \frac{4}{I_B} = \frac{4\text{ V}}{0.16\text{mA}} = 25 \text{ k}\Omega \quad \textbf{Ans.}$$

$$\text{Stability factor, } S = \frac{1+\beta}{1+\beta\frac{R_C}{R_B+R_C}} = \frac{1+100}{1+100\times\frac{1}{25+1}} = 20.84 \textbf{ Ans.}$$

Example 4.32. In a common emitter circuit an N-P-N transistor having a value of β = 50 is used with $V_{CC} = 10$ V and $R_C = 2$ kΩ. If a 100 kΩ resistor is connected between collector and base and $V_{BE} = 0$, determine. *(i)* the position of quiescent point and *(ii)* stability factor S.

Solution : Applying KVL to base-emitter circuit we have

$$V_{CC} - I'_C R_C - I_B R_B - V_{BE} = 0$$

$$\text{or } V_{CC} - (I_B + I_C)R_C - I_B R_B = 0 \quad \because V_{BE} = 0$$

$$\text{or } V_{CC} - (I_B + \beta I_B)R_C - I_B R_B = 0 \quad \because I_C = \beta I_B$$

or Base current, $I_B = \frac{V_{CC}}{R_B + (1+\beta)R_C} = \frac{10\,V}{100\,k\Omega + (1+50)\times 2\,k\Omega} = 49.5\,\mu A$

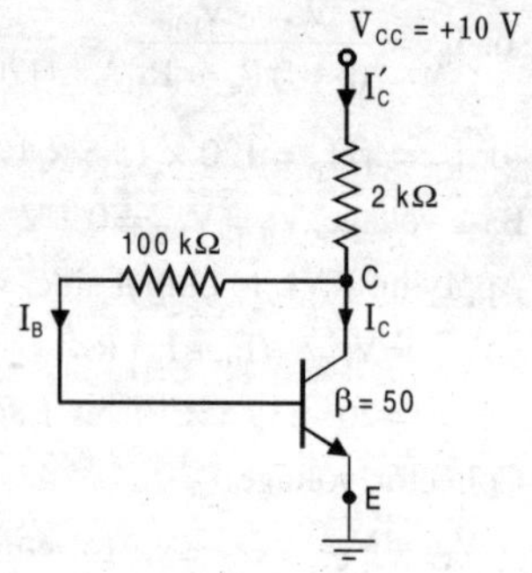

Fig. 4.33

Collector current, $I_C = \beta I_B = 50 \times 49.5\,\mu A = 2.475\,mA$

Applying KVL to collector-emitter circuit we have

$$V_{CE} = V_{CC} - (I_C + I_B)\,R_C$$
$$= 10 - (2.475\,mA + 49.5\,\mu A) \times 2\,k\Omega$$
$$= 4.95\,V$$

Hence Q point lies at

$$I_{CQ} = 2.475\,mA \text{ and } V_{CEQ} = 4.95\,V \quad \textbf{Ans.}$$

$$\text{Stability factor, } S = \frac{1+\beta}{1+\beta\frac{R_C}{R_B+R_C}} = \frac{50+1}{1+50\times\frac{2}{2+100}}$$
$$= 25.75 \quad \textbf{Ans.}$$

Example 4.33. Calculate the maximum and minimum collector current in the circuit of Fig. 4.34. [*U.P. Technical Univ. Electronics Engineering Dec. 2004*]

Solution : Maximum value of β *i.e.,* $\beta_{max} = 180$

Minimum value of β *i.e.,* $\beta_{min} = 60$

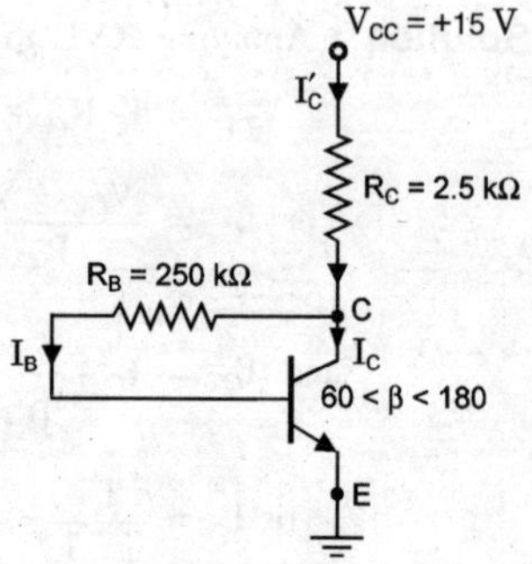

Fig. 4.34

Applying KVL to base-emitter loop we have

$$V_{CC} - (I_C + I_B)\,R_C - I_B\,R_B - V_{BE} = 0$$

$$\text{or} \quad \left(I_C + \frac{I_C}{\beta}\right)R_C + \frac{I_C}{\beta}R_B = V_{CC} - V_{BE} = 15 - 0.7 = 14.3$$

$$\text{or } I_C = \frac{14.3}{R_C + \frac{R_C}{\beta} + \frac{R_B}{\beta}} \qquad ...(i)$$

Collector current will be maximum for maximum value of β. Substituting β = 180 in above equation we have

$$I_{C\,max} = \frac{14.3}{2.5\times10^3 + \frac{2.5\times10^3}{180} + \frac{250\times10^3}{180}} = 3.664\,mA \quad \textbf{Ans.}$$

Collector current will be minimum for minimum value of β. Substituting β = 60 in Eq. (*i*), we have

$$I_{C\,min} = \frac{14.3}{2.5\times10^3 + \frac{2.5\times10^3}{60} + \frac{250\times10^3}{60}} = 2.13\,mA \quad \textbf{Ans.}$$

Example 4.34. Determine the resistor R_B for a fixed bias and collector-to-base bias and compare the two stability factors.

$V_{CC} = 12$ V, $R_L = 33$ kΩ, $I_B = 0.3$ mA, β = 100, $V_{CE} = 6$ V.

[*Pb. Technical Univ. Electronic Devices and Circuits May 2006*]

Solution : For a fixed biasing circuit (Referring to Fig. 4.2), we have

$$I_B = \frac{V_{CC} - V_{BE}}{R_B}$$

$$\text{or } R_B = \frac{V_{CC} - V_{BE}}{I_B} = \frac{12 - 0}{0.3\times10^{-3}} = 40\,k\Omega \quad \text{neglecting } V_{BE}$$

Stability factor, S = β + 1 = 1 + 100 = 101

For collector-to-base bias (referring to Fig. 4.30)

Applying KVL to the output loop, we have

$$V_{CC} = (I_B + I_C)R_C + V_{CE}$$

or Collector resistance, $$R_C = \frac{V_{CC} - V_{CE}}{I_B + I_C} = \frac{V_{CC} - V_{CE}}{I_B + \beta I_B} = \frac{V_{CC} - V_{CE}}{(\beta + 1)I_B}$$

$$= \frac{12 - 6}{(100 + 1) \times 0.3 \times 10^{-3}} = 198\ \Omega$$

Applying KVL to the input circuit, we have

$$V_{CC} = (I_B + I_C)\ R_C + I_B\ R_B + V_{BE}$$

$$\text{or } R_B = \frac{(V_{CC} - V_{BE} - I_C R_C)\beta}{I_C} - R_C \qquad \because\ \beta = \frac{I_C}{\beta}$$

$$= \frac{(12 - 0.3 \times 10^{-3} \times 100 \times 198) \times 100}{100 \times 0.3 \times 10^{-3}} - 198 \quad \text{Neglecting } V_{BE}$$

$= 20{,}002\ \Omega$ or $20\ k\Omega$ **Ans.**

Stability factor, $$S = \frac{1+\beta}{1 + \dfrac{\beta R_C}{R_B + R_C}} = \frac{1 + 100}{1 + \dfrac{100 \times 198}{20{,}000 + 198}} = \frac{101}{1.98} = 51 \textbf{ Ans.}$$

Stability factor with collector-to-base bias is lesser than that with fixed bias, so collector-to-base bias provides improved stability as compared to that of a fixed bias circuit.

4.5.6. Base Bias Circuit With Collector and Emitter Feedbacks

In this circuit, as obvious from its name, both collector and emitter feedbacks are used in an attempt to reduce circuit sensitivity to changes in β. Increase in β causes increase in emitter voltage but decreases collector voltage. As a result voltage across base resistor R_B is reduced causing I_B to fall thereby partially compensating the increase in β.

In this circuit a current flows in resistance R_C from the supply V_{CC}. At point A this current divides itself in two parts, one flowing in the collector of transistor, I_C and the second part flows in the base resistor R_B, I_B. Thus current flowing in resistor R_C will be equal to $I_B + I_C$.

No current flows to the output terminals since coupling capacitor C_C offers infinite impedance to it (open circuit for dc).

Let the potential of point A be V_C.

Applying Kirchhoff's second law from point A to ground via R_B, we have

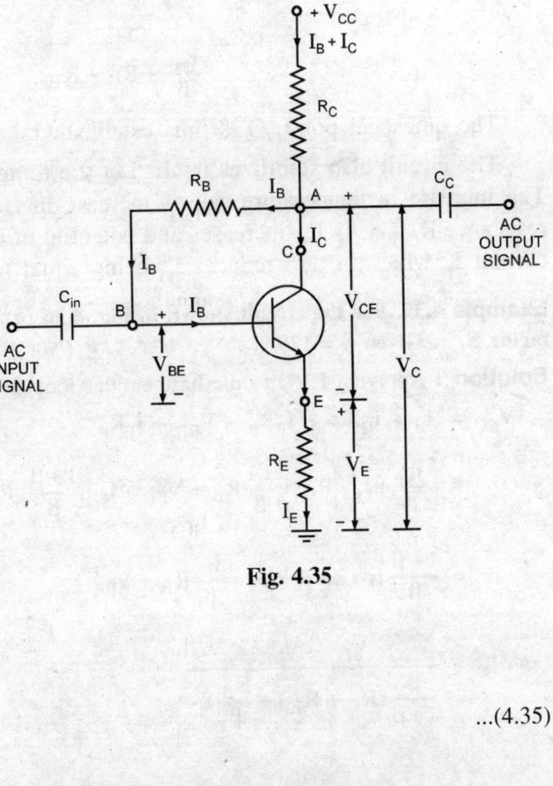

Fig. 4.35

$$V_C - I_B\ R_B - V_{BE} - I_E\ R_E = 0$$

$$\text{or } I_B = \frac{V_C - V_{BE} - I_E\ R_E}{R_B} \qquad \text{...(4.35)}$$

Neglecting V_{BE} in comparison to collector potential V_C which is usually very small.

$$I_B \approx \frac{V_C - I_E R_E}{R_B} \qquad ...(4.36)$$

Now applying Kirchhoff's second law for circuit starting from V_{CC} through collector to emitter terminal, we have

$$V_{CE} = V_{CC} - R_C\,(I_B + I_C) - I_E\,R_E$$
$$\simeq V_{CC} - I_C\,R_C - I_C\,R_E \simeq V_{CC} - I_C\,(R_C + R_E) \qquad ...(4.37)$$

neglecting I_B being too small and I_E being almost equal to I_C.

Applying Kirchhoff's second law to the circuit starting from collector point A to ground via R_B

$$V_{CE} = I_B\,R_B + V_{BE} \qquad ...(4.38)$$

Comparing Eqs. (4.37) and (4.38) we have

$$I_B\,R_B + V_{BE} = V_{CC} - I_C\,R_C - I_C\,R_E$$

or $$\frac{I_C}{\beta}R_B + V_{BE} = V_{CC} - I_C\,(R_C + R_E)$$

or $$I_C\left[\frac{R_B}{\beta} + R_C + R_E\right] = V_{CC} - V_{BE}$$

or $$I_C = \frac{V_{CC} - V_{BE}}{\dfrac{R_B}{\beta} + R_C + R_E} \qquad ...(4.39)$$

Neglecting V_{BE}, being very small in comparison to supply voltage V_{CC},

$$I_C = \frac{V_{CC}}{\dfrac{R_B}{\beta} + R_C + R_E} \qquad ...(4.40)$$

The quiescent point, Q is thus established.

The circuit also stabilizes itself. Let the temperature of transistor junction rise when it is loaded. The increase in temperature causes increase in I_C and β. When I_C increases voltage drop in collector resistance R_C *i.e.*, $I_C\,R_C$ increases and potential of collector drops. This causes reduction in value of base current I_B which in turn reduces I_C being equal to βI_B.

Example 4.35. For the circuit shown in Fig. 4.36 (*a*), determine (*i*) V_C (*ii*) stability factor $S_{I_{CO}}$. Given $\beta = 120$. [*G.G.S.I.P. Univ. Analog Electronics December 2010*]

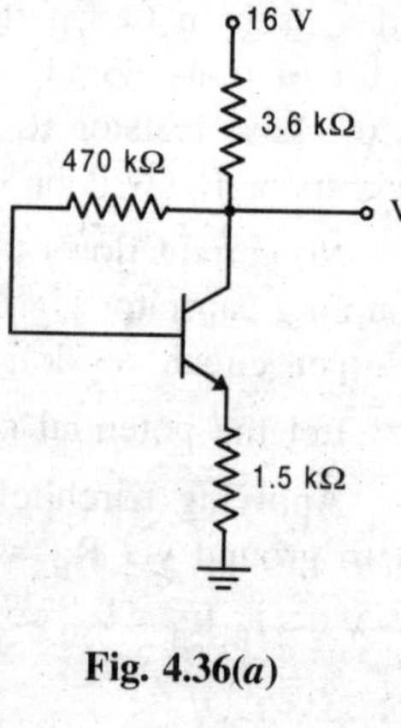

Fig. 4.36(*a*)

Solution : Applying KVL to the base-emitter loop in Fig. 4.36 (*b*), we have

$$V_{CC} = (I_B + I_C)R_C + I_B R_B + V_{BE} + I_E R_E$$

$$= \left(\frac{I_C}{\beta} + I_C\right)R_C + \frac{I_C}{\beta}R_B + V_{BE} + I_C\left(\frac{1+\beta}{\beta}\right)R_E$$

$$= \frac{1+\beta}{\beta}I_C\,(R_C + R_E) + \frac{I_C}{\beta}R_B + V_{BE}$$

or $$I_C = \frac{V_{CC} - V_{BE}}{\dfrac{\beta}{1+\beta}(R_C + R_E) + \dfrac{R_B}{\beta}I_C}$$

$$= \frac{16 - 0.7}{\frac{120}{121}(3.6 + 1.5) \times 10^3 + \frac{470 \times 10^3}{120}}$$

$$= \frac{15.3}{(5.058 + 3.917) \times 10^3} = 1.7 \text{ mA}$$

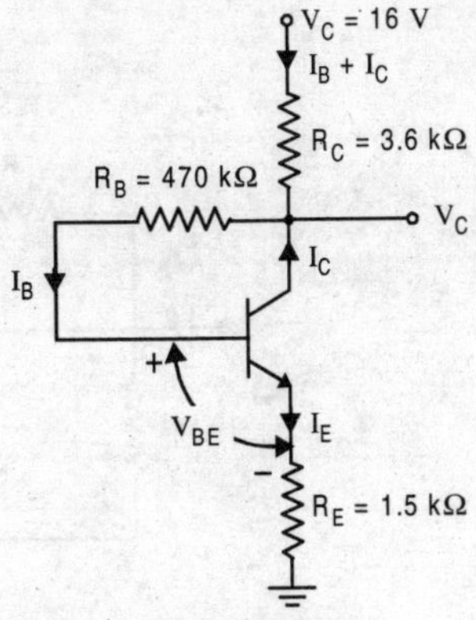

Fig. 4.36(*b*)

Collector voltage,

$V_C = V_{CC} - I_C R_C = 16 - 1.7 \times 10^{-3} \times 3.6 \times 10^3 = 9.88$ V **Ans.**

Stability factor,

$$S = \frac{1+\beta}{1+\beta\frac{R_C + R_E}{R_B + R_C + R_E}} = \frac{1+120}{1+120 \times \frac{3.6+1.5}{470+3.6+1.5}}$$

$$= \frac{121}{1+120 \times \frac{5.1}{475.1}} = 52.88 \text{ Ans.}$$

Example 4.36. For the circuit shown in Fig. 4.37, I_E = 1 mA, β = 99 and V_{BE} = 0.7 V. Determine (*a*) the current through R_1 and R_C (*b*) the output voltage V_0 and (*c*) the value of R_F. [*GATE 2002*]

Solution : Emitter current, I_E = 1 mA $\qquad \because I_E = (\beta + 1)I_B$

$\therefore$ Base current, $I_B = \frac{I_E}{\beta+1} = \frac{1 \text{ mA}}{99+1} = 0.01$ mA

Collector current, $I_C = \beta I_B = 99 \times 0.01 = 0.99$ mA

Applying KVL to base-emitter loop we have

$V_{BE} + I_E R_E = I_1 R_1$ where I_1 is current through R_1

or Current through R_1, $I_1 = \frac{V_{BE} + I_E R_E}{R_1}$

$$= \frac{0.7 + 1 \times 10^3 \times 1 \times 10^{-3}}{17} = 0.1 \text{ mA} \quad \textbf{Ans.}$$

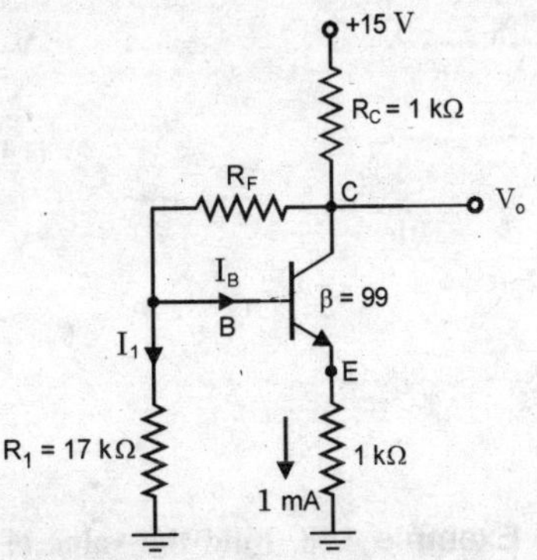

Fig. 4.37

(*i*) Current through R_F, I_{R_F} = Current through R_1 + I_B = 0.1 + 0.01 = 0.11 mA **Ans.**

Current through R_C, $I'_C = I_C$ + Current through R_F = 0.99 + 0.11 = 1.1 mA **Ans.**

(*ii*) Output voltage, $V_0 = V_{CC} - I'_C R_C = 15 - 1.1 \times 10^{-3} \times 1 \times 10^3 = 13.9$ V **Ans.**

$$(iii)\ R_F = \frac{V_0 - V_{BE} - I_E R_E}{I_{R_F}} = \frac{13.9 - 0.7 - 1 \times 10^{-3} \times 1 \times 10^3}{0.11 \times 10^{-3}} = 110.9 \text{ k}\Omega \quad \textbf{Ans.}$$

Example 4.37. Find the value of R_1 in the circuit shown in Fig. 4.38 (*a*). Assume β = 49, I_E = 2 mA, V_{BE} = 0.7 V. [*G.G.S.I.P. Univ. Analog Electronics December 2010*]

Solution : The given circuit is redrawn as shown in Fig. 4.38 (*b*).

Emitter current, I_E = 2 mA

Base current, $I_B = \frac{I_E}{\beta+1} = \frac{2 \text{ mA}}{49+1} = 0.04$ mA

Collector current, $I_C = I_E - I_B = 2 - 0.04 = 1.96$ A

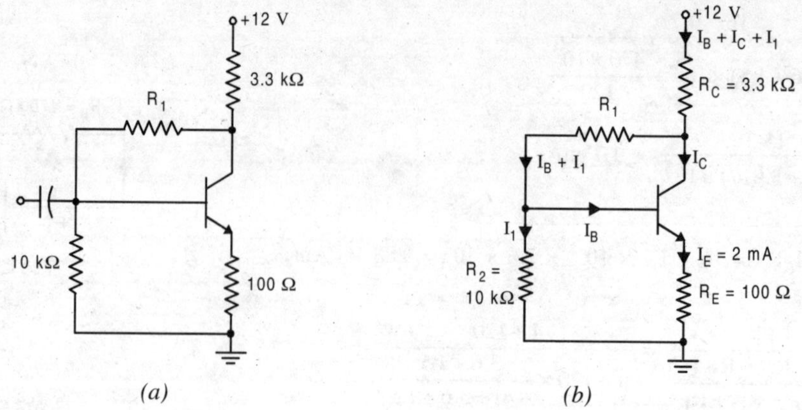

Fig. 4.38

Current through R_2, $I_1 = \dfrac{V_{BE} + I_E R_E}{R_2} = \dfrac{0.7 + 2 \times 10^{-3} \times 100}{10 \times 10^3} = 0.09$ mA

Applying KVL to input side, we have

$$V_{CC} = (I_B + I_C + I_1)R_C + (I_B + I_1)R_1 + I_1 R_2$$

$$\text{or } R_1 = \frac{V_{CC} - (I_B + I_C + I_1)R_C - I_1 R_2}{I_B + I_1}$$

$$= \frac{12 - (0.04 + 1.96 + 0.09) \times 10^{-3} \times 3.3 \times 10^3 - 0.09 \times 10^{-3} \times 10 \times 10^3}{(0.04 + 0.09) \times 10^{-3}}$$

$$= \frac{12 - 6.897 - 0.9}{0.13 \times 10^{-3}} = 32.33 \text{ k}\Omega \quad \textbf{Ans.}$$

Example 4.38. Find the value of stability factor S with respect to I_{CO} for the circuit shown in Fig. 4.39 (*a*).

[*G.G.S.I.P. Univ. Analog Electronics December 2010*]

Solution : Drawing dc equivalent circuit, we have circuit as shown in Fig. 4.39 (*b*).

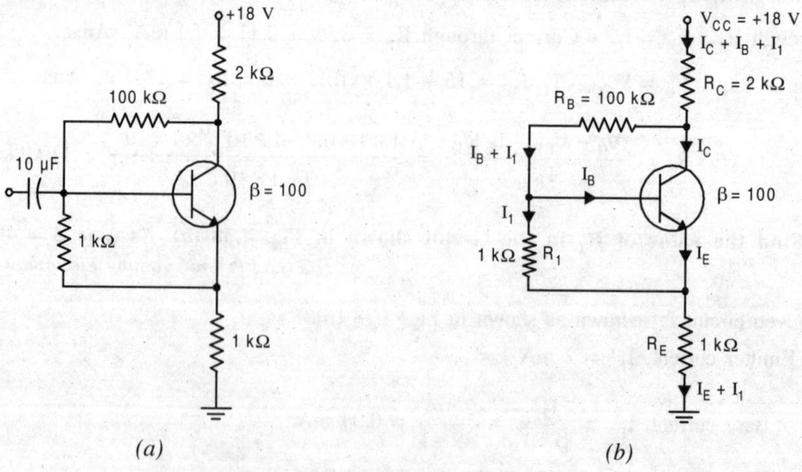

Fig. 4.39

Applying KVL to the input side of the circuit shown in Fig. 4.39 (*b*), we have

$$V_{CC} = (I_C + I_B + I_1)R_C + (I_B + I_1) R_B + V_{BE} + (I_E + I_1)R_E$$

$$= \left(I_C + I_B + \frac{V_{BE}}{R_1}\right)R_C + \left(I_B + \frac{V_{BE}}{R_1}\right)R_B + V_{BE} + \left(I_C + I_B + \frac{V_{BE}}{R_1}\right)R_E$$

$$\text{or } V_{CC} = V_{BE}\left(1 + \frac{R_C + R_B + R_E}{R_1}\right) + I_C(R_C + R_E) + I_B(R_C + R_B + R_E)$$

Differentiating above equation w.r.t. I_C we have

$$0 = 0 + R_C + R_E + \frac{dI_B}{dI_C}(R_C + R_B + R_E)$$

$$\text{or } \frac{dI_B}{dI_C} = -\frac{R_C + R_E}{R_C + R_B + R_E} = -\frac{2+1}{2+100+1} = \frac{-3}{103}$$

Substituting the value of $\frac{dI_B}{dI_C}$ and β in Eq. (4.5), we have

$$\text{Stability factor, } S = \frac{1+\beta}{1-\beta\frac{dI_B}{dI_C}} = \frac{1+100}{1-100\times\left(\frac{-3}{103}\right)} = \frac{101}{\frac{403}{103}} = 25.8 \textbf{ Ans.}$$

4.6. GUIDELINES FOR DESIGN OF TRANSISTOR BIASING CIRCUITS

In practice, for low power transistors, the following guidelines may be followed for designing the transistor biasing circuits.

1. In common practice R_E is taken to be between 0.5 kΩ to 1.0 kΩ. From the point of view of stabilization larger the value of R_E, better it is but larger value of R_E develops larger voltage drop across it and, therefore, output is reduced. Hence a compromise has to be made in the selection of the value of the emitter resistor R_E.

2. The zero signal collector current I_C is chosen so that the operating point lies at the middle of the dc load line. However, in initial stages of most transistor amplifiers zero signal I_C is taken to be 1 mA. This value of I_C is considered to be most suitable because of the following advantages.

(*i*) The output impedance of a transistor is very high for I_C = 1 mA. So higher voltage gain is obtained.

(*ii*) There is little danger of overheating of collector junction because of small value of I_C.

It is not advisable to operate the transistor below zero signal I_C = 1 mA because of nonlinear characteristics of transistor.

3. Emitter voltage V_E is usually maintained one-tenth of supply voltage V_{CC} *i.e.*, $V_E = 0.1\ V_{CC}$. The voltage across collector-emitter terminals V_{CE} will be 0.5 V_{CC} so as to locate operating point Q at the middle of the dc load line. The remaining voltage 0.4 V_{CC} appears across R_C.

$$\text{Thus } R_C = 4\ R_E$$

4. The values of resistors R_1 and R_2 are selected so as to give current flowing through them 10 times the base current *i.e.*, $I_1 \geq 10\ I_B$. This condition is for providing better stabilisation.

5. The zero signal collector current I_C should be slightly more (say 20%) than the maximum collector current swing due to signal otherwise a part of negative half cycle of a signal may be cutoff.

For example if collector current swing is expected to be of 2.5 mA due to signal, then zero signal I_C may be taken equal to 3.0 mA. Value of zero signal I_C higher than this value (say 10.0 mA) may unnecessarily overheat the transistor and cause wastage of supply power. Moreover, a higher zero signal I_C will reduce the value of collector resistance R_C for the given V_{CC} and so reduction in voltage gain.

Highlights

1. No matter what type of configuration of a transistor is used in, the basic relationships between the currents are always the same, and the base-emitter voltage V_{BE} is the threshold value if the transistor is in the **"on"** state.
2. The operating point defines where the transistor will operate on its characteristics curves under dc conditions. For linear amplification (amplification with minimum distortion), the operating point should not be too close to the maximum power, voltage, or current rating besides avoiding the regions of cutoff and saturation.
3. The variation in a transistor's dc collector current I_C with temperature in a CE amplifier depends upon β, I_B and I_{CO}, all of which increase with temperature.

   β is very sensitive to temperature, and V_{BE} decreases about 2.5 mV for each 1°C rise in temperature. The reverse saturation current typically doubles for every 10°C rise in temperature.
4. The resulting increase in collector current and shift in operating point may cause clipping and distortion of the input signal, or it may cause thermal runaway in worst condition.
5. The stability factor S is defined as the rate of change of collector current with respect to leakage current I_{CO}, with V_{BE} and β both constant.

   The noteworthy point is that the networks that are the most stable and least sensitive to temperature variations have the smallest stability factors. The larger the value of S, the more likely the circuit to exhibit thermal instability.
6. The fixed bias configuration is the simplest of transistor biasing arrangements, but it is quite unstable due to its sensitivity to β at the operating point.

   The stability factor S for a CE circuit using fixed bias is (β + 1) for either a silicon or germanium transistor.
7. The saturation (maximum) collector current can usually be determined, for any configuration, by superimposing an imaginary short circuit between the collector and emitter terminals of the transistor. The resulting current through the short is then the saturation current.
8. The equation for the load line of a transistor network can be determined by applying Kirchhoff's voltage law to the output or collector network. The Q-point is then determined by finding the intersection between the base current and the load line drawn on the device characteristics.
9. The stabilization methods involve the reduction of forward bias with temperature rise in order to reduce base current I_B and hence collector current I_C by some means of negative feedback, in different biasing arrangements.
10. The voltage divider bias circuit is the most commonly used biasing arrangement. Its popularity is primarily due to its low sensitivity to variations in β from one transistor to another of the same lot (with the same transistor label). This circuit provides bias stabilization with

$$S = 1+\beta\frac{1+\frac{R_{Th}}{R_E}}{1+\beta+\frac{R_{Th}}{R_E}}$$

11. Collector-to-base bias makes the use of a resistor R_B connected from collector to base and provides bias stabilization with

$$S = \frac{1+\beta}{1+\beta\frac{R_C}{R_B+R_C}}$$

This arrangement may not provide much overall increase in stability in a silicon transistor circuit since the reduced value of R_B will allow a large increase in I_B with temperature than fixed bias.

This arrangement also reduces the voltage gain of the amplifier, an undesirable feature.

Exercises

1. What is transistor biasing ? What are the basic conditions which are to be necessarily fulfilled for achieving faithful amplification of input signal in transistor amplifiers.
2. Taking the example of CE amplifier, explain the criterion for selection of a suitable operating point and factors affecting its stability. Hence define stability factor. **[U.P.S.C. I.E.S. Elec. Engineering-II, 1994]**
3. Explain how operating point is selected for amplification in CE mode using graphical method only. **[U.P. Technical Univ. Electronics Engg. First Semester 2005-06]**
4. What do you understand by transistor biasing ? Give its importance. **[R.G. Technical Univ. Basic Electrical and Electronics Engineering Dec. 2010]**
5. Explain the need for stabilization of any BJT. List the types of stability factors. **[G.G.S.I.P. Univ. Analog Electronics December 2010]**
6. What is need for stabilization ? List the various factors responsible for shift in Q-point in a transistor. Explain the effect of each parameter on stability of Q-point. **[G.G.S.I.P. Univ. Analog Electronics December 2009]**
7. What is the significance of 'stability factor' ? Define $S_{(ICO)}$, $S_{(\beta)}$ and $S_{(VBE)}$. **[G.G.S.I.P. Univ. Analog Electronics December 2009]**
8. What is bias stabilization ? Determine the general stability factors for all configurations. **[Pb. Technical Univ. Electronic Devices & Circuits May 2005]**
9. Define stability factor and obtain an expression for a fixed bias circuit. **[M.D. Univ. Rohtak Analog Electronics May 2007]**
10. Explain the working of emitter bias circuit. **[P.T.U. Electronic Devices & Circuits December 2008]**
11. Draw the circuit diagram of a BJT emitter bias and derive the expression of Quotient point. Write its advantage over BJT fixed bias circuit. Also define bias stabilization and stability factors. **[G.B. Technical Univ. Electronics Engineering First Semester 2011-12]**
12. Sketch a voltage divider bias circuit using an npn transistor. Show all voltage polarities and current directions. Explain the operation of the circuit and write the approximate equations for V_B, I_E, I_C and V_{CE}. **[U.P. Technical Univ. Electronics Engg. First Semester 2008-09]**
13. Draw a self bias circuit. Explain why such circuit is an improvement over fixed circuit. **[Rajasthan Technical Univ. 2007]**
14. Define stability factor. Explain self bias circuit with diagrams and derive expression for stability factor. **[Pb. Technical Univ. Electronic Devices & Circuits May 2008]**
15. Why is transistor biasing required ? Describe different schemes of transistor biasing in CE N-P-N transistor circuit. State their advantages. **[Mahamaya Technical Univ. Electronics Engineering Second Semester 2011-12]**
16. Derive expressions of stability factors with respect to I_{CO}, β and V_{BE} for emitter bias circuit. Compare the features of fixed bias, self bias and emitter bias circuits. **[G.G.S.I.P. Univ. Delhi Analog Electronics 2011-12]**

Short Answer Type Questions With Answers

Q. 1. What is operating point ? Explain its physical significance.

[Pb. Technical Univ. Electronic Devices & Circuits Dec. 2006]

Ans. The zero signal values of I_C and V_{CE} are known as the operating point. It is called operating point because the variations of I_C and V_{CE} take place about this point when signal is applied. It is also known as quiescent point or Q point.

The operating point defines where the transistor will operate on its characteristics curves under dc conditions. For linear amplification (amplification with minimum distortion), the operating point should not be too close to the maximum power, voltage, or current rating besides avoiding the regions of cutoff and saturation.

Q. 2. What is an ac load line ? **[P.T.U. Analog Electronics December 2004]**

Ans. This is the line on the output characteristics of a transistor circuit which gives the values of V_{CE} and I_C when an ac signal is applied. The slope of the ac load line is given as $\frac{-1}{R_C \parallel R_L}$ where $R_C \parallel R_L$ represents ac load resistance.

Q. 3. What is faithful amplification ?

Ans. The process of raising the strength of a weak signal without any change in its general shape is referred to as faithful amplification.

Q. 4. What is operating point ? Why is it necessary to stabilize operating point of a transistor amplifier ?

[Pb. Technical Univ. Electronic Devices & Circuits Dec. 2007]

Ans. The zero signal values of I_C and V_{CE} are known as the operating point. It defines where the transistor will operate on its characteristics curves under dc conditions.

The stabilization of operating point is necessary because of (*i*) temperature dependence of collector current I_C, (*ii*) individual variations and (*iii*) thermal runaway.

Q. 5. How does I_{CBO} vary with the temperature ? **[Pb. Technical Univ. Electronic Devices & Circuits Dec. 2003]**

Ans. I_{CBO} is exactly like the reverse saturation current of a reverse biased diode, which, theoretically, increases by 8 per cent per °C for silicon and 11 per cent per °C for germanium. But from experimental data it is found that the reverse saturation current increases 7 per cent for both silicon and germanium. This is because a surface leakage current component of reverse saturation current is independent of temperature. Since $(1.07)^{10} = 2.0$, the reverse saturation current approximately doubles for every 10°C rise in temperature.

Q. 6. What do you understand by thermal runaway ? **[Pb. Technical Univ. December 2008]**

Ans. The collector current I_C, being equal to $\beta I_B + (1 + \beta) I_{CO}$, increases with the increase in temperature. This leads to increased power dissipation with further increase in temperature. Being a cumulative process it can lead to *thermal runaway* resulting in burn-out of the transistor. The self destruction of an unstabilized transistor is called the *thermal runaway*.

Q. 7. What is bias stabilization ? **[Pb. Univ. Dec. 1993]**

Ans. The maintenance of the operating or quiescent point stable (independent of temperature variations or variations in transistor parameters) is known as bias stabilization.

Q. 8. State the condition for thermal stability of operating point of a BJT amplifier.

Ans. The operating point should not shift with temperature variations.

Q. 9. What is meant by midpoint biasing ?

Ans. When an amplifier circuit is so designed that the operating point Q lies at the centre of the dc load line, the amplifier is said to be midpoint biased.

Q. 10. Which two methods make the collector current constant ? Explain them.

[Pb. Technical Univ. December 2005]

Ans. Operating point stability may be achieved by adopting either stabilization techniques or compensation techniques.

Stabilization technique makes use of a resistive biasing circuit that permits such a variation of base current (biasing current) I_B as to maintain collector current I_C almost constant inspite of variations in reverse saturation current I_{CO}, base-emitter voltage V_{BE} and β.

Compensation techniques make use of temperature sensitive devices such as diodes, transistors, thermistors etc. Such devices produce compensating voltages and currents so as to make operating point stable.

Q. 11. What is the requirement of good biasing circuit ? **[Pb. Technical Univ. May 2007]**

Ans. *Requirements of Biasing Circuit*. A biasing network associated with a transistor should fulfill the following requirements :

1. Establish the operating point in the middle of the active region of the characteristics, so that on applying the input signal the instantaneous operating point does not move either to the cutoff region or to the saturation region, even at the extreme values of the input signal.
2. Stabilize the collector current I_C against temperature variations.
3. Make the operating point independent of transistor parameters so that replacement of transistor by another transistor of the same type in the circuit does not shift the operating point.

Q. 12. Define stability factor. **[P.T.U. December 2004]**

Ans. Stability factor is defined as the rate of change of collector current I_C w.r.t. reverse saturation current I_{CO} with V_{BE} and β constant.

Q. 13. What is the significance of stability factor in transistor operation ?

Ans. The stability factor indicates the change in collector current I_C due to change in collector leakage current I_{CO}. Thus a stability factor of high value say 50 means that I_C changes 50 times as much as any change in I_{CO}. In order to achieve greater thermal stability, it is desirable to have as low stability factor as possible. Thus importance of stability factor is with regard to stability of the transistor.

Q. 14. Why fixed bias circuit, inspite of its simplicity, is not much used in practice ?

[Pb. Technical Univ. May 2009]

Ans. Rise in temperature causes increase in leakage current I_{CO}, increase in current amplification factor β and decrease in base-emitter voltage V_{BE}, so operating point is not stabilized. This is the reason that fixed bias circuit is not commonly used.

Q. 15. Explain why collector-to-base bias circuit is seldom used ?

[U.P. Technical Univ. Electronics Engineering December 2004]

Ans. The base resistor R_B not only provides a dc feedback for the stabilization of the operating point, but it also causes an ac feedback in a collector-to-base bias circuit. This is the reason that this circuit is not much used.

Problems

1. In a fixed biasing circuit Fig. 1, R_B = 1 MΩ and R_C = 5 kΩ, V_{CC} = 6 V, β = 100. Determine the Q-point.

[Ans. 3 V, 0.6 mA**]**

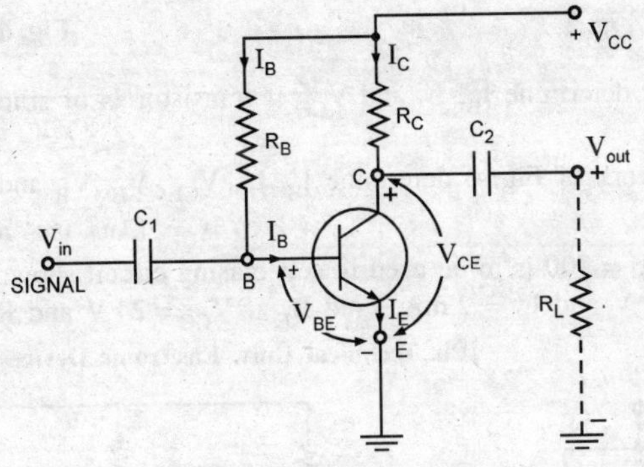

Fig. 1

2. Draw the load line for the following Fig. 2. What is I_C at saturation point ? Find V_{CE} at cutoff point.

[U.P. Technical Univ. Basic Electronics First/Second Semes. 2001-02]

[Ans. 6.06 mA; 20 V**]**

3. A silicon transistor with $V_{BE\ Sat}$ = 0.8 V, $\beta = h_{FE}$ = 100, $V_{CE\ Sat}$ = 0.2 V is used in circuit shown in Fig. 3. Find the maximum value of R_C for which the transistor remains in saturation.

[U.P. Technical Univ. Basic Electronics Second Semester 2002-03]

[Ans. 4.667 kΩ**]**

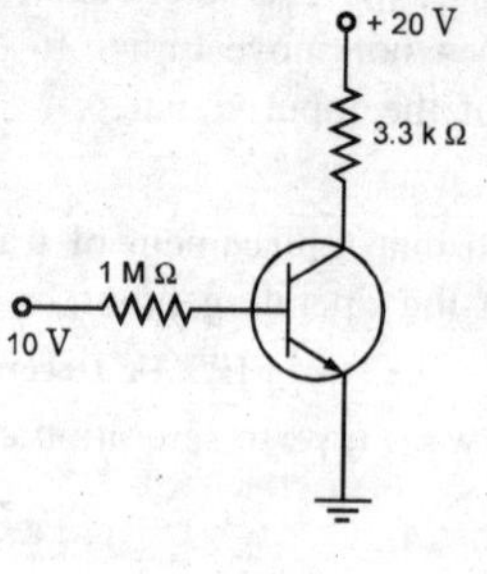

Fig. 2

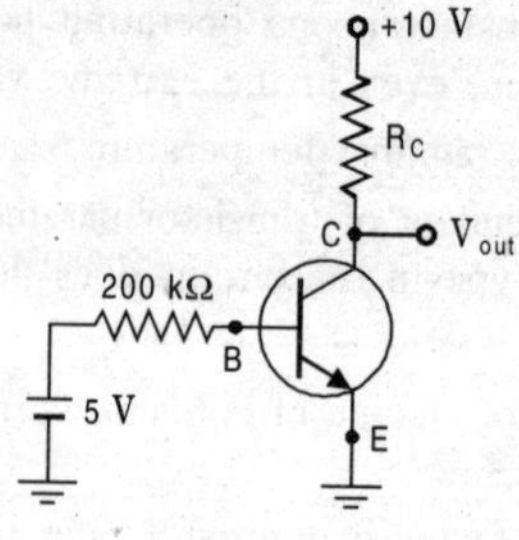

Fig. 3

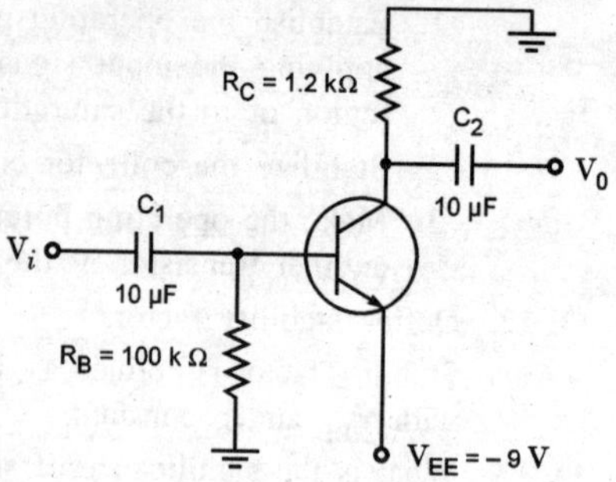

Fig. 4

4. Determine V_C and V_B for the network given in Fig. 4 with $\beta = 45$ and $V_{BE} = 0.7$ V.

[U.P. Technical Univ. Feb. 2002]

[Ans. $V_C = -4.482$ V; $V_B = -8.3$ V]

5. Figure 5 shows the circuit of fixed bias using a silicon transistor with $\beta = 100$. Determine (*i*) base current, (*ii*) collector current (*iii*) V_C, V_B and V_{CB} (*iv*) operating point and (*v*) stability factor S.

[Ans. (*i*) 10 μA, (*ii*) 1 mA, (*iii*) 8 V, 0.7 V, 7.3 V, (*iv*) (8 V, 1 mA), (*v*) 101]

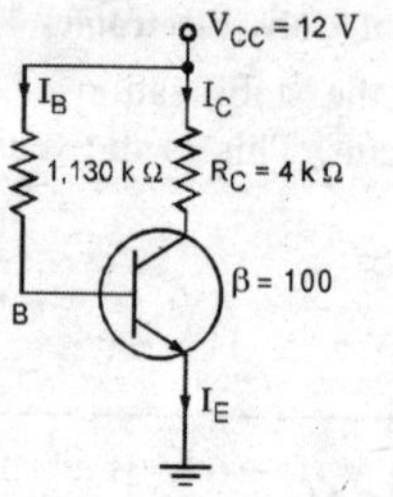

Fig. 5

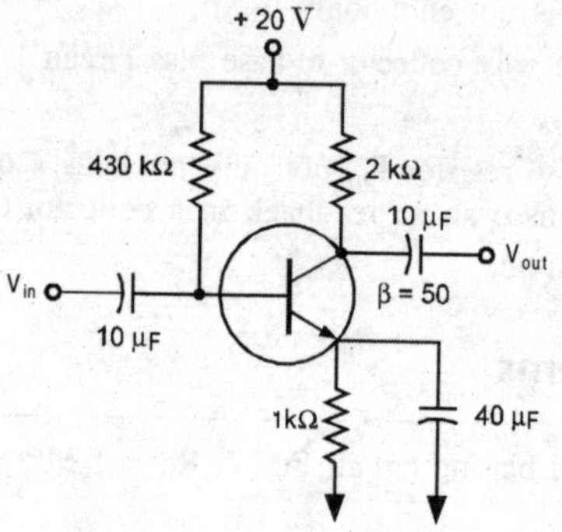

Fig. 6

6. In a fixed biasing circuit determine I_B, I_C and V_{CE} if transistor is of silicon, $V_{CC} = 10$ V, $R_B = 2.5$ MΩ, $R_C = 15$ kΩ and $\beta = 90$

[Ans. 3.72 μA; 0.335 mA; 4.978 V]

7. For the emitter bias network of Fig. 6 determine I_B, I_C, V_{CE}, V_E, V_B and V_C.

[Ans. 0.04 mA; 2 mA; 14 V, 2 V, 2.8 V, 16 V]

8. A silicon transistor with $\beta = 100$ is to be used in self biasing circuit shown in Fig. 7, such that the Q-point corresponds to $V_{CE} = 12$ V and $I_C = 2$ mA. Find R_E if $V_{CC} = 24$ V and $R_C = 5$ kΩ.

[Pb. Technical Univ. Electronic Devices and Circuits 2005] **[Ans.** 1 kΩ]

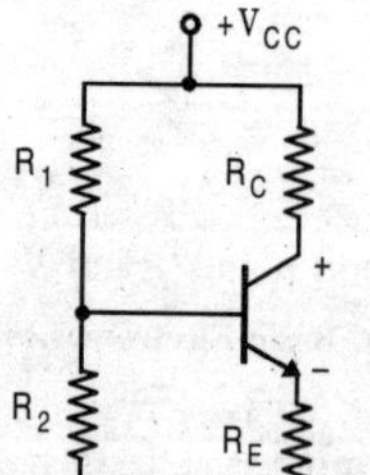

Fig. 7

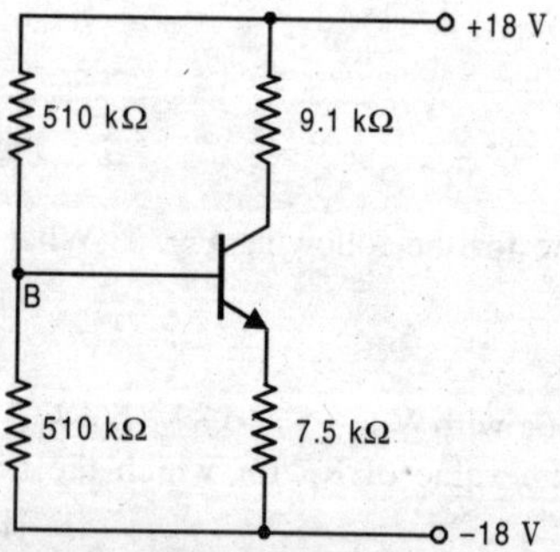

Fig. 8

9. Calculate I_B, I_C, V_E and V_{CE} for the circuit given in Fig. 8 if $\beta = 130$.

[U.P. Technical Univ. Electronics Engg. First Semester 2006-07]

[**Ans.** 14 μA, 1.82 mA, –4.24 V, 1.44 V]

10. A transistor biased by potential divider and emitter biasing has its operation point fixed at 2 mA, 6 V. If $V_{CC} = 15$ V, $R_E = 1.0$ kΩ, $R_2 = 10$ kΩ, $V_{BE} = 0.3$ V, find the values of R_C and R_1.

[M.D. Univ. Rohtak Analog Electronics Dec. 2007]

[**Ans.** 3.5 kΩ; 65 kΩ]

11. A silicon transistor having $\beta = 40$, $V_{BE} = 2$ V, $V_{CC} = 24$ V, $R_C = 4$ kΩ is used in CE amplifier. The operating point is required to be at $V_{CE} = 10$ V and $I_C = 1.8$ mA with stability factor not exceeding 4. Find values of R_1, R_2 and R_E. **[M.D. Univ. Rohtak Analog Electronics May 2008]**

[**Ans.** $R_E = 3.78$ kΩ; 29.94 kΩ; 19.6 kΩ]

12. Determine the quiscent current and the collector-to-emitter voltage for a silicon transistor with $\beta = 50$ in a self biasing arrangement shown in Fig. 7.

The circuit component values are :

$V_{CC} = 20$ V, $R_C = 2$ K, $R_1 = 100$ K

$R_E = 0.1$ K, $R_2 = 5$ K. **[M.D. Univ. Rohtak Analog Electronics May 2006]** [**Ans.** 1.28 mA, 17.312 V]

13. Design a self bias circuit for a Si transistor having $V_{BE} = 0.6$ V, $\beta = 99$. Desired operating point $V_{CE} = 5$ V and $I_C = 1$ mA. Assume $V_{CC} = 10$ V, $R_C = 3$ kΩ and stability factor S = 5.

[M.D. Univ. Rohtak Analog Electronics Dec. 2008]

[**Ans.** 2 kΩ; 31 kΩ, 11.4 kΩ]

14. In a self-bias circuit shown in Fig. 9 a silicon transistor is used with $\beta = 50$, $V_{BE} = 0.6$ V, $V_{CC} = 18$ V and $R_C = 4.3$ kΩ. It is desired to establish a quiescent point at $I_C = 1.5$ mA, $V_{CE} = 10$ V and a stability factor of $S \leq 4$. Find R_E, R_1 and R_2. **[U.P.S.C. I.E.S. Electrical Engineering-II, 1993]**

[**Ans.** 1 kΩ; 26.3 kΩ; 3.7 kΩ]

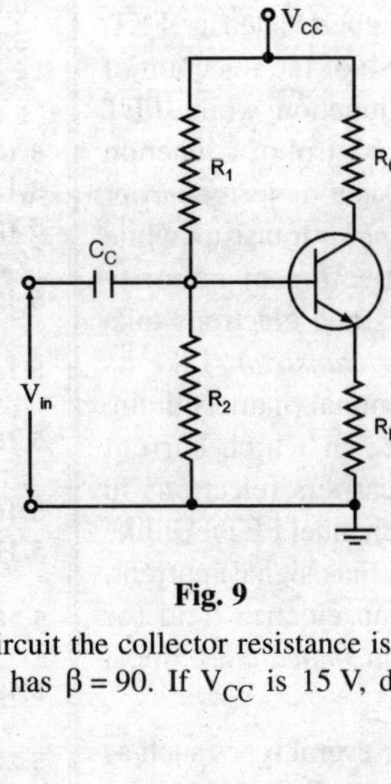

Fig. 9

15. For a collector feedback biasing circuit the collector resistance is 1 kΩ and the base resistance is 90 kΩ. The transistor used is silicon one and has $\beta = 90$. If V_{CC} is 15 V, determine Q-point. [**Ans.** 7.15 mA; 7.85 V]

❑❑❑

CHAPTER

5

FIELD-EFFECT TRANSISTORS (FETs) AND UJT

5.1. INTRODUCTION

INSIDE THIS CHAPTER

We know that P-N junctions have two dominant features *viz.* the injection of minority carriers with forward bias and a variation of the depletion width W with reverse bias. These two P-N junction properties are used in two important types of transistors *viz. bipolar junction* transistor abbreviated as BJT and the *junction field-effect transistor* abbreviated as JFET. The BJT discussed in chapter 3 makes use of the injection of minority carriers across a forward biased junction, while JFET discussed in this chapter depends on the control of a junction depletion width under reverse bias. FET is a majority carrier device, and is, therefore, often called a *unipolar* transistor while BJT operates by the injection and collection of minority carriers. Since in this device, both holes and electrons take part in conduction, it is called a *bipolar transistor*. Like its bipolar counterpart, the FET is a three-terminal (namely drain, source and gate) semiconductor device in which current conduction is by one type of majority carriers (electrons in case of an N-channel FET or holes in a P-channel FET). Unlike the BJT, a FET needs virtually no input (bias signal) current. Field-effect devices are controlled by an electric field (or voltage) and give an extremely *high input impedance*—most important advantage over BJT devices.

Field-effect transistors (FETs) are of several types such as junction field-effect transistors (JFETs), metal semiconductor field-effect transistors (MESFETs), and metal-insulator semiconductor field-effect transistors (MISFETs).

In a junction field-effect transistor (JFET) the gate voltage controls the depletion width of a reversed biased P-N junction. A metal semiconductor field-effect transistor

(MESFET) is the same basic device except that the P-N junction is replaced by a Schottky barrier rectifying junction. Metal-insulator-semiconductor field-effect transistor (MISFET) is one in which the metal gate electrode may be separated from the semiconductor by an insulator. The metal-oxide semiconductor field-effect transistor (MOSFET) is a special type of MISFET which makes use of an oxide layer as insulator.

The current in a JFET is through a semiconductor region called the *channel,* with ohmic contacts at each end. The basic transistor action is the modulation of the channel conductance by the electric field that is perpendicular to the channel. The modulating electric field is induced in the space charge region of a reverse-biased P-N junction or a reverse-biased Schottky barrier junction and is, therefore, a function of voltage applied to the gate.

Although the bipolar junction transistors (BJTs) reigned supreme in the early days of semiconductor integrated electronics, it has been gradually supplanted in most applications by the silicon MOSFETs. The main reason is, unlike BJTs, the different types of FETs are characterised by high input impedance, as the control voltage is applied to a reverse-biased junction or Schottky barrier, or across an insulator. Such devices are particularly suitable for controlled switching between a conducting state and a nonconducting state and that is why they are quite useful in digital circuits. They are also well suited for integration of many devices on a single chip. The MOSFET, is, without doubt, the core of the integrated circuit (IC) design at present time.

5.2. JUNCTION FIELD-EFFECT TRANSISTORS

Depending upon the majority carriers, JFETs have been classified into two types, namely (*i*) N-channel JFETs with electrons as the majority carriers and (*ii*) P-channel JFETs with holes as the majority charge carriers. N-channel JFETs are preferred because of mobility of electrons higher than that of holes. N-channel and P-channel FETs are shown in Figs. 5.1 (*a*) and 5.1 (*b*) respectively.

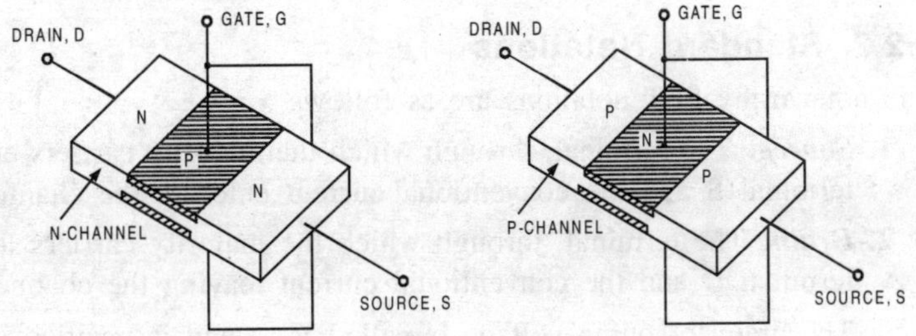

(*a*) *N-Channel JFET* (*b*) *P-Channel JFET*

Fig. 5.1. ***JFETs***

5.2.1. Basic Construction

Its structure is quite simple. In an N-channel JFET an N-type silicon bar, referred to as the *channel*, has two smaller pieces of P-type silicon material diffused on the opposite sides of its middle part, forming P-N junctions, as illustrated in Fig. 5.1 (*a*). The two P-N junctions forming diodes or gates are connected internally and a common terminal,* called the *gate terminal*, is brought out. Ohmic contacts (direct electrical connections) are made at the two ends of the channel—one lead is called the *source terminal* S and the other *drain terminal* D.

* When a manufacturer connects a separate lead to each gate, the device is called a dual-gate JFET. The main use of a dual-gate JFET is with a mixer, a special circuit used in communication equipment.

The silicon bar behaves like a resistor between its two terminals D and S. The gate terminal is analogous to the base of an ordinary transistor (BJT). It is used to control the flow of current from source to drain. Thus, source and drain terminals are analogous to emitter and collector terminals respectively of a BJT.

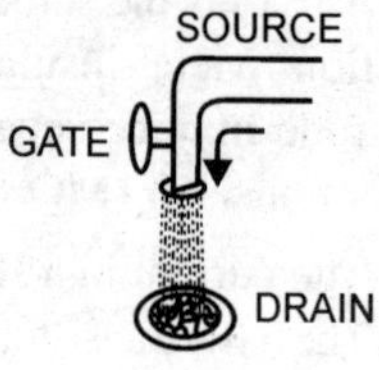

Fig. 5.2. ***Water Analogy For The JFET Control Mechanism***

Analogies are seldom perfect and at times can be misleading, but the water analogy of Fig. 5.2 does provide a sense for the JFET control at the gate terminal and the appropriateness of the terminology applied to the terminals of the device. The source of water pressure can be likened to the applied voltage from drain to source, which results in a flow of water (electrons) from the spigot (source). The "gate", through an applied signal (potential) controls the flow of water (charge) to the "drain", the drain and source terminals are at opposite ends of the N-channel as introduced in Fig. 5.1 because the terminology is defined for electron flow.

In Fig. 5.1 (*a*), the gate is P-region, while the source and the drain are N-regions. Because of this, a JFET is similar to two diodes. The gate and the source form one of the diodes, and the gate and the drain form the other diode. These two diodes are usually referred to as the *gate-source diode* and the *gate-drain diode*. Since JFET is a silicon device, it takes only 0.7 V for forward bias to get significant current in either diode.

With the gate terminal not connected, and a potential applied (+ve at the drain and –ve at the source), a current called the *drain current*, I_D flows through the channel located between the two P-regions. This current consists of only majority carriers–electrons in this case.

P-channel JFET is similar in construction to N-channel JFET except that P-type semiconductor material is sandwiched between two N-type junctions, as shown in Fig. 5.1 (*b*). In this case majority carriers are holes.

5.2.2. Standard Notations

The noteworthy FET notations are as follows :

1. ***Source.*** The terminal, through which the majority carriers enter the channel, is called the *source* terminal S and the conventional current entering the channel at S is designated as I_S.
2. ***Drain.*** The terminal, through which the majority carriers leave the channel, is called the *drain* terminal D and the conventional current leaving the channel at D is designated as I_D.
   The drain-to-source voltage is called V_{DS}, and is positive if D is more positive than S.
3. ***Gate.*** There are two internally connected heavily doped impurity regions formed by alloying, by diffusion, or by any other method available to create two P-N junctions. These impurity regions are called the gate G. A voltage V_{GS} is applied between the gate and source in the direction to reverse bias the P-N junction. Conventional current entering the channel at G is designated as I_G.
4. ***Channel.*** The region between the source and drain, sandwiched between the two gates is called the *channel* and the majority carriers move from source to drain through this channel.

5.2.3. Schematic Symbols

The schematic symbols for N-type and P-type JFETs are shown in Figs. 5.3 (*a*) and 5.3 (*b*) respectively. The vertical line in the symbol may be thought as channel and source S and drain D connected to the line.

Note that the direction of the arrow at the gate indicates the direction in which the gate current flows when the gate junction is forward biased. Thus for the N-channel JFET, the arrow at the gate junction points into the device and in P-channel JFET, it is away from the device.

An alternative symbol for an N-channel JFET is shown in Fig. 5.3 (*c*). This symbol with its offset gate is preferred by many engineers and technicians. The offset gate points to the source end of the device, a definite advantage in complicated multistage circuit.

The symbol shown in Fig. 5.3 (*d*) is used where the terminals of the two gate regions are provided with separate connecting leads. In this case device is referred to as a **tetrode-connected FET.**

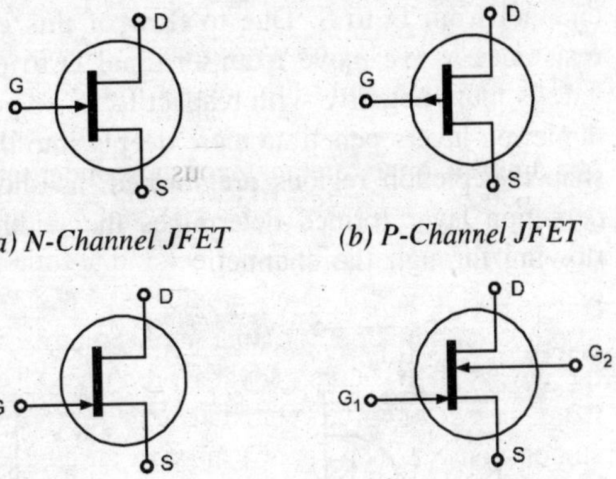

(a) N-Channel JFET *(b) P-Channel JFET*

(c) Offset Gate Symbol For an N-Channel JFET *(d) Tetrode-Connected JFET*

Fig. 5.3. ***Schematic Symbols For JFETs***

5.2.4. Polarity Conventions

The polarities for N-channel and P-channel JFETs are shown in Figs. 5.4 (*a*) and 5.4 (*b*) respectively. In both of the cases the voltage between gate and source is such that the gate is reverse biased. This is the normal method of connection of JFETs. The drain and source terminals are interchangeable* *i.e.,* either end can be used as a source and the other end as a drain. The source terminal is always connected to that end of the drain voltage supply which provides the necessary charge carriers *i.e.,* in an N-channel JFET source terminal, S is connected to the –ve end of the drain voltage supply for obtaining electrons whereas in P-channel JFET source terminal, S is connected to the +ve end of the drain voltage supply for getting holes.

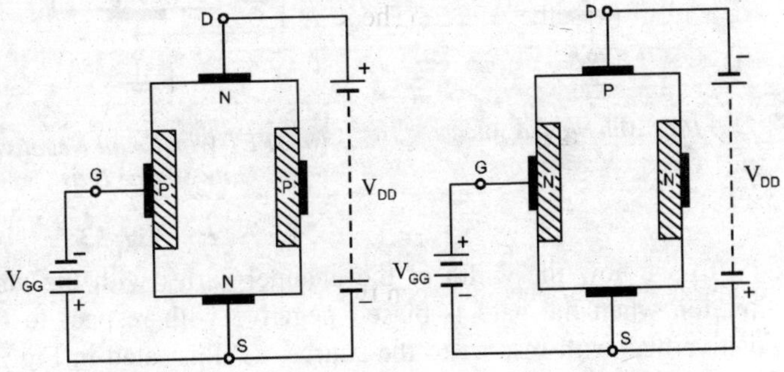

(a) N-Channel JFET *(b) P-Channel JFET*

Fig. 5.4. ***Polarity Conventions For JFETs***

Other noteworthy points are as follows :

(*i*) For N-channel JFET, I_D and V_{DD} are positive while V_{GG} is negative.

(*ii*) For P-cannel JFET, I_D and V_{DD} are negative while V_{GG} is positive.

(*iii*) Assumed positive directions of I_S, I_D and I_G are into the JFET.

From the above statements it may be concluded that the action of a P-channel JFET is complementary, that is, all voltages and currents are reversed.

5.2.5. Operation

Let us consider an N-channel JFET for discussing its operation.

(*i*) When neither any bias is applied to the gate (*i.e.,* when $V_{GS} = 0$) nor any voltage to the drain w.r.t. source (*i.e.,* when $V_{DS} = 0$), the depletion regions around the P-N junctions are of equal thickness and symmetrical.

(*ii*) When positive voltage is applied to the drain terminal D w.r.t. source terminal S without connecting gate terminal G to supply, as illustrated in Fig. 5.5 (*a*), the electrons (which are the majority carriers) flow from terminal S to terminal D whereas conventional drain current I_D flows through the

---

* The drain and source terminals are not interchangeable at high frequencies. Almost always, the manufacturer minimizes the internal capacitance on the drain side of the JFET *i.e.,* the capacitance between the gate and the drain is smaller than the capacitance between the gate and the source.

channel from D to S. Due to flow of this current, there is a uniform voltage drop across the channel resistance as we move from terminal D to terminal S. This voltage drop reverse biases the diode. The gate is more negative with respect to those points in the channel which are nearer to D than to S. Hence, depletion layers penetrate more deeply into the channel at points lying closer to D than to S. Thus wedge-shaped depletion regions are formed, as shown in Fig. 5.5 (*a*), when V_{DS} is applied. The size of the depletion layer formed determines the width of the channel and hence the magnitude of current I_D flowing through the channel.

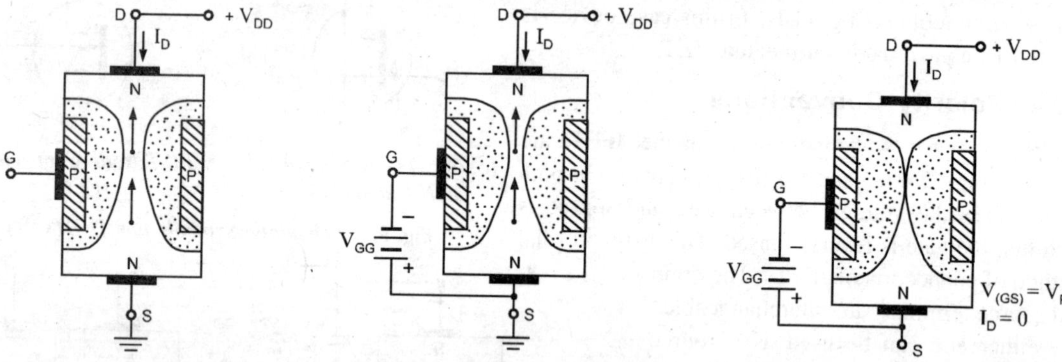

(*a*) *JFET With No Bias Voltage*   (*b*) *JFET With Small Negative Gate Source Bias*   (*c*)

Fig. 5.5

To see how the width of the channel varies with the variation in gate voltage, let us consider the situation when the gate is biased negative with respect to the source while the drain is applied with positive bias with respect to the source, as illustrated in Fig. 5.5 (*b*). Now the P-N junctions are reverse biased and depletion regions are formed. P-regions are heavily doped compared to the N-channel, so the depletion regions penetrate deeply into the channel. Since a depletion region is a region depleted of the charge carriers, it behaves as an insulator. The result is that the channel is narrowed, the resistance is increased and drain current I_D is reduced. If the negative voltage at the gate is further increased, depletion layers meet at the centre and the drain current I_D is cut off completely. On the other hand, if the negative bias to the gate is reduced, the width of the depletion layers gets reduced causing decrease in resistance and, therefore, increase in drain current I_D. The gate-source voltage V_{GS} at which drain current I_D is cut off completely (pinched off), is called the *pinch-off voltage* V_P. It is also to be noted that

1. The amount of reverse bias is not the same throughout the length of the P-N junction. When the drain current flows through the channel, there is a voltage drop along its length. The result is that the reverse bias at the drain end is more than that at the source end making the width of depletion layer more at the drain end than that at the source end. Thus the channel becomes narrower at the drain end in comparison to that at source end, as shown in Fig. 5.5 (*b*).
2. The channel is not completely closed at the drain end. Because in that case there will be no drain current, so there will be no voltage drop along the channel length and amount of reverse bias will become uniform and the wedge shaped depletion region will become rectangular one. The channel will open and the drain current will flow.

   However, at pinch-off voltage, the channel width is reduced to a constant minimum value to allow the flow of drain current.
3. The N-channel JFET behaves as a vacuum tube triode. The drain and source perform the same functions as the plate and cathode, respectively and, like the grid of a triode, the JFET gate

controls the drain current. As is also the case with a grid, gate current is to be avoided, so the gate channel junctions are normally never forward biased.

The device is called the *field-effect transistor* (FET) because the drain current (output current) is controlled by the effect of the extension of the field associated with the depletion region developed by the reverse bias at the gate.

P-channel JFET operates in the same manner as an N-channel JFET except that channel current carriers will be the holes in place of electrons and the polarities of V_{GS} and V_{DS} are reversed.

5.3. CHARACTERISTICS OF JFETs

There are two types of static characteristics viz. (1) output or drain characteristic and (2) transfer characteristic.

1. *Output or Drain Characteristic*. The curve drawn between drain current I_D and drain-source voltage V_{DS} with gate-to-source voltage V_{GS} as the parameter is called the *drain* or *output characteristic*. This characteristic is analogous to collector characteristic of a BJT.

***(a)* Drain Characteristic With Shorted Gate.** The circuit diagram for determining the drain characteristic with shorted gate for an N-channel JFET is given in Fig. 5.6 (*a*) and the drain characteristic with shorted gate is shown in Fig. 5.6 (*b*).

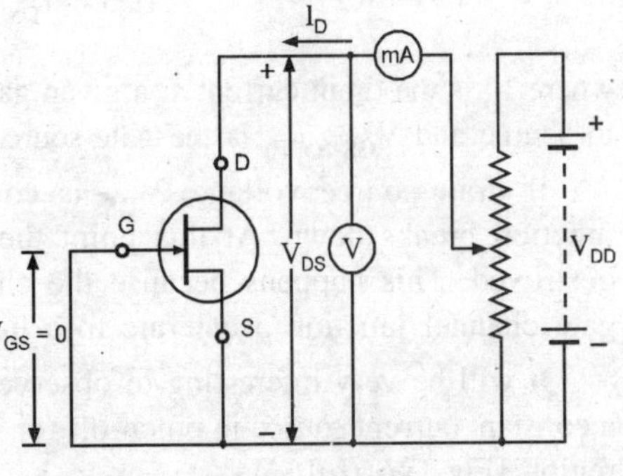

(a) Circuit Diagram For Determining Drain Characteristic With Shorted Gate For an N-Channel JFET

Initially when drain-source voltage V_{DS} is zero, there is no attracting potential at the drain, so no current flows inspite of the fact that the channel is fully open. This gives drain current $I_D = 0$. For small applied voltage V_{DS}, the N-type bar acts as a simple semiconductor resistor, and the drain current I_D increases linearly with the increase in V_{DS}, up to the knee point. This region (to the left of the knee point) of the curve is called the *channel ohmic region*, because in this region the FET behaves like an ordinary resistor.

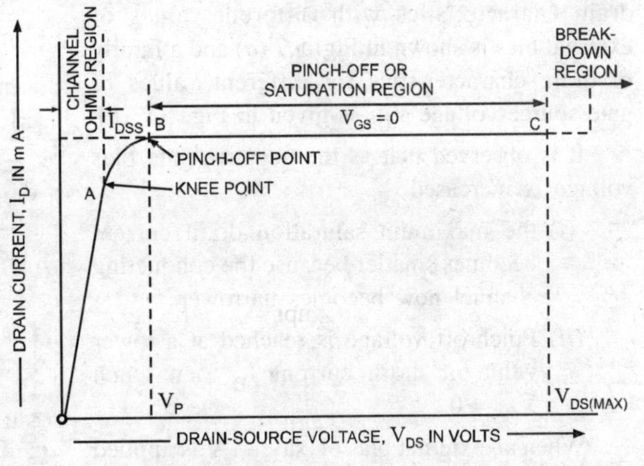

(b) JFET Drain Characteristic With Shorted Gate

Fig. 5.6

With the increase in drain current I_D, the ohmic voltage drop between the source and channel region reverse biases the gate junction. The reverse biasing of the gate junction is not uniform throughout. The reverse bias is more at the drain end than that at the source end of the channel, so with the increase in V_{DS}, the conducting portion of the channel begins to constric more at the drain end. Eventually a voltage V_{DS} is reached at which the channel is *pinched off*. The drain current I_D no longer increases with the increase in V_{DS}. It approaches a constant saturation value. The value of voltage V_{DS} at which the channel is pinched off (*i.e.,* all the free charges from the channel get removed), is called the *pinch-off voltage* V_P. The pinch-off voltage V_P, not too sharply defined on the curve, where the drain current I_D begins to level off and attains a constant value. From point A (knee point) to the point B (pinch-off point) the drain current I_D increases with the increase in voltage V_{DS} following a reverse square law. The region of the characteristic in which drain current I_D remains fairly constant is called the *pinch-off region*. It is also sometimes called the *saturation region* or *amplifier region*. In this region the JFET operates as a *constant current device* since drain current (or output current) remains almost constant. It is the normal operating region of the JFET when used as an amplifier. The drain current in the pinch-off region with $V_{GS} = 0$ is referred to as the *drain-source saturation current*, I_{DSS}.

It is to be noted that in the pinch-off (or saturation) region the channel resistance increases in proportion to increase in V_{DS} and so keeps the drain current almost constant and the reverse bias required by the gate channel junction is supplied entirely by the voltage drop across the channel resistance due to flow of I_{DSS} and not by the external bias because $V_{GS} = 0$.

Drain current in the pinch-off region is given by Shockley's equation

$$I_D = I_{DSS}\left(1 - \frac{V_{GS}}{V_P}\right)^2 = I_{DSS}\left(1 - \frac{V_{GS}}{V_{GS\,(OFF)}}\right)^2 \qquad ...(5.1)$$

where I_D is the drain current at a given gate-source voltage V_{GS}, I_{DSS} is the drain current with gate shorted to source and $V_{GS\,(OFF)}$ is the gate-source cutoff voltage.

If drain-source voltage, V_{DS} is continuously increased, a stage comes when the gate channel junction breaks down. At this point the drain current increases very rapidly, and the JFET may get destroyed. This happens because the charge carriers making up the reverse saturation current at the gate channel junction accelerate to a high velocity and produce an *avalanche effect*.

It will be very interesting to observe that JFET behaves as an ordinary resistor in ohmic region, as a constant current source in pinch-off (or saturation) region and as a constant voltage source in breakdown region [Fig. 5.6 (*b*)].

(*b*) Drain Characteristics With External Bias. The circuit diagram for determining the drain characteristics with different values of external bias is shown in Fig. 5.7 (*a*) and a family of drain characteristics for different values of gate-source voltage V_{GS} is given in Fig. 5.7 (*b*).

It is observed that as the negative gate bias voltage is increased

(*i*) the maximum saturation drain current becomes smaller because the conducting channel now becomes narrower.

(*ii*) Pinch-off voltage is reached at a lower value of drain current I_D than when $V_{GS} = 0$.

When an external bias of, say, –1 V is applied between the gate and the source, the gate channel junctions are reverse biased even when drain current, I_D is zero. Hence the depletion regions are already penetrating the channel to a certain extent when drain-source voltage, V_{DS} is zero. Due to this reason, a smaller voltage drop along the channel (*i.e.*, smaller than that for $V_{GS} = 0$) will increase the depletion regions to the point where they pinch-off the current. Consequently, the pinch-off voltage V_P is reached at a lower drain current, I_D when $V_{GS} = 0$.

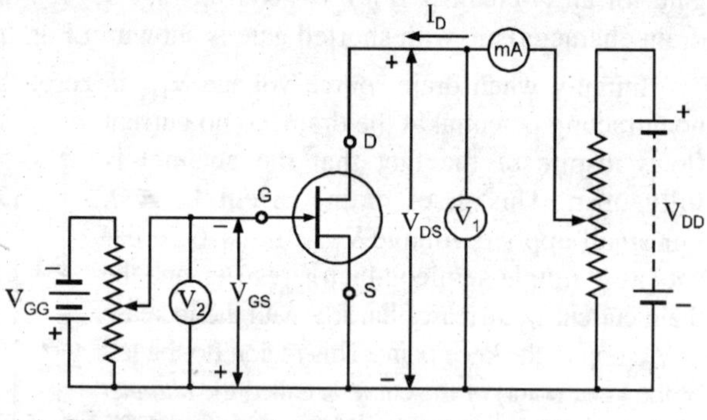

(*a*) *Circuit Diagram For Determining Drain Characteristic With External Bias For an N-Channel JFET*

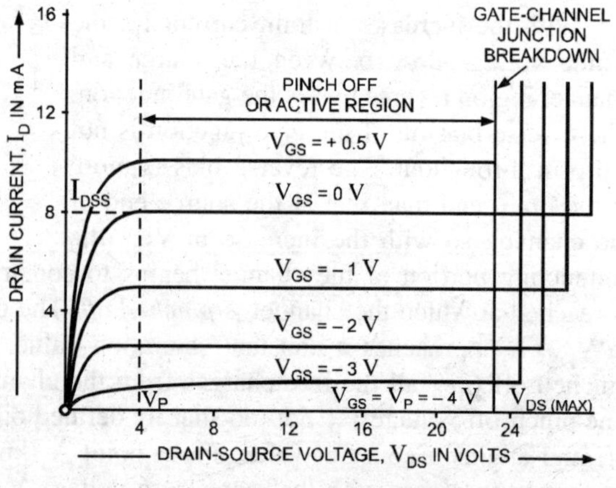

(*b*) *JFET Drain-Characteristics With External Bias*

Fig. 5.7

(*iii*) The ohmic region portion decreases.

(*iv*) Value of drain-source voltage V_{DS} for the avalanche breakdown of the gate junction is reduced.

Value of drain-source voltage, V_{DS} for breakdown with the increase in negative bias voltage is reduced simply due to the fact that gate-source voltage, V_{GS} keeps adding to the reverse bias at the junction produced by current flow. Thus the maximum value of V_{DS} that can be applied to a FET is the lowest voltage which causes avalanche breakdown.

It is also observed that with $V_{GS} = 0$, I_D saturates at I_{DSS} and the characteristic shows $V_P = 4$ V. When an external bias of –1 V is applied, the gate channel junctions still require –4 V to achieve pinch-off. It means that a 3 V drop is now required along the channel instead of the previous 4.0 V. Obviously, this drop of 3 V can be achieved with a lower value of drain current. Similarly, when $V_{GS} = -2$ V and – 3 V, pinch-off is achieved with 2 V and 1 V respectively, along the channel. These drops of 2 V and 1 V are, of course, achieved with further reduced values of drain current, I_D. It is further observed that when the gate-source bias is numerically equal to pinch-off voltage, V_P (–4 V in this case), no channel drop is required and, therefore, drain current, I_D is zero. The gate-source bias voltage required to reduce drain current I_D to zero is designated the gate-source cutoff voltage, $V_{GS\,(OFF)}$ and as explained,

$$V_{GS\,(OFF)} = V_P$$

Hence for working of JFET in the pinch-off or active region it is necessary that the following conditions be fulfilled.

$$V_P < V_{DS} < V_{DS\,(MAX)} \qquad ...(5.2)$$

$$V_{GS\,(OFF)} < V_{GS} < 0 \qquad ...(5.3)$$

$$0 < I_D < I_{DSS} \qquad ...(5.4)$$

2. *Transfer Characteristics*. The transfer characteristics for a JFET can be determined experimentally, keeping drain-source voltage, V_{DS} constant and determining drain current, I_D for various values of gate-source voltage, V_{GS}. The circuit diagram is shown in Fig. 5.7 (*a*). The curve is plotted between gate-source voltage, V_{GS} and drain current, I_D, as illustrated in Fig. 5.8. It is similar to the transconductance characteristic of a vacuum tube or a transistor. It is observed that

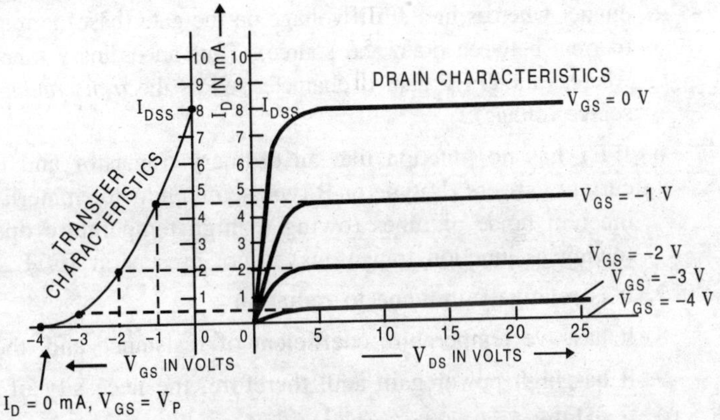

Fig. 5.8. ***Derivation of Transfer Characteristics From Drain Characteristics***

(*i*) drain current decreases with the increase in –ve gate-source bias,

(*ii*) drain current, $I_D = I_{DSS}$ when $V_{GS} = 0$

(*iii*) drain current, $I_D = 0$ when $V_{GS} = V_P$

The transfer characteristic follows Eq. (5.1)

The transfer characteristic can also be derived from the drain characteristic by noting values of drain current, I_D corresponding to various values of gate-source voltage, V_{GS} for a constant drain-source voltage and plotting them.

It may be noted that a P-channel JFET operates in the same way and have the similar characteristics as an N-channel JFET except that channel carriers are holes instead of electrons and the polarities of V_{GS} and V_{DS} are reversed.

5.4. MERITS AND DEMERITS OF JFETs

Junction field-effect transistors combine several merits of both conventional (or bipolar) transistors and vacuum tubes. Some of these are enumerated below :

1. Its operation depends upon the flow of majority carriers only. It is, therefore, a *unipolar* (one type of carrier) device. On the other hand in an ordinary transistor both majority and minority carriers take part in conduction and, therefore, ordinary transistor is sometimes called the *bipolar* transistor. The vacuum tube is another example of a unipolar device.
2. It is simpler to fabricate, smaller in size, rugged in construction and has longer life and higher efficiency. Simpler to fabricate in IC form and space requirement is also lesser.
3. It has high input impedance (of the order of 100 MΩ), because its input circuit (gate to source) is reverse biased, and so permits high degree of isolation between the input and output circuits. However, the input circuit of an ordinary transistor is forward biased and, therefore, ordinary transistor has low input impedance.
4. It carries very small current because of reverse-biased gate and, therefore, it operates just like a vacuum tube where control grid (corresponding to gate in JFET) carries extremely small current and input voltage controls the output current. This is the reason that JFET is essentially a *voltage driven* device. Ordinary transistor is a current operated device since input current controls the output current.
5. An ordinary transistor uses a current into its base for controlling a large current between collector and emitter whereas in a JFET voltage on the gate (base) terminal is used for controlling the drain current (current between drain and source). Thus an ordinary transistor gain is characterised by current gain whereas the JFET gain is characterised as the *transconductance* (the ratio of drain current and gate-source voltage).
6. JFET has no junction like an ordinary transistor and the conduction is through bulk material current carriers (N-type or P-type semiconductor material) that do not cross junctions. Hence the inherent noise of tubes (owing to high temperature operation) and that of ordinary transistors (owing to junction transitions) is not present in JFET.
7. It is relatively immune to radiation.
8. It has –ve temperature coefficient of resistance and, therefore, has better thermal stability.
9. It has high power gain and, therefore, the necessity of employing driver stages is eliminated.
10. It exhibits no offset voltage at zero drain current and, therefore, makes an excellent signal chopper.
11. It has square-law characteristics and, therefore, it is very useful in the tuners of radio and TV receivers.
12. It has got high frequency response.

The main drawback of JFET is its relative small gain-bandwidth product in comparison with that of a conventional transistor and greater susceptibility to damage in its handling. Other disadvantages of JFET are low voltage gains because of small transconductance and higher cost than that of BJTs.

5.5. FET CONFIGURATIONS

FETs (JFETs and MOSFETs) have three configurations just as BJTs have. The three configurations of FETs—the common gate, common source, and common drain configurations have been given in Fig. 5.9. Input and output terminals of different configurations have been given in Table 5.1.

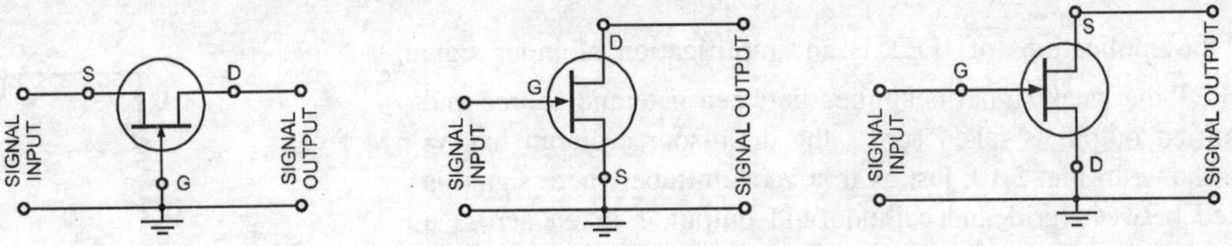

(a) Common Gate Configuration *(b) Common Source Configuration* *(c) Common Drain Configuration*

Fig. 5.9. ***N-Channel FET Configurations***

TABLE 5.1. ***FET Configurations***

Configuration	*Input Terminal*	*Output Terminal*
Common gate	Source	Drain
Common source	Gate	Drain
Common drain	Gate	Source

TABLE 5.2. ***FET-BJT Correspondence***

FET Configuration	*BJT Configuration*
Common gate	Common base
Common source	Common emitter
Common drain	Common collector

The three FET configurations have characteristics that are similar to the three corresponding BJT configurations, as given in Table 5.2.

5.6. JFET TEMPERATURE EFFECTS

The JFET, like other solid state devices, is subject to the effect of temperature. However, it is possible to bias the JFET such that it will exhibit a *zero temperature coefficient i.e.,* drain current I_D will be independent of temperature. This is explained below :

There are two primary mechanisms underlying the temperature sensitivity of the conductivity of a JFET channel.

1. Decrease in the depletion region width at the channel-gate P-N junction with the increase in temperature. This results in increase of channel thickness.
2. Decrease in carrier mobility with the increase in temperature.

Increase in channel thickness with the increase in temperature makes drain current I_D to increase. Another way of looking into the situation is to note that $V_{GS\,(OFF)}$ increases in magnitude with the increase in temperature. So far a fixed value of V_{GS}, I_D will increase with the temperature. $V_{GS\,(OFF)}$ [V_P] has a positive temperature coefficient of about 2.2 mV/°C.

The second factor [decrease in carrier mobility with the increase in temperature] makes the channel conductivity to decrease with the increase in temperature. The result is that drain current I_D decreases (assuming V_{DS} constant) with the increase in temperature. Thus we see that I_{DSS} tends to fall as the temperature rises.

Thus we have two distinct mechanisms affecting drain current I_D as a function of temperature. Carrier mobility gives I_D a negative temperature coefficient while the effect of $V_{GS\,(OFF)}$ is to give I_D a positive temperature coefficient. Since both of these mechanisms occur simultaneously, it is possible to bias the JFET so as to exhibit a zero temperature coefficient. Thus JFETs have better thermal stability.

The value of V_{GS} that will give drain current I_D for both N- and P-channel JFETs a zero temperature coefficient is given as

$$|V_{GS}| \simeq |V_P| - 0.63 \text{ V}. \quad ...(5.5)$$

5.7. JFET AS AN AMPLIFIER

One of the applications of JFET is an amplification of input signal.

In FET the weak signal is applied between gate and source and an amplified output is taken across the drain-source circuit having load, as shown in Fig. 5.10, just as in a vacuum tube where signal is connected between grid and cathode and output is taken across a load connected in plate circuit.

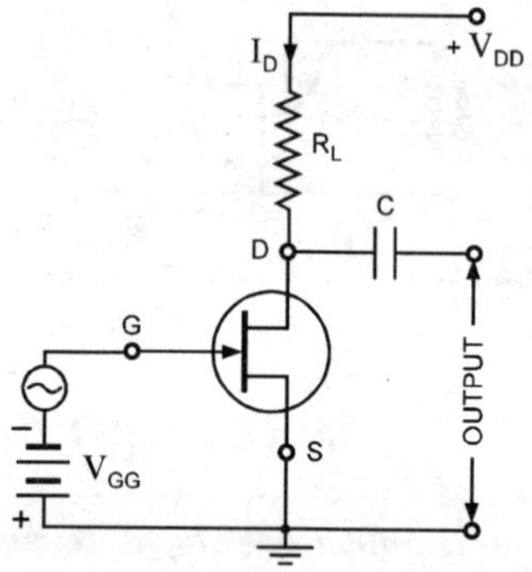

Fig. 5.10. *JFET As An Amplifier*

For proper working of JFET it is necessary that the gate is made negative w.r.t. source (*i.e.*, the input circuit is always reverse biased). This is achieved either by inserting a battery V_{GG} in the gate circuit with the polarity shown in Fig. 5.10 or by providing a biasing circuit.

On application of an alternating weak signal, the fixed reverse bias of the gate changes and a small change in reverse bias on the gate causes a large change in drain current owing to change in width of the channel. This fact makes the FET capable of amplifying the strength of weak signals. During the positive half cycle of the input signal, the reverse bias on the gate decreases causing increase in channel width and hence the drain current. During the negative half cycle of the input signal, the reverse bias on the gate increases causing drain current to decrease. The result is that the small change in voltage at the gate produces a large change in the drain current. These large variations in drain current give large output across the load R_L and thus, JFET operates as an amplifier.

5.8. JFET PARAMETERS

JFET has certain parameters which determine the performance. Such parameters are (*i*) ac drain resistance, (*ii*) transconductance, (*iii*) amplification factor, and (*iv*) dc drain resistance, as explained below :

1. *AC Drain Resistance*. It is defined as the ratio of change in drain-source voltage to change in drain current at constant gate-source voltage and is denoted by r_d.

i.e., AC drain resistance, $$r_d = \frac{\Delta V_{DS}}{\Delta I_D} \text{ at constant } V_{GS} \quad ...(5.6)$$

It is also called the *dynamic drain resistance*.

Referring to the output or drain characteristic shown in Fig. 5.7 (*b*), it is clear that in the active region the change in drain current, I_D is very small for change in drain-source voltage, V_{DS} because the characteristic curves are almost flat. Hence ac drain resistance of a JFET is very large ranging from 10 kΩ to 1 MΩ.

2. *Transconductance*. The control that the gate-source voltage has over the drain current, I_D is measured by transconductance. It is denoted by g_{fs}. It may be defined as the ratio of change in drain current to the change in gate-source voltage at constant drain-source voltage.

i.e., Transconductance, $$g_m = \frac{\Delta I_D}{\Delta V_{GS}} \text{ at constant } V_{DS} \quad ...(5.7)$$

It is also called the *forward transconductance* (g_{fs}) or forward transadmittance (Y_{fs}). It is measured in mA/volt or millisiemens.

The transconductance measured at I_{DSS} is denoted by g_{mo}.

Mathematical Expression For Transconductance. Differentiating both sides of Shockley equation

$$I_D = I_{DSS}\left(1 - \frac{V_{GS}}{V_P}\right)^2, \text{ we get}$$

$$dI_D = 2I_{DSS}\left(1 - \frac{V_{GS}}{V_P}\right)\left(-\frac{1}{V_P}\right) dV_{GS}$$

$$\text{or} \quad \frac{dI_D}{dV_{GS}} = 2I_{DSS}\left(1 - \frac{V_{GS}}{V_P}\right)\left(-\frac{1}{V_P}\right)$$

$$\text{or} \quad g_m = -\frac{2I_{DSS}}{V_P}\left(1 - \frac{V_{GS}}{V_P}\right) \quad ...(5.8)$$

Substituting $V_{GS} = 0$ in above equation, we get

$$g_{mo} = -\frac{2I_{DSS}}{V_P} \quad ...(5.9)$$

From Eqs. (5.8) and (5.9) we have

$$g_m = g_{mo}\left(1 - \frac{V_{GS}}{V_P}\right) \quad ...(5.10)$$

3. *Amplification Factor*. It is defined as the ratio of change in drain-source voltage to the change in gate-source voltage at constant drain current and is denoted by μ.

i.e., Amplification factor, $\mu = \frac{\Delta V_{DS}}{\Delta V_{GS}}$ at constant I_D

Amplification factor of a JFET indicates how much more control the gate-source voltage has over drain current in comparison to the drain-source voltage.

$$\text{Amplification factor, } \mu = \frac{\Delta V_{DS}}{\Delta V_{GS}} = \frac{\Delta V_{DS}}{\Delta I_D} \times \frac{\Delta I_D}{\Delta V_{GS}}$$

$$= r_d \times g_m$$

$$= \text{ac drain resistance} \times \text{transconductance} \quad ...(5.11)$$

Amplification factor μ of a FET may be as high as 100.

4. *DC Drain Resistance*. It is also called the static or ohmic resistance of the channel and is defined as ratio of drain-source voltage and drain current. It is denoted by R_{DS}

$$\textit{i.e.,}\ R_{DS}^* = \frac{V_{DS}}{I_D} \quad ...(5.12)$$

Example 5.1. When a reverse voltage of 10 V is applied between gate and source of JFET the gate current is 0.1 μA. Determine resistance between gate and source.

Solution : Gate-source voltage, $V_{GS} = 10.0$ V

* Drain resistance r_d is not to be confused with the drain-source resistance, R_{DS}. r_d is a dynamic or ac resistance whereas R_{DS} is the dc resistance of the channel when the depletion regions are absent *i.e.*, when the device is biased on in the channel ohmic region of the characteristics.

Gate current, $I_G = 0.1\ \mu A = 1 \times 10^{-7}$ A

$$\text{Gate-to-source resistance, } R_{GS} = \frac{V_{GS}}{I_G} = \frac{10}{1\times10^{-7}} = 100\ M\Omega \textbf{ Ans.}$$

Example 5.2. The readings obtained experimentally from a JFET are as follows :

Drain-source voltage, V_{DS} in volts	**5**	**12**	**12**
Gate-source voltage, V_{GS} in volts	**0**	**0**	**– 0.25**
Drain current, I_D in mA	**8**	**8.2**	**7.5**

Determine (*i*) ac drain resistance (*ii*) transconductance and (*iii*) amplification factor.

Solution : (*i*) The drain current increases from 8.0 mA to 8.2 mA when drain-source voltage is increased from 5 V to 12 V, keeping gate-source voltage constant at 0 V.

$$\therefore \text{ AC drain resistance, } r_d = \frac{\Delta V_{DS}}{\Delta I_D} = \frac{(12-5)\text{ V}}{(8.2-8)\text{ mA}} = \frac{7}{0.2\times10^{-3}} = 35\ k\Omega \textbf{ Ans.}$$

(*ii*) The drain current changes from 8.2 mA to 7.5 mA when gate-source voltage is changed from 0 to –0.25 V, keeping drain-source voltage constant at 12 V.

$$\therefore \text{ Transconductance, } g_m = \frac{\Delta I_D}{\Delta V_{GS}} = \frac{(7.5-8.2)\text{ mA}}{(-0.25-0)\text{ V}} = 2.8\text{ mA/V} \textbf{ Ans.}$$

(*iii*) Amplification factor, $\mu = r_d \times g_m = 35 \times 1{,}000 \times 2.8 \times 10^{-3} = 98$ **Ans.**

Example 5.3. Calculate the minimum value of V_{DS} required for an NMOSFET to operate in pinch-off condition when $V_{GS} = 1$ V, $V_P = -2$ V, $I_{DSS} = 10$ mA. Also find the corresponding drain current.

[*U.P.S.E. I.E.S. Electronics & Telecommunication Engineering-I, 2003*]

Solution : Drain current in pinch-off region

$$I_D = I_{DSS}\left(1-\frac{V_{GS}}{V_P}\right)^2 = 10\times10^{-3}\left(1-\frac{1}{(-2)}\right)^2 = 10\times10^{-3}\times\left(1\frac{1}{2}\right)^2 = 22.5\times10^{-3}\text{ A or } 22.5\text{ mA} \textbf{ Ans.}$$

The minimum value of V_{DS} required for the MOSFET to operate in pinch-off condition is equal to V_P *i.e.*, $V_{DS\,(min)} = V_P = -2$ V **Ans.**

Example 5.4. Given $I_{DSS} = 9$ mA and $V_P = -3.5$ V, determine I_D when $V_{GS} = 0$ V and $V_{GS} = -2$ V.

[*Mahamaya Technical Univ. Electronics Engineering Second Semester 2011-12*]

Solution : $I_{DSS} = 9$ mA

Pinch-off voltage, $V_P = -3.5$ V

When $V_{GS} = 0$ V

$$I_D = I_{DSS}\left(1-\frac{V_{GS}}{V_P}\right)^2 = 9\times10^{-3}\left(1-\frac{0}{V_P}\right)^2 = 9\text{ mA} \textbf{ Ans.}$$

When $V_{GS} = -2$ V

$$I_D = I_{DSS}\left(1-\frac{V_{GS}}{V_P}\right)^2 = 9\times10^{-3}\left(1-\frac{-2}{-3.5}\right)^2 = 1.653\text{ mA} \textbf{ Ans.}$$

Example 5.5. A JFET has drain current of 15 mA. If $I_{DSS} = 25$ mA and $V_P = 5$ V, find V_{GS}.

[*M.D. Univ. Rohtak Analog Electronics May 2008*]

Solution : Drain current, $I_D = 15$ mA

Drain-source saturation current, $I_{DSS} = 25$ mA

Pinch-off voltage, $V_P = 5$ V

From Eq. (5.1) $$I_D = I_{DSS}\left(1 - \frac{V_{GS}}{V_P}\right)^2$$

or $$V_{GS} = V_P\left(1 - \sqrt{\frac{I_D}{I_{DSS}}}\right) = 5\left(1 - \sqrt{\frac{15}{25}}\right) = 5(1 - \sqrt{0.6}) = 1.127 \text{ V} \quad \textbf{Ans.}$$

Example 5.6. A JFET has $V_P = -4.5$ V, $I_{DSS} = 10$ mA and $I_{DS} = 2.5$ mA. Determine the transconductance.

Solution : Drain-source saturation current, $I_{DSS} = 10$ mA

Pinch-off voltage, $V_P = -4.5$ V

Drain-source current, $I_{DS} = 2.5$ mA

From Shockley's equation, drain-source current, $$I_{DS} = I_{DSS}\left[1 - \frac{V_{GS}}{V_P}\right]^2$$

or $$V_{GS} = V_P\left[1 - \sqrt{\frac{I_{DS}}{I_{DSS}}}\right] = -4.5\left[1 - \sqrt{\frac{2.5}{10}}\right] = -2.25 \text{ V}$$

From Eq. (5.8), transconductance, $$g_m = \frac{-2I_{DSS}}{V_P}\left[1 - \frac{V_{GS}}{V_P}\right] = \frac{-2 \times 10 \times 10^{-3}}{-4.5}\left[1 - \frac{-2.25}{-4.5}\right]$$

$$= 2.22 \text{ mA/V} \quad \textbf{Ans.}$$

Example 5.7. For an N-channel JFET drain current with gate shorted = 8.7 mA, pinch-off voltage = –1 V. Find the value of (*i*) drain current and (*ii*) transconductance.

[*Pb. Technical Univ. Electronic Devices and Circuits Dec. 2009*]

Solution : $I_{DSS} = 8.7$ mA

$V_P = -1$ V

Assume $V_{GS} = -0.7$ V

(*i*) Drain current, $$I_D = I_{DSS}\left[1 - \frac{V_{GS}}{V_P}\right]^2 = 8.7 \times 10^{-3}\left[1 - \frac{-0.7}{-1}\right]^2 = 4.437 \text{ mA} \quad \textbf{Ans.}$$

(*ii*) Transconductance, $$g_m = \frac{-2I_{DSS}}{V_P}\left[1 - \frac{V_{GS}}{V_P}\right] = \frac{-2 \times 8.7 \times 10^{-3}}{-1}\left[1 - \frac{-0.7}{-1}\right] = 5.22 \text{ mA/V or } 5.22 \text{ mS} \quad \textbf{Ans.}$$

Example 5.8. A FET has a drain current of 4 mA. If $I_{DSS} = 6$ mA and $V_{GS\,(off)} = -6$ V, find the values of V_{GS} and V_P.

Solution : Drain-source current, $I_D = 4$ mA

Drain-source saturation current, $I_{DSS} = 6$ mA

Gate-to-source voltage, $$V_{GS} = V_{GS(off)}\left[1 - \sqrt{\frac{I_D}{I_{DSS}}}\right] = -6\left[1 - \sqrt{\frac{4 \text{ mA}}{6 \text{ mA}}}\right] = -1.1 \text{ V} \quad \textbf{Ans.}$$

Pinch-off voltage, $V_P = V_{GS(off)} = -6$ V **Ans.**

Example 5.9. A JFET has $g_m = 10$ mS, $I_{DSS} = 10$ μA. Calculate $V_{GS\,(OFF)}$.

Solution : Transconductance, $g_m = 10 \text{ mS} = 10 \times 10^{-3}$ S

Drain-source saturation current, $I_{DSS} = 10\ \mu\text{A} = 10 \times 10^{-6}$ A

$$V_{GS\,(OFF)} \rightarrow V_{GS} = V_P$$

$$g_m = g_{mo} = \frac{-2I_{DSS}}{V_P} = \frac{-2I_{DSS}}{V_{GS(OFF)}}$$

$$\text{or} \quad 10 \times 10^{-3} = \frac{-2 \times 10 \times 10^{-6}}{V_{GS(OFF)}}$$

$$\text{or} \quad V_{GS\,(OFF)} = \frac{-2 \times 10 \times 10^{-6}}{10 \times 10^{-3}} = -2 \text{ mV} \textbf{ Ans.}$$

Example 5.10. An N-channel JFET has I_{DSS} = 10 mA and V_P = –4 V. Determine the minimum value of V_{DS} for pinch-off region and drain current I_D for V_{GS} = – 2 V in pinch-off region.

Solution : Pinch-off voltage, $V_P = -4$ V

Gate-source voltage, $V_{GS} = -2$ V

Drain-source saturation current, $I_{DSS} = 10 \text{ mA} = 10 \times 10^{-3}$ A

$$\text{Drain current, } I_D = I_{DSS}\left[1 - \frac{V_{GS}}{V_P}\right]^2 = 10 \times 10^{-3}\left[1 - \frac{-2}{-4}\right]^2 = 2.5 \text{ mA } \textbf{Ans.}$$

The minimum value of V_{DS} for pinch-off region is equal to V_P. Thus the minimum value of V_{DS}

$$V_{DS\,(min)} = V_P = -4 \text{ V } \textbf{Ans.}$$

Example 5.11. For an N-channel JFET I_{DSS} = 8.7 mA; V_P = –3 V; V_{GS} = – 1 V.
Find the values of (*i*) I_D (*ii*) g_{mo} (*iii*) g_m. [*Bombay Univ. Basic Electronics 1985*]

Solution : Drain-source saturation current, I_{DSS} = 8.7 mA = 0.0087 A

Pinch-off voltage, $V_P = -3$ V

Gate-source voltage, $V_{GS} = -1$V

$$(i) \text{ Drain current, } I_D = I_{DSS}\left(1 - \frac{V_{GS}}{V_P}\right)^2 = 0.0087\left(1 - \frac{-1}{-3}\right)^2 = 3.8667 \text{ mA } \textbf{Ans.}$$

$$(ii) \text{ Transconductance for } V_{GS} = 0,\ g_{mo} = \frac{-2I_{DSS}}{V_P} = \frac{-2 \times 8.7 \text{ mA}}{-3} = 5.8 \text{ mA/V or } 5.8 \text{ mS } \textbf{Ans.}$$

$$(iii) \text{ Transconductance, } g_m = g_{mo}\left(1 - \frac{V_{GS}}{V_P}\right) = 5.8\left(1 - \frac{-1}{-3}\right) = 3.867 \text{ mA/V or } 3.867 \text{ mS } \textbf{Ans.}$$

Example 5.12. An N-channel JFET has a pinch-off voltage of – 4.5 V and I_{DSS} = 9 mA. At what value of V_{GS} will I_{DS} be equal to 3 mA? What is its g_m at this I_{DS} ?

Solution : Pinch-off voltage, $V_P = -4.5$ V

Drain-source saturation current, $I_{DSS} = 9 \text{ mA} = 9 \times 10^{-3}$ A

Drain-source current, $I_{DS} = 3 \text{ mA} = 3 \times 10^{-3}$ A

From Shockley's equation (Eq. 5.1)

$$V_{GS} = V_P\left[1 - \sqrt{\frac{I_{DS}}{I_{DSS}}}\right] = -4.5 \times \left[1 - \sqrt{\frac{3 \times 10^{-3}}{9 \times 10^{-3}}}\right] = -1.902 \text{ V } \textbf{Ans.}$$

Transconductance g_m for I_{DS} = 3 mA for which

$$V_{GS} = -1.902 \text{ V}$$

$$g_m = \frac{-2I_{DSS}}{V_P}\left[1 - \frac{V_{GS}}{V_P}\right] = \frac{-2 \times 9 \times 10^{-3}}{-4.5}\left(1 - \frac{-1.902}{-4.5}\right) = 2.31 \text{ mA/V or } 2.31 \text{ mS } \textbf{Ans.}$$

5.9. JFET DATA SHEET

JFET data sheets are similar to BJT data sheets—starting with a device type number and a brief description of the device to indicate the most important applications. These datas are followed by maximum ratings, dc characteristics, ac characteristics, mechanical data etc. Some of the important parameters specified are :

1. ***Saturation Current and Pinch-off Voltage.*** The drain-source saturation current I_{DSS} and the pinch-off voltage V_P have already been discussed in Art. 5.3. I_{DSS} is sometimes termed the *pinch-off current* and is referred to as I_{DP} at $V_{GS} = 0$. I_{DP} may also be specified at values of V_{GS} other than zero, and in this case V_{GS} is also specified.
2. ***Transconductance.*** It has already been discussed in Art. 5.8. It is measured in siemens (S), formerly in mho (℧). Its typical value for a FET may lie between 1 mS and 10 mS.
3. ***Drain Resistance.*** It has already been discussed in Art. 5.8.
4. ***Drain-Source ON Resistance.*** The drain resistance r_d is not to be confused with the drain-source ON resistance R_{DS}, also designated $R_{D\,(on)}$. It has already been discussed in Art. 5.8. It is measured at a specified gate-source voltage V_{GS} and drain current I_D and is important when using the JFET as a switch. When the JFET is biased in its saturation or ohmic region of operation it exhibits a resistance between drain and source ranging from 10 to a few hundred ohms.
5. ***Gate Cutoff Current and Input Resistance.*** The gate channel junction in a JFET is a P-N junction, and since it is normally reverse biased, a minority charge carrier current flows. This is the gate-source cutoff current I_{GSS}, also called the *gate reverse current.* I_{GSS} is very small current of the order of few nanoamperes. The device *input resistance* (resistance of the reverse biased gate channel junctions) is inversely proportional to I_{GSS} and its typical values for a JFET are $10^9\ \Omega$ at 25°C and $10^7\ \Omega$ at 100°C.
6. ***Breakdown Voltage.*** There are several ways in which the FET breakdown voltage may be specified. BV_{DGO} is the drain-gate breakdown voltage with the source open circuited. BV_{GSS} is the gate-source breakdown voltage with the drain shorted to the source. Both are a measure of the voltage at which the reverse-biased gate channel junctions break down. Typical values for each are of the order of 25 V.
7. ***Noise Figure.*** FET usually has much lower thermal noise as compared to that in case of BJT. This is because, unlike the BJT, there are very few charge carriers crossing a junction in the FET. As in the case of bipolar device, the FET noise figure NF is specified as a *spot noise figure* at a particular frequency and bias conditions, and for a given value of bias resistance.
8. ***Capacitances.*** FET capacitances may be specified as *gate-drain capacitance* C_{gd}, *gate-source capacitance* C_{gs}, and *drain-source capacitance* C_{ds}. Instead, the capacitance is sometimes specified as the *common source input capacitance*, C_{iss} or C_{gss}. This is the gate-source capacitance measured with the drain shorted to the source. In this case a reverse transfer capacitance C_{rss} is also specified. C_{rss} being another term for C_{gd}. These quantities are very important for FET high-frequency and switching circuits.

5.10. DC LOAD LINE AND BIAS POINT

The dc load line for a FET circuit is drawn upon the output characteristics of the device in exactly the same way as for a BJT circuit.

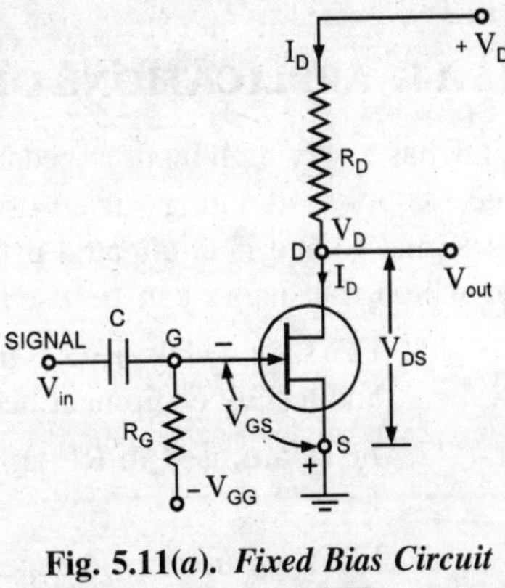

Fig. 5.11(*a*). *Fixed Bias Circuit For JFET*

Consider the FET circuit and characteristics shown in Figs. 5.11 (*a*) and 5.11 (*b*) respectively. The drain-source voltage is given as

$$V_{DS} = V_{DD} - I_D R_D \qquad ...(5.13)$$

where V_{DD} is the supply voltage and $I_D R_D$ is the voltage drop across R_D.

Consider the following two particular situations :

(*i*) When $I_D = 0$, $V_{DS} = V_{DD}$ ...cutoff point B

(*ii*) When $V_{DS} = 0$, $I_D = \frac{V_{DD}}{R_D}$ ...saturation point A

By joining these two points A and B, dc load line is obtained.

A *dc bias point* or *quiescent point* (Q-point) is selected on the FET dc load line. This point defines the dc conditions that exist in a circuit when no input signal is applied. The bias point is selected to give maximum possible variations in output (drain-source) voltage when the drain current is varied by an input signal. In this case the positive-going and negative-going output voltage swings should be as closely equal as possible. Where maximum output voltage swing is not required, the bias point may be at any convenient position on the load line.

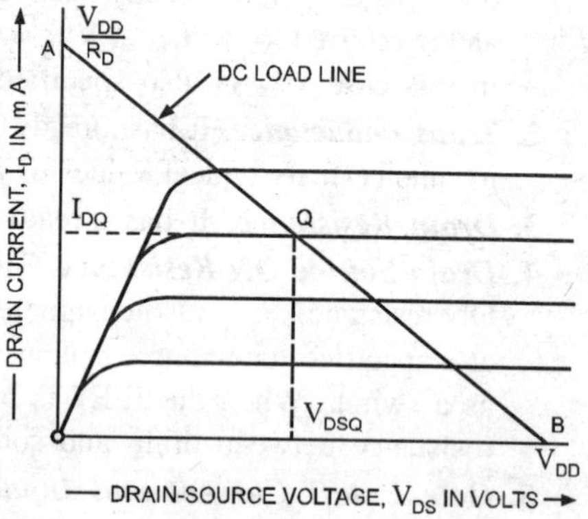

Fig. 5.11(*b*)

For the load line in Fig. 5.11 (*b*), the bias conditions are $\frac{V_{DD}}{2}$ and $\frac{V_{DD}}{2R_D}$.

For a FET amplifier, V_{DS} must remain in the pinch-off region of the characteristics. This means that it must not be allowed to go below the level of the pinch-off voltage [V_P in Fig. 5.7 (*b*)]. Thus, in the design of a FET bias circuit, the drain source voltage should always be a minimum of (V_P + 1) volt.

$$\textit{i.e.,} \quad V_{DS\,(min)} = V_P + 1 \text{ volts} \qquad ...(5.14)$$

When an external bias voltage V_{GS} is included, the pinch-off voltage for that bias level on the drain characteristics is $V_P - V_{GS}$, where V_P is the pinch-off voltage at $V_{GS} = 0$ [Fig. 5.7 (*b*)]. Consequently, the minimum drain-source voltage may be reduced to

$$V_{DS\,(min)} = V_P - V_{GS} + 1 \text{ volts} \qquad ...(5.15)$$

For the operation of FET as a switch, the device will either be biased *off* or *on* into the channel ohmic region of the drain characteristics.

5.11. APPLICATIONS OF FETs

FET has a very high input impedance (100 MΩ in case of JFETs and 10^4 to 10^9 MΩ in case of MOSFETs), the major shortcomings of an ordinary transistor *i.e.,* low input impedance with consequent of loading of signal source is eliminated in FET. Hence FET is an ideal device for use in almost every application in which transistors can be used.

1. FETs are widely used as input amplifiers in oscilloscopes, electronic voltmeters and other measuring and testing equipment because of their high input impedance.
2. FETs are used in RF amplifiers in FM tuners and communication equipments for the low noise level.
3. FETs are used as voltage variable resistors (VVRs) in operational amplifiers and tone controllers etc. because it is a voltage controlled device.

4. FETs are used in mixer circuits in FM and TV receivers, and communication equipments because of their low intermodulation distortion.
5. FETs are used in low frequency amplifiers in hearing aids and inductive transducers because of the small coupling capacitors.
6. FETs are used in digital circuits in computers, LSI and memory circuits because of very small size.

5.12. FET AS A VOLTAGE VARIABLE RESISTOR (VVR)

FET is a device that is usually operated in the constant-current portion of its output characteristics. But if it is operated in the region prior to pinch-off (*i.e.,* where V_{DS} is small, say below 100 mV), it will behave as a voltage-variable resistor (VVR). It is due to the fact that in this region drain-to-source resistance R_{DS} can be controlled by varying the gate bias voltage V_{GS}. In such applications the FET is also referred to as a *voltage-variable resistor* or *voltage dependent resistor*.

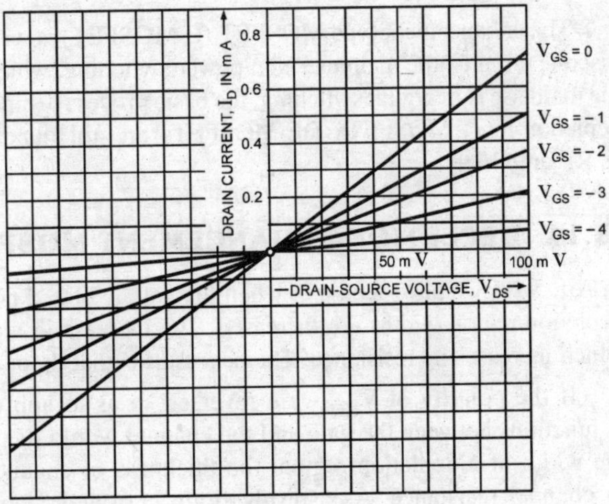

Fig. 5.12. ***FET Low-Level Drain Characteristics For 2N5951***

Figure 5.12 shows the drain characteristic curves for a 2N5951 in the ohmic region (*i.e.,* for low V_{DS}). From the characteristic curve it can be seen that R_{DS} varies with V_{GS}. For example, when $V_{GS} = 0$, $R_{DS} = 133\ \Omega$ and when $V_{GS} = -2$ V, $R_{DS} = 250\ \Omega$. Because of this a JFET operating in the ohmic region with small ac signals acts as a voltage-controlled resistance.

Note that the drain curves shown in Fig. 5.12 extend on both sides of the origin. This means that a JFET can be employed as a *voltage-variable resistor* for small ac signals, typically those less than 100 mV. When it is employed in this way, it does not require a dc drain voltage from the supply. All that is required is an ac input signal.

5.13. METAL-INSULATOR-SEMICONDUCTOR FIELD-EFFECT TRANSISTORS (MISFETs)

The metal-insulator-semiconductor field-effect transistor (MISFET) is one of the most widely used electronic devices, particularly in digital circuits, because of relatively small size— thousands of devices can be fabricated in a single integrated circuit (IC). Since it is constructed with the gate terminal insulated from the channel, it may be referred to generally as an *insulated-gate field-effect transistor* (IGFET). However, since most of such devices are made using silicon for the semiconductor, SiO_2 for the insulator, and metal or heavily doped polysilicon for the gate electrode, the term *metal-oxide-semiconductor field-effect* transistor (MOSFET) or simply MOS is commonly used. MISFET is the more general term, where the insulator is not necessarily silicon dioxide (SiO_2) and the semiconductor is not necessarily silicon (Si). Here discussion will be about MOSFETs although the same physics applies to the MISFETs.

Like, a JFET, a MOSFET is also a three terminal (source, gate and drain) device and drain current in it is also controlled by gate bias. The operation of MOSFET is similar to that of JFET. It can be employed in any of the circuits covered for the JFET and, therefore, all the equations apply equally well to the MOSFET and

JFET in amplifier connections. However, MOSFET has lower capacitance and input impedance much more than that of a JFET owing to small leakage current. In case of a MOSFET the positive voltage may be applied to the gate and still the gate current remains zero.

MOSFETs are of two types namely (*i*) *enhancement type* MOSFET or E-MOSFET and (*ii*) depletion enhancement MOSFET or DE-MOSFET. In the depletion-mode construction a channel is physically constructed and a current between drain and source is due to voltage applied across the drain-source terminals. The enhancement MOSFET structure has no channel formed during its construction. Voltage is applied to the gate, in this case, to develop a channel of charge carriers so that a current results when a voltage is applied across the drain-source terminals.

The enhancement type MOSFET (E-MOSFET) is widely used in both discrete and integrated circuits. In discrete circuits, the main use is in power switching, which means turning largest currents *on* and *off*. In ICs, the main use is in digital switching, the basic process behind modern computers. Although their use is declined, depletion-mode MOSFETs (DE-MOSFETs) are still found in high-frequency front-end communication circuits as RF amplifiers.

5.14. DEPLETION-ENHANCEMENT MOSFETs (DE-MOSFETs)

In Art. 5.2.5 we have seen that when the gate is biased negative w.r.t. the source in an N-channel JFET, the depletion-region widths are increased. The increase in the depletion regions reduces the channel thickness, which increases its resistance. The net result is that drain current I_D is reduced.

If the polarity of V_{GG} were reversed so as to apply a positive bias to the gate w.r.t. source, the P-N junctions between the gate and the channel would then be forward biased. Since a forward bias reduces the width of a depletion region, the thickness of channel would increase with a corresponding decrease in channel resistance. As a result, drain current I_D would increase beyond the JFET's I_{DSS} value.

The normal operation of a JFET is in its depletion mode of operation. However, as discussed above, it is also possible to *enhance* the conductivity of the JFET channel. However, the forward bias of the silicon P-N junction is usually restricted to a maximum of 0.5 V (more conservative limit is 0.2 V) so as to limit the gate current.

As we have seen in Art. 5.8, the greater the I_D is compared to I_{DSS}, the greater the transconductance g_m will be. Since voltage gain is directly proportional to g_m in general, the higher the g_m, the better it is. This is one of the advantages of being able to enhance the channel.

As its name suggests, the *depletion-enhancement* MOSFET (DE-MOSFET) was developed to be used in either or both the depletion and enhancement modes.

5.14.1. Construction

Figure 5.13 shows the construction of an N-channel DE-MOSFET. It consists of a lightly doped P-type substrate into which two blocks of heavily doped N-type material are diffused forming the source and drain. An N-channel is formed by diffusion between the source and drain. The type of impurity for the channel is the same as for the source and drain. Now a thin layer of SiO_2 dielectric is grown over the entire surface and holes are cut through the SiO_2 (silicon dioxide) layer to make contact with the N-type blocks (Source and Drain). Metal is deposited through the holes to provide drain and source terminals, and on the surface area between drain and source, a metal plate is deposited. This layer

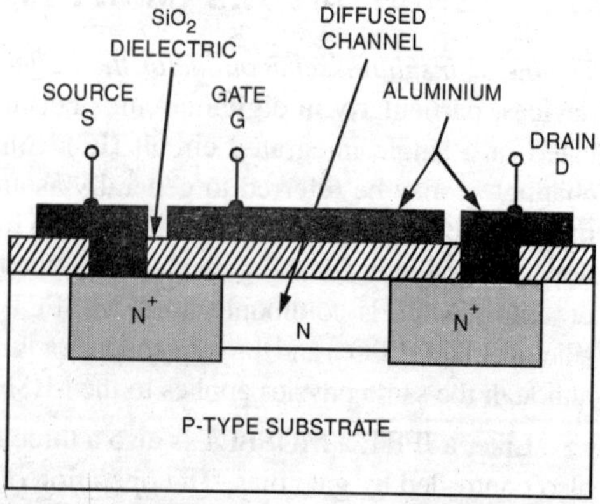

Fig. 5.13. ***N-Channel DE-MOSFET Structure***

constitutes the gate. SiO_2 layer results in an extremely high input impedance of the order of 10^{10} to 10^{15} Ω for this device. The chip area of a MOSFET is typically 0.003 μm^2 or less which is about only 5% of the area required by a BJT.

A P-channel DE-MOSFET is constructed like an N-channel DE-MOSFET, starting with an N-type substrate and diffusing P-type drain and source blocks and connecting them internally by a P-dopped channel region.

In some cases the substrate is internally connected to the source terminal S and in others an additional terminal labelled SS is provided resulting in a four terminal device.

Silicon dioxide (SiO_2) is a type of insulator referred to as a *dielectric,* which develops opposing (as indicated by the prefix di-) electric fields within the dielectric when exposed to an externally applied field. The fact that the SiO_2 layer is an insulating layer means that there is no direct electrical connection between the gate terminal G and the channel of a MOSFET. It is the insulating layer of SiO_2 in the MOSFET construction that accounts for the very desirable high input impedance of the device.

In fact, the input resistance of a MOSFET is usually much higher than that of a typically JFET, even though the input impedance of most JFETs is sufficiently high for applications. Because of very high input impedance, the gate current I_G is essentially zero for dc-biased configurations.

5.14.2. Operation

It is obvious from Fig. 5.13 that there is no P-N junction between gate and channel. Here the diffused channel N, the insulating dielectric SiO_2 layer and the metal layer of the gate forms a parallel plate capacitor. DE-MOSFET can be operated with either a positive or a negative gate. When gate is positive with respect to the source it operates in the enhancement—or E-mode and when the gate is negative w.r.t. the source, as illustrated in Fig. 5.14, it operates in depletion-mode.

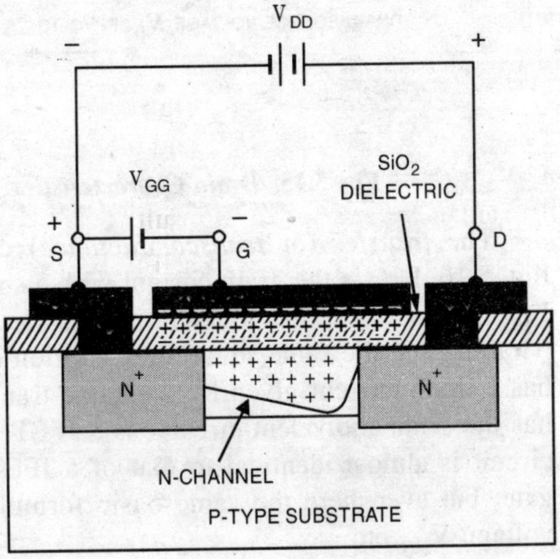

Fig. 5.14. ***Depletion Mode Operation***

When the drain is made + ve with respect to source, a drain current will flow, even with zero gate potential and the MOSFET is said to be operating in E-mode. In this mode of operation gate attracts the – ve charge carriers from the P-substrate to the N-channel and thus reduces the channel resistance and increases the drain current. The more positive the gate is made, the more drain current flows.

On the other hand when the gate is made –ve with respect to the substrate, the gate repells some of the –ve charge carriers out of the N-channel. This creates a depletion region in the channel, as illustrated in Fig. 5.14, and, therefore, increases the channel resistance and reduces the drain current. The more –ve the gate, the less the drain current. In this mode of operation the device is referred to as a *depletion-mode MOSFET*. Here too much negative gate voltage can pinch-off the channel. Thus operation is similar to that of JFET.

5.14.3. Characteristics

Typical *drain characteristics*, for various levels of gate-source voltage, of an N-channel MOSFET are shown in Fig. 5.15. The upper curves are for positive V_{GS} and the lower curves are for negative V_{GS}. The bottom drain curve is for $V_{GS} = V_{GS\,(OFF)}$. For a specified drain-source voltage V_{DS}, $V_{GS\,(OFF)}$ is the gate-source voltage at which drain current reduces to a certain specified negligibly small value, as shown in Fig. 5.15. This voltage corresponds to the pinch-off voltage V_P of JFET. For V_{GS} between $V_{GS\,(OFF)}$ and zero, the device operates in *depletion-mode* while for V_{GS} exceeding zero the device operates in *enhancement mode*. These drain curves again display an ohmic region, a constant current source region and a cutoff region. MOSFET has two major applications: a constant current source and a voltage variable resistor.

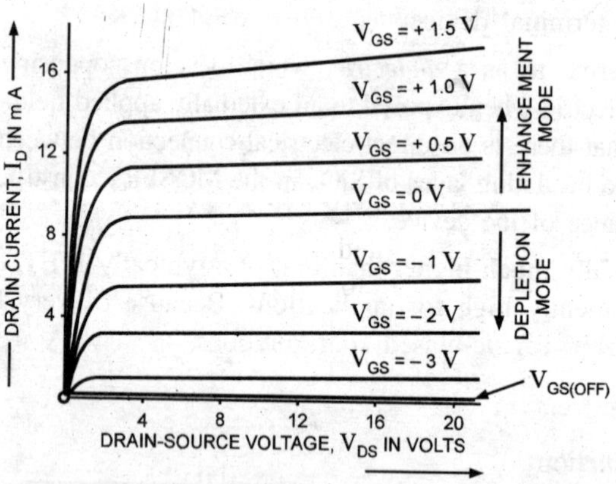

Fig. 5.15. *Drain Characteristics*

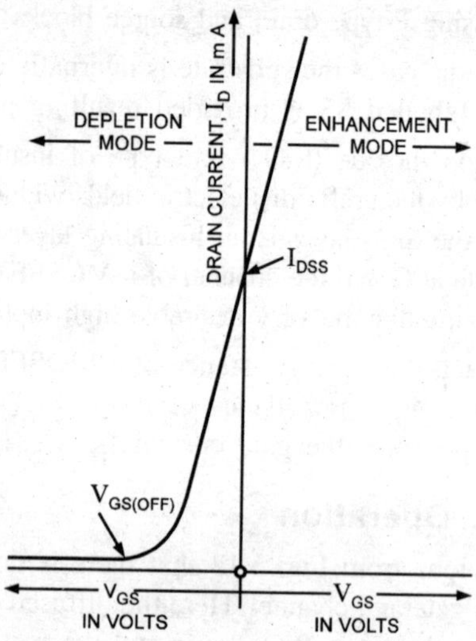

Fig. 5.16. *Transfer Characteristic*

The *transfer* (or *transconductance*) characteristic for an N-channel DE-MOSFET is shown in Fig. 5.16. I_{DSS} is the drain current with a shorted gate. Since the curve extends to the right of the origin, I_{DSS} is no longer the maximum possible drain current. Mathematically, the curve is still part of a parabola and the same square-law relation exists as with a JFET. In fact, the depletion-mode MOSFET has a drain current given by the same transconductance equation as before, Eq. (5.1). Furthermore, it has the same equivalent circuits as a JFET. Because of this, the analysis of a depletion-mode MOSFET circuit is almost identical to that of a JFET circuit. The only difference is the analysis for a positive gate, but even here the same basic formulas are used to determine the drain current I_D, gate-source voltage V_{GS} etc.

The foregoing discussion is applicable in principle also to the P-channel DE-MOSFET. For such a device the sign of all currents and voltages in the characteristics [Figs. 5.15 and 5.16] must be reversed.

5.14.4. Schematic Symbols

Figure 5.17 (*a*) shows the schematic symbol for a DE-MOSFET. Just to the right of the gate is the thin vertical line representing the channel. The drain lead comes out from the top of the channel and

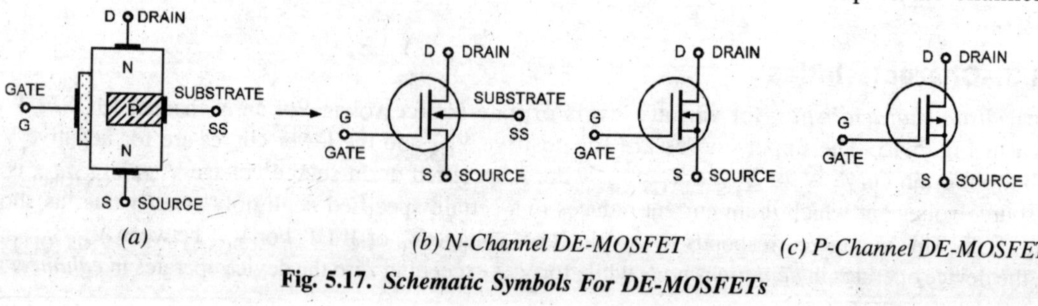

Fig. 5.17. *Schematic Symbols For DE-MOSFETs*

the source lead connects to the bottom. The arrow is on the P-substrate and points to the N-material. In some applications, a voltage can be applied to the substrate for added control of drain current. For this reason, some DE-MOSFETs have four terminal leads. But in most applications, the substrate is connected to the source. Usually the substrate is connected to the source internally by the manufacturer. This results in a three terminal device whose schematic symbol is shown in Fig. 5.17 (*b*).

Schematic symbol for a three terminal P-channel DE-MOSFET device is shown in Fig. 5.17 (*c*). The schematic symbol of a P-channel DE-MOSFET is similar to that of an N-channel DE-MOSFET, except that the arrow points outward.

5.14.5. DE-MOSFET Amplifiers

DE-MOSFET can be operated either with positive or negative gate, so its operating or quiescent point can be set at $V_{GS} = 0$ V, as illustrated in Fig. 5.18 (*b*). In Fig. 5.18 (*a*) the gate G is at ground potential for dc since ac signal source appears to be short circuited for dc and sets Q-point at $V_{GS} = 0$ V. The ac input signal causes the gate-source voltage V_{gs} to vary above and below zero developing drain current I_D and drain voltage V_{out}. The biasing circuit given in Fig. 5.18 (*a*) has $V_{gs} = 0$ V, $I_D = I_{DSS}$ and dc drain voltage

$$V_{DS} = V_{DD} - I_{DSS}R_D \quad ...(5.16)$$

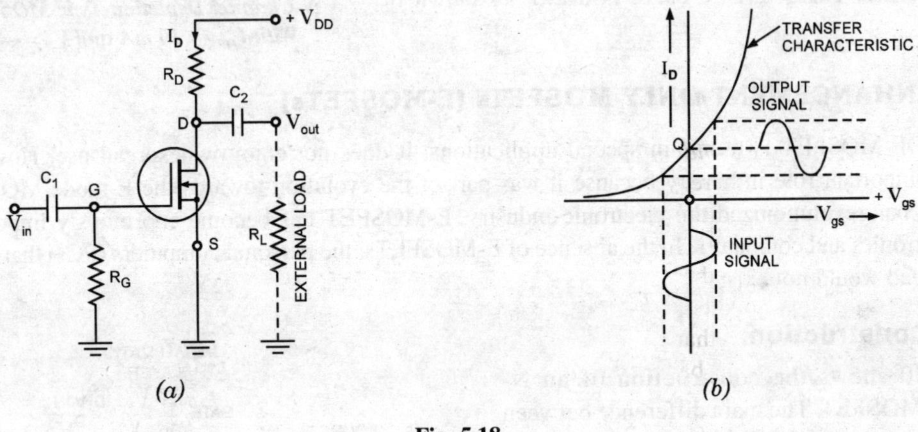

Fig. 5.18

The zero bias of Fig. 5.18 (*a*) is unique with depletion-mode MOSFETs. Although any of the JFET biasing method will work with a depletion-mode MOSFET, the zero-bias method shown in Fig. 5.18 (*a*) is preferred as it does the job simply and adequately.

After biasing the depletion-mode MOSFET to a Q-point, it can amplify small signals. The JFET formulas for voltage gain can be applied directly to a MOSFET amplifier. But like the JFET, the depletion-mode MOSFET has a relatively low voltage gain. MOSFETs have excellent low-noise properties, a definite advantage for any stage near the front end of a system where the signal is weak, such as a TV receiver. As with a JFET, the transconductance g_m in case of MOSFET can be controlled by varying the gate-source voltage V_{gs}. Due to the fact that a DE-MOSFET is normally an *on* device, it is also possible to use self-bias by adding a source resistor. The operation becomes the same as a self-biased JFET circuit.

Example 5.13. Sketch the transfer characteristic for an N-channel depletion type MOSFET with I_{DSS} = 10 mA and V_P = – 4 V.

Solution : From Eq. (5.1)

At $V_{GS} = 0$, $I_D = I_{DSS} = 10$ mA

At $V_{GS} = 1$, $I_D = 10 \text{ mA} \left(1 - \frac{+1\text{ V}}{-4\text{ V}}\right)^2 = 15.625$ mA

At $V_{GS} = -1$ V, $I_D = 10 \text{ mA} \left(1 - \frac{1}{4}\right)^2 = 5.625$ mA

At $V_{GS} = -2$ V, $I_D = 10 \text{ mA} \left(1 - \frac{2}{4}\right)^2 = 2.5$ mA

At $V_{GS} = -4$ V $I_D = 10 \text{ mA} \left(1 - \frac{4}{4}\right)^2 = 0$ mA

The transfer characteristic curve is drawn as shown in Fig. 5.19.

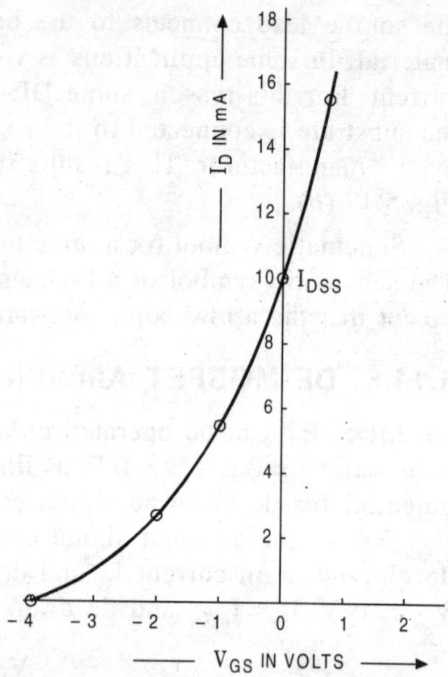

Fig. 5.19. ***Transfer Characteristic For An N-Channel Depletion Type MOSFET With I_{DSS} = 10 mA and V_P = –4 V***

5.15. ENHANCEMENT-ONLY MOSFETs (E-MOSFETs)

Although DE-MOSFET is useful in special applications, it does not enjoy widespread use. However, it played an important role in history because it was part of the evolution towards the E-mode MOSFET, a device that has revolutionized the electronic industry. E-MOSFET has become enormously important in digital electronics and computers. In the absence of E-MOSFETs, the personal computers (PCs) that are now so widespread would not exist.

5.15.1. Construction

Figure 5.20 shows the construction of an N-channel E-MOSFET. The main difference between the construction of DE-MOSFET and that of E-MOSFET, as we see from Figs. 5.13 and 5.20 respectively, is that in E-MOSFET substrate extends all the way to the silicon dioxide (SiO_2) and no channels are doped between the source and the drain. Channels are electrically induced in these MOSFETs, when a positive gate-source voltage V_{GS} is applied to it.

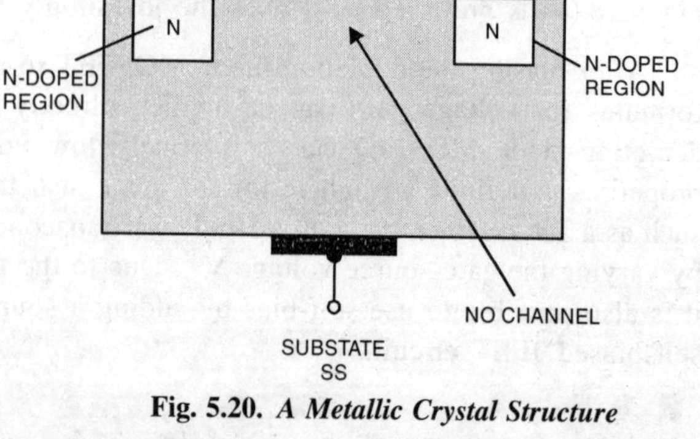

Fig. 5.20. ***A Metallic Crystal Structure***

5.15.2. Operation

As its name indicates, this MOSFET operates only in the *enhancement mode* and has no depletion mode. It operates with large positive gate voltage only. It does not conduct when gate-source voltage V_{GS} = 0. This is the reason that it is called *normally-off* MOSFET. In these MOSFETs drain current I_D flows only when

V_{GS} exceeds V_{GST} [gate-to-source threshold voltage].

When drain is applied with +ve voltage with respect to source and no potential is applied to the gate, the two N-regions and one P-substrate form two P-N junctions connected back to back with a resistance of the P-substrate. Both the junctions cannot be forward biased at the same time, so only an extremely small drain current *i.e.,* a reverse leakage current flows. If the P-type substrate is now connected to the source terminal, there is zero voltage across the source-substrate junction, and the drain-substrate junction remains reverse biased. When the gate is made positive with respect to the source and the substrate, negative (*i.e.,* minority) charge carriers within the substrate are attracted to the +ve gate and accumulate close to the surface of the substrate. As the gate voltage is increased, more and more electrons accumulate under the gate. Since these electrons cannot flow across the insulated layer of silicon dioxide to the gate, they accumulate at the surface of the substrate just below the gate. These accumulated minority charge carriers make an N-type channel stretching from drain to source. When this occurs, a channel is induced by forming what is termed an *inversion layer* (N-type). Now a drain current starts flowing. The strength of the drain current depends upon the channel resistance, which in turn depends on the number of charge carriers attracted to the positive gate. Thus drain current is controlled by the gate potential.

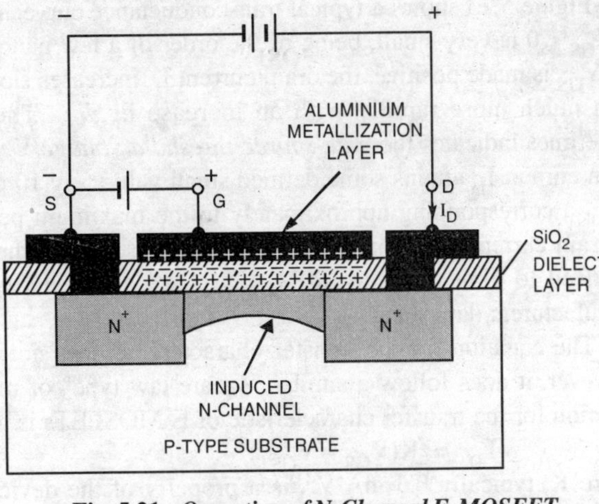

Fig. 5.21. ***Operation of N-Channel E-MOSFET***

Since the conductivity of the channel is enhanced by the +ve bias on the gate, this device is called the *enhancement MOSFET* or E-MOSFET.

The minimum value of gate-to-source voltage V_{GS} that is required to form the inversion layer (N-type) is termed the *gate-to-source threshold voltage* V_{GST}. For V_{GS} below V_{GST}, the drain current $I_D = 0$. But for V_{GS} exceeding V_{GST} an N-type inversion layer connects the source to drain and the drain current I_D is large. Depending upon the device being used, V_{GST} may vary from less than 1 V to more than 5 V.

JFETs and DE-MOSFETs are classified as the depletion-mode devices because their conductivity depends on the action of depletion layers. E-MOSFET is classified as an enhancement-mode device because its conductivity depends on the action of the inversion layer. Depletion-mode devices are normally ON when the gate-source voltage $V_{GS} = 0$, whereas the enhancement-mode devices are normally OFF when $V_{GS} = 0$.

5.15.3. Characteristics

Drain characteristics of an N-channel E-MOSFET are shown in Fig. 5.22. The lowest curve is the V_{GST} curve. When V_{GS} is lesser than V_{GST}, I_D is approximately zero. When V_{GS} is greater than V_{GST}, the device *turns on* and the drain current I_D is controlled by the gate voltage. The characteristic curves have almost vertical and almost horizontal parts. The almost vertical components of the curves correspond to the ohmic region, and the horizontal components correspond to the constant current region. Thus E-MOSFET can be operated in either of these regions *i.e.,* it can be used as a variable voltage resistor (VVR) or as a constant current source.

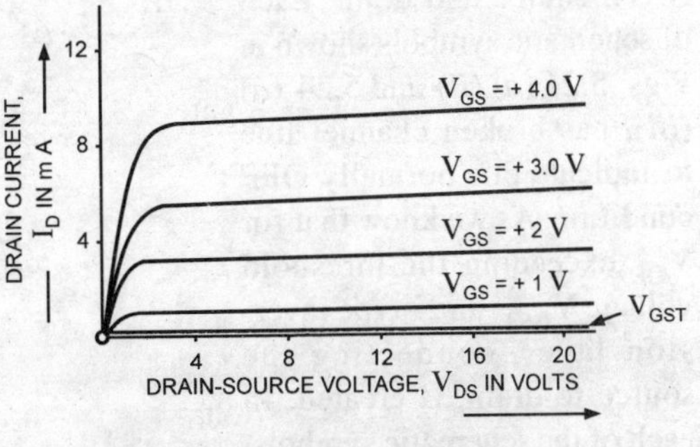

Fig. 5.22. ***Drain Characteristics***

Figure 5.23 shows a typical transconductance curve. The current I_{DSS} at $V_{GS} \leq 0$ is very small, being of the order of a few nanoamperes. When the V_{GS} is made positive, the drain current I_D increases slowly at first, and then much more rapidly with an increase in V_{GS}. The manufacturer sometimes indicates the *gate-source threshold voltage* V_{GST} at which the drain current I_D attains some defined small value, say 10 μA. The current $I_{D\,(ON)}$, corresponding approximately to the maximum permissible value of drain current given on the drain characteristics and the values of V_{GS} required to give this current $V_{GS\,(ON)}$ are also usually given on the manufacturers data sheet.

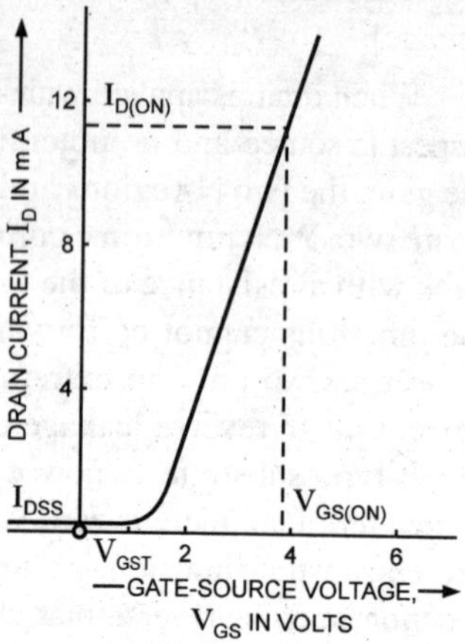

Fig. 5.23. ***Transfer Characteristic***

The equation for the transfer characteristic does not obey Eq. (5.1). However, it does follow a similar "square-law type" of relationship. The equation for the transfer characteristic of E-MOSFETs is given as

$$I_D = K(V_{GS} - V_{GST})^2 \qquad ...(5.17)$$

where K, typically 0.3 mA/V², is a property of the device construction. Note that I_{DSS} cannot be associated with an E-MOSFET, because no drain current flows with $V_{GS} = 0$ volt.

With JFETs and DE-MOSFETs, the value of I_{DSS} and $V_{GS\,(OFF)}$ are the key quantities required for analysis. With E-MOSFETs, the key quantities are $I_{D\,(ON)}$, V_{GST}, and $V_{GS\,(ON)}$ indicated in Fig. 5.23. These three quantities are the first items to look for on a data sheet. By substituting these quantities in Eq. (5.17), the equation can be arranged in a more useful form:

$$I_D = KI_{D\,(ON)} \qquad ...(5.18)$$

$$\text{where } K = \left(\frac{V_{GS} - V_{GST}}{V_{GS\,(ON)} - V_{GST}}\right)^2 \qquad ...(5.19)$$

The input impedance of the E-MOSFET is very high (of the order of 10^{15} Ω) as the gate of the E-MOSFET is insulated from the channel. Transconductance for this device typically ranges from 1,000 μS to 2,000 μS (or 1 to 2 mA per volt).

5.15.4. Schematic Symbols

Schematic symbols for an N-channel E-MOSFET are shown in Fig. 5.24 (*a*). For zero value of V_{GS}, the E-MOSFET is OFF because there is no conducting channel between source and drain. Each of schematic symbols shown in Figs. 5.24 (*a*) (*ii*) and 5.24 (*a*) (*iii*), has broken channel line to indicate this normally OFF condition. As we know that for V_{GS} exceeding the threshold voltage V_{GST}, an N-type inversion layer, connecting the source to drain, is created. In each of the schematic symbols,

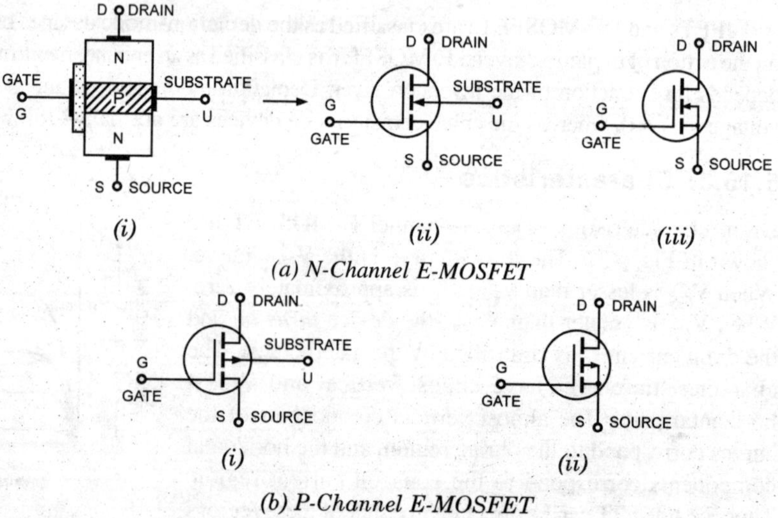

Fig. 5.24. ***Schematic Symbols For E-MOSFETs***

the arrow points to this inversion layer, which acts like an N-channel when the device is conducting. In each case, the fact that the device has an insulated gate is indicated by the gate not making direct contact with the channel. The schematic symbol shown in Fig. 5.24 (*a*) (*iii*) shows the source and substrate internally connected, while the other symbol shown in Fig. 5.24 (*a*) (*ii*) shows the substrate connection brought out separately from the source.

The schematic symbols for a P-channel E-MOSFET are shown in Fig. 5.24 (*b*). In these cases the arrow points outwards.

5.15.5. E-MOSFET Data Sheets

As already mentioned above, there are three key quantities [$I_{D\,(ON)}$, V_{GST} and $V_{GS\,(ON)}$] with E-MOSFETs. Data sheets may vary from one manufacturer to another one and one may contain some informations somewhat different from those given in another manufacturer's data sheet. But almost all data sheets include information regarding V_{GST}, $I_{D\,(ON)}$ and most include information regarding $V_{GS\,(ON)}$. Some also include information on the drain resistance R_{DS}. This resistance may be symbolised R_{DS} or it may appear as $r_{DS\,(ON)}$, $r_{ds\,(ON)}$ etc.

5.16. MOSFET RESISTOR

The MOSFET can be employed as a resistor. DE-MOSFET can be employed as a resistor by connecting the gate to the source, as shown in Fig. 5.25 (*a*), so that gate-to-source voltage V_{GS} is zero. E-MOSFET may also be employed as a resistor by connecting the gate to the drain, as shown in Fig. 5.25 (*b*) so that gate-to-source voltage V_{GS} becomes equal to drain-to-source voltage V_{DS}.

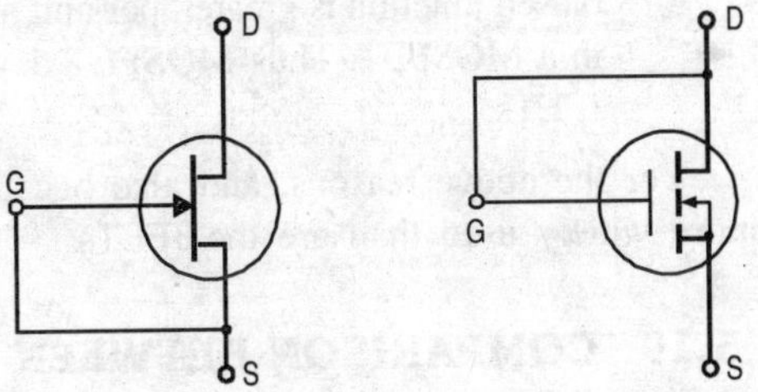

(*a*) DE-MOSFET (*b*) E-MOSFET

Fig. 5.25. *MOSFET Resistor*

5.17. COMPARISON OF N-CHANNEL FETs WITH P-CHANNEL FETs

S. No.	*Particulars*	*N-Channel FETs*	*P-Channel FETs*
1.	Current carriers	Electrons	Holes
2.	Mobility	Large	Poor
3.	Noise	Less	More
4.	Transconductance	Larger	Lesser

5.18. COMPARISON OF JFETs AND MOSFETs

JFETs and MOSFETs are quite similar in their operating principles and in their electrical characteristics. However, they differ in some aspects, as detailed below :

1. In MOSFETs (DE-MOSFETs as well as E-MOSFETs), the transverse electric field induced across an insulating layer deposited on the semiconductor material controls the conductivity of the channel whereas in the JFETs the transverse electric field across the reverse-biased PN junction controls the conductivity of the channel.
2. JFETs can only be operated in the depletion mode whereas MOSFETs can be operated in either depletion or in enhancement mode. In a JFET, if the gate is forward biased, excess-carrier injunction occurs and the gate current is substantial. Thus channel conductance is enhanced to some degree

due to excess carriers but the device is never operated with gate forward biased because gate current is undesirable.

3. MOSFETs have input impedance much higher than that of JFETs because of small leakage currents. The input impedance of the MOSFET is of the order of 10^{10} to 10^{15} Ω whereas the input impedance of JFET is of the order of 10^8 Ω.
4. JFETs have characteristic curves more flatter than those of MOSFETs indicating a higher drain resistance.

   The drain resistance of a JFET is of the order of 10^5 to 10^6 Ω whereas the drain resistance of a MOSFET is of the order of 1 to 50 kΩ.
5. MOSFET is very susceptible to overload voltage and needs special handling during installation. It may get damaged easily if not properly handled.
6. When JFET is operated with a reverse bias on the junction, the gate current I_G is larger than it would be in a comparable MOSFET. The current caused by minority carrier extraction across a reverse biased junction is greater, per unit area, than the leakage current that is supported by the oxide layer in a MOSFET. Thus MOSFET devices are more useful in electrometer applications than are the JFETs.

For the above reasons, and also because MOSFETs are somewhat easier to manufacture, they are more widely used than are the JFETs.

5.19. COMPARISON BETWEEN NMOS AND PMOS

1. P-channel is much easier and cheaper to produce than the N-channel device.
2. The drain resistance of P-channel MOSFET is three times higher than that of an identical N-channel device.
3. The N-channel MOSFET is smaller for the same complexity than the P-channel MOSFET.
4. The N-channel MOSFET has the higher packing density which makes it faster in switching applications due to the smaller junction areas and low inherent capacitances.
5. A P-channel MOSFET occupies a larger area than the N-channel MOSFET for the given drain current rating. This is because the electron mobility is 2.5 times of that of an hole.
6. The N-channel MOSFET has higher false turn-on possibility than the P-channel device because of the positively charged contaminants.

5.20. UNIJUNCTION TRANSISTOR (UJT)

Unijunction transistor (abbreviated as UJT), also called the *double-base diode,* is a 2 layer, 3 terminal solid-state (silicon) switching device. The device has a unique characteristic that when it is triggered, its emitter current increases regeneratively (due to negative resistance characteristic) until it is restricted by emitter power supply. The low cost per unit, combined with its unique characteristics, have warranted its use in a wide variety of applications. A few include oscillators, pulse generators, sawtooth generators, trigger circuits, phase control, timing circuits, and voltage- or current-regulated supplies. The device is, in general, a low-power absorbing device under normal operating conditions and provides tremendous aid in the continual effort to design relatively efficient systems.

5.20.1. Construction

The basic structure of a unijunction transistor is shown in Fig. 5.26 *(a)*. It essentially consists of a lightly-doped N-type silicon bar with a small piece of heavily doped P-type material alloyed to its one side to produce single P-N junction. The single P-N junction accounts for the terminology *unijunction*. The silicon bar, at its ends, has two ohmic contacts designated as base-1 (B_1) and base-2 (B_2), as shown and the P-type region is termed the emitter (E). The emitter junction is usually located closer to base-2 (B_2) than base-1 (B_1) so that the device is not symmetrical, because symmetrical unit does not provide optimum electrical characteristics for most of the applications.

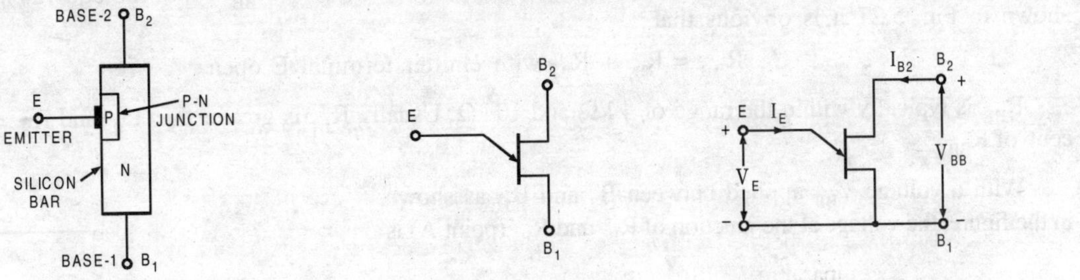

(a) Basic Structure *(b) Schematic Symbol* *(c) Basic Arrangement*

Fig. 5.26. ***Unijunction Transistor***

The symbol for unijunction transistor is shown in Fig. 5.26 (*b*). The emitter leg is drawn at an angle to the vertical line representing the N-type material slab and the arrowhead points in the direction of *conventional current* when the device is forward biased, active or in the conducting state. The basic arrangement for the UJT is shown in Fig. 5.26 (*c*).

A complementary UJT is formed by diffusing an N-type emitter terminal on a P-type base. Except for the polarities of voltage and current, the characteristics of a complementary UJT are exactly the same as those of a conventional UJT.

The noteworthy points about UJT are given below :

1. The device has only one junction, so it is called the *unijunction* device.
2. The device, because of one P-N junction, is quite similar to a diode but it differs from an ordinary diode that it has three terminals.
3. The structure of a UJT is quite similar to that of an N-channel JFET. The main difference is that P-type (gate) material surrounds the N-type (channel) material in case of JFET and the gate surface of the JFET is much larger than emitter junction of UJT.
4. In a unijunction transistor the emitter is heavily doped while the N-region is lightly doped, so the resistance between the base terminals is relatively high, typically 4 to 10 kΩ when the emitter is open.
5. The N-type silicon bar has a high resistance and the resistance between emitter and base-1 is larger than that between emitter and base-2. It is because emitter is closer to base-2 than base-1.
6. UJT is operated with emitter junction forward biased while the JFET is normally operated with the gate junction reverse biased.
7. UJT does not have ability to amplify but it has the ability to control a large ac power with a small signal.
8. It exhibits a negative resistance characteristic and so it can be employed as an oscillator.

5.20.2. Equivalent Circuit

The equivalent circuit of a UJT is shown in Fig. 5.27. The P-N junction is represented in the emitter by a diode.

Silicon bar being lightly doped has a high resistance and can be represented as two resistors connected in series–R_{B_1} from base-1 to point A and R_{B_2} from base-2 to point A, as shown in Fig. 5.27. The resistance R_{B_1} is shown as a variable resistor since its magnitude varies with the emitter current I_E. In fact, for a typical UJT, R_{B_1} may vary from 5 kΩ down to 50 Ω for a corresponding change of emitter current I_E from 0 to 50 μA. The total resistance of silicon bar from one end to the other end *i.e.,* the resistance between B_2 and B_1 with emitter terminal E open is called the *interbase resistance* R_{BB}. From the equivalent circuit shown in Fig. 5.27 it is obvious that

$$R_{BB} = R_{B_1} + R_{B_2} \text{ with emitter terminal E open} \quad ...(5.20)$$

R_{BB} is typically within the range of 4 kΩ and 10 kΩ. Usually R_{B_1} is greater than R_{B_2} and R_{B_1} = 60 per cent of R_{BB}.

With a voltage V_{BB} applied between B_2 and B_1, as shown in the figure, the voltage at the junction of R_{B_1} and R_{B_2} (point A) is

$$V_A = V_{BB} \times \frac{R_{B_1}}{R_{B_1} + R_{B_2}} = \frac{R_{B_1}}{R_{BB}} V_{BB} \quad ...(5.21)$$

$$= \eta V_{BB} \quad ...(5.22)$$

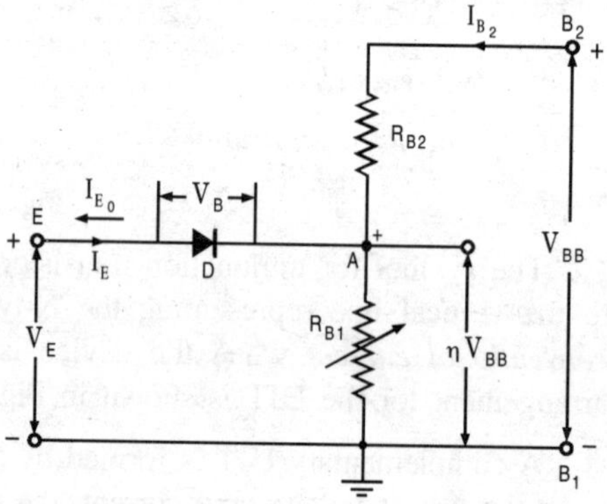

Fig. 5.27. *Equivalent Circuit of a UJT*

The Greek letter η (eta) is called the *intrinsic stand-off* ratio of the device and is defined as

$$\eta = \frac{R_{B_1}}{R_{B_1} + R_{B_2}} \text{ with terminal E open}$$

$$= \frac{R_{B_1}}{R_{BB}} \quad ...(5.23)$$

The intrinsic stand-off ratio is the property of a UJT and is always less than unity. The typical range of η is from 0.5 to 0.8.

Voltage V_A is also the voltage at the cathode of the diode representing the P-N junction. With the emitter terminal open circuited, the only current flowing is

$$I_{B_2} = \frac{V_{BB}}{R_{BB}} \quad ...(5.24)$$

If the emitter terminal is grounded, the P-N junction is reverse biased and a small *emitter reverse current* I_{E_0} flows.

5.20.3. Operation

Imagine that the emitter supply voltage is turned down to zero. Then the intrinsic stand-off voltage reverse biases the emitter diode, as mentioned above. If V_B is the barrier voltage of the emitter diode, then the total reverse bias voltage is $V_A + V_B = \eta V_{BB} + V_B$. For silicon V_B = 0.7 V.

Now let the emitter supply voltage V_E be slowly increased. When V_E becomes equal to ηV_{BB}, I_{E_0} will be reduced to zero. With equal voltage levels on each side of the diode, neither reverse nor forward current will flow. When emitter supply voltage is further increased, the diode becomes forward biased as soon as it exceeds the total reverse bias voltage $(\eta V_{BB} + V_B)$. This value of emitter voltage V_E is called the *peak point voltage* and is denoted by V_P. When $V_E = V_P$, emitter current I_E starts to flow through R_{B_1} to ground (*i.e.,* B_1). This is the minimum current that is required to trigger the UJT. This is called the *peak point emitter current* and denoted by I_P. I_P is inversely proportional to the interbase voltage, V_{BB}. Now when the emitter diode starts conducting, charge carriers are injected into the R_{B_1} region of the bar. Since the resistance of a semiconductor material depends upon doping, the resistance of region R_{B_1} decreases rapidly due to additional charge carriers (holes). With this decrease in resistance, the voltage drop across R_{B_1} also decreases, causes the emitter diode to be more heavily forward biased. This, in turn, results in larger forward current, and consequently more charge carriers are injected causing still further reduction in the resistance of the R_{B_1} region. Thus the emitter current goes on increasing until it is limited by the emitter power supply. Since V_A decreases with the increase in emitter current, the UJT is said to have *negative resistance characteristic*.

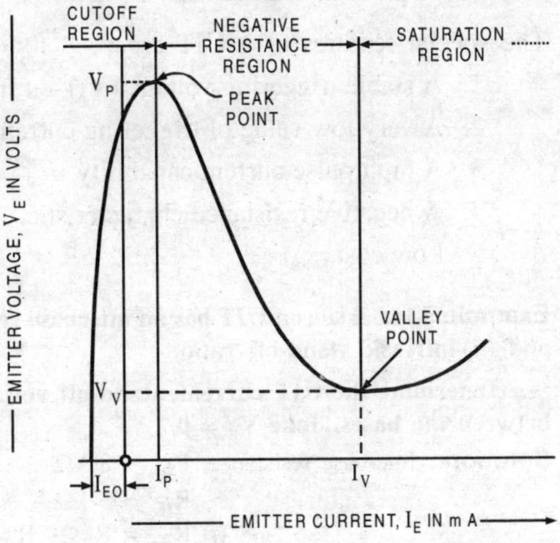

Fig. 5.28. *Static Emitter-Characteristic For a UJT*

It is seen that the base-2 (B_2) is used only for applying external voltage V_{BB} across it. Terminals E and B_1 are the active terminals. UJT is usually triggered into conduction by applying a suitable positive pulse to the emitter. It can be turned off by applying a negative trigger pulse.

5.20.4. Characteristics

The static emitter characteristic (a curve showing the relation between emitter voltage V_E and emitter current I_E) of a UJT at a given interbase voltage V_{BB} is shown in Fig. 5.28.* From Fig. 5.28 it is noted that for emitter potentials to the left of peak point, emitter current I_E never exceeds I_{EO} (measured in microamperes). The current I_{EO} corresponds very closely to the reverse leakage current I_{CO} of the conventional BJT. This region, as shown in the figure, is called the *cutoff* region. Once conduction is established at $V_E = V_P$ the emitter potential V_E starts decreasing with the increase in emitter current I_E. This corresponds exactly to the decrease in resistance R_{B_1} for increasing current I_E. This device, therefore, has a *negative resistance region* which is stable enough to be used with a great deal of reliability in the areas of applications listed earlier. Eventually, the valley point reaches, and any further increase in emitter current I_E places the device in the saturation region, as shown in the figure. In this region characteristic approaches that of a semiconductor diode in the equivalent circuit shown in Fig. 5.27. Three other important parameters for the UJT are I_P, V_V and I_V (shown in the Fig. 5.28) and are defined below :

Peak Point Emitter Current I_P. It is the emitter current at the peak point. It represents the minimum current that is required to trigger the device (UJT). It is inversely proportional to the interbase voltage V_{BB}.

Valley Point Voltage V_V. The valley point voltage is the emitter voltage at the valley point. The valley voltage increases with the increase in interbase voltage V_{BB}.

Valley Point Current I_V. The valley point current is the emitter current at the valley point. It increases with the increase in interbase voltage V_{BB}.

* V-I characteristic of UJT is one of the very few curves, in which current I is taken on X-axis and voltage V is taken on Y-axis, rather say, it is a I-V characteristic curve.

5.20.5. Special Features of UJT

The special features of a UJT are :

1. A stable triggering voltage (V_p)—a fixed fraction of applied interbase voltage V_{BB}.
2. A very low value of triggering current.
3. A high pulse current capability.
4. A negative resistance characteristic.
5. Low cost.

Example 5.14. A silicon UJT has an interbase resistance of 8 kΩ and R_{B_1} of 5 kΩ with $I_E = 0$. Determine (*i*) R_{B_2} and (*ii*) intrinsic stand-off ratio.

Determine the UJT current, stand-off voltage and peak-point voltage when a voltage of 15 V is applied between the bases. Take $V_B = 0.7$ V.

Solution : Interbase resistance, $R_{BB} = 8 \text{ k}\Omega$

$$R_{B_1} = 5 \text{ k}\Omega$$

(*i*) $R_{B_2} = R_{BB} - R_{B_1} = 8 - 5 = 3 \text{ k}\Omega$ **Ans.**

(*ii*) Intrinsic stand-off ratio, $\eta = \dfrac{R_{B_1}}{R_{BB}} = \dfrac{5}{8} = 0.625$ **Ans.**

Assuming V_E less than peak point voltage

$$\text{UJT current, } I_{B_1} = I_{B_2} = \frac{V_{BB}}{R_{BB}} = \frac{15}{8 \text{ k}\Omega} = 1.875 \text{ mA } \textbf{Ans.}$$

$$\text{Stand-off voltage} = \eta V_{BB} = 0.625 \times 15 = 9.375 \text{ V } \textbf{Ans.}$$

$$\text{Peak point voltage, } V_P = \eta V_{BB} + V_B = 9.375 + 0.7 = 10.075 \text{ V } \textbf{Ans.}$$

5.21. UJT TRIGGERING OF AN SCR

One common application of the unijunction transistor is the triggering of the other devices such as the SCR, triac etc. The basic elements of such a triggering circuit are shown in Fig. 5.29. The resistor R_E is chosen so that the load line determined by R_E passes through the device characteristic in the negative resistance region *i.e.*, to the right of the peak point but to the left of the valley point, as shown in Fig. 5.30. If the load line does not pass to the right of the peak point P, the device cannot turn *on*.

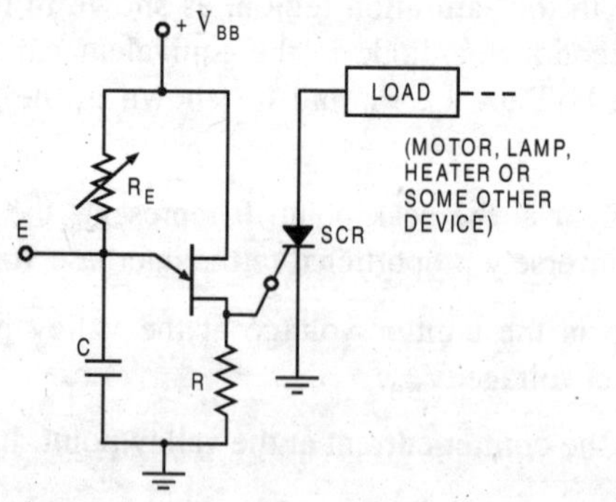

Fig. 5.29. ***UJT Triggering of an SCR***

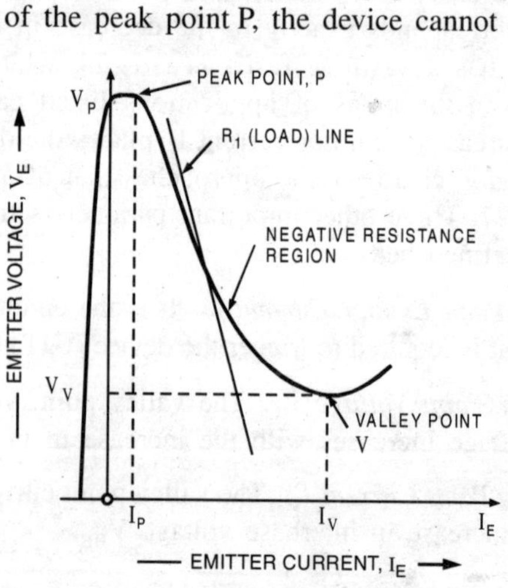

Fig. 5.30

For ensuring turn-on of UJT $R_E < \dfrac{V_{BB} - V_P}{I_P}$ ...(5.25)

This can be established as below :

Consider the peak point at which $I_{R_E} = I_P$ and $V_E = V_P$ (the equality $I_{R_E} = I_P$ is valid because the charging current of capacitor at this instant is zero *i.e.*, the capacitor, at this particular instant, is changing from a charging state to a discharging state). Then $V_E = V_{BB} - I_{R_E} R_E$.

$$\text{So } R_{E\,(max)} = \frac{V_{BB} - V_E}{I_{R_E}} = \frac{V_{BB} - V_P}{I_P} \text{ at the peak point.}$$

$$\text{or } R_E < \frac{V_{BB} - V_P}{I_P}$$

At the valley point, V $I_E = I_V$ and $V_E = V_V$ so that

$$V_E = V_{BB} - I_{R_E} R_E$$

$$\text{So } R_{E\,(min)} = \frac{V_{BB} - V_E}{I_{R_E}} = \frac{V_{BB} - V_V}{I_V}$$

$$\text{or for ensuring turn-off } R_E > \frac{V_{BB} - V_V}{I_V} \quad \text{...(5.26)}$$

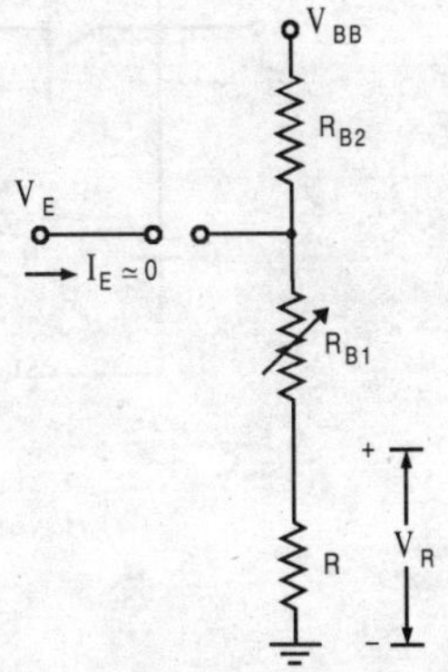

Fig. 5.31

So the range of resistor R_E is given as

$$\frac{V_{BB} - V_P}{I_P} > R_E > \frac{V_{BB} - V_V}{I_V}$$

The resistor R is chosen small enough so as to ensure that SCR is not turned *on* by voltage V_R when emitter terminal E is open or $I_E = 0$ (Fig. 5.31).

$$\text{The voltage } V_R = \frac{RV_{BB}}{R + R_{BB}} \text{ for open-emitter terminal.}$$

The capacitor C determines the time interval between triggering pulses and the time duration of each pulse.

By varying R_E, we can change the time constant $R_E C$ and alter the point at which the UJT fires. This allows us to control the conduction angle of the SCR, which means the control of load current.

5.22. UJT RELAXATION OSCILLATOR

The relaxation oscillator shown in Fig. 5.32 (*a*) consists of UJT and a capacitor C which is charged through resistor R_E when interbase voltage V_{BB} is switched *on*. During the charging period, the voltage across the capacitor increases exponentially until it attains the peak point voltage V_P. When the capacitor voltage attains voltage V_P, the UJT switches *on* and the capacitor C rapidly discharges via B_1. The resulting current through the external resistor R develops a voltage spike, as illustrated in Fig. 5.32 (*b*) and the capacitor voltage drops to the value V_V. The device then cuts off and the capacitor commences charging again. The cycle is repeated continually generating a sawtooth waveform across capacitor C. The resulting waveforms of capacitor voltage V_C and the voltage across resistor $R(V_R)$ are shown in Fig. 5.32 (*b*). The frequency of the output sawtooth wave can be varied by varying the value of resistor R_E as it controls the time constant ($T = R_E C$) of the capacitor charging circuit. The discharge time t_2 is difficult to calculate because the UJT is in its negative resistance region and its resistance is continually changing. However, t_2 is normally very much less than t_1 and can be neglected for approximation.

For satisfactory operation of the above oscillator, the following two conditions for the turn-on and turn-off of the UJT must be met, as derived in Art. 5.21 {Eqs. (5.25) and (5.26)}.

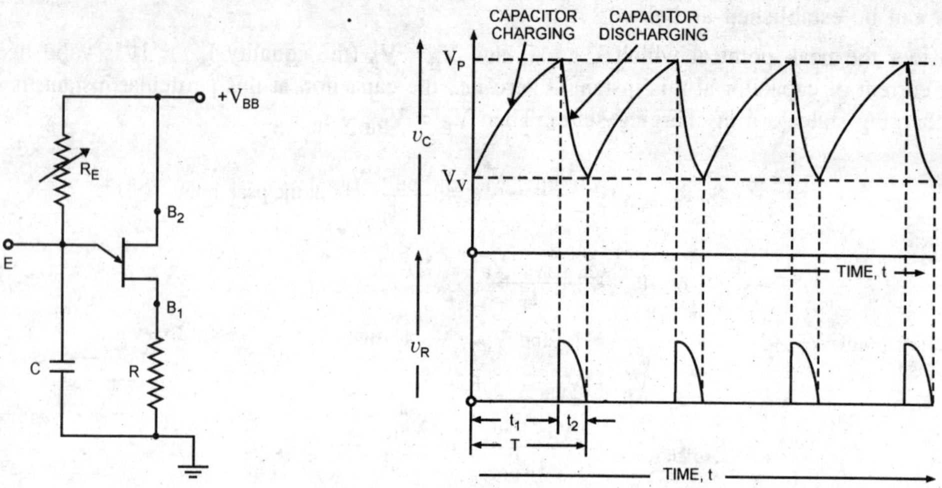

(a) Basic Circuit (b) Voltage Waveforms across C and R

Fig. 5.32. ***UJT Relaxation Oscillator***

$$R_E < \frac{V_{BB} - V_P}{I_P} \quad \text{and} \quad R_E > \frac{V_{BB} - V_V}{I_V}$$

i.e., the range of resistor R_E should be as given below

$$\frac{V_{BB} - V_P}{I_P} > R_E > \frac{V_{BB} - V_V}{I_V}$$

The time period and, therefore, frequency of oscillation can be derived as below.

During charging of capacitor, the voltage across the capacitor is given as

$$v_C = V_{BB}\ (1 - e^{-t/R_EC})$$

where R_EC is the time constant of the capacitor charging circuit and t is the time from the commencement of the charging.

The discharge of the capacitor commences at the end of charging period t_1, when the voltage across the capacitor v_C becomes equal to V_P *i.e.*, $(\eta\ V_{BB} + V_B)$

$$V_P = \eta V_{BB} + V_B = V_{BB}\,(1 - e^{-t_1/R_EC})$$

Neglecting V_B in comparison to ηV_{BB} we have

$$\eta V_{BB} = V_{BB}\,(1 - e^{-t_1/R_EC})$$

$$\text{or} \quad e^{-t_1/R_EC} = 1 - \eta$$

$$\text{or} \quad e^{t_1/R_EC} = \frac{1}{1-\eta}$$

$$\text{or} \quad t_1 = R_EC \log_e \frac{1}{1-\eta}$$

So charging time period, $t_1 = 2.3 R_EC \log_{10} \dfrac{1}{1-\eta}$

Since discharging time duration t_2 is negligibly small as compared to charging time duration t_1, so taking time period of the wave, $T = t_1$.

Time period of the sawtooth wave,

$$T \cong 2.3R_EC \log_{10} \frac{1}{1-\eta}$$

and frequency of oscillation $f = \dfrac{1}{T} \simeq \dfrac{1}{2.3R_EC \log_{10} \dfrac{1}{1-\eta}}$ ...(5.27)

By including a small resistor in each base circuit, three useful outputs (sawtooth waves, positive triggers, and negative triggers), as shown in Fig. 5.33, can be obtained.

When the UJT fires, the sudden surge of current through B_1 causes a voltage drop across R_1 and produces the positive going spikes. Also, at the UJT firing time, the fall of V_{EB_1} causes I_{B_2} to rise rapidly and generate the negative going spikes across R_2, as shown in Fig. 5.33. R_1 and R_2 should be much smaller than R_{BB} to avoid altering the firing voltage of the UJT. A wide range of oscillation frequencies can be achieved by making R_E adjustable and including a switch to select different values of capacitance, as illustrated.

As already mentioned in Art. 5.21, there is upper and lower limits to the signal source resistance R_E for the satisfactory operation of the UJT [refer to Eqs. (5.25) and (5.26)].

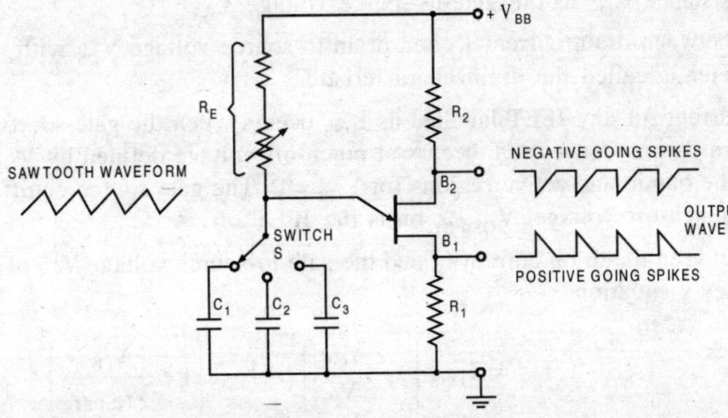

Fig. 5.33. ***Variable Frequency UJT Relaxation Oscillator***

Example 5.15. A UJT relaxation oscillator circuit shown in Fig. 5.32 *(a)* has R_E = 60 kΩ and C = 0.25 µF. Determine the pulse repetition frequency. Take intrinsic stand-off ratio to be 0.65.

Solution : From Eq. (5.27),

$$\text{Pulse repetition frequency, } f = \frac{1}{2.3\, R_EC \log_{10} \dfrac{1}{1-\eta}} = \frac{1}{2.3 \times 60 \times 10^3 \times 0.25 \times 10^{-6} \log_{10} \dfrac{1}{1-0.65}}$$

$$= 63.5 \text{ Hz } \textbf{Ans.}$$

5.23. UJT AS OVERVOLTAGE DETECTOR

A simple dc overvoltage detector circuit is given in Fig. 5.34. It operates on the fact that the device remains switched *off* as long as the input voltage remains less than the peak point voltage V_P of the UJT and as soon as input voltage exceeds V_P, the device is switched *on*.

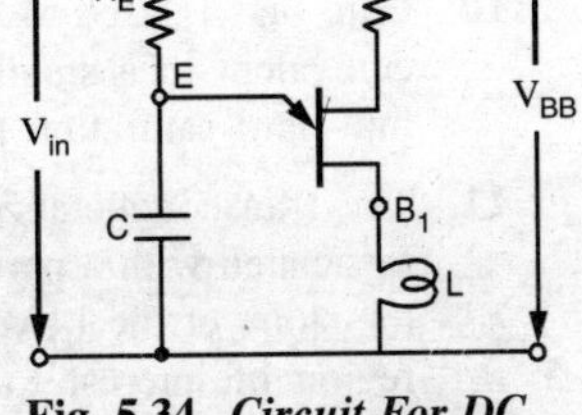

Fig. 5.34. ***Circuit For DC Voltage Detector***

Let the input voltage V_{in} to be kept constant. For it V_{BB} is to be chosen in such a way that this is slightly lesser than V_{in}. As soon as V_{in} exceeds V_P, then UJT is switched *on*, and consequently the capacitor commences discharging through low resistance path E and B_1. The current flowing in pilot lamp L lights it, which is an indication of an overvoltage across the circuit.

Highlights

1. The FET is a majority carrier device, and is, therefore, often called a unipolar transistor. Like its bipolar counterpart, the FET is a three-terminal (namely drain, source and gate) semiconductor device in which current conduction is by one type of majority carriers (either electrons or holes).
2. Field-effect devices are controlled by an electric field (or voltage) and give an extremely high input impedance—most important advantage over BJT devices. FETs are of several types such as JFETs, MESFETs and MISFETs.
3. The JFET is a voltage controlled device and has two diodes; the gate-source diode and the gate-drain diode. For normal operation, the gate-source diode is reverse biased. Then the gate voltage controls the drain current.
4. The JFET has a high input impedance (of the order of 100 MΩ), a high degree of isolation between input and output, and a low noise figure. The main disadvantage of a JFET is its relatively low gain-bandwidth product, due to the junction capacitive effects.
5. The JFET can actually be used as a voltage-controlled resistor because of unique sensitivity of the drain-to-source resistance R_{DS} to the gate-to-source voltage V_{GS}.
6. The curve drawn between drain current I_D and drain-to-source voltage V_{DS} with gate-to-source voltage V_{GS} as the parameter is called the drain characteristic.

   Maximum drain current for any JFET labelled as I_{DSS} occurs when the gate-source voltage V_{GS} is zero. The minimum drain current for a JFET occurs at pinch-off voltage defined by $V_{GS} = V_P$. The pinch-off voltage separates the ohmic and active regions for $V_{GS} = 0$. The gate-source cutoff voltage has the same magnitude as the pinch-off voltage. $V_{GS(off)}$ turns the JFET off.
7. The relationship between the drain current I_D and the gate-to-source voltage V_{GS} of a JFET is a nonlinear defined by Shockley's equation

$$I_D = I_{DSS}\left[1-\frac{V_{GS}}{V_P}\right]^2 = I_{DSS}\left[1-\frac{V_{GS}}{V_{GS(off)}}\right]^2$$

   As the current level approaches I_{DSS}, the sensitivity of I_D to variations in V_{GS} increases significantly.
8. The curve drawn between drain current I_D and gate-to-source voltage V_{GS} for a given value of drain-to-source voltage V_{DS} is called the transfer characteristic. The transfer characteristics are the characteristics of the device itself and are not sensitive to the circuit in which the JFET is used. The drain current increases more rapidly as V_{GS} approaches zero. Because the drain current equation contains a squared quantity, JFETs are referred to as square-law devices. The normalized transconductance curve indicates that I_D equals one-quarter of the maximum when V_{GS} equals half of the cutoff $\left(i.e.\, I_D = \frac{I_{DSS}}{4} \text{ when } V_{GS} = \frac{V_P}{2}\right)$ and at a point where $I_D = \frac{I_{DSS}}{2}$, $V_{GS} \cong 0.3$ V.
9. Maximum operating conditions are determined by the product of the V_{DS} and I_D.
10. Gate bias is used to bias a JFET in the ohmic region. When it operates in the ohmic region, a JFET is equivalent to a small resistance of R_{DS}. To ensure operation in the ohmic region, the JFET is driven into hard saturation by using $V_{GS} = 0$ and $I_{D(sat)} \ll I_{DSS}$.
11. The transconductance parameter g_m is determined by the ratio of the change in drain current ΔI_D associated with a particular change in gate-to-source voltage ΔV_{GS} in the region of interest. The steeper the slope of the I_D versus V_{GS} curve, the greater is the value of g_m. In addition, the closer the point or region of interest to the saturation current I_{DSS}, the greater is the transconductance parameter. The quantity g_m increases as V_{GS} approaches zero. Data sheets may list g_m or g_{fs} in siemens.

12. JFETs are mainly small-signal devices because most JFETs have a power rating of less than 1 W. When reading data sheets, start with the maximum ratings. Sometimes data sheets omit the minimum $V_{GS\ (off)}$ or other parameters. The large spread in JFET parameters justifies using ideal approximations for preliminary analysis and troubleshooting.

13. The operation of a metal-semiconductor field-effect transistor (MOSFET) is identical to that of a JFET. In the MOSFET, however, a metal-semiconductor rectifying contact, instead of a P-N junction, is used for the gate electrode.

14. The MOSFET is a field-effect transistor in which the metal gate is insulated by a very thin oxide layer from the semiconductor channel. It is acually a four-terminal device, with the substrate or body being the fourth terminal.

    There are two types of MOSFETs—the N-channel in which current is due to the flow of electrons in the inversion layer, and the P-channel in which current is due to the flow of holes in the inversion layer. Each of these devices can be either enhancement mode (E-mode) in which device is normally "off" and is turned on by applying a gate voltage, or depletion mode (DE-mode) in which the device is normally "on" and is turned off by applying a gate voltage.

15. The current in a MOSFET is due to the flow of carriers in the inversion layer between the source and drain terminals. The inversion layer charge density and channel conductance are controlled by the gate voltage, which means that the channel current is also controlled by the gate voltage.

16. The depletion-mode MOSFET, has a source, gate, and drain. The gate is insulated from the channel. Because of this, the input resistance is very high. The DE-MOSFET has limited use, mainly in RF circuits.

17. The drain characteristic curves for a DE-MOSFET are similar to those of a JFET when operating in depletion mode. Unlike JFETs, DE-MOSFETs can also operate in the enhancement mode. When operating in enhancement mode, the drain current is greater than I_{DSS} (*i.e.*, $I_D > I_{DSS}$). DE-MOSFETs are mainly employed as RF amplifiers. They have good high frequency response, generate low levels of electrical noise, and maintain high input impedance values when V_{GS} is negative or positive.

18. The enhancement type MOSFET is normally off. When the gate voltage is equal to the threshold voltage (V_{GST}), an N-type inversion layer connects the source to the drain. When the gate voltage exceeds the V_{GST}, the device conducts heavily. Because of the thin insulating layer, MOSFETs are easily destroyed unless precautions are taken in their handling.

19. Since the E-MOSFET is primarily a switching device, it usually operates between cutoff and saturation. When it is biased in the ohmic region, it acts like a small resistance.

20. Unijunction transistor (UJT) is a 2 layer, 3 terminal solid state (silicon) switching device. One common application of the UJT is the triggering of other devices like SCR, triac etc. Other applications include oscillators, pulse generators, sawtooth generators, phase control, timing circuits, and voltage- or current-regulated supplies.

Exercises

1. With the help of neat diagram, describe the construction and operation of a JFET.
   [G.G.S.I.P. Univ. Analog Electronics December 2009]
2. Draw the basic structure of an N-channel JFET. Draw and explain the drain characteristics qualitatively.
   [G.G.S.I.P. Univ. Analog Electronics December 2009]
3. Explain the principle of operation of an N-channel JFET. Draw a family of drain characteristics and mark the various regions of operation. **[G.G.S.I.P. Univ. Analog Electronics December 2010]**
4. Describe construction, working and characteristics of N-channel JFET.
   [Pb. Technical Univ. Electronic Devices and Circuits May 2009]

5. Sketch the basic structure of an N-channel JFET and draw the volt-ampere (V/I) characteristics and explain about each region of the volt-ampere (V/I) characteristics qualitatively.
[U.P. Technical Univ. Solid State Devices & Circuits 2009-10]

6. Explain the construction, operation and volt-ampere characteristics of a JFET.
[G.G.S.I.P. Univ. Analog Electronics-I, 2011-12]

7. Why JFET is called voltage controlled device ? Draw the structure and output characteristics of P-channel JFET. Indicate different regions in the characteristic and explain its significance.
[G.B. Technical Univ. Electronics Engineering Second Semester 2010-11]

8. Define transconductance (g_m), output resistance (r_p) and gain of a FET. How these parameters are determined graphically ? **[U.P. Technical Univ. Electronics Engineering Second Semester 2007-08]**

9. Compare the features of BJT and FET. **[G.G.S.I.P. Univ. Analog Electronics, 2011-12]**

10. Prove that the transconductance g_m of a JFET is given by

$$g_m = \frac{2}{|V_P|}\sqrt{I_{DS}\, I_{DSS}}$$

where V_P = pinch-off voltage, I_{DS} = drain current, I_{DSS} = maximum, I_{DS} corresponding to V_{GS} = 0 volt.

11. Explain the working of a JFET. Define the parameters of a JFET and develop its equivalent circuit.
[M.D. Univ. Rohtak Analog Electronics December 2008]

12. Define transconductance (g_m), output resistance (r_p) and gain of a FET. How these parameters are determined graphically. **[U.P. Technical Univ. Electronics Engg. Second Semester 2007-08]**

13. Explain the meaning of pinch-off voltage in a JFET. How does the current flow in a JFET after pinch-off ? Explain the meaning of the terminologies (*i*) transconductance, g_m (*ii*) drain resistance, r_d and (*iii*) amplification factor, μ of a JFET. **[U.P. Technical Univ. Solid State Devices & Circuits 2008-09]**

14. Explain the characteristics of N-channel depletion type MOSFET.
[G.G.S.I.P. Univ. Analog Electronics December 2010]

15. Draw the structure of an N-channel depletion type MOSFET. Explain its working with the help of output drain characteristics and transfer characteristics. **[G.G.S.I.P. Univ. Analog Electronics December 2010]**

16. Draw the structure of an N-channel JFET and explain its principle of operation. Also draw its drain and tarnsfer characteristics with the help of suitable circuit. Describe how an FET can be used as a voltage-controlled resistor. **[G.B. Technical Univ. Electronics Engineering First Semester 2011-12]**

17. Explain the MOSFET structure and operation with neat sketch. Also show that how it can be used as voltage controlled linear resistor. **[U.P. Technical Univ. Solid State Devices & Circuits 2007-08]**

18. Draw a schematic diagram of a MOSFET and explain its mechanism of operation. Sketch the drain current-drain voltage characteristics of the device and explain the meaning of threshold voltage.
[U.P. Technical Univ. Solid State Devices & Circuits 2008-09]

19. Draw the structure and transfer curve for enhancement MOSFET and depletion MOSFET. Choose P or N type. **[G.G.S.I.P. Univ. Analog Electronics December 2009]**

20. Draw the cross section of a P-channel enhancement MOSFET and explain its characteristics.
[G.G.S.I.P. Univ. Analog Electronics-I December 2004]

21. What is the significant difference between the construction of an Enhancement-type MOSFET and depletion type MOSFET ? Sketch the basic construction of a P-channel depletion type MOSFET.
[Mahamaya Technical Univ. Electronics Engineering Second Semester 2010-11]

22. Draw a neat sketch of the terminal drain current versus drain voltage characteristics of an N-channel enhancement mode MOSFET and explain the characteristics. **[U.P.S.C. I.E.S. E.&T.E.-I, 2007]**

23. Draw and label cross-sectional view of depletion and enhancement type MOSFETS. Draw the drain characteristics of both. **[M.D. Univ. Rohtak Analog Electronics May 2008]**

24. Explain the transfer characteristics for N-channel and P-channel MOSFETs alongwith their output characteristics of an N-channel enhancement MOSFET. **[U.P.S.C. I.E.S. Electrical Engineering-II 2006]**

25. Draw a neat cross-sectional view of an N-channel MOSFET and explain the modes of operation of this device with typical current-voltage characteristics. What is threshold voltage ? What techniques are usually followed for lowering this parameter for a MOSFET ?

26. What is the significant difference between the construction of an enhancement type MOSFET and depletion type MOSFET ? Explain with suitable diagrams.

[Mahamaya Technical Univ. Electronics Engineering First Semester 2011-12]

27. Explain the difference between the enhancement mode and depletion mode MOSFETs.

[U.P.S.C. I.E.S Elec. Engineering-II, 1997]

28. Sketch the cross-sectional view of an enhancement mode MOSFET. Explain its operation and characteristics.

[U.P.S.C. I.E.S. Electrical Engineering II, 1998]

29. Draw diagrams showing constructional details of N-channel MOSFET. Explain the operation of N-channel MOSFET in depletion mode. **[Pb. Technical Univ. Electronic Devices & Circuits May 2008]**

30. Explain the output and transfer characteristics of enhancement and depletion type MOSFETs.

[G.G.S.I.P. Univ. Analog Electronics 2011-12]

31. Distinguish between JFET and MOSFET. **[U.P.S.C. I.E.S. Elec. Engineering-II, 1995]**

32. Give a comparison between JFET and MOSFET. **[G.G.S.I.P. Univ. Analog Electronics-I, December 2004]**

33. Write short notes on the following :

(*i*) JFET. **[M.D. Univ. Analog Electronics May 2005]**

(*ii*) FET as a VVR. **[M.D. Univ. Analog Electronics May 2007]**

34. Describe the construction and working of UJT ? Also draw its characteristics.

[Pb. Technical Univ. Electronic Devices & Circuits Dec. 2008]

35. Describe the principle of operation of UJT.

Describe the intrinsic stand-off ratio with regard to UJT. **[Rajasthan Technical Univ. 2008]**

36. Explain the working principle of UJT. Also explain why UJT has a negative resistance region.

[Rajasthan Univ. 2006]

Short Answer Type Questions With Answers

Q. 1. What is the major difference between a bipolar and a unipolar device ? Explain with example.

[U.P. Technical Univ. Second Semester 2009-10]

Ans. In a bipolar device, current conduction is by both majority carriers and minority carriers (both by holes as well as electrons), such as in a case of BJT. On the other hand, in a unipolar device, the current conduction is by only one type of majority carriers (either by electrons or by holes), such as in case of FET.

Q. 2. Explain why BJTs are called bipolar devices while FETs are called unipolar devices.

Ans. In ordinary transistors, current conduction is by both majority carriers and minority carriers (both by holes as well as electrons) and, therefore, these are called bipolar transistors. In field-effect transistors current conduction is by only one type of majority carriers (either by electrons or by holes) and, therefore, these are called unipolar transistors.

Q. 3. Define "current-controlled device" and "voltage-controlled device".

Ans. A current-controlled device is one in which a current defines the operating conditions of the device, whereas a voltage-controlled device is one in which a particular voltage defines the operating conditions.

Q. 4. Why is terminology "field effect" appropriate for this terminal device ? **[Pb. Technical Univ. May 2005]**

Ans. The device is called field-effect transistor (FET) because the drain current (output current) is controlled by the effect of extension of the field associated with the depletion region developed by the reverse bias at the gate.

Q. 5. Why I_G is effectively zero ampere for a JFET transistor ? **[Pb. Technical Univ. May 2005]**

Ans. The gate channel junction in a JFET is a P-N junction and since it is normally reverse biased, a minority charge carrier current flows. This is gate-source cutoff current, known as gate reverse current. So the gate current, I_G can be said to be effectively zero ampere.

Q. 6. What do you understand by terms channel and drain in JFET ? **[Pb. Technical Univ. May 2008]**

Ans. ***Channel.*** The region between the source and drain, sandwiched between the two gates is called the *channel* and the majority carriers move from source to drain through this channel.

Drain. The terminal, through which the majority carriers leave the channel, is called the *drain* terminal D and the conventional current leaving the channel at D is designated as I_D.

Q. 7. Why the channel of a JFET is never completely closed at the drain end ?

Ans. If the channel is completely closed in JFET, then there will be no drain current, so there will be no voltage drop along the channel length and amount of reverse bias will become uniform and wedge-shaped depletion region will become rectangular one.

Q. 8. How is drain current controlled in a JFET ?

Ans. In a JFET drain current is controlled by controlling the reverse bias given to its gate (*i.e.,* V_{GS}).

Q. 9. What is meant by drain characteristic of FETs ?

Ans. The curve drawn between drain current and drain-source voltage with gate-to-source voltage as the parameter is called the drain characteristic.

Q. 10. What happens to the drain current of a P-channel JFET when a positive voltage is applied on its gate ? **[Pb. Technical Univ. December 2008]**

Ans. Whenever the positive voltage is applied to the gate of P-channel JFET, the depletion region width increases because of which channel is narrowed, the resistance is increased and drain current is reduced.

Q. 11. What is meant by transfer characteristic of FETs ?

Ans. The curve drawn between drain current and gate-source voltage for a given value of drain-source voltage is called the transfer characteristic.

Q. 12. No current flows through the channel of a JFET when $V_{GS} = 0$ and $V_{DS} = 0$, though channel is fully open. Why ?

Ans. When drain-source voltage is zero, there is no attracting potential at the drain, so no current flows inspite of the fact that the channel is fully open.

Q. 13. Define pinch-off voltage for a JFET. **[U.P.S.C. I.E.S. 2002]**

Ans. The value of drain-source voltage at which channel is pinched off (*i.e.,* all the free charges from the channel get removed) is called the pinch-off voltage.

Q. 14. What are the parameters that control the pinch-off voltage of JFET ?

Ans. Parameters that control the pinch-off voltage of JFET are electron charge, donor/acceptor concentration density, permittivity of channel material and half width of channel bar.

Q. 15. Why FET is called a voltage controlled device ? **[Pb. Univ. April 1993]**

Ans. In a FET, drain current is controlled by the effect of the extension of the field associated with the depletion region developed by the reverse bias on the gate, so it is called a voltage controlled device.

Q. 16. How does the FET behave (*i*) for small values of $|V_{DS}|$ and (*ii*) for large values of $|V_{DS}|$? **[Pb. Univ. May 1994; U.P. Technical Univ. 2005-06]**

Ans. (*i*)FET behaves as an ordinary resistor for small value of $|V_{DS}|$ *i.e.,* in ohmic region.

(*ii*)FET behaves as a constant current source for large values of $|V_{DS}|$ till breakdown occurs.

Q. 17. How does the current vary with the gate voltage in the saturation region ? **[Pb. Univ. May 1994]**

Ans. Drain current decreases with the increase in gate-source bias $|V_{GS}|$. When $V_{GS} = 0$; drain current $I_D = I_{DSS}$, drain-source saturation current and when $V_{GS} = V_P$; drain current $I_D = 0$.

Q. 18. How does the transconductance vary with drain current ? **[Pb. Univ. May 1994]**

Ans. Transconductance g_m of a FET varies with the drain current I_D as given by the following equation

$$g_m = \frac{2}{|V_P|}\sqrt{I_{DS} \cdot I_{DSS}}$$

Q. 19. What is Shockley equation ? **[Pb. Technical Univ. May 2005]**

Ans. Drain current in the pinch-off region of JFET is given by the equation, known as Shockley equation,

$$I_D = I_{DSS}\left[1-\frac{V_{GS}}{V_P}\right]^2 = I_{DSS}\left[1-\frac{V_{GS}}{V_{GS(OFF)}}\right]^2$$

where I_D is the drain current at a given gate-source voltage V_{GS}, I_{DSS} is the drain current with gate shorted to source and V_{GS} is the gate-source cutoff voltage.

Q. 20. Explain how the mutual conductance of a FET varies with I_D and V_G. **[U.P. Technical Univ. 2006-07]**

Ans. Mutual conductance g_m of a FET is determined by the ratio of the change in drain current ΔI_D associated with a particular change in gate-to-source voltage ΔV_{GS} in the region of interest. The steeper the slope of the I_D versus V_{GS} curve, the greater is the value of g_m. The quantity g_m increases as V_{GS} approaches zero.

Q. 21. What is meant by gate-source cutoff voltage ?

Ans. The gate-source bias voltage required to reduce the drain current to zero is designated the gate-source cutoff voltage $V_{GS\ (OFF)}$. It is equal to pinch-off voltage V_P.

Q. 22. What is meant by saturation region ?

Ans. The region of drain characteristic of a FET in which drain current remains fairly constant is called the saturation or pinch-off region.

Q. 23. What is meant by drain-source saturation current I_{DSS} ?

Ans. The drain current in pinch-off or saturation region with zero gate-source voltage ($V_{GS} = 0$) is referred to the drain-source saturation current I_{DSS}.

Q. 24. What is physical significance of high input impedance of an FET ? **[Pb. Technical Univ. December 2006]**

Ans. Input circuit of an FET (gate to source) is reverse biased and the input gate current is very small (of the order of few nA), so the input impedance is very high. It permits high degree of isolation between the input and output circuits.

Q. 25. What is dynamic resistance of a JFET ?

Ans. The ratio of change in drain-source voltage to change in drain current at a given gate-source voltage is known as ac drain resistance or dynamic resistance r_d

$$i.e.,\ r_d = \frac{\Delta V_{DS}}{\Delta I_D} \text{ at constant } V_{GS}.$$

Q. 26. What is meant by transconductance with reference to JFET ?

Ans. The control that gate-source voltage has over the drain current is measured by the transconductance of a JFET. It may be defined as the ratio of change in drain current to the change in gate-source voltage for a given value of drain-source voltage *i.e.,*

$$g_m = \frac{\Delta I_D}{\Delta V_{GS}} \text{ at constant } V_{DS}.$$

Q. 27. Why is MOSFET called sometimes IGFET ?

Ans. MOSFET is constructed with the gate terminal insulated from the channel so it is sometimes called insulated gate FET (or IGFET).

Q. 28. What is meant by gate-to-source threshold voltage in E-MOSFET ?

Ans. The minimum value of gate-to-source voltage V_{GS} that is required to form the inversion layer is termed the gate-to-source threshold voltage V_{GST}.

Q. 29. Name the factors which make the JFET superior to BJT ?

Ans. The high input impedance, low output impedance and low noise level make JFET far superior to the BJT.

Q. 30. Why N-channel MOSFETs are preferred over P-channel ones in integrated circuits ? **[U.P.S.C. I.E.S. E.& T.E.-I 2008]**

Ans. The hole mobility in silicon at normal field intensities is approximately 0.05 m^2/V-s and the same for electron is about 0.13 m^2/V-s. Thus the P-channel device will have more than twice the ON resistance of an equivalent N-channel of the same geometry and under the same working conditions. In other words, P-channel device must have more than twice the area of the N-channel device to achieve the same resistance. So, N-channel MOS circuits can be smaller for the same complexities than P-channel devices. The higher packing density of the N-channel MOS also makes it faster in switching applications due to the smaller junction areas. The operating speed is limited primarily by the internal time constants (RC), and the capacitance is directly proportional to the junction cross-sections. For all these reasons, N-channel MOS are preferred over P-channel MOS.

Q. 31. Why enhancement mode MOSFETs are preferred over the depletion mode ones in the fabrication of integrated circuits ? **[U.P.S.C. I.E.S. E. & T.E.-I, 2008]**

Ans. Since the conductivity of the channel is enhanced by the positive bias given to the gate in case of enhancement mode MOSFETs, the E-MOSFETs are preferred over the DE-MOSFETs in the fabrication of ICs.

Q. 32. What does UJT stand for ? Justify the name UJT. **[P.T.U. December 2004]**

Ans. UJT stands for unijunction transistor. It essentially consists of a lightly doped N-type silicon bar with a small piece of heavily doped P-type material alloyed to its one side to produce single P-N junction. The single P-N junction justifies the name UJT.

Q. 33. What is negative resistance region in UJT ? Explain its physical significance.

[Pb. Technical Univ. December 2006]

Ans. Once conduction is established at emitter voltage $V_E = V_P$, peak value of emitter voltage, the emitter potential V_E starts decreasing with the increase in emitter current I_E. This corresponds to a negative resistance region which is stable enough to be used with a great deal of reliability in the area of applications—oscillator.

Problems

1. When gate-source voltage is –12 V and gate current is 0.1 μA, determine the resistance between the gate and source of a given JFET. **[Ans.** 120 MΩ**]**
2. In case of a JFET the variation in drain current is of 100 mA when drain-source voltage is varied by 1.25 V, keeping gate-source voltage V_{GS} constant, determine the dynamic resistance of the JFET.

   [Ans. 12.5 kΩ**]**
3. In a JFET the drain current is changed by 0.25 mA when the gate-source voltage is changed by 0.125 V, keeping drain-source voltage constant. Calculate the transconductance of the given JFET. **[Ans.** 2 mS**]**
4. A JFET has drain current of 10 mA. If I_{DSS} = 20 mA and V_P = 4 V, find V_{GS}.

   [M.D. Univ. Analog Electronics May 2006] **[Ans.** 1.172 V**]**
5. The intrinsic stand-off ratio for a UJT is determined to be 0.65. If the interbase resistance is 7.5 kΩ, determine R_{B_1} and R_{B_2}. **[Ans.** 4.875 kΩ; 2.625 kΩ**]**
6. A silicon UJT has an interbase resistance of 10 kΩ and R_{B_1} of 6 kΩ with I_E = 0. Voltage of 20 V is applied between the bases. Determine (*i*) R_{B_2} (*ii*) intrinsic stand-off ratio (*iii*) UJT current (*iv*) stand-off voltage and (*v*) peak-point voltage. Take V_B = 0.7 V. **[Ans.** (*i*) 4 kΩ (*ii*) 0.6 (*iii*) 2 mA (*iv*) 12 V (*v*) 12.7 V**]**

❑❑❑

Unit-III

Large Signal Amplifiers

Contents of This Unit

UNIT-III : LARGE SIGNAL AMPLIFIERS : Class A Direct Coupled With Resistive Load; Transformer Coupled With Resistive Load; Harmonic Distortion; Variation of Output Power With Load; Push-Pull Amplifiers; Operation of Class-B Push-Pull Amplifier; Crossover Distortion; Transistor Phase Inverter; Complementary-Symmetry Amplifier.

CHAPTER 6. Large Signal Amplifiers.

CHAPTER

6

LARGE SIGNAL AMPLIFIERS

6.1. INTRODUCTION

INSIDE THIS CHAPTER

An amplifier system consists of signal pickup transducer, followed by small signal amplifier(s), a large signal amplifier and an output transducer. A *transducer* is used to convert one form of energy into another type. For example a microphone is used to convert acoustical energy into electrical energy. Conversely, a loudspeaker is used to convert electrical energy into acoustical energy.

The input transducer produces small electrical (typically voltage) signal, that needs sufficient amplification to operate some output device such as a loudspeaker, a servomotor, a solenoid or relay. The factors of prime interest in small signal voltage amplifiers are usually linearity and gain. Since the signal voltage and current from the input transducer is usually very small, the amount of power handling capacity and power efficiency are of slight concern. The functions of voltage amplifiers are to present a high resistance to the input transducer to minimize loading effects and to provide a large enough voltage signal to the large signal amplifier stages to operate such output devices (loudspeaker, servomotor etc.). A large signal amplifier must operate efficiently and be capable of handling large amounts of power–typically, a few watts to hundred of watts. Large signal amplifier that drives the output transducer demands even more consideration than the small signal voltage amplifiers that we have focused so far. The factors of greatest concern to the large signal power amplifiers are the power efficiency of the circuit, the maximum amount of power that the circuit is capable of handling, and impedance matching to the output device.

Power amplifier is meant to raise the power level of the input signal. In order to get large power at the output, it is necessary that the input-signal voltage is large. That is why, in an electronic system, a voltage amplifier always precedes the power amplifier, as illustrated in the block diagram (Fig. 6.1). Also, that is why power amplifiers are called large signal amplifiers.

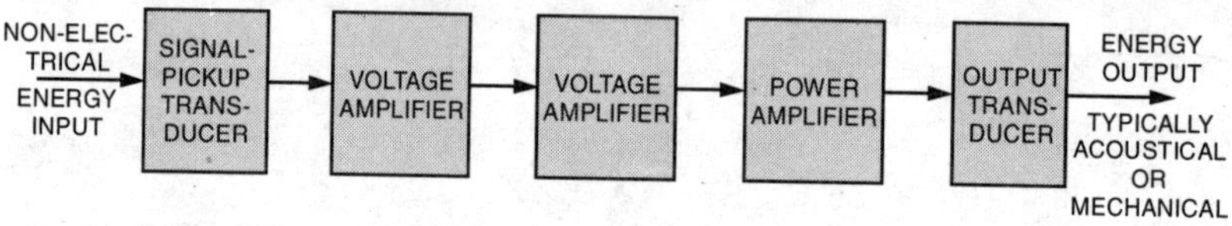

Fig. 6.1. ***Amplifier System***

In fact, power amplifier does not amplify power.* What a power amplifier actually does is that it draws power from dc supply connected to the output circuit and converts it into useful ac signal power. The type of ac power available at the output terminals of the power amplifier is controlled by the input signal. Thus a *power amplifier may be defined as a device that converts dc power into ac power and whose action is controlled by the input signal.*

The transistors employed in power amplifiers are called **power transistors**. They differ from other transistors in the following respects :

(*i*) The base is made thicker to handle large currents *i.e.,* in power amplifiers, transistors with comparatively smaller β are used.

(*ii*) The area of collector region of a power transistor is made considerably larger in order to dissipate the heat developed in the transistor during operation. Moreover, heat sinks are used for improving the heat dissipation.

(*iii*) The emitter and base layers are heavily doped. The contact area between the base layers and base leads is in ring like form so that the area is increased. By doing so ohmic resistance between emitter and base is reduced and due to low resistance, small power is required at input.

6.2. DIFFERENCE BETWEEN VOLTAGE AND POWER AMPLIFIERS

The primary function of the *voltage amplifier* is to raise the voltage level of the signal. It is designed to achieve the largest possible gain. Only very little power can be drawn from its output.

The voltage gain of an amplifier is given by the expression $A_v = \beta R_C/R_{in}$. So in order to achieve high voltage amplification (*i*) transistors having thin base *i.e.,* the transistors with high β (exceeding 100) are used (*ii*) input resistance R_{in} is kept quite low in comparison to collector load R_C and (*iii*) to permit high collector load, which is necessary to give high voltage gain, the voltage amplifiers are always operated at low collector current (of the order of 1 mA). R-C coupling is preferred for coupling various stages of voltage amplifiers because it is smaller in size, smaller in weight, cheaper in cost and it occupies less space.

A power amplifier is required to deliver a large amount of power and as such it has to handle large currents. So in order to achieve high power amplification, (*i*) large sized power transistors are used in order to dissipate the heat produced in transistor during operation, (*ii*) transistor having thick base (*i.e.,* transistors with comparatively smaller β) are used in order to handle large currents (*iii*) transformer coupling is used for impedance matching and (*iv*) collector resistance is made low. If its value is high, power losses in the collector resistor will be increased. Moreover, voltage drop across it will increase and, therefore, operating voltage will be reduced. This is the reason that low collector load resistance is used.

The noteworthy point is that we cannot obtain a power amplifier by simply replacing a transistor of low power dissipation in a voltage amplifier circuit by another transistor with high power dissipation.

* No device can amplify power because the idea of amplification of power will be against the basic principles of physics.

Let us consider a voltage divider biased common emitter voltage amplifier circuit, with replaced transistor, shown in Fig. 6.2. Let I_{CQ} and V_{CEQ} be the collector current and collector-emitter voltage respectively at the operating point.

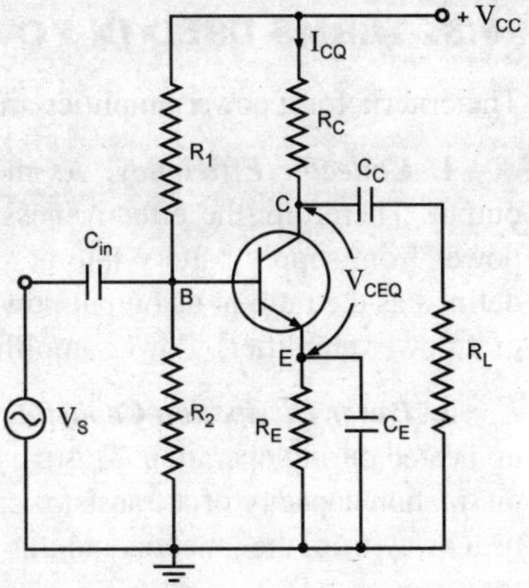

Fig. 6.2. *Simple Amplifier Circuit*

Now collector–emitter voltage,

$$V_{CEQ} = V_{CC} - I_{CQ} (R_C + R_E) \quad ...(6.1)$$

Total dc power supplied to the amplifier from collector supply

$$= V_{CC} I_{CQ} \quad ...(6.2)$$

DC power input to transistor = $V_{CEQ} I_{CQ}$ ...(6.3)

Out of total dc power of $V_{CC} I_{CQ}$ drawn from the dc supply, $V_{CEQ} I_{CQ}$ is only the power that is converted into useful ac power and rest is wasted.

i.e., Power wasted = $V_{CC} I_{CQ} - V_{CEQ} I_{CQ}$

$$= I_{CQ} [V_{CC} - V_{CEQ}]$$

$$= I_{CQ} [I_{CQ} (R_C + R_E)] = I_{CQ}^2 (R_C + R_E) \quad ...(6.4)$$

$\because$ from Eq. (6.1) $V_{CC} - V_{CEQ} = I_{CQ} (R_C + R_E)$

The $I^2_{CQ}(R_C + R_E)$ is the power that gets wasted unnecessarily in heating of resistors. This is the reason that a voltage amplifier cannot be used as a power amplifier.

Power wastage can be reduced by short circuiting collector and emitter resistors, R_C and R_E, but it is not possible. Emitter resistor R_E cannot be eliminated because it is part of biasing network. In the absence of R_E, the stabilization of the operating point is adversely affected. Collector resistor R_C also cannot be short circuited because in that case load resistor R_L will be short circuited (from ac point of view) and there will be no transfer of power to the load, R_L.

However, collector resistor R_C can be replaced by an inductor whose dc resistance is zero (acts as a short circuit for dc) and ac impedance is high. By doing so, two things are achieved. Power loss in the amplifier circuit is reduced because dc power loss in the choke is almost zero. In addition we can afford to use lower voltage supply V_{CC} for the same amplifier because no dc voltage drop occurs across the choke. This is the reason that transformer coupling is preferred over R-C coupling in power amplifiers.

Still more improvements can be made in the above circuit so as to make it more efficient in operation and will be discussed in this unit.

The comparison between voltage and power amplifiers is tabulated as below :

S. No.	Type of Amplifier	Voltage Amplifier	Power Amplifier
1.	Current gain, β	High, exceeding 100	Low, 20 – 50
2.	Collector load, R_C	High, about 10 kΩ	Low, 5 – 20 Ω
3.	Input voltage	Low, a few mV	High, 2 – 4 volts
4.	Collector current, I_C	Low, about 1 mA	High, exceeding 100 mA
5.	Power output	Low	High, a few watts to hundred of watts
6.	Power dissipation capacity	Less than 0.5 W	More than 0.5 W
7.	Output impedance	High, about 10 kΩ	Low, about 200 Ω
8.	Coupling	Usually R-C coupling	Invariably transformer coupling or tuned circuit

6.3. TERMS USED IN POWER AMPLIFIERS

The criteria for a power amplifier are collector efficiency, power dissipation capacity and distortion.

1. *Collector Efficiency*. As mentioned earlier, power amplifier converts dc power into ac power at output. Therefore, the effectiveness of a power amplifier is measured by its capability of converting dc power from supply battery into ac power at the output. This is known as *collector efficiency* and may be defined as the ratio of ac output power to the zero signal power (or dc power drawn from the supply battery of a power amplifier). Power amplifiers are designed to provide maximum collector efficiency.

2. *Power Dissipation Capacity*. As mentioned earlier, a power transistor handles large currents and gets up heated during operation. This may cause the junction temperature to rise to a dangerous level. The power dissipation capacity of a transistor can be increased by making a suitable provision for rapid conduction of heat away from the junction and this is achieved by attaching a metal case, called the *heat sink*, to the power transistor case.

Power dissipation capacity of a power transistor is defined as its ability to dissipate the heat developed in it.

3. *Distortion* is defined as the change of output waveshape from the input waveshape of the amplifier. Since power amplifier handles large signals and for large input signals, the transistor operation becomes nonlinear, the problem of distortion arises. Lesser the distortion, better is the amplifier.

6.4. CLASSIFICATION OF POWER AMPLIFIERS

The power amplifiers are primarily divided into two categories viz. *audio power amplifiers* and *radio power amplifiers*.

Audio power amplifiers, also called the *small signal power amplifiers*, raise the power levels of signals that have audio frequency range (20 Hz–20 kHz).

Radio power amplifiers, also called *large signal power amplifiers,* raise the power level of signals that have radio frequency range. They amplify a specific frequency or narrowband of frequencies while rejecting all other frequencies.

Classification According To Mode of Operation. Transistor power amplifiers handle large signals. Many of them are driven so hard by the input large signal that collector current is either cutoff or is in saturation region during a large portion of the input cycle. So such amplifiers are generally classified according to their mode of operation. This classification is based on the amount of transistor bias and amplitude of the input signal. It takes into account the portion of the cycle for which the transistor conducts. They are classified as below :

1. *Class A Power Amplifiers*. In this case, transistor is so biased that the output current flows for the entire cycle of the input signal. Thus the operating point is so selected that the transistor operates only over the linear region of its load line. So such an amplifier can amplify input signal of small amplitude. As the transistor operates over the linear portion of load line, the output waveform is exactly similar to input waveform. So class A amplifiers are characterised by a high fidelity of the output. Such amplifiers are used where freedom from distortion is prime aim. Operation is restricted only over a small central region of the load line so such amplifiers can be used for amplifying signals of small amplitude. Also ac power output per transistor is small. The maximum possible overall efficiency with resistive load is 25%. The maximum possible collector efficiency with resistive load is 50%. In case an output transformer is used, both of these efficiencies are 50%.

2. *Class B Power Amplifiers*. In this case, the transistor bias and signal amplitude are such that output current flows only during positive half cycle of the input signal. At zero signal, the collector current is zero and no biasing system is required in class B amplifiers. The operating point is selected at collector cutoff voltage. Because of total absence of negative half cycle from the output the signal distortion is high. Zero signal input represents the best condition for class B amplifiers because of zero collector current. The transistor dissipates more power with increase in signal strength. In comparison to class A amplifiers average current is less, power dissipation is less. So overall efficiency is increased. The theoretical efficiency in class B operation is about 78.5% while it is only 50% in class A operation.

3. *Class AB Power Amplifiers.* An amplifier may be biased at a dc level above the zero base current level of class B power amplifiers and above one-half the supply voltage level of class A; this bias condition is class AB. Class AB operation still needs a push-pull connection to achieve a full output cycle, but the dc bias level is usually closer to zero base current level for better power efficiency. For class AB operation the output signal swing occurs between 180º and 360º and is neither class A nor class B operation.

4. *Class C Power Amplifiers.* A class C power amplifier is biased for operation for less than 180º of the input signal cycle and will operate only with a tuned or resonant circuit which provides a full cycle of operation for the tuned or resonant frequency. Such power amplifiers are, therefore, employed in special areas of tuned circuits, such as radio or communications.

5. *Class D Power Amplifiers.* Class D power amplifiers are designed to operate with digital or pulse type signals. Using digital techniques makes it possible to have a signal that varies over the entire cycle (using sample-and-hold circuitry) to recreate the output from many pieces of input signal. The main advantage of class D power amplifiers is that it is *on* (using power) only for short intervals and the overall efficiency can practically be very high (above 90%).

Classification Based on Deriving Output. Transistor amplifiers, according to the method of deriving output, are classified as single-ended, double-ended (or push-pull) and complementary symmetry push-pull power amplifiers.

Single-ended power amplifier uses single transistor and derives output power w.r.t. one end permanently grounded.

Double-ended or push-pull amplifier uses two transistors in a single stage. It consists of two loops in which the transistor collector currents flow in opposite directions but add in the load.

Complementary symmetry push-pull power amplifier uses two transistors having complementary symmetry (one N-P-N and another P-N-P). The term complementary arises from the fact that one transistor is the N-P-N type and the other is P-N-P type. They have symmetry as they are made with the same material and technology and are of same maximum ratings.

Classification Based on Use. The amplifiers, according to use, may be classified as voltage, power, current or general purpose amplifiers. In general, the load of an amplifier is impedance. The two most important special cases are the idealized resistive load and the tuned circuit operating near its resonant frequency. Class AB and class B operation are used with untuned power amplifiers whereas class C operation is employed with tuned radio frequency amplifiers. Many important waveshaping functions may be accomplished by class B or class C overdriven amplifiers.

6.5. PRACTICAL POWER AMPLIFIER STAGES

Power amplifier is meant to amplify a weak signal until sufficient power is available to operate an output device such as a loudspeaker, a solenoid or a relay. Power amplifier, to provide the desired power amplification, has generally 3 stages (voltage amplification stage, driver stage and output stage), as shown by a block diagram in Fig. 6.3.

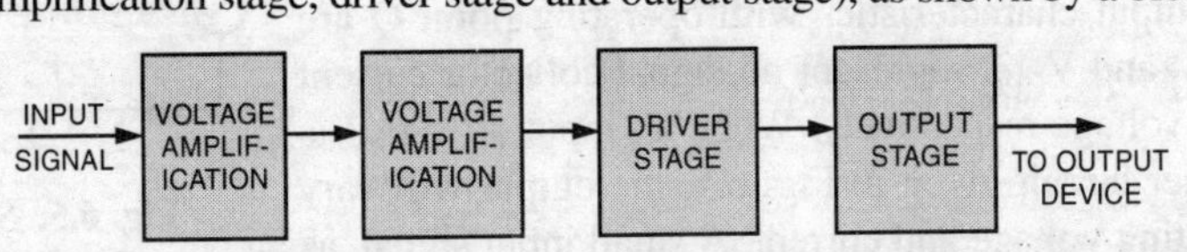

Fig. 6.3. ***Stages in a Practical Power Amplifier***

1. *Voltage Amplification Stage*. As already mentioned in Art. 6.1, the signal developed by the input transducer is very small (of the order of few millivolts) and needs sufficient amplification so as to operate the output device. Therefore, for raising the level of the weak input signal, it is amplified in two or more stages, R-C coupling is usually employed.

2. *Driver Stage*. The stage that immediately precedes the output stage is called the *driver stage*. The output from the last voltage amplification stage is fed to the driver stage and output from the driver stage is fed to the output stage. The driver stage renders power amplification as usual. Here, concentrated effort is made to provide maximum power gain and so transformer coupled class A power amplifier is employed in this stage. The driver transformer is usually a step-down transformer and facilitates impedance matching.

3. *Output Stage*. The output stage essentially consists of a power amplifier and is meant for transferring maximum power to the output device. In order to transfer maximum power at high efficiency, push-pull arrangement is employed in the output stage (if a single transistor is used as a class A amplifier in the output stage to provide faithful amplification, the operation efficiency will be quite low). In this arrangement two transistors are used in class B operation and are fed from the centre-tapped secondary of the transformer whose primary forms the collector load for the driver stage.

6.6. CLASS A POWER AMPLIFIERS (DIRECT COUPLED WITH RESISTIVE LOAD)

A class A power amplifier is defined as a power amplifier in which output current flows for the entire cycle (360°) of the input signal, as illustrated in Fig. 6.4. In other words, the transistor remains forward biased throughout the input cycle.

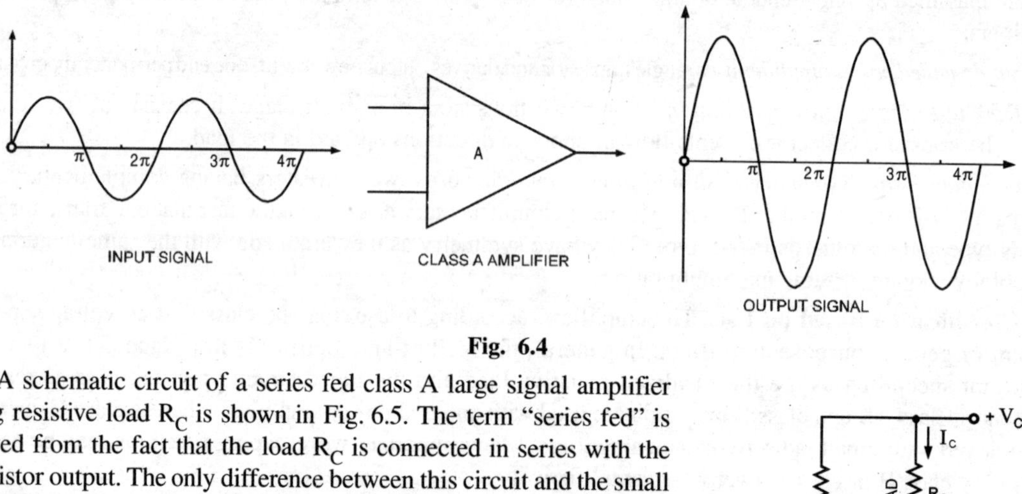

Fig. 6.4

A schematic circuit of a series fed class A large signal amplifier using resistive load R_C is shown in Fig. 6.5. The term "series fed" is derived from the fact that the load R_C is connected in series with the transistor output. The only difference between this circuit and the small signal amplifier circuits considered previously is that the signals handled by the large signal circuit are in the range of volts and the transistor used is a power transistor capable of operating in the range of few watts. This circuit is seldom used for power amplification because of its poor collector efficiency but will give clear understanding of class A operation to the readers. The output characteristics with operating point Q are shown in Fig. 6.6. I_{CQ} and V_{CEQ} represent no signal collector current and collector-emitter voltage respectively. When an input ac signal is applied to the amplifier circuit given in Fig. 6.5, the output will vary from its dc bias operating voltage and current. A small input signal, as indicated in Fig. 6.6, will cause the base current to vary above and

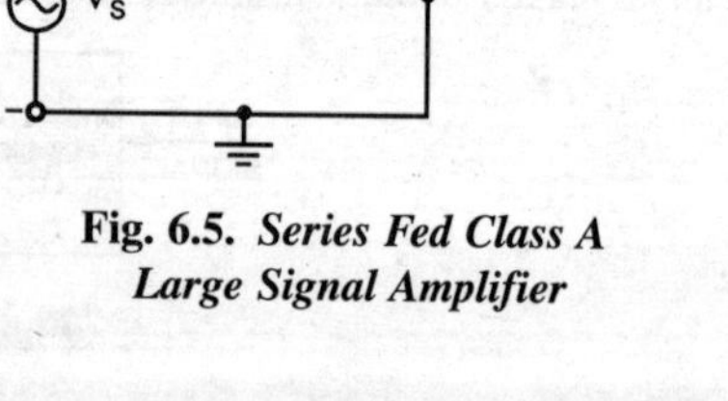

Fig. 6.5. ***Series Fed Class A Large Signal Amplifier***

below the dc bias point, which will then cause the collector current (output) I_c to vary from the dc bias point set as well as the collector-emitter voltage V_{ce} to vary around its dc bias value. With the strengthening of the input signal, the output will vary further around the established dc bias point until either the current or voltage attains a limiting condition. For the current this limiting condition is either zero current at the lower end or $\frac{V_{CC}}{R_C}$ at the upper end. For the collector-emitter voltage this limit is either 0 V or the supply voltage V_{CC}. Let the output current vary between limits $I_{c\,min}$ and $I_{c\,max}$ and similarly the collector-emitter voltage between limits $V_{ce\,min}$ and $V_{ce\,max}$.

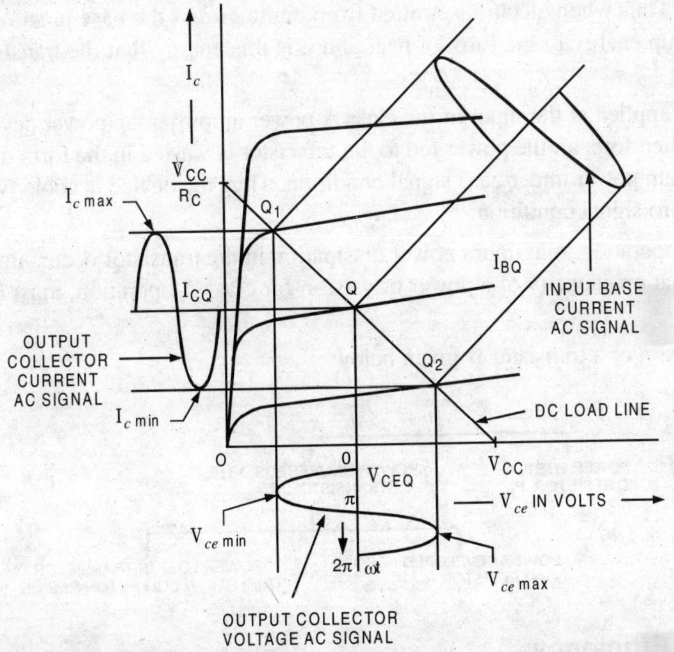

Fig. 6.6. ***Transistor Output Characteristics Showing DC Load Line and AC Signals***

6.6.1. Power Distribution

Input power from the collector supply V_{CC},

$$P_{in\,(dc)} = V_{CC}\, I_{CQ} \quad \text{...(6.5)}$$

The power drawn from the collector supply is used in the following two components

(*i*) Power dissipated in collector load as heat,

$$P_{RC\,(dc)} = I_{CQ}^2\, R_C \quad \text{...(6.6)}$$

(*ii*) Power supplied to the transistor,

$$P_{tr\,(dc)} = P_{in\,(dc)} - P_{RC\,(dc)} = V_{CC}\, I_{CQ} - I_{CQ}^2\, R_C \quad \text{...(6.7)}$$

Power supplied to the transistor, $P_{tr\,(dc)}$ is further subdivided into

(*a*) ac power developed across the load resistor constituting ac power output and is given as

$$P_{out\,(ac)} = I_c^2\, R_C = I_c V_{ce} \quad \text{...(6.8 a)}$$

where I_c and V_{ce} are the rms values of collector current and collector-emitter voltage

Note : RMS value = $\frac{1}{2}\left[\frac{\text{peak-to-peak}}{\sqrt{2}}\right] = \frac{1}{2\sqrt{2}}$ peak-to-peak.

$$= \frac{1}{2} \frac{I_{c\,(\text{peak-to-peak})}}{\sqrt{2}} \times \frac{1}{2} \frac{V_{ce\,(\text{peak-to-peak})}}{\sqrt{2}} \qquad \text{...(6.8 b)}$$

$$= \frac{I_{c\,(\text{peak-to-peak})} \times V_{ce\,(\text{peak-to-peak})}}{8} \qquad \text{...(6.8 c)}$$

(*b*) Power dissipated, in the form of heat, by the transistor itself. The cause of power dissipation in transistor is explained below :

Consider an N-P-N transistor. The potential difference across the depletion layer formed near the collector junction is called the barrier potential. This potential gives the P-region (base) slightly more energy than N-region (collector). Thus when electrons emitted from emitter cross the base junction and enter the collector region, they give up energy in the form of heat and it is this energy that the transistor has to dissipate to the surrounding.

With zero signal applied at the input of the class A power amplifier, ac power developed across the load reduces to zero and, therefore, all the power fed to the transistor is wasted in the form of heat. Thus, a *transistor dissipates* maximum power under zero signal condition. Thus the device is cooler when delivering power to a load than with zero signal condition.

Since in class A operation, maximum power dissipation in the transistor occurs under zero signal condition, the power dissipation capacity of a power transistor, for class A operation, must be at least equal to the zero signal rating.

Power flow diagram of a transistor is given below :

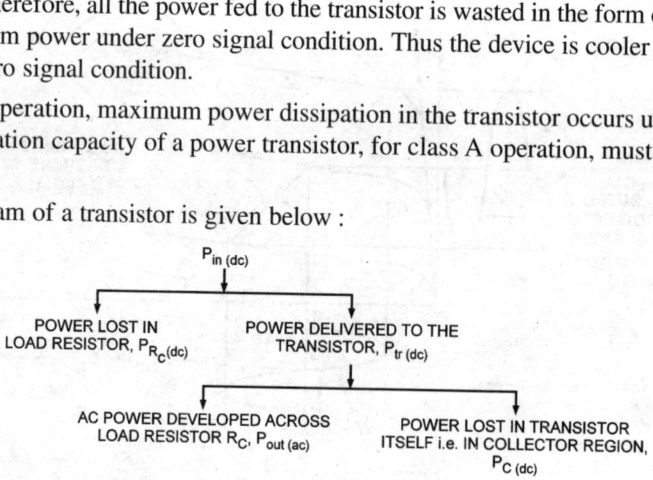

6.6.2. Collector Efficiency

The collector efficiency of a transistor is given as

$$\eta_{\text{collector}} = \frac{\text{Average ac power output, } P_{out\,(ac)}}{\text{Average dc power input to the transistor, } P_{tr\,(dc)}}$$

6.6.3. Power Efficiency

A measure of the ability of an active device to convert the dc power of supply into the ac (signal) power delivered to the load is called the *power* or *conversion* or *theoretical efficiency*. By definition the efficiency is

$$\eta = \frac{\text{AC power delivered to the load, } P_{out\,(ac)}}{\text{Total power drawn from dc supply, } P_{in\,(dc)}}$$

Now ac power delivered to the load,

$$P_{out\,(ac)} = \frac{I_{c\,(\text{peak-to-peak})} \times V_{ce\,(\text{peak-to-peak})}}{8} \qquad \text{...Refer to Eq. (6.8 c)}$$

$$= \frac{(I_{c\,max} - I_{c\,min}) \times (V_{ce\,max} - V_{ce\,min})}{8}$$

and from Eq. (6.5), total power drawn from dc supply,

$$P_{in\,(dc)} = V_{CC}\, I_{CQ}$$

So power efficiency, $$\eta = \frac{P_{out\,(ac)}}{P_{in\,(dc)}} = \frac{(I_{c\,max} - I_{c\,min}) \times (V_{ce\,max} - V_{ce\,min})}{8\, V_{CC}\, I_{CQ}} \qquad ...(6.9)$$

6.6.4. Maximum Power and Efficiency

If the operating point Q is set at the midpoint of the maximum signal swing, the resulting maximum power condition may be achieved. For the circuit shown in Fig. 6.5, this would be determined using

$$\text{Maximum } V_{ce\,(peak\text{-}to\text{-}peak)} = V_{CC}$$

$$\text{Maximum } I_{c\,(peak\text{-}to\text{-}peak)} = \frac{V_{CC}}{R_C}$$

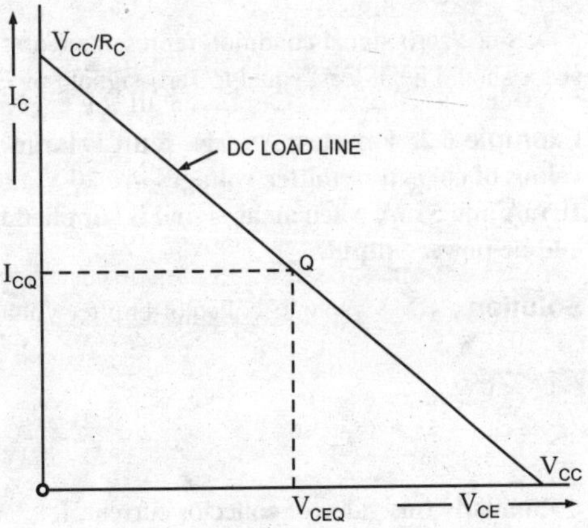

Fig. 6.7

From Eq. (6.8 *c*) we have maximum ac power developed across the load resistor,

$$P_{out\,(ac)\,max} = \frac{1}{8} \cdot \frac{V_{CC}}{R_C} \times V_{CC} = \frac{V_{CC}^2}{8\,R_C} \qquad ...(6.10)$$

For the quiescent point Q,

$$I_{CQ} = \frac{V_{CC}/R_C}{2} \qquad ...(6.11)$$

and dc power drawn from dc supply,

$$P_{in\,(dc)\,max} = V_{CC}\, I_{CQ} = \frac{V_{CC} \times \dfrac{V_{CC}}{R_C}}{2} = \frac{V_{CC}^2}{2\,R_C} \qquad ...(6.12)$$

Power supplied to the transistor,

$$P_{tr}(dc) = V_{CC} I_{CQ} - I_{CQ}^2 R_C \qquad \text{...Refer to Eq. (6.7)}$$

$$= V_{CC} \cdot \frac{V_{CC}}{2\,R_C} - \left(\frac{V_{CC}}{2\,R_C}\right)^2 R_C = \frac{V_{CC}^2}{2R_C} - \frac{V_{CC}^2}{4R_C} = \frac{V_{CC}^2}{4R_C} \qquad ...(6.13)$$

So maximum efficiency of an amplifier (class A power) is given as

$$\eta_{max} = \frac{P_{out\,(ac)\,max}}{P_{in\,(dc)\,max}} = \frac{V_{CC}^2/8R_C}{V_{CC}^2/2R_C} = \frac{2R_C}{8R_C} = 0.25 \text{ or } 25\% \qquad ...(6.14)$$

This is the maximum per cent efficiency for a series-fed class A power amplifier. Since this maximum efficiency will occur only under ideal conditions and for the maximum ac signal swings, most series-fed class A power amplifiers have power efficiencies much less than 25%.

From Eqs. (6.10) and (6.13) we have

$$\frac{P_{out\,(ac)\,max}}{P_{tr\,(dc)}} = \frac{V_{CC}^2/8R_C}{V_{CC}^2/4R_C} = \frac{1}{2} \qquad ...(6.15)$$

Thus, we see that a class A power amplifier with a direct coupled resistive load can supply only one-half the amount of ac power to the load that the transistor can dissipate and the dc power input is four times the amount of ac signal power in the load.

Class A power amplifiers have got advantages of simple construction and distortionless output voltage.

The drawbacks of class A power amplifiers are very low efficiency and large power dissipation in the power transistors.

Example 6.1. A power transistor working in class A operation has zero signal power dissipation of 10 watts. If the ac output power is 4 watts, find (*i*) collector efficiency (*ii*) power rating of the transistor.

[*Pb. Technical Univ. Analog Electronics Dec. 2005*]

Solution : AC power output, $P_{out\,(ac)} = 4\text{ W}$

Power delivered to the transistor, $P_{tr\,(ac)}$ = Zero-signal power dissipation = 10 W

(*i*) Collector efficiency, $\eta_{collector} = \dfrac{P_{out\,(ac)}}{P_{tr\,(ac)}} \times 100 = \dfrac{4}{10} \times 100 = 40\%$ **Ans.**

Since zero-signal condition represents worst condition for the transistor, the power dissipation capacity of the transistor should be at least equal to zero-signal power dissipation *i.e.,* 10 W **Ans.**

Example 6.2. For class A series fed CE large signal amplifier using resistance load, the maximum and minimum values of collector-emitter voltages are 20 V and 10 V and maximum and minimum values of collector current are 10 mA and 5 mA when an ac signal is supplied to it. Determine rms values of collector voltage and collector current and ac power output.

Solution : RMS value of collector-emitter voltage,

$$V_{ce\,rms} = \frac{V_{ce\,(peak\text{-}to\text{-}peak)}}{2\sqrt{2}} = \frac{V_{ce\,max} - V_{ce\,min}}{2\sqrt{2}} = \frac{20-10}{2\sqrt{2}} = 3.54\text{ V } \textbf{Ans.}$$

Similarly rms value of collector current, $I_{c\,rms} = \dfrac{I_{c\,max} - I_{c\,min}}{2\sqrt{2}} = \dfrac{10-5}{2\sqrt{2}} = 1.77\text{ mA}$ **Ans.**

Hence ac power output, $P_{out\,(ac)} = V_{ce\,rms} \times I_{c\,rms} = 3.54 \times 1.77 \times 10^{-3} = 6.25\text{ mW}$ **Ans.**

Example 6.3. Calculate the input power, output power delivered to the transistor, power lost in transistor, collector efficiency and overall efficiency of the amplifier shown in Fig. 6.8 for an input voltage resulting in base current of 10 mA peak. Take β = 25 and V_{BE} = 0.7 V.

Solution : Saturation current,

$$I_{C\,(sat)} = \frac{V_{CC}}{R_C} = \frac{20}{20} = 1\text{ A or 1,000 mA}$$

$$V_{CE} = V_{CC} = 20\text{ V}$$

Now dc load line is drawn joining points (20 V, 0) and (0, 1000 mA), as shown in Fig. 6.9.

For operating point Q

Base current, $I_B = \dfrac{V_{CC} - V_{BE}}{R_B} = \dfrac{20-0.7}{1\text{ k}\Omega} = 19.3\text{ mA}$

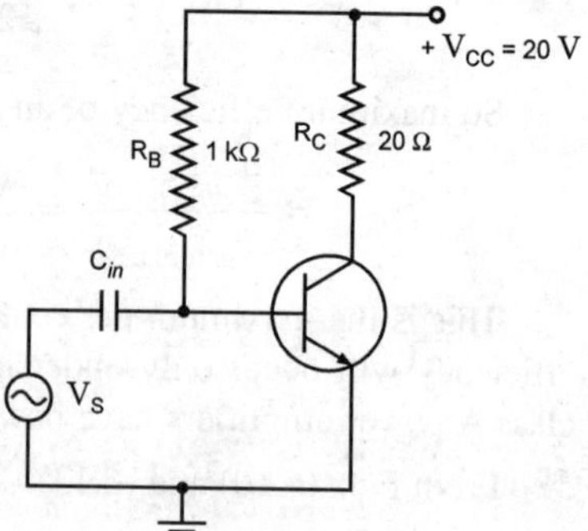

Fig. 6.8

Collector current, $I_{CQ} = \beta I_B = 25 \times 19.3 = 482.5\text{ mA}$

Collector-emitter voltage,

$$V_{CEQ} = V_{CC} - I_{CQ} R_C = 20 - 482.5 \times 10^{-3} \times 20 = 10.35\text{ V}$$

The operating point Q (10.35 V, 482.5 mA) is marked on the dc load line.

Peak value of collector current,

$$I_{c\,peak} = \beta I_{b\,peak} = 25 \times 10\text{ mA peak} = 250\text{ mA}$$

Input power, $P_{in\,(dc)} = V_{CC} I_{CQ} = 20 \times 482.5 \times 10^{-3} = 9.65$ W **Ans.**

$$\text{Output power, } P_{out\,(ac)} = \left(\frac{I_{c\,peak}}{\sqrt{2}}\right)^2 R_C = \left(\frac{250 \times 10^{-3}}{\sqrt{2}}\right)^2 \times 20$$

$$= 0.625 \text{ W } \textbf{Ans.}$$

Power delivered to the transistor,

$$P_{tr\,(dc)} = V_{CC} I_{CQ} - I^2_{CQ} R_C$$
$$= 20 \times 482.5 \times 10^{-3} - (482.5 \times 10^{-3})^2 \times 20$$
$$= 9.65 - 4.66 = 4.99 \text{ W } \textbf{Ans.}$$

Power lost in transistor

$$= P_{tr\,(dc)} - P_{out\,(ac)} = 4.99 - 0.625 = 4.365 \text{ W } \textbf{Ans.}$$

Collector efficiency,

$$\eta_{collector} = \frac{P_{out\,(ac)}}{P_{tr\,(dc)}} \times 100 = \frac{0.625}{4.99} \times 100 = 12.53\% \textbf{ Ans.}$$

Overall efficiency

$$\eta_{overall} = \frac{P_{out\,(ac)}}{P_{in\,(dc)}} = \frac{0.625}{9.65} \times 100 = 6.48\% \textbf{ Ans.}$$

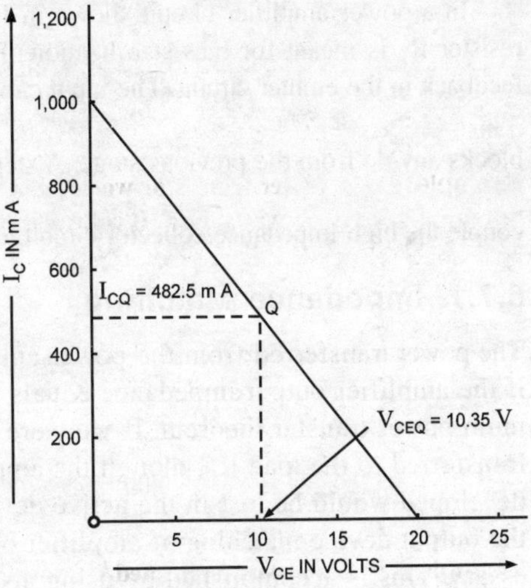

Fig. 6.9

Power rating of transistor = Zero-signal power dissipation = $V_{CEQ} I_{CQ} = 10.35 \times 482.5 \times 10^{-3} = 5$ W **Ans.**

6.7. CLASS A POWER AMPLIFIERS (TRANSFORMER COUPLED WITH RESISTIVE LOAD)

This is also sometimes referred to as *single ended power amplifier*. The term "single ended" (denoting only one transistor) is used to distinguish it from the push-pull amplifier using two transistors.

In case of a direct coupled class A power amplifier shown in Fig. 6.5, the quiescent current flows through the collector resistive load and causes large wastage of dc power in it. This dc power dissipated in the load resistor does not contribute to the useful ac output power. Furthermore, it is generally inadvisable to pass the dc through the output device such as in a voice coil of a loudspeaker. For these reasons an arrangement using a suitable transformer for coupling the load (say, a loudspeaker) to the amplifier is usually employed, as shown in Fig. 6.10. This arrangement enhances the conversion efficiency by a factor of 2 by eliminating the dc power dissipation in the load. Since the primary of the transformer has negligible dc resistance, there is negligible dc power lost at this point. The ac power is, however, coupled magnetically across the transformer into the load R_L. This coupling method also prevents a large dc current from flowing through the load, which otherwise could be harmful if the load were a loudspeaker, since it would cause saturation of the magnetic circuit and impair the reproduction of the audio signal.

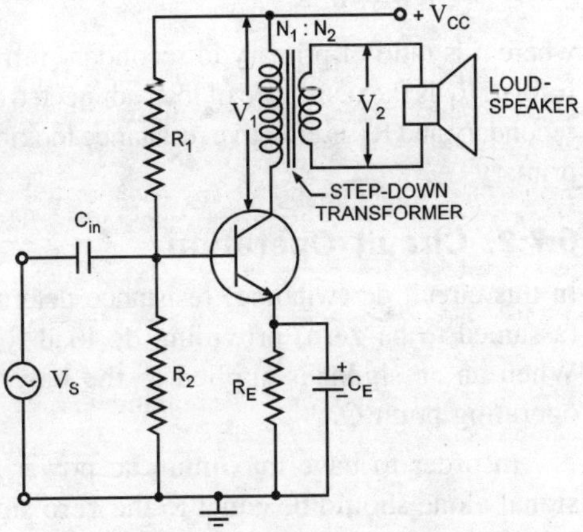

Fig. 6.10. *Transformer Coupled Class A Power Amplifier*

This arrangement also permits impedance matching.

In a power amplifier circuit shown in Fig. 6.10, R_1 and R_2 provide potential divider biasing and emitter resistor R_E is meant for bias stabilization. The emitter bypass capacitor C_E is meant for preventing negative feedback in the emitter circuit. The input capacitor C_{in} couples ac signal voltage to the base of the transistor but blocks any dc from the previous stage. A step-down transformer of suitable turn ratio $\left(a = \frac{N_1}{N_2}\right)$ is provided to couple the high impedance collector circuit to low impedance load.

6.7.1. Impedance Matching

The power transferred from the power amplifier to the load (say a loudspeaker) will be maximum only if the amplifier output impedance equals the load impedance R_L. This is in accordance with the maximum power transfer theorem. If we were not able to achieve the above condition, lesser power will be transferred to the load R_L, though the amplifier is capable of delivering more power, and rest of power developed would be lost in the active device. Hence for transfer of maximum power from amplifier to the output device matching of amplifier output impedance with the impedance of output device is necessary. This is accomplished by using a step-down transformer of suitable turn ratio. The transformer impedance matching circuit is shown separately in Fig. 6.11, where R_L' is the resistance looking into the primary of the transformer and is given as

$$\frac{R'_L}{R_L} = \frac{V_1/I_1}{V_2/I_2} = \frac{V_1}{V_2} \cdot \frac{I_2}{I_1} = \frac{N_1}{N_2} \cdot \frac{N_1}{N_2} = \left(\frac{N_1}{N_2}\right)^2 \qquad \text{where } \frac{V_1}{V_2} = \frac{N_1}{N_2} \text{ and } \frac{I_2}{I_1} = \frac{N_1}{N_2}$$

Thus the ratio of the transformer input and output resistances varies directly as the square of the transformer turn ratio :

$$\frac{R'_L}{R_L} = \left(\frac{N_1}{N_2}\right)^2 = a^2$$

$$\text{or} \quad R'_L = a^2 R_L \qquad \text{...(6.16)}$$

where a is ratio of primary to secondary turns of step-down transformer, R_L is the resistance of load connected across the transformer secondary and R'_L is effective resistance looking into the transformer primary.

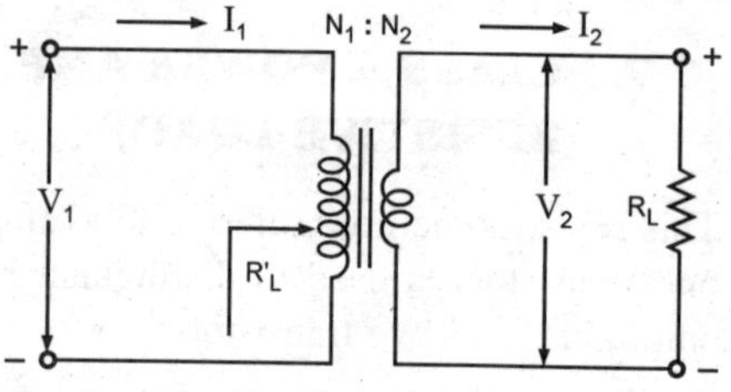

Fig. 6.11

6.7.2. Circuit Operation

In this circuit dc (winding) resistance determines the dc load line. Typically, this resistance is quite small (assumed to be zero) providing dc load to be a vertical line rising from V_{CC}, as shown in Fig. 6.12. When an ac signal is applied to the base of the transistor the collector current will vary around the operating point Q.

In order to have maximum ac power output, the peak value of collector current due to input ac signal alone should be equal to the zero signal collector current. To achieve this, the operating point Q is located at the centre of the load line. This is achieved by adjusting the biasing circuit (R_1, R_2 and R_E).

When ac signal is applied, collector current fluctuates from maximum to minimum (zero), and operating point Q moves up and down the load line. At the peak of the positive half cycle of the input signal, the total collector current $I_{c\,max} = 2\,I_{CQ}$ and collector-emitter voltage $V_{ce\,min} = 0$ while at the peak of the negative half cycle of the input signal, the collector current $I_{c\,min} = 0$ and collector-emitter voltage $V_{ce\,max} = 2\,V_{CC}$. Thus collector-emitter voltage varies in opposite phase to the collector current. The variation of collector voltage appears across primary of the transformer. Now ac voltage is induced in the transformer secondary which in turn develops ac power and supplies to the load.

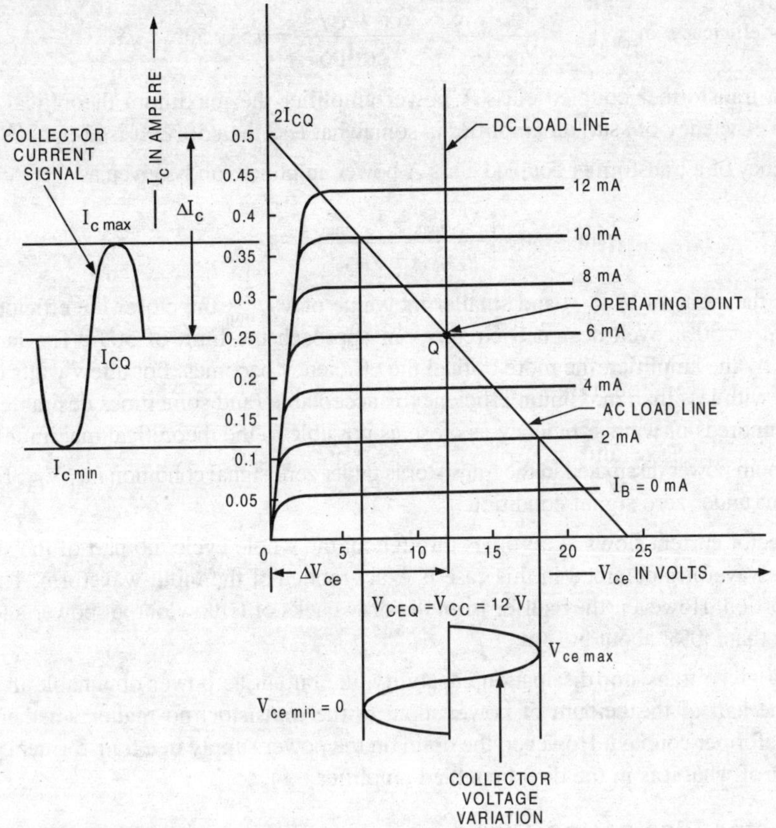

Fig. 6.12. ***Graphical Representation of Transformer Coupled Class A Audio Power Amplifier***

6.7.3. Circuit Analysis

In an ideal transformer, there is no voltage drop in primary so $V_{CEQ} = V_{CC}$ and power input to the transistor becomes equal to the dc power drawn from the collector supply V_{CC} as power lost in the primary is negligibly small.

Thus power input to the transistor, P_{tr} = Power drawn from collector supply, $P_{in\,(dc)} = V_{CC}\, I_{CQ}$

and overall efficiency becomes equal to collector efficiency and

$$= \frac{P_{out\,(ac)}}{V_{CC}\, I_{CQ}}$$

Under condition of development of maximum ac power, voltage swings from $V_{ce\,max}$ to zero and collector current from $I_{c\,max}$ to zero. So

$$V_{rms} = \frac{1}{\sqrt{2}}\left[\frac{V_{ce\,max} - V_{ce\,min}}{2}\right] = \frac{V_{ce\,max}}{2\sqrt{2}} = \frac{2\,V_{CC}}{2\sqrt{2}} = \frac{V_{CC}}{\sqrt{2}}$$

$$\text{and } I_{rms} = \frac{1}{\sqrt{2}}\left[\frac{I_{c\,max} - I_{c\,min}}{2}\right] = \frac{I_{c\,max}}{2\sqrt{2}} = \frac{2I_{CQ}}{2\sqrt{2}} = \frac{I_{CQ}}{\sqrt{2}}$$

AC power developed across the load,

$$P_{out\,(ac)} = V_{rms}\, I_{rms} = \frac{V_{CC}}{\sqrt{2}} \cdot \frac{I_{CQ}}{\sqrt{2}} = \frac{V_{CC}\, I_{CQ}}{2}$$

Collector efficiency, $\eta_{collector} = \frac{P_{out\,(ac)}}{V_{CC}\,I_{CQ}} = \frac{V_{CC}\,I_{CQ}/2}{V_{CC}\,I_{CQ}} = 0.5$ or 50% ...(6.17)

Thus for a transformer coupled class A power amplifier the maximum theoretical efficiency is 50%. In practice, the efficiency of such an amplifier is somewhat less than 50%. It is about 30%.

The efficiency of a transformer coupled class A power amplifier can be given as

$$\%\eta = 50 \times \left(\frac{V_{ce\,max} - V_{ce\,min}}{V_{ce\,max} + V_{ce\,min}}\right)^2 \% \quad ...(6.18)$$

The larger the value of $V_{ce\,max}$ and smaller the value of $V_{ce\,min}$ the closer the efficiency approaches the theoretical limit of 50%. Well-designed circuits can approach the limit of 50%. The larger the amount of power handled by the amplifier, the more critical the efficiency becomes. For a few watts of power a simpler, cheaper circuit with less than maximum efficiency is acceptable (and sometimes desirable). For power levels in the tens to hundreds of watts efficiency as close as possible to the theoretical maximum would be desired.

The maximum power dissipated in the transistor is under zero signal condition *i.e.,* $V_{CC}\,I_{CQ}$ where I_{CQ} is the collector current under zero signal condition.

Since collector current flows at all times throughout the whole cycle, no part of the signal is cut off and thus the output waveform obtained in this case is exact replica of the input waveform. Thus such amplifiers have least distortion. However, they suffer from the drawbacks of (*i*) low output power and (*ii*) low collector efficiency (less than 50%; about 30%).

Thus, for a given transistor dissipation capability, the output ac power obtainable in a load is the same (*i.e.,* ideally one half of the amount of power input to the transistor) no matter whether the load is direct coupled or transformer coupled. However, the drain on the power supply in a transformer coupled amplifier is ideally one half of what it is in the direct coupled amplifier.

6.7.4. Collector Dissipation Curve

In a power amplifier, the main source of power supply is collector dc supply and some of it is converted into ac power and appears in the load and rest of it is dissipated in the form of heat. Thus in a power amplifier

$$P_{in\,(dc)} = P_{out\,(ac)} + P_d, \quad \text{neglecting transformer losses being very small}$$

or Power dissipation, $P_d = P_{in\,(dc)} - P_{out\,(ac)} = V_{CC}\,I_{CQ} - P_{out\,(ac)}$ ...(6.19)

From above Eq. (6.19) it is obvious that power dissipated by the transistor (P_d) is the difference of average dc power input from the collector supply battery, $P_{in\,(dc)}$ (which is constant for a given battery and operating point) and ac output power, $P_{out\,(ac)}$ supplied to the load. For zero output (*i.e.,* when no signal is applied), the transistor is required to dissipate the maximum power *i.e.,* $P_d = V_{CC}\,I_{CQ}$. This is the worst case for a transistor. When the signal is applied, the transistor has to dissipate less power. Hence in designing a power amplifier, the important factor that is to be kept in mind is the power dissipation rating of the transistor.

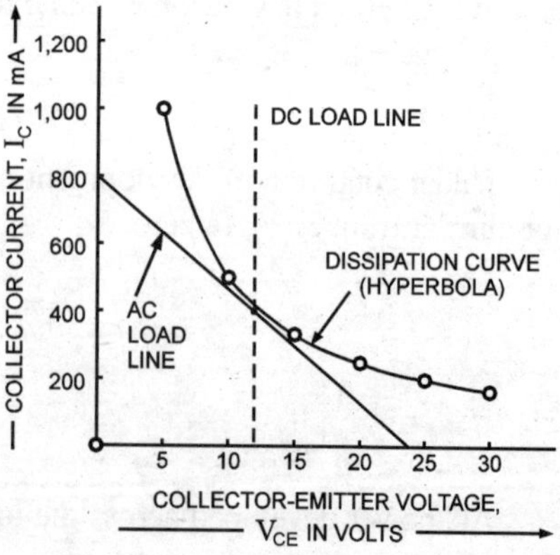

Fig. 6.13

Let the power dissipation capacity of a transistor be 5 W. For designing a power amplifier, first of all its *collector dissipation curve* is plotted. For plotting the collector dissipation curve of a power amplifier collector current I_C is determined corresponding to some arbitrary values of collector-emitter voltage V_{CE} from the relation $P_d = V_{CE}\,I_C = 5$ W, the power dissipation capacity of the transistor. Some sample calculations are given below in tabular form. The curve obtained from these values is a hyperbola, as shown in Fig. 6.13.

S.No.	V_{CE} in volts	I_C in mA	P_d in watts
1	5	1,000	5
2	10	500	5
3	15	333	5
4	20	250	5
5	25	200	5
6	30	167	5

For using this transistor in a power amplifier, its operating point must be below the dissipation curve, otherwise transistor will get damaged. This is the importance of dissipation curve.

6.7.5. Drawbacks of Single Ended Transformer Coupled Amplifier

A single ended transformer coupled amplifier has the following drawbacks :

1. The total harmonic distortion is unacceptably high for most audio applications except in cheaper amplifiers. The second harmonic distortion is usually the greatest offender.
2. Maximum output power is delivered at a load resistance that is different from the value for minimum harmonic distortion. Consequently, the amplifier is designed to operate into a load that provides slightly less than maximum output power, but a significantly lower total distortion.
3. The output transformer is subject to saturation problems due to flow of direct current in transformer primary.

6.7.6. Output Transformer Saturation

For the proper operation of a transformer, the necessary requirement is that the flux in the transformer core varies in proportion with the variation in primary current. The transformer core is said to be saturated if the core flux does not change in proportion with the change in primary current. The effects of transformer core saturation are as follows:

1. With the increase in primary current of transformer, it is possible that there may not be a correct corresponding increase in magnetic flux to induce the proper voltage in the secondary.
2. The primary current may increase to a very high value due to sudden reduction in the primary impedance because the saturation of the core will reduce the primary inductance to a very low value.

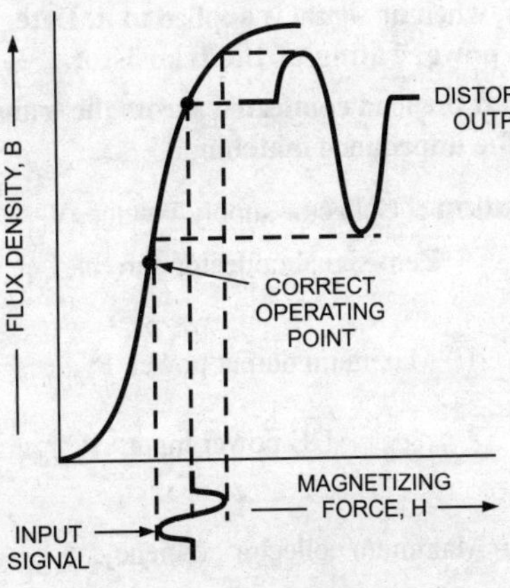

Fig. 6.14

In a single ended transformer coupled amplifier, the relatively large dc collector current flowing under zero signal condition may cause a large magnetic flux density in the transformer core, that may result in saturation of transformer core. Then, when the collector current increases due to increase in input signal, it is possible that magnetic flux may not increase in proportion to the collector current and so proper voltage may not be induced in the secondary. On the negative swing, however, the collector current would be able to fall and reduce the flux by the correct amount, inducing in the secondary a voltage larger than was obtained on the positive half-cycle, as shown in Fig. 6.14. Thus saturation of output transformer core may result in distorted output.

Example 6.4. A transformer coupled class A large signal amplifier has maximum and minimum values of collector-emitter voltage of 25 V and 2.5 V. Determine its collector efficiency.

Solution :

$$V_{ce\,max} = 25\text{ V}$$
$$V_{ce\,min} = 2.5\text{ V}$$

From Eq. (6.18)

$$\text{Collector efficiency, } \eta_{collector} = 50 \times \left(\frac{V_{ce\,max} - V_{ce\,min}}{V_{ce\,max} + V_{ce\,min}}\right)^2 \% = 50 \times \left(\frac{25 - 2.5}{25 + 2.5}\right)^2 = 33.5\% \textbf{ Ans.}$$

Example 6.5. A class A transformer-coupled amplifier uses 25:1 transformer to derive 4 Ω load. Calculate the effective load (seen by the transistor connected to the large turns side of the transformer).

[*Rajasthan Technical Univ. Analog Electronics 2009*]

Solution :

$$\text{Load resistance, } R_L = 4\ \Omega$$
$$\text{Turn ratio, } a = 25$$

Load as seen by transistor connected to large turns side of the transformer,

$$R'_L = a^2 \times R_L = 25^2 \times 4 = 2{,}500\ \Omega \text{ or } 2.5\text{ k}\Omega \textbf{ Ans.}$$

Example 6.6. A class A power amplifier has a transformer as the load. If the transformer has a turn ratio of 10 and the secondary load is 100 Ω, find the maximum ac power output. Given that zero-signal collector current is 100 mA.

[*Pb. Technical Univ. Analog Electronics Dec. 2005*]

Solution :

$$\text{Load resistance, } R_L = 100\ \Omega$$
$$\text{Turn ratio, } a = 10$$

Load, as seen by the transformer primary,

$$R_L' = a^2 R_L = (10)^2 \times 100 = 10{,}000\ \Omega$$

$$\text{For maximum power output, } I_{max} = 2 I_{CQ} \text{ and } I_{min} = 0$$

$$\text{So rms value of current, } I_{rms} = \frac{I_{max} - I_{min}}{2} \times \frac{1}{\sqrt{2}} = \frac{1}{2\sqrt{2}} \times 2I_{CQ} = \frac{I_{CQ}}{\sqrt{2}} = \frac{100}{\sqrt{2}}\text{ mA}$$

$$\text{Maximum power output, } P_{out\,(ac)} = I_{rms}^2\ R_L' = \frac{I_{CQ}^2}{2}\ R_L' = \frac{(100 \times 10^{-3})^2}{2} \times 10{,}000 = 50\text{ W} \textbf{ Ans.}$$

Example 6.7. A transformer coupled class A power amplifier draws a current of 200 mA from a collector supply of 10 V, when no signal is applied to it. Determine (*i*) maximum output power (*ii*) maximum collector efficiency and (*iii*) power rating of the transistor.

If the load connected across the transformer secondary is of 2 Ω and transformer turn ratio is 5 : 1, comment on the impedance matching.

Solution :

$$\text{Collector supply voltage, } V_{CC} = 10\text{ V}$$
$$\text{Zero-signal collector current, } I_{CQ} = 200\text{ mA} = 0.2\text{ A}$$

$$(i) \text{ Maximum output power, } P_{out\,(ac)} = \frac{V_{CC}\ I_{CQ}}{2} = \frac{10 \times 0.2}{2} = 1\text{ W} \textbf{ Ans.}$$

$$\text{DC power input, } P_{in\,(dc)} = V_{CC}\ I_{CQ} = 10 \times 0.2 = 2\text{ W}$$

$$(ii) \text{ Maximum collector efficiency, } \eta_{max} = \frac{P_{out\,max}}{P_{in\,(dc)}} \times 100 = \frac{1}{2} \times 100 = 50\% \textbf{ Ans.}$$

(*iii*) Power rating of the transistor = Zero-signal power dissipated = $V_{CC}\ I_{CQ} = 10 \times 0.2 = 2$ W **Ans.**

$$\text{Load, } R_L = 2\ \Omega$$

Load, as seen by the transformer primary, $R_L' = a^2 R_L = (5)^2 \times 2 = 50\ \Omega$, which is also the ratio of $\frac{V_{CC}}{I_{CQ}}\left(\frac{10}{0.2}\right)$. Thus the load is properly matched.

Example 6.8. A power transistor is to be used as a Class A transformer coupled amplifier and is to deliver a maximum of 5 W to a 4 Ω load. Operating point is adjusted for symmetrical clipping with collector supply voltage of 20 V. Assume ideal characteristics with V_{min} = 0 V. Calculate :

(*i*) Transformer turns ratio.

(*ii*) Peak collector current.

(*iii*) Operating point values of I_{CEQ} and V_{CEQ}.

(*iv*) Power dissipation rating of transistor.

(*v*) Collector circuit efficiency. [*M.D. Univ. Analog Electronics May, 2009*]

Solution : (*iii*) $V_{CEQ} = V_{CC} = 20$ V **Ans.**

$$I_{CQ} = \frac{2 \times P_{out\,(ac)}}{V_{CC}} = \frac{2 \times 5}{20} = 0.5 \text{ A } \textbf{Ans.}$$

(*ii*) Peak collector current, $I_{CQ\,max} = 2I_{CQ} = 2 \times 0.5 = 1$ A

$$(iii)\ R_L' = \frac{2 \times P_{out\,(ac)}}{I_{CQ}^2} = \frac{2 \times 5}{0.5^2} = 40\ \Omega$$

$$(i) \text{ Transformer turns ratio, } a = \sqrt{\frac{R_L'}{R_L}} = \sqrt{\frac{40}{4}} = 3.16 \textbf{ Ans.}$$

(*iv*) Power dissipation rating of transistor $= V_{CC}I_{CQ} = 20 \times 0.5 = 10$ W **Ans.**

$$(v) \text{ Maximum collector efficiency, } \eta = \frac{P_{out\,max}}{P_{in\,(dc)}} = \frac{5}{20 \times 0.5} = 50\% \textbf{ Ans.}$$

Example 6.9. A transistor rated for a maximum collector dissipation of 100 W operates a single ended class A output stage from a 10 V supply. Calculate the approximate values of (*i*) maximum undistorted ac power output, (*ii*) the quiescent current and (*iii*) turns ratio of the output transformer.

Assume load resistance of 16 Ω and overall collector efficiency as 0.5.

Solution : Maximum collector dissipation rating of transistor,

$$P_{in} = V_{CC}I_{CQ} = 100 \text{ W}$$

$$\text{Overall collector efficiency of, } \eta = 0.5$$

(*i*) Maximum undistorted ac power output, $P_{out(ac)} = V_{CC}I_{CQ} \times$ overall collector efficiency $= 100 \times 0.5 = 50$ W **Ans.**

$$(ii) \text{ Quiescent current, } I_{CQ} = \frac{P_{in\,(dc)}}{V_{CC}} = \frac{P_{in}}{V_{CC}} = \frac{100}{10} = 10 \text{ A } \textbf{Ans.}$$

$$\text{Load seen by the transformer primary, } R_L' = \frac{V_{CC}}{I_{CQ}} = \frac{10}{10} = 1\ \Omega$$

$$\text{Load resistance, } R_L = 16\ \Omega$$

(*iii*) Transformer turns-ratio (primary to secondary)

$$a = \sqrt{\frac{R_L'}{R_L}} = \sqrt{\frac{1}{16}} = \frac{1}{4} \textbf{ Ans.}$$

Example 6.10. A transformer coupled class A audio power amplifier driving an 8 Ω speaker is shown in Fig. 6.15. If the circuit component values result in a dc base current of 6 mA and the input signal V_{in} results in a peak base swing of 4 mA, determine (*i*) $V_{ce\ max}$ (*ii*) $V_{ce\ min}$ (*iii*) $I_{C\ max}$ (*iv*) $I_{c\ min}$ (*v*) rms values of load current and voltage (*vi*) ac power developed across the load (*vii*) power dissipated by the transistor and (*viii*) the efficiency of the amplifier circuit. Transistor output characteristics are shown in Fig. 6.16.

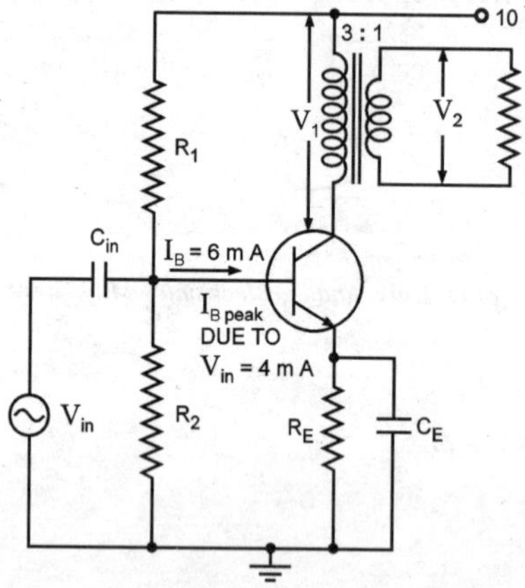

Fig. 6.15

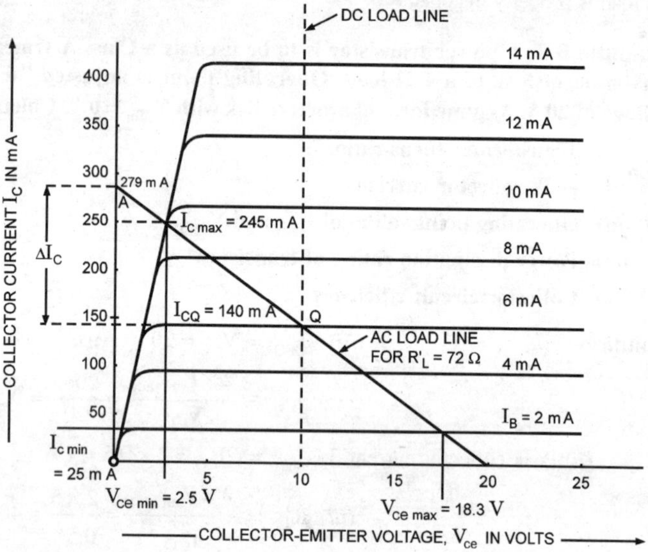

Fig. 6.16

Solution : $V_{CEQ} = V_{CC} = 10$ V, so dc load line drawn vertically from the voltage point $V_{CEQ} = 10$ V.

For base current $I_B = 6$ mA, and operatinng point Q in Fig. 6.16 gives

$$V_{CEQ} = 10\text{ V and } I_{CQ} = 140\text{ mA}$$

Effective ac resistance, $R_L' = a^2 R_L = (3)^2 \times 8 = 72\ \Omega$

$$\text{Current swing, } \Delta I_C = \frac{\Delta V_{CEQ}}{R_L'} = \frac{10\text{ V}}{72\ \Omega} = 139\text{ mA}$$

Now point A is marked corresponding to collector current of 140 + 139 = 279 mA on the current axis. AC load line is now drawn joining points A and Q.

For the given peak base current swing of 4 mA, from Fig. 6.16

(*i*) $V_{ce\ max} = 18.3$ V **Ans.**

(*ii*) $V_{ce\ min} = 2.5$ V **Ans.**

(*iii*) $I_{c\ max} = 245$ mA **Ans.**

(*iv*) $I_{c\ min} = 25$ mA **Ans.**

(*vi*) AC power developed across transformer primary,

$$P_{out\ (ac)} = \frac{(V_{ce\ max} - V_{ce\ min})(I_{c\ max} - I_{c\ min})}{8} = \frac{(18.3 - 2.5)(245 - 25) \times 10^{-3}}{8} = 0.434\text{ W } \textbf{Ans.}$$

(*v*) RMS value of voltage across primary,

$$V_{1\ rms} = \frac{V_{ce\ max} - V_{ce\ min}}{2\sqrt{2}} = \frac{18.3 - 2.5}{2\sqrt{2}} = 5.59\text{ V}$$

RMS value of voltage across load,

$$V_{L\ rms} = V_{2\ rms} = \left(\frac{1}{a}\right) V_{1\ rms} = \left(\frac{1}{3}\right) \times 5.59 = 1.86\text{ V } \textbf{Ans.}$$

RMS value of load current, $I_{rms} = \frac{V_{2\,rms}}{R_L} = \frac{1.86}{8} = 232.5$ mA **Ans.**

Power input, $P_{in\,(dc)} = V_{CC} \times I_{CQ} = 10 \times 140 \times 10^{-3} = 1.4$ W

(*vii*) Power dissipated in the transistor, $P_d = P_{in\,(dc)} - P_{out\,(ac)} = 1.4 - 0.434$ W $= 0.966$ W **Ans.**

(*viii*) Amplifier circuit efficiency, $\eta = \frac{P_{out\,(ac)}}{P_{in\,(dc)}} \times 100 = \frac{0.434}{1.4} \times 100 = 31\%$ **Ans.**

6.8. CLASS B POWER AMPLIFIERS

In class B operation the transistor is so biased that zero signal collector current is zero. Hence class B operation does not need any biasing system. The operating point is set at cutoff, as illustrated in Fig. 6.17. It remains forward biased for only half cycle of the input signal *i.e.,* its conduction angle is 180º.

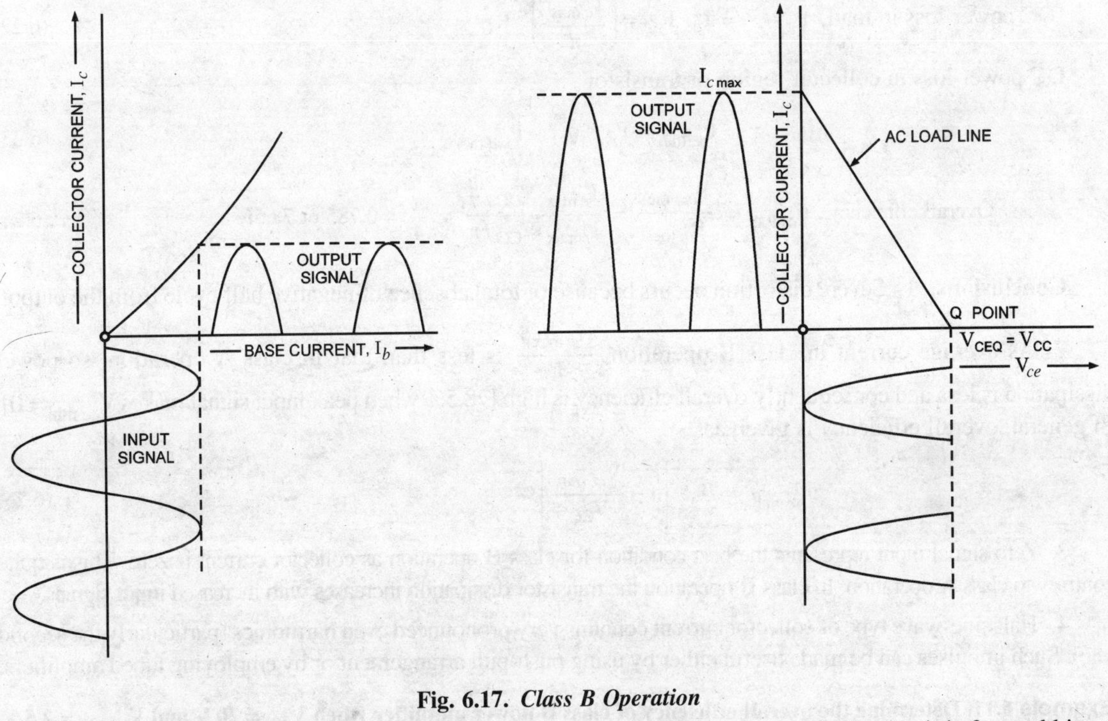

Fig. 6.17. ***Class B Operation***

As illustrated in Fig. 6.17, during the positive half cycle of the input ac signal, the circuit is forward biased and, therefore, collector current flows. On the other hand, during negative half cycle of the input ac signal, the circuit is reverse biased and no collector current flows.

6.8.1. Power and Efficiency Calculations

Input dc power, $P_{in\,(dc)} = V_{CC}\, I_{dc}$ where I_{dc} is the average or direct current taken from the collector supply.

If $I_{c\,max}$ is the maximum or peak value of collector or output current, then

$$I_{dc} = \frac{1}{2\pi}\int_0^{\pi} I_{c\,max} \sin\theta\, d\theta = \frac{I_{c\,max}}{2\pi}\left[-\cos\theta\right]_0^{\pi} = \frac{I_{c\,max}}{\pi} \qquad \because i_c = I_{c\,max} \sin\theta \text{ for } 0 < \theta < \pi$$

$$\therefore P_{in\,(dc)} = \frac{V_{CC}\, I_{c\,max}}{\pi} \qquad ...(6.20)$$

RMS value of collector current, $I_{c\,rms} = \frac{I_{c\,max}}{\sqrt{2}}$

RMS value of output voltage, $V_{rms} = \frac{V_{CC}}{\sqrt{2}}$

Hence output power during half cycle,

$$P_{out\,(ac)} = \frac{1}{2}\frac{I_{c\,max}}{\sqrt{2}} \times \frac{V_{CC}}{\sqrt{2}} = \frac{I_{c\,max}\, V_{CC}}{4} \qquad ...(6.21)$$

In above equation factor $\frac{1}{2}$ is used because power is developed during one half cycle only.

$$\text{DC power loss in load, } P_{R_{C(dc)}} = I_{dc}^2\, R_C = \left(\frac{I_{c\,max}}{\pi}\right)^2 R_C \qquad ...(6.22)$$

DC power loss in collector region or transistor

$$= P_{in\,(dc)} - P_{R_C\,(dc)} - P_{out\,(ac)} \qquad ...(6.23)$$

$$\text{Overall efficiency, } \eta_{overall} = \frac{P_{out\,(ac)}}{P_{in\,(dc)}} = \frac{I_{c\,max}\, V_{CC}/4}{I_{c\,max}\, V_{CC}/\pi} = \frac{\pi}{4} = 0.785 \text{ or } 78.5\% \qquad ...(6.24)$$

Conclusions. 1. Severe distortion occurs because of total absence of negative half cycle from the output.

2. As average current in class B operation $\frac{(I_{c\,max})}{\pi}$ is less than that in class A operation, so power dissipation is less and consequently overall efficiency is high [78.5% when peak input signal makes $V_{ce\,min} = 0$]. In general, overall efficiency is given as

$$\eta_{overall} = 78.5\left[1 - \frac{V_{ce\,min}}{V_{CC}}\right]\% \qquad ...(6.25)$$

3. Zero signal input represents the best condition for class B operation as collector current is zero. This is quite contrary to class A operation. In class B operation the transistor dissipation increases with increased input signals.

4. Half sine-wave type of collector current contains very pronounced even harmonics (particularly the second one). Such impulses can be made useful either by using push-pull arrangement or by employing tuned amplifiers.

Example 6.11. Determine the overall efficiency of class B power amplifier when V_{CC} = 20 V and $V_{ce\,min}$ = 2.5 V.

Solution : From Eq. (6.25)

$$\text{Overall efficiency, } \eta_{overall} = 78.5\left[1 - \frac{V_{ce\,min}}{V_{CC}}\right]\% = 78.5\left(1 - \frac{2.5}{20}\right) = 68.69\% \textbf{ Ans.}$$

Example 6.12. For a class B amplifier providing a 20 V peak signal to 16 Ω load (speaker) and a power supply of V_{CC} = 30 V, determine the input power, output power and circuit efficiency.

Solution : Maximum value of output signal, $V_{ce\,max}$ = 20 V

Maximum (or peak) value of collector or output current,

$$I_{c\,max} = \frac{V_{ce\,max}}{R_L} = \frac{20}{16} = 1.25 \text{ A}$$

$$\text{Input power, } P_{in(dc)} = \frac{V_{CC}\, I_{c\,max}}{\pi} = \frac{30 \times 1.25}{\pi} = 11.94 \text{ W} \quad \textbf{Ans.}$$

$$\text{Output power, } P_{out(ac)} = \frac{V_{CC}\, I_{c\,max}}{4} = \frac{30 \times 1.25}{4} = 9.375 \text{ W} \quad \textbf{Ans.}$$

$$\text{Circuit efficiency, } \eta = \frac{P_{out(ac)}}{P_{in(dc)}} \times 100 = \frac{9.375}{11.94} \times 100 = 78.5\% \quad \textbf{Ans.}$$

6.9. HARMONIC DISTORTION IN POWER AMPLIFIERS

Output signal variations of less than 360° of the signal cycle are considered to have distortion. This means that the output signal is no longer just an amplified version of the input signal but in some way a distorted or changed from that of the input. The poor quality of music coming from a radio or hi-fi system with the music or voice no longer sounding like that which was originally recorded or transmitted is the result of distortion.

A good amplifier should not only give an enlarged version of the input signal at the output, but should also provide a faithful reproduction of the input waveforms. One type of distortion that is common to most of the amplifiers is called the *amplitude* or *harmonic distortion*, which is caused due to nonlinearity of the active device employed for amplification. The active device (semiconductor device or tube) used for amplification may not increase equally all portions of the input signal over its positive and negative excursions. Harmonic distortion increases as we go from class A operation to class C operation. Even in class A operation some distortion occurs.

When nonlinear distortion is present, the output waveform contains components of frequencies which are harmonics (integer multiples) of the input signal frequency. There may be present second, third and higher order harmonic components, but the most important in terms of the amount of distortion for the classes of operation we will consider is the second harmonic. For a signal occurring in class AB or class B, the distortion may be mainly even harmonics of which the second harmonic component is the greatest.

An instrument such as a spectrum analyzer permits the measurement of the harmonics present in the signal by providing a display of the fundamental component of signal and a number of harmonics on a CRT screen. Similarly, a wave analyzer instrument allows a more precise measurement of the harmonic components of a distorted signal by filtering out each of these components and providing a reading of these components, one at a time. In any case the technique of considering any distorted signal as containing a fundamental component and harmonic components is practical and useful.

6.9.1. Determination of Second Harmonic Distortion

A collector current waveform is shown in Fig. 6.18, with the quiescent, minimum and maximum, signal levels, and the time they occur marked on the waveform. An equation which approximately describes the distorted signal waveform is

$$i_C \simeq I_{CQ} + I_0 + I_1 \cos(\omega t) + I_2 \cos(2\omega t) \qquad ...(6.26)$$

The given current waveform contains the original quiescent current I_{CQ}, that occurs with zero input signal, an additional dc current I_0, due to the nonzero average of the distorted signal; the fundamental component of the distorted ac signal I_1; and a second harmonic component I_2, at twice the fundamental frequency. Although other harmonics are present, only the second is

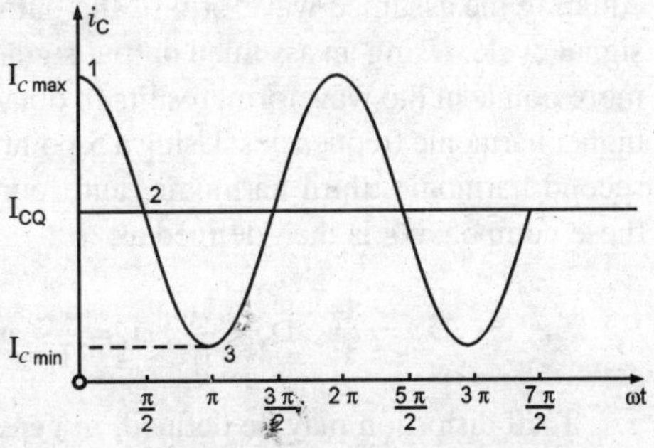

Fig. 6.18. *Waveform For Obtaining Second Harmonic Distortion*

considered here. Equating the resultant current from Eq. (6.26) at a few points in the cycle to that shown on the current waveform we get

At point 1 *i.e.*, when $\omega t = 0$

$$i_C = I_{c\,max} = I_{CQ} + I_0 + I_1 \cos(0) + I_2 \cos(0) = I_{CQ} + I_0 + I_1 + I_2 \quad ...(6.27)$$

At point 2 *i.e.*, when $\omega t = \frac{\pi}{2}$

$$i_c = I_{CQ} = I_{CQ} + I_0 + I_1 \cos\left(\frac{\pi}{2}\right) + I_2 \cos\left(\frac{2\pi}{2}\right) = I_{CQ} + I_0 - I_2 \quad ...(6.28)$$

$$\text{or} \quad I_0 = I_2 \quad ...(6.29)$$

At point 3 *i.e.*, when $\omega t = \pi$

$$I_c = I_{c\,min} = I_{CQ} + I_0 + I_1 \cos(\pi) + I_2 \cos(2\pi) = I_{CQ} + I_0 - I_1 + I_2 \quad ...(6.30)$$

Solving Eqs. (6.27), (6.29) and (6.30), we get

$$I_0 = I_2 = \frac{I_{c\,max} + I_{c\,min} - 2I_{CQ}}{4} \quad ...(6.31)$$

$$\text{and } I_1 = \frac{I_{c\,max} - I_{c\,min}}{2} \quad ...(6.32)$$

By definition the per cent of the second harmonic distortion is given as

$$D_2 = \frac{I_2}{I_1} \times 100\% = \frac{\frac{1}{2}(I_{c\,max} + I_{c\,min}) - I_{CQ}}{I_{c\,max} - I_{c\,min}} \times 100 \quad ...(6.33)$$

The second harmonic distortion is the percent of the second harmonic component present in the output current waveform with respect to the amount of the fundamental component. Obviously, 0% distortion is the ideal condition of no distortion.

In a similar way, the per cent of second harmonic distortion present in the output voltage waveform w.r.t. the amount of the fundamental component can be determined and is given as

$$D_2 = \frac{V_2}{V_1} \times 100\,\% = \frac{\frac{1}{2}(V_{ce\,max} + V_{ce\,min}) - V_{CEQ}}{V_{ce\,max} - V_{ce\,min}} \times 100 \quad ...(6.34)$$

6.9.2. Five Point Method of Determination of Harmonic Distortion in Power Amplifiers

The method of determination of second harmonic distortion is called the *3-point method* as it involved equating the assumed waveform of the output current/voltage to the measured voltage at three points in the signal cycle. Using an assumed output signal equation containing more harmonic terms, alongwith choosing more points in the waveform, results in obtaining relations for the magnitude of the harmonic components at higher harmonic frequencies. Using a 5-point method provides the dc component, first harmonic (fundamental), second harmonic, third harmonic, and fourth harmonic components. The harmonic distortion for each of these components is then defined as

$$D_2 = \frac{I_2}{I_1};\ D_3 = \frac{I_3}{I_1};\ D_4 = \frac{I_4}{I_1} \text{ and so on.} \quad ...(6.35)$$

Total distortion may be defined, in general, using the individual distortion components

$$D = \sqrt{D_2^2 + D_3^2 + D_4^2 + \cdots} \quad ...(6.36)$$

When distortion occurs, the output power computed on the basis of non-distortion is no longer correct. For example, the power computed from Eq. (6.8*c*) gives output power only for the non-distorted case. In case of a distorted output, the output power due to the fundamental component of the distorted signal is given as

$$P_1 = \frac{I_1^2 R_C}{2}$$

Similarly $P_2 = \frac{I_2^2 R_C}{2}$; $P_3 = \frac{I_3^2 R_C}{2}$ and so on.

Total power due to all harmonic components of the distorted signal is given as

$$P = (I_1^2 + I_2^2 + I_3^2 + ...) \frac{R_C}{2}$$

$$= \left[1 + \left(\frac{I_2}{I_1}\right)^2 + \left(\frac{I_3}{I_1}\right)^2 + ...\right] \frac{I_1^2 R_C}{2}$$

$$= [1 + D_2^2 + D_3^2 + ...] P_1 = (1 + D^2) P_1 \qquad ...(6.37)$$

If the total distortion is 10% *i.e.*, 0.1, then the total power is $[1 + (0.1)^2] P_1 = 1.01 P_1$

$$\text{Percentage increase in power} = \frac{1.01P_1 - P_1}{P_1} \times 100 = 1.0\%$$

i.e., a 10% distortion represents a power increase of only 1% of the fundamental. Thus, only a small error is introduced in using only the fundamental term P_1 for determination of the output power.

Example 6.13. An output waveform displayed on an oscilloscope provided the following measured values.

(*a*) $V_{CE\,min} = 1.2$ V, $V_{CE\,max} = 22$ V, $V_{CEQ} = 10$ V

(*b*) $V_{CE\,min} = 2$ V, $V_{CE\,max} = 18$ V, $V_{CEQ} = 10$ V

Determine the per cent second harmonic distortion in each case.

[*Pb. Technical Univ. Electronic Devices and Circuits Dec. 2009*]

Solution : (*a*) According to Eq. (6.34)

$$D_2 = \frac{\frac{1}{2}(V_{CE\,max} + V_{CE\,min}) - V_{CEQ}}{V_{CE\,max} - V_{CE\,min}} \times 100 = \frac{\frac{1}{2}(22 + 1.2) - 10}{22 - 1.2} \times 100 = 7.7\% \textbf{ Ans.}$$

(*b*) $D_2 = \frac{\frac{1}{2}(18 + 2) - 10}{18 - 2} \times 100 = 0\%$ *i.e.*, no distortion

Example 6.14. A sinusoidal signal $V_s = 1.95 \sin 400t$ is applied to a power amplifier. The resulting current is

$$i_0 = 12 \sin 400\,t + 1.2 \sin 800\,t + 0.9 \sin 1{,}200\,t + 0.4 \sin 1{,}600\,t$$

Calculate (*i*) the total harmonic distortion (*ii*) the percentage increase in power because of distortion.

Solution : Harmonic distortion of each component is given as

$$D_2 = \frac{V_2}{V_1} = \frac{1.2}{12} = 0.1$$

$$D_3 = \frac{V_3}{V_1} = \frac{0.9}{12} = 0.075$$

$$D_4 = \frac{V_4}{V_1} = \frac{0.4}{12} = 0.0333$$

$$\text{Total distortion, } D = \sqrt{D_2^{\,2} + D_3^{\,2} + D_4^{\,2}} = \sqrt{(0.1)^2 + (0.075)^2 + (0.0333)^2} = 0.1294 \text{ or } 12.94\% \textbf{ Ans.}$$

$$\text{Total power, } P = (1 + D^2)\,P_1 = [1 + (0.1294)^2]P_1 = 1.01674P_1$$

Percentage increase in power because of distortion

$$= \frac{P - P_1}{P_1} \times 100 = \frac{(1.01674 - 1)\,P_1}{P_1} \times 100 = 1.674\% \textbf{ Ans.}$$

Example 6.15. For a given input current $i_i = 50 \times 10^{-3}$ sin $2\omega t$, the output current given by a class A amplifier is $i_0 = 2 \times 10^{-2} + 50 \times 10^{-2}$ sin $200t + 10^{-4}$ sin $400t + 3 \times 10^{-6}$ cos $600t$. Determine the distortion factor for it.

[Rajasthan Technical Univ. Analog Electronics 2008]

Solution : Considering the magnitude of fundamental component, second and third harmonic components as

$I_0 = 2 \times 10^{-2}$ dc component

$I_1 = 50 \times 10^{-2}$ fundamental component

$I_2 = 10^{-4}$ second harmonic component

$I_3 = 3 \times 10^{-6}$ third harmonic component

Percentage harmonic distortion of various components is given by

$$D_2 = \frac{I_2}{I_1} \times 100 = \frac{10^{-4}}{50 \times 10^{-2}} \times 100 = 0.02\%$$

$$D_3 = \frac{I_3}{I_1} \times 100 = \frac{3 \times 10^{-6}}{50 \times 10^{-2}} \times 100 = 0.0006\%$$

$$\text{Distortion factor, } D = \sqrt{D_2^2 + D_3^2} = \sqrt{(0.02)^2 + (0.0006)^2} = 0.02\% \textbf{ Ans.}$$

6.10. VARIATION OF OUTPUT POWER IN TRANSFORMER COUPLED POWER AMPLIFIER

The variations of output power and the distortion with the reflected load R'_L, for a typical transistor amplifier, are shown in Fig. 6.19.

The curves shown in Fig. 6.19 reveal that the load that provides minimum total harmonic distortion is different from the load that gives maximum power output. Fortunately, the power curve is fairly flat in the region of maximum power so that a well-designed amplifier would work into a load that provides power slightly less than the maximum output power, but a significantly lower distortion.

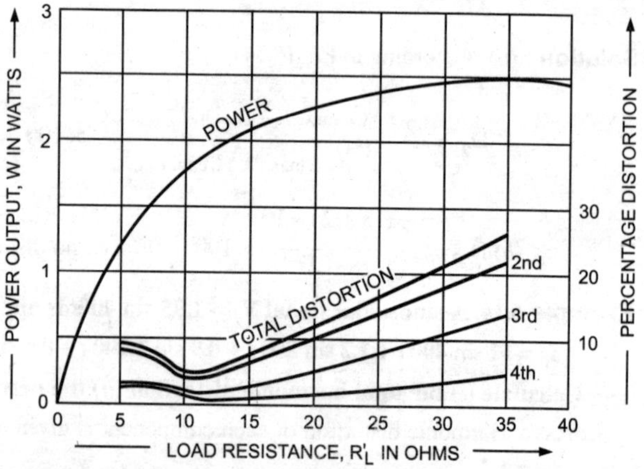

Fig. 6.19. ***Power Output and Distortion For a Typical Transistor Amplifier as a Function of Load Resistance, R'_L***

6.11. PUSH-PULL AMPLIFIERS

As already mentioned in Art. 6.4, *double-ended or push-pull* amplifier makes use of two identical transistors in a single stage. It consists of two loops in which the transistor collector currents flow in opposite directions but add in the load. Push-pull amplifiers eliminate all of the three drawbacks listed in Art. 6.7.5.

The push-pull amplifier is a power amplifier and is frequently used in the output stages of electronic circuits. It is employed whenever high output at high efficiency is required. The audio power amplifiers used in transistor receivers, tape recorders, record-players, public address systems etc. make use of such circuits. These systems are usually operated by batteries or cells and, therefore, amplifier efficiency is of prime importance.

6.11.1. Basic Principle of Working of Push-Pull Amplifier

The basic principle on which a push-pull amplifier operates is that the input signal is converted before amplification, into two separate signals, which are identical except for a 180° phase difference. Each of these signals is applied as the input to one of the transistors of the push-pull amplifier. Since the inputs to the two transistors are 180° out of phase, so the output of the transistors. The output transformer is connected in the collector circuits of the transistors in such a way that the collector currents of the two transistors combine to provide an overall signal having the same waveform as the original input signal. The magnitude of this combined output signal is larger than the input signal. Push-pull amplifiers can be operated in class A, class AB or class B mode.

6.12. CLASS A PUSH-PULL AMPLIFIER

A class A push-pull amplifier circuit is shown in Fig. 6.20. By class A push-pull amplifier means that current flows in the output of the active device (each transistor) for the entire of the input cycle.

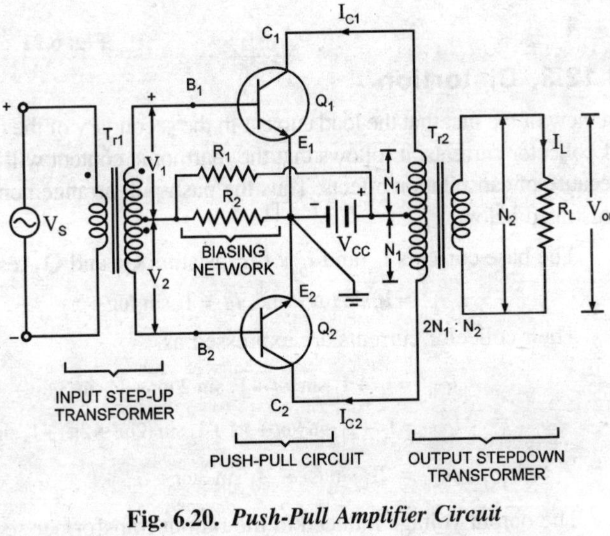

Fig. 6.20. ***Push-Pull Amplifier Circuit***

6.12.1. Circuit Arrangement

The circuitry of a typical push-pull amplifier is shown in Fig. 6.20. As already mentioned, push-pull amplifier uses two identical transistors, say Q_1 and Q_2. The emitter terminals of the two transistors Q_1 and Q_2 connected together. The input signal is applied to the inputs of two transistors through centre tapped step-up transformer T_{r_1} which provides opposite polarity signals to the two transistor inputs. The collector terminals of both the transistors are connected to end terminals of centre tapped primary of output transformer T_{r_2}. The power supply V_{CC} is connected between the emitter terminals and the centre tap of primary of output transformer. Resistors R_1 and R_2 are used to provide the proper bias for the circuit. The load R_L is connected across the secondary of the output transformer T_{r_2}. The turn ratio ($2N_1 : N_2$) of the output transformer is chosen so as to match the load with the output impedance of the amplifier and, therefore, transfer maximum power. The quiescent currents of the two transistors, which are equal in magnitude, flow in opposite directions through each half of primary of the output transformer T_{r_2}, so no saturation of the magnetic core occurs.

6.12.2. Circuit Operation

When the base current of one transistor is being driven positive with respect to the quiescent point Q, the collector current increases, thus causing a decrease in collector potential relative to ground. At the same time, however, a reverse action takes place in the base circuit of the second transistor, *i.e.,* base current decreases causing a drop in the collector current with a consequent rise in collector potential w.r.t. ground. This means that the ac current flowing through the transformer primary winding is in the same direction. As i_{C_1} increases (*i.e.,* pulls), the current i_{C_2} decreases (*i.e.,* pushes). Hence the name *push-pull amplifier.* The overall operation results in ac voltage induced in the secondary of the output transformer and thus ac power is delivered to the load. The difference of two collector currents is illustrated in Fig. 6.21.

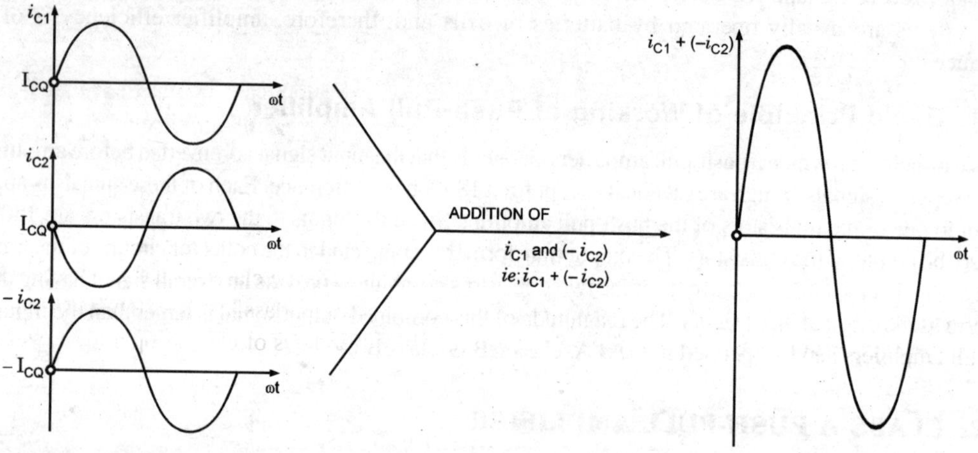

Fig. 6.21

6.12.3. Distortion

In view of the fact that the load current in the secondary of the output transformer is proportional to the difference of collector currents, it follows that the harmonic content will be less than it is in either of the collector currents because of cancellation effects. Thus the push-pull arrangement yields much less distortion in the output. This is illustrated below.

The base currents i_{b_1} and i_{b_2} of transistors Q_1 and Q_2 respectively are expressed as

$$i_{b_1} = I_b \sin(\omega t) \text{ and } i_{b_2} = I_b \sin(\omega t + \pi)$$

Their collector currents are expressed as

$$i_{C_1} = I_C + I_1 \sin \omega t + I_2 \sin 2\omega t + I_3 \sin 3\omega t + ...$$

$$i_{C_2} = I_C + I_1 \sin(\omega t + \pi) + I_2 \sin(2\omega t + 2\pi) + I_3 \sin(3\omega t + 3\pi) = I_C - I_1 \sin \omega t + I_2 \sin 2\omega t - I_3 \sin 3\omega t + ...$$

$$\text{and } i_{C_1} - i_{C_2} = 2I_1 \sin \omega t + 2I_3 \sin 3\omega t + ...$$

The output voltage induced in the output transformer secondary, being proportional to $(i_{C_1} - i_{C_2})$,

$$v_{out} = K(i_{C_1} - i_{C_2}) = 2K[I_1 \sin \omega t + I_3 \sin 3\omega t + I_5 \sin 5\omega t + ...]$$

From the above equation for output voltage v_{out} it is obvious that all even harmonics are eliminated. This conclusion was reached on the assumption that the two transistors are identical. If their characteristics differ appreciably, the appearance of even harmonics is not ruled out.

6.12.4. Advantages

1. Because of absence of even harmonics in the output of the push-pull amplifier, such a circuit gives more output per active device for a given amount of distortion or less distortion for a given power output per transistor.
2. As already mentioned earlier, the dc components of the collector current oppose each other magnetically in the transformer core. This eliminates any tendency toward core saturation and consequent nonlinear distortion that may arise from the curvature of the transformer magnetization curve.
3. Another advantage of this system is that the effects of ripple voltage that may be contained in the power supply because of inadequate filtering will be balanced out. This cancellation results because the currents

produced by the ripple voltage are in opposite directions in the transformer winding, and so will not appear in the load. Of course, the power supply hum will also act on the voltage amplifier stages, and so will be part of the input to the power stage. This hum will not be eliminated by the push-pull circuit.

6.12.5. Disadvantages

The drawbacks of push-pull amplifiers are (*i*) requirement of two identical* or matched pair transistors (*ii*) need of use of driver stage to furnish two equal and opposite voltages at the input and (*iii*) need of bulky and expensive transformers.

6.13. CLASS B PUSH-PULL AMPLIFIER

The circuitry for the class B push-pull operation is the same as that for the class A operation except that the devices are biased at cutoff. The transistor circuit of Fig. 6.20 operates as class B if $R_2 = 0$ because a transistor is essentially at cutoff if the base is shorted to the emitter. The advantages of class B operation over class A operation are given below :

1. It is possible to obtain greater power output—class B push-pull amplifier provides practically 4-times the power supplied by a single ended amplifier provided load resistance remains the same.
2. It gives higher operating efficiency (theoretical 78.5%). It is primarily due to the fact that no power is drawn by the circuit under zero signal condition.

Because of above advantages, a push-pull class B transistor circuit is preferred in systems where the power supply is limited, such as those operating from solar cells or a battery. The automatic cancellation of all even-order harmonics from the output makes class B push-pull amplifiers highly desirable for communication sound equipment.

The drawbacks of class B push-pull amplifiers are that the harmonic distortion is higher, self-bias cannot be used, and the supply voltages must have good regulation.

Class B push-pull amplifiers are widely employed for audio work in portable record-players, as stereo amplifiers and in high-fidelity radio receivers.

6.13.1. Circuit Arrangement

Circuitry for a class B push-pull amplifier is shown in Fig. 6.22. As illustrated in the figure, the bases of the two CE-connected identical transistors Q_1 and Q_2 have been connected to the opposite ends of the secondary

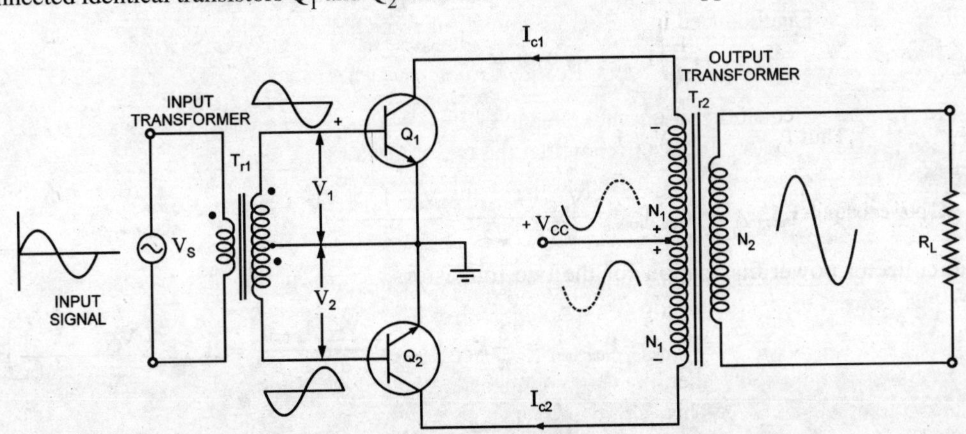

Fig. 6.22. ***Class B Push-Pull Amplifier***

* If the parameters of the two transistors used in the circuit are not identical, there will be unequal amplification of the two halves of input signal.

of the input transformer T_{r_1}. Load resistance R_L is coupled to the collector terminals of the two transistors through output transformer T_{r_2}. The emitter terminals of the transistors are connected to the centre tap of the secondary of the input transformer T_{r_1} while the supply V_{CC} is connected to the centre tap on the primary of output transformer. This is done in order to have balanced circuits. Since zero bias is required for cutoff, the two bases of the two transistors have been grounded.

6.13.2. Circuit Operation

When the input signal is applied, the centre tapped secondary of the input transformer develops two signals which are identical but in phase opposition. The transistors, Q_1 and Q_2 are driven by these two signals. Thus when V_1 is going positive, V_2 is going negative, so that transistor Q_2 is being biased further off when transistor Q_1 is being biased on. As the collector current in Q_1 increases from zero, it produces a half sine wave of voltage across the upper half of the primary of the output transformer, as shown. When the positive half cycle of the input signal to Q_1 begins to go negative, the signal at Q_2 base is commencing to go positive. Thus, as Q_1 becomes biased off again, Q_2 is biased on and a half cycle of voltage waveform is generated across the lower primary winding of output transformer.

The effect of the two half cycles in separate halves of the primary of output transformer is to produce a magnetic flux in the transformer core, which flows first in one direction and then in the opposite direction. This flux links with the secondary of the output transformer and generates a complete sine wave output, which is passed on to the load. In the class B circuit, the two output transistors are said to be operating in push-pull.

As explained above, a full sine wave is developed at the output of the transformer. When no signal is present, both transistors remain off, and thus no current is drawn from dc supply source V_{CC}. Hence there is no dissipation of power with zero signal. Some power is dissipated in each transistor only while the transistor is conducting.

The load resistance (usually a loudspeaker) is connected across the secondary of the output transformer. The turn-ratio $2N_1 : N_2$ of the transformer is chosen so that load resistance R_L is matched with the output impedance of the transistor. Undermatched conditions, maximum power is delivered to the load by the amplifier.

6.13.3. Conversion Efficiency of Class B Amplifier

The power provided to the load R_L coupled to the amplifier is drawn from the dc power supply V_{CC} and is considered an input dc power. The input dc power is given as

$$P_{in\,(dc)} = V_{CC} I_{dc} \quad \text{where } I_{dc} \text{ is the average or direct current taken from the dc supply } V_{CC}$$

In class B operation the current drawn from a single power supply is a full-wave rectified signal, thus

$$I_{dc} = \frac{1}{\pi}\int_0^{\pi} I_{c\,max} \sin\theta \, d\theta = \frac{2I_{c\,max}}{\pi}$$

$$\text{Thus } P_{in\,(dc)} = \frac{2}{\pi} V_{CC} I_{c\,max} \qquad \text{...(6.38)}$$

$$\text{AC power output, } P_{out\,(ac)} = V_{rms} I_{rms} = \frac{V_{CC}}{\sqrt{2}} \cdot \frac{I_{c\,max}}{\sqrt{2}} = \frac{V_{CC} I_{c\,max}}{2} \qquad \text{...(6.39)}$$

Total collector power dissipation for the two transistors,

$$= P_{in\,(dc)} - P_{out\,(ac)} = \frac{2}{\pi} V_{CC} I_{c\,max} - \frac{V_{CC} I_{c\,max}}{2} = 2I_{c\,max}\left(\frac{V_{CC}}{\pi} - \frac{V_{CC}}{4}\right)$$

$$\text{Overall efficiency, } \eta_{overall} = \frac{P_{out\,(ac)}}{P_{in\,(dc)}} = \frac{V_{CC}\dfrac{I_{c\,max}}{2}}{2V_{CC}\dfrac{I_{c\,max}}{\pi}} = \frac{\pi}{4} = 0.785 \text{ or } 78.5\% \qquad \text{...(6.40)}$$

Example 6.16. Determine the input power, output power and efficiency resulting in a class-B push-pull amplifier providing a signal of 20 V peak to a 16 Ω load, using a single supply of V_{CC} = 30 V.

[U.P.S.C. I.E.S. Electronics & Telecommunication Engineering-II, 2006]

Solution : Maximum value of collector current,

$$I_{c\,max} = \frac{V_{CC}}{R'_L} = \frac{30}{16} = 1.875 \text{ A}$$

$$\text{Input power, } P_{in(dc)} = \frac{2}{\pi} V_{CC} I_{c\,max} = \frac{2}{\pi} \times 30 \times 1.875 = 35.8 \text{ watts} \quad \textbf{Ans.}$$ ...Refer to Eq. (6.38)

$$\text{AC power output, } P_{out\,(ac)} = \frac{V_{CC} I_{c\,max}}{2} = \frac{30 \times 1.875}{2} = 28.125 \text{ watts} \quad \textbf{Ans.}$$ ...Refer to Eq. (6.39)

$$\text{Overall efficiency, } \eta = \frac{P_{out\,(ac)}}{P_{in\,(dc)}} \times 100 = \frac{28.125}{35.8} \times 100 = 78.5\% \quad \textbf{Ans.}$$

Example 6.17. A class B push-pull amplifier must deliver 10 W of audio power to the output load. (*i*) If the output transformer is 80% efficient, what is the minimum power drain on the power supply under optimum conditions and (*ii*) what is the minimum average dissipation rating required for each transistor ?

Solution : Power delivered to load, P_L = 10 W

$$\text{AC power output of amplifier, } P_{out\,(ac)} = \frac{P_L}{\text{Output transformer efficiency}} = \frac{10}{0.8} = 12.5 \text{ W}$$

Efficiency of a class B push-pull amplifier under optimum conditions, η = 78.5% = 0.785

(*i*) Minimum power drain on the power supply $= P_{in\,(dc)} = \dfrac{P_{out\,(ac)}}{\eta} = \dfrac{12.5}{0.785} = 15.9 \text{ W}$ **Ans.**

(*ii*) Minimum average power dissipation, $P_d = P_{in\,(dc)} - P_{out\,(ac)} = 15.9 - 12.5 = 3.4 \text{ W}$

Minimum average dissipation rating required for each transistor $= \dfrac{P_d}{2} = \dfrac{3.4}{2} = 1.7 \text{ W}$ **Ans.**

Example 6.18. A class B push-pull power amplifier is supplied with V_{CC} = 50 V. The signal swings the collector voltage down to V_{min} = 5 V. The total dissipation in both transistors is 40 W. Find the total power and conversion efficiency. *[Pb. Technical Univ. Analog Electronics Dec. 2006; M.D. Univ. Analog Electronic Circuits May 2007]*

Solution : DC supply voltage, V_{CC} = 50 V

V_{min} = 5 V

Total power dissipation, P_d = 40 W

Also total power dissipation, $P_d = P_{in\,(dc)} - P_{out\,(ac)}$

$$\therefore 40 = \frac{2}{\pi} V_{CC} I_{c\,max} - \frac{I_{c\,max}}{2} (V_{CC} - V_{c\,min}) = I_{c\,max}\left[\frac{2V_{CC}}{\pi} - \frac{V_{CC} - V_{c\,min}}{2}\right]$$

$$= I_{c\,max}\left[\frac{2 \times 50}{\pi} - \frac{50 - 5}{2}\right] = 9.33 I_{c\,max}$$

$$\text{or } I_{c\,max} = \frac{40}{9.33} = 4.287 \text{ A}$$

$$\text{Total power input, } P_{in\,(dc)} = \frac{2}{\pi} V_{CC} I_{c\,max} = \frac{2}{\pi} \times 50 \times 4.287 = 136.45 \text{ W} \quad \textbf{Ans.}$$

$$\text{AC power output, } P_{out}\,(ac) = \frac{I_{c\,max}}{2} (V_{CC} - V_{c\,min}) = \frac{4.287}{2} \times (50 - 5) = 96.46 \text{ W}$$

$$\text{Conversion efficiency, } \eta = \frac{P_{out\,(ac)}}{P_{in\,(dc)}} \times 100 = \frac{96.46}{136.45} \times 100 = 70.7\% \quad \textbf{Ans.}$$

Example 6.19. Consider the idealised push-pull class B power amplifier shown in Fig. 6.20. Given that $R_2 = 0$, $V_{CC} = 20$ V, $N_2 = 2N_1$, $R_L = 20\ \Omega$ and the transistors have $h_{FE} = 20$. The input is sinusoid. For the maximum output signal at $V_{max} = V_{CC}$ determine (*i*) the output signal power, (*ii*) the collector dissipation in each transistor.

[*U.P.S.C. I.E.S. Electrical Engineering II 2006*]

Solution : Load seen by the transformer primary,

$$R'_L = R_L \times \left(\frac{N_1}{N_2}\right)^2 = R_L \left(\frac{N_1}{2N_1}\right)^2 = \frac{20}{4}\ \Omega = 5\ \Omega$$

Peak value of ac output current, $I_{max} = \dfrac{V_{max}}{R'_L} = \dfrac{V_{CC}}{R'_L}$

(*i*) Thus ac power output, $P_{out(ac)} = I^2_{rms}\, R'_L = \dfrac{I^2_{max}}{2} R'_L = \dfrac{1}{2}\dfrac{V^2{}_{CC}}{R'_L} = \dfrac{V^2_{CC}}{2R'_L} = \dfrac{(20)^2}{2 \times 5} = 40$ W **Ans.**

Power input, $P_{in\ (dc)} = \dfrac{2}{\pi} V_{CC}\, I_{c\ max} = \dfrac{2}{\pi} V_{CC} \times \dfrac{V_{CC}}{R'_L} = \dfrac{2V^2_{CC}}{\pi R'_L} = \dfrac{2 \times 20^2}{\pi \times 5} = 50.93$ W.

Total collector power dissipation for the two transistors

$$= P_{in(dc)} - P_{out(ac)} = 50.93 - 40 = 10.93 \text{ W}$$

(*ii*) Collector dissipation in each resistor $= \dfrac{10.93}{2} = 5.465$ W. **Ans.**

6.14. CROSSOVER DISTORTION

In addition to the distortion introduced due to the nonlinearity of the collector characteristics and due to non-matching of the two transistors, there is one more source of distortion, that is caused by nonlinearity of the input characteristic.

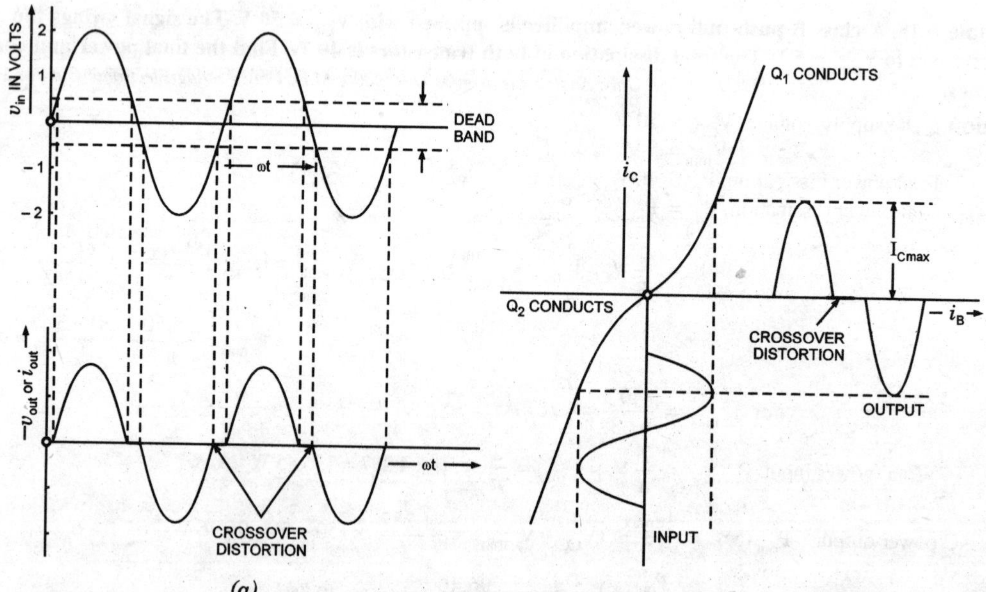

(*a*) (*b*)

Fig. 6.23. ***Crossover Distortion***

Recall that silicon transistors must have at least 0.5 V to 0.6 V of forward base-emitter bias before they will go into conduction. Since in class B push-pull amplifiers the forward bias is produced by the input signal, both of the transistors will be non-conducting (or off), when the input signal is approximately ± 0.5 V. This forms a "dead band" in the input and produces *crossover distortion* in the output. In simple words, crossover distortion occurs as a result of one transistor cutting off before the other begins conducting, as illustrated in Fig. 6.23 (*a*). The distortion introduced is called the crossover distortion because it occurs during the time operation crossover from one transistor to the other in the push-pull amplifier. The same is shown in Fig. 6.23 (*b*) by using transfer characteristics of the two transistors.

To eliminate crossover distortion, it is necessary to add a small amount of forward bias to take the transistors to the average of conduction or slightly beyond. This does slightly lowers the efficiency of the circuit and there is a waste of standby power, but it alleviates the crossover distortion problem. Technically the operation of transistors lies between class B and class A mode. Therefore, the circuit operation is often referred to as being class AB operation (refer to Art. 6.15).

6.15. CLASS AB PUSH-PULL AMPLIFIERS

The basic circuit of class A B push-pull amplifier is the same as that of class A push-pull amplifier shown in Fig. 6.20 except that the voltage drop across resistor R_2 is so adjusted that it is approximately equal to cutin voltage (0.5 V for *Si* and 0.1 V for *Ge*). Now operation becomes AB type *i.e.*, the collector current flows for more than half the cycle of the input signal but less than the complete cycle. The distortion introduced in class AB push-pull amplifier is less than that in class B push-pull amplifier but more than in class A push-pull amplifier. The drawbacks of class AB push-pull amplifiers in comparison to class B push-pull amplifiers are lower conversion efficiency and wastage of standby power.

6.16. PHASE-SPLITTER CIRCUIT FOR PUSH-PULL AMPLIFIER [OR PHASE INVERTER]

In a push-pull amplifier circuit an input transformer is used to provide the polarity inversion between the two input signals to the push-pull circuit, but the transformer is quite expensive and bulky. The same result may be achieved by using a circuit called the *phase-splitter* or *phase invertor*. Such a circuit is shown in Fig. 6.24.

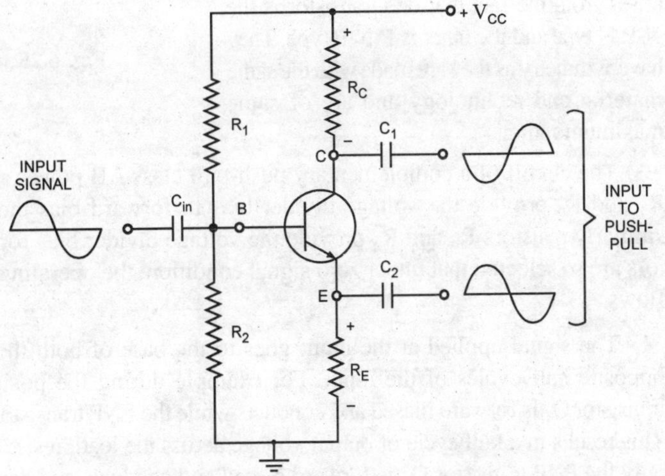

Fig. 6.24. ***Phase-Splitter Circuit For Push-Pull Amplifier***

The signal appearing at the collector is of polarity opposite to that of the input signal. During positive half cycle of input voltage, base current, and therefore, collector current increases causing increase in $I_c R_C$ drop and consequently decrease in collector-emitter voltage V_{ce}. Thus we get negative half cycle of V_{ce} *i.e.*, phase of signal at collector of a CE transistor is reversed. The output from the emitter is of the same polarity as the input. Thus the two output signals obtained from this circuit are of opposite polarity. Values of resistors R_C, R_E and β are chosen so as to make the voltage gain for the output signal at collector to be unity. The gain for signal at emitter is 1. Thus the circuit gives two signals of equal magnitude but of opposite polarities. The use of input transformer can thus be eliminated.

However, the circuit explained above has three shortcomings as follows :

1. Since this circuit is of the emitter-follower type, there is no voltage gain.
2. Since the emitter and collector currents are not equal, the output voltages will have some imbalance.
3. This phase inverter could not be used to drive a class B push-pull amplifier because output voltages are only equal if driving a resistive load over the entire cycle.

This circuit can be used on a class A push-pull circuit, where the load is constant on each output.

6.17. COMPLEMENTARY SYMMETRY PUSH-PULL AMPLIFIERS

The use of transformers, at input as well as at output ends, in the push-pull amplifier shown in Fig. 6.20 makes it bulky and expensive especially in this age of the integrated circuits. Another drawback of the circuit given in Fig. 6.20 is that it needs two out-of-phase signals which necessitates an input tapped transformer or phase inverter, and thus makes the driver circuitry quite complicated.

The above two drawbacks of an ordinary push-pull amplifier have been overcome in the complementary symmetry push-pull amplifier.

This arrangement uses two transistors having complementary symmetry (one N-P-N and another P-N-P). The term complementary arises from the fact that one transistor is the N-P-N type and the other is P-N-P type. They have symmetry as they are made with the same material and technology and are of same maximum rating.

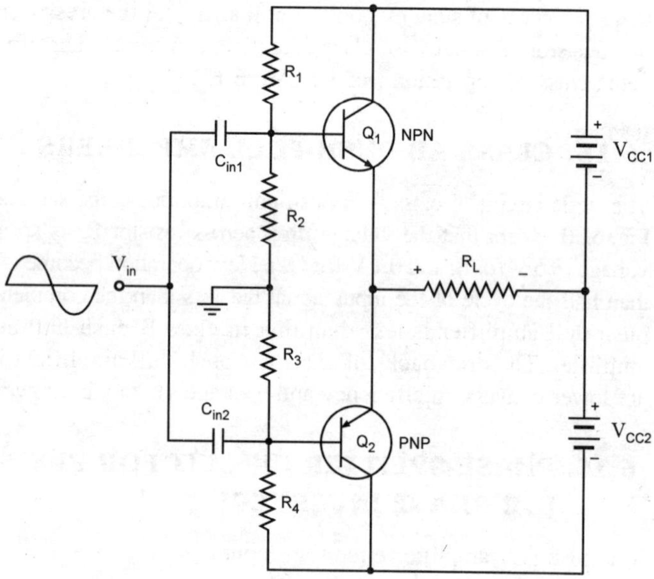

Fig. 6.25. ***Complementary Symmetry Push-Pull Class B Power Amplifier***

The circuit of a complementary push-pull class AB power amplifier is shown in Fig. 6.25. The resistors R_1 and R_2 provide the voltage divider bias to forward bias the emitter-base junction of transistor Q_1 and similarly resistors R_3 and R_4 provide the voltage divider bias for emitter junction of transistor Q_2. The resistors are so selected that under zero signal condition, the operating point is at cutoff and so no collector current flows.

The signal applied at the input goes to the base of both the transistors. The two transistors conduct in opposite half cycles of the input. For example during the positive half cycle of the input signal, the NPN transistor Q_1 is forward biased and conducts while the PNP transistor Q_2 is reverse biased and so does not conduct. This results in a half cycle of output voltage across the load, resistor R_L. Similarly during the negative half cycle only the PNP transistor Q_2 is forward biased and conducts and develops second half cycle of the output voltage across the load resistor R_L. Transistor Q_1 being reverse biased does not conduct during the negative half cycle of the input signal. Thus, during a complete cycle of input, a complete cycle of the output signal is developed.

Since collector current from each transistor flows through the load during alternate half cycles of the input signal, no centre tapped output transformer is required. However, for impedance matching an output transformer (without centre tapping) can be used.

One obvious drawback of this circuit arrangement is the requirement for two supply voltages. Another drawback of this circuit is difficulty of obtaining matched complementary transistors. If there is an unbalance in the characteristics of the two transistors, even harmonics will no longer be cancelled and considerable distortion will be introduced. Very often negative feedback is used in power amplifiers to reduce nonlinear distortion.

Highlights

1. Power amplifier is meant to raise the power level of the input signal. In order to get large power at the output, it is necessary that the input signal voltage is large. That is why, in an electronic system, a voltage amplifier always precedes the power amplifier, as illustrated in the block diagram (Fig. 6.1). Also, that is why power amplifiers are called large signal amplifiers.
2. A power amplifier is required to deliver a large amount of power and as such it has to handle large current. So in order to achieve high power amplification, (*i*) large sized power transistors are used in order to dissipate the heat produced in transistor during operation, (*ii*) transistor having thick base (*i.e.*, transistors with comparatively smaller β) are used in order to handle large currents (*iii*) transformer coupling is used for impedance matching and (*iv*) collector resistance is made low. If its value is high, power losses in the collector resistor will be increased. Moreover, voltage drop across it will increase and, therefore, operating voltage will be reduced. This is the reason that low collector load resistance is used.
3. The power amplifiers are primarily divided into two categories viz. *audio power amplifiers* and *radio power amplifiers*.

   Audio power amplifiers, also called the *small signal power amplifiers*, raise the power levels of signals that have audio frequency range (20 Hz – 20 kHz).

   Radio power amplifiers, also called *large signal power amplifiers* raise the power level of signals that have radio frequency range. They amplify a specific frequency or narrowband of frequencies while rejecting all other frequencies.
4. Important features of large signal amplifiers are : Necessity of impedance matching; Use of power transistors; Presence of harmonic distortion in the output; Capability of handling a large power; Bulky in size.
5. A class A power amplifier is defined as a power amplifier in which output current flows for the full cycle duration of the input ac signal.
6. In a class A amplifier utilising a direct coupled resistive load, the maximum conversion efficiency is given as

$$\eta_{max} = \frac{P_{out\,(ac)\,max}}{P_{in\,(dc)}} \times 100 = 25\%$$

7. In a class A power amplifier, maximum power dissipation occurs in the transistor under zero signal condition. The power dissipation capacity of a power transistor, for class A operation, must be at least equal to the zero signal rating.
8. When a signal is applied to class A power amplifier, the power dissipated in the transistor reduces by an amount equal to the ac power delivered to the load.
9. A class A power amplifier with direct coupled resistive load can supply only one-half the amount of ac power to the load that the transistor can dissipate and the dc power input is four times the amount of ac signal power in the load.
10. Class A power amplifiers have the advantages of simple construction and distortionless output voltage and disadvantages of very low efficiency and large power dissipation in power transistors.
11. Transformer coupled class A power amplifier is sometimes referred to a single ended power amplifier. The term "single ended" (denoting only one transistor) is used to distinguish it from the push-pull amplifier using two transistors.
12. For transfer of maximum power from amplifier to the output device, matching of amplifier output impedance with the impedance of output device is necessary. This is accomplished by using step-down transformer of suitable turn ratio.

The ratio of the transformer input and output resistances varies directly as the square of the transformer turn ratio:

$$\frac{R'_L}{R_L} = \left(\frac{N_1}{N_2}\right)^2 = a^2,$$

where R_L is the resistance of load connected across the transformer secondary and R'_L is the effective resistance looking into the transformer primary.

13. A transformer coupled class A power amplifier has a maximum theoretical efficiency of 50%, but its maximum ac power delivering capability is still one-half of its maximum dissipation capability.
14. The operating point of a transformer coupled power amplifier is located at,

$$I_{CQ} = \frac{2P'_{out\,(ac)}}{V_{CC}} \text{ and } V_{CEQ} = V_{CC}$$

ignoring collector and emitter resistances

where $P'_{out(ac)}$ is the power delivered to the primary of the output transformer.

15. A graph of power output and total harmonic distortion against load resistance indicates occurrence of maximum output power at a higher value of load resistance than that which provides minimum distortion.
16. A single ended transformer coupled amplifier has two additional disadvantages. The total harmonic distortion is unacceptably high with the second harmonic contributing the most; dc current flow in the primary of the transformer causes saturation problem.
17. In class B operation, no biasing system is required; the operating point is set at cutoff. Conduction angle is 180°. Less power dissipation resulting in higher overall efficiency; Severe distortion; Increase of transistor power dissipation with increase in input signal.
18. In class AB operation, the biasing circuit is so adjusted that the operating point Q lies near the cutoff voltage. During a small portion (less than half a cycle) of the negative cycle, no collector current flows in class AB operation. Class AB operation needs a push-pull arrangement to achieve a full output cycle.
19. A class A push-pull amplifier consists of two active devices (transistors), driven by two signals having a phase difference of 180°, and a centre-tapped output transformer.
20. The operation of a push-pull amplifier is such that all the even harmonics are eliminated from the output without changing the magnitudes of the fundamental and odd harmonics.
21. A pure class B push-pull amplifier has both of its active devices biased at cutoff, with centre tapped transformers used both at the input and output. Thus, no power is drawn under zero signal condition.
22. Crossover distortion occurs as a result of one transistor cutting off before the other begins conducting in class B push-pull amplifier. The distortion introduced is called the crossover distortion because it occurs during the time of operation crossover from one transistor to the other in the push-pull amplifier.
23. Class AB operation eliminates the crossover distortion by slightly forward biasing both transistors to permit a small idling current to flow at all times.
24. The maximum conversion efficiency in case of a class B push-pull amplifier is 78.5% while in case of class AB operation it is 70%.
25. The relation between maximum output power $P'_{out\,(ac)}$ and reflected load resistance R'_L is given by

$$R'_L = \frac{V_{CC}^2}{2P'_{out\,(ac)}}$$

26. The transistor dissipation in a class B amplifier is zero under quiescent or zero signal condition but goes through a maximum power dissipation 2 $P_{(ac)\,max}$ twice for each cycle of the input signal, having an average power dissipation of $P_{(ac)\,av}$ over a complete cycle.
27. The current flowing from the power supply in a class B push-pull amplifier flows in full-wave rectifier pulses, so the power supply must have good regulation.

28. Use of high power transistors makes possible to dispense with the output transformer by placing the load so that each collector current pulse passes directly through the load, but in opposite directions, thus providing the advantage of push-pull.

29. Complementary symmetry amplifiers do not depend upon either an input or an output transformer. They make use of opposite types of transistors, NPN and PNP, so that no phase-splitting device is required.

30. The output transformer is selected on the basis of the collector-to-collector resistance, $R_{CC} = 4\,R'_L$.

Exercises

1. Explain the difference between a voltage and a power amplifier.
[Pb. Technical Univ. Analog Electronics May 2005]
2. Differentiate between voltage and power amplifiers, Give the classification of power amplifiers and the use of each class.
3. Distinguish amongst class A, class B and class C operation of amplifiers.
4. Explain how the position of Q-point varies in case of classes A, B and C operation.
5. What do you understand by class A, class B and class C power amplifiers ? Define and explain the following terms as applied to power amplifiers :
(*a*) Collector efficiency, (*b*) Distortion, (*c*) Power dissipation capability.
[Pb. Technical Univ. Analog Electronics May 2005]
6. Draw the circuit of a simple class A amplifier which can provide power amplification and explain its working.
7. Explain how ac power is developed across a loudspeaker in a single-ended power amplifier. Explain why a power amplifier is also known as the large-signal amplifier ?
8. Prove that a class A power amplifier is cooler in the presence of signal than in the absence of signal.
9. Draw the circuit diagram of a class A transformer coupled power amplifier and explain its operation. Derive an expression for its maximum efficiency.
10. Show that the maximum collector efficiency of class A transformer coupled power amplifier is 50%.
[Pb. Technical Univ. Analog Electronics December 2005]
11. Show that the maximum efficiency of class B operation will not exceed 78.5 per cent.
[Pb. Technical Univ. Analog Electronics December 2006]
12. What are the various types of distortion present in power amplifier ? How do you graphically determine the second harmonic distortion ?
13. Distinguish between power amplifiers and voltage amplifiers. Discuss a method of determining the total harmonic distortion of a power amplifier.
14. Discuss five point method of computing harmonic distortion.
[Pb. Technical Univ. Electronic Devices and Circuits 2005-06]
15. What is the difference between a voltage amplifier and a power amplifier ?
Derive the conversion efficiency of a class B power amplifier and clearly explain the reasons. Why are most of the power amplifiers used in practice designed to operate in class AB stage ?
16. How are transformer-coupled power amplifiers designed ? Explain with example.
17. Explain how the position of Q point varies in case of class A, B and C operation.
18. Explain the push-pull amplifier circuit in detail with a neat diagram.
[Pb. Technical Univ. Analog Electronics May 2005]
19. Draw and explain the working of a push-pull amplifier circuit.
[Pb. Technical Univ. Analog Electronics May 2007]
20. Explain the functioning of a push-pull amplifier. **[Pb. Technical Univ. Analog Electronics May 2006]**
21. Draw the circuit and explain the operation of a class A push-pull amplifier.
22. Discuss crossover distortion. **[Pb. Technical Univ. Analog Electronics May 2006]**

23. List advantages of push-pull amplifiers. **[Pb. Technical Univ. Electronic Devices & Circuits 2005-06]**

24. Draw the configuration of push-pull amplifier and explain its working.

[Pb. Technical Univ. Electronic Devices & Circuits 2003-04]

25. Explain the origin and magnitude of harmonic distortion in power amplifiers. Explain how even harmonics are eliminated in push-pull amplifiers.

26. Explain why even harmonics are not present in a push-pull amplifier.

[U.P. Technical Univ. Solid State Devices and Circuits 2005-06]

27. Draw the circuit of a class B push-pull amplifier and explain its operation. What is crossover distortion ? How is it minimised ? Explain why even harmonics are not present in a push-pull amplifier.

28. What type of bias is used in a true class B push-pull amplifier ? Explain how it is obtained in a transistor amplifier. **[U.P. Technical Univ. Solid State Devices & Circuits February 2002]**

29. Draw and explain the working of push-pull class B amplifier. What are its advantages and disadvantages ?

[Pb. Technical Univ. Analog Electronics December 2007]

30. Prove that the maximum transistor power dissipation of a class B push-pull amplifier is given by

$$\text{Pull}_{D(max)} = 0.2P_{P(max)}$$

[Pb. Technical Univ. Analog Electronics December 2004]

31. Show that maximum theoretical efficiency of a class B push-pull amplifier is 78.5%.

[U.P. Technical Univ. Solid State Devices & Circuits February 2002]

32. Explain crossover distortion, its causes and remedy.
What penalty is paid for biasing an amplifier into class AB operation ?

[U.P. Technical Univ. Solid State Devices and Circuits February 2002]

33. Draw the circuit diagram of a push-pull amplifier and explain its working. Explain crossover distortion.

[U.P. Technical Univ. Solid State Devices and Circuits February 2006-07]

34. Explain the origin of crossover distortion and describe a method to minimise this distortion.

[U.P. Technical Univ. Solid State Devices and Circuits February 2006-07]

35. Explain how two complementary transistors in a class B push-pull amplifier act simultaneously as phase inverters and an output push-pull pair.

[U.P. Technical Univ. Solid State Devices and Circuits February 2002]

36. Write short notes on the following :

(*i*) Classification of power amplifiers.

(*ii*) Harmonic distortion.

(*iii*) Push-pull amplifiers. **[Rajasthan Univ. 2002]**

(*iv*) Need of class AB push-pull amplifier. **[Rajasthan Univ. 2004]**

(*v*) Crossover distortion. **[M.D. Univ. May 2008, May 2009]**

(*vi*) Complementary symmetry power amplifiers. **[Rajasthan Univ. 2004, 2005, 2007; Dec. 2009; Dec. 2010]**

Short Answer Type Questions With Answers

Q. 1. How do you classify amplifiers ? What are small signal amplifiers ? How are they different from large signal amplifier ? **[R.G. Technical Univ. June 2009]**

Ans. When the input signal is so weak as to produce small fluctuations in the collector current compared to its quiescent value, the amplifier is known as *small signal amplifier.* On the other hand, when the fluctuations in collector current are large *i.e.* beyond the linear portion of its characteristics, the amplifier is called the large signal amplifier. In case of a small signal amplifier, the transistor can be represented by a linear circuit model, and a simple analytical method can be used for determination of various performance characteristics of the amplifier. But in case of a large signal power amplifier, the equivalent linear circuit model of the transistor and analytical method cannot be applied and graphical method is preferred because of involvement of certain non-linear operation in it. Voltage amplifiers and audio power amplifiers are the small signal power amplifiers while radio power amplifiers are the large signal power amplifiers.

Q. 2. Why a power amplifier is always preceded by a voltage amplifier.

[P.T.U. A.E. December 2006; M.D.U. December 2005]

Ans. Signal voltage at the input is quite small but a large voltage input signal is necessitated so as to provide large power at the output. So voltage amplifier is required for raising the voltage level of the weak input signal before it is fed to the power amplifier.

Q. 3. Why is a power amplifier called a large signal amplifier ? Explain. **[Rajasthan Technical Univ. 2009]**

Ans. Power amplifiers are required to handle large voltage signals so as to deliver large power at the output. That is why power amplifiers are called large signal amplifiers.

Q. 4. What is the collector junction temperature for *Ge* and *Si* in case of transistor power amplifier ?

Ans. The collector junction temperature for *Ge* lies in the range of 150ºC to 225ºC while in case of silicon it ranges 60ºC to 100ºC.

Q. 5. Why a class A power amplifier is cooler in the presence of signal than in the absence of signal ?

Ans. With the zero signal applied at the input of the class A power amplifier, ac power developed across the load reduces to zero and, therefore, all the power drawn from collector supply V_{CC} is wasted in the form of heat. Thus, a power transistor dissipates maximum power under zero signal condition. Thus the device is cooler in the presence of signal than in the absence of signal.

Q. 6. What is the maximum collector circuit efficiency of transformer coupled class A power amplifier?

Ans. 50 per cent.

Q. 7. Why do we use transformer in the output stage of an amplifier ? **[P.T.U. Analog Electronics Dec. 2005]**

Ans. For transfer of maximum power from amplifier to the output device matching of amplifier output impedance with the impedance of the output device is necessary and this is accomplished by using a step-down transformer of suitable turn ratio at the output.

Q. 8. What principal factor contributes to the doubling of the conversion efficiency in a transformer coupled load amplifier compared with an amplifier with a direct coupled resistive load ?

Ans. Negligible power loss in the transformer primary (in the collector load) is the major factor that contributes to the doubling of the conversion efficiency in a transformer coupled load amplifier compared with an amplifier with a direct coupled resistive load.

Q. 9. Under what conditions would the dc load line in a transformer coupled amplifier not be drawn vertically on the output characteristics ? Which way would it slope ?

Ans. If the resistance of the transformer primary is significant and biasing stabilizing resistance is included, the dc load line will slope slightly to the left.

Q. 10. What equations specify the operating point in terms of the output power delivered to the transformer primary ?

Ans. The operating point of a transformer coupled amplifier is specified by the following equations:

$$I_{CQ} = \frac{2P'_{out(ac)}}{V_{CC}}$$

where $P'_{out(ac)}$ is the power delivered to the transformer primary

and $V_{CEQ} = V_{CC}$

where V_{CC} is the collector supply voltage in volts.

Q. 11. What should be the effective ac resistance reflected into the primary of the output transformer in a transformer coupled power amplifier ?

Ans. Effective ac resistance reflected into the primary of the output transformer in a transformer coupled power amplifier is given by

$$R'_L = \frac{V_{CC}^2}{2P'_{out(ac)}}$$

where V_{CC} is the collector supply voltage in volts and $P'_{out\ (ac)}$ is the power delivered to the transformer primary, in watts.

Q. 12. Why may an amplifier be operated with a load resistance lower than the value that would provide maximum output power ?

Ans. From the power output and distortion load resistance curves, it is seen that the load providing minimum total harmonic distortion is lower than the load that provides maximum power output. That is why, an amplifier is operated with a load resistance lower than the value that would provide maximum output power so that output with least distortion is obtained.

Q. 13. Comment on the maximum efficiency of class B operation. **[A.M.I.E. Sec. B. Summer 1996]**

Ans. The maximum possible value of efficiency for class B operation is 78.5%. The efficiency of class B operation exceeds that of class A operation because with zero excitation there is no current in class B operation whereas power is drawn from supply source in class A operation.

Q. 14. Although a class-B single ended power amplifier has high efficiency, yet it is less used in practical circuits. Explain why ? **[P.T.U. Analog Electronics May 2007]**

Ans. Severe distortion occurs, because of total absence of negative half cycles from the output in a class-B single ended power amplifier so it is less used in practical circuits.

Q. 15. What is class A push-pull amplifier ?

Ans. Class A push-pull amplifier is that in which current flows in the output of each transistor for the whole of the input cycle.

Q. 16. Why does the output stage employ push-pull arrangement ? **[P.T.U. Analog Electronics December 2005]**

Ans. Push-pull arrangement is used in the output stage in order to have high output power at high efficiency with less distortion.

Q. 17. Draw the circuit of push-pull amplifier. Explain why is it called push-pull ? **[P.T.U. Analog Electronics May 2006]**

Ans. Refer to Fig. 6.20. The circuit is called a push-pull arrangement because during one-half cycle one transistor is in conducting (pushing) state and the other is in non-conducting (pulling) state and in another half cycle, the states of the two transistors are reversed. Moreover, by doing so, the transistors also amplify the signal, hence, the name push-pull amplifier.

Q. 18. What is crossover distortion ? How can it be minimized ? **[Pb. Technical Univ. 2005-06]**

Ans. Crossover distortion occurs in class B push-pull amplifiers as a result of one transistor cutting off before the other begins conducting. The distortion introduced is called the crossover distortion because it occurs during the time operation crossover from one transistor to the other in the push-pull amplifier.

For elimination of crossover distortion, it is necessary to add a small amount of forward bias to take the transistors to the average of conduction or slightly beyond. Technically the operation of transistors will lie between class B and class A mode (*i.e.,* class AB operation).

Q. 19. The complementary symmetry push-pull amplifiers are called so. Why ?

Ans. The complementary symmetry push-pull amplifier uses two transistors having complementary symmetry (one N-P-N and another P-N-P). The term complementary arises from the fact that one transistor is the N-P-N and the other is P-N-P. They have the symmetry as they are made with the same material and technology and are of same maximum ratings.

Q. 20. What is the maximum collector circuit efficiency of a complementary push-pull class B power amplifier ?

Ans. Maximum collector circuit efficiency of a class B complementary push-pull power amplifier is 78.5%.

Q. 21. What are the drawbacks of complementary symmetry push-pull class B power amplifier ?

Ans. One obvious drawback of this circuit arrangement is the requirement for two supply voltages. Another drawback of this circuit is difficulty of obtaining matched complementary transistors. If there is an unbalance in the characteristics of the two transistors, even harmonics will no longer be cancelled and considerable distortion will be introduced. Very often negative feedback is used in power amplifiers to reduce nonlinear distortion.

Problems

1. A power transistor working in class A operation has zero signal power dissipation of 10 watts. If the ac output power is 4 watts, find (*i*) collector efficiency (*ii*) power rating of the transistor. **[Ans.** 40%; 10 W**]**
2. For a class A power amplifier, the maximum and minimum values of collector-emitter voltage are 20 V and 4 V respectively. V_{CC} = 25 V. Determine overall efficiency of the amplifier. **[Ans.** 16%**]**

3. For class A, CE large signal amplifier shown in Fig. 6.5, V_{CC} = 25 V, V_{CEQ} = 12.5 V, I_{CQ} = 0.5 A, R_C = 25 Ω. If the output current varies by ± 0.25 A when an input signal is applied to the base, determine (*i*) input power (*ii*) output power (*iii*) power delivered to the transistor (*iv*) power lost in transistor, (*v*) collector efficiency and (*vi*) overall efficiency. [**Ans.** (*i*) 12.5 W (*ii*) 0.78125 W (*iii*) 6.25 W (*iv*) 5.47 W (*v*) 12.5% (*vi*) 6.25%]

4. A class A amplifier shown in Fig. 6.10 operates from V_{CC} = 20 V, draws a no-signal current of 5 A and feeds a load of 40 Ω through a step-up transformer of N_2/N_1 = 3.16. Find (*a*) whether the amplifier is properly matched for maximum power transfer (*b*) maximum ac signal power output (*c*) maximum dc power input (*d*) conversion efficiency at maximum signal input. **[Gauhati Univ. 1986]** [**Ans.** (*a*) properly matched (*b*) 50 W (*c*) 100 W (*d*) 50%]

5. A transformer coupled class A transistor power amplifier uses a transistor with β = 100 and supplies to load of 8 Ω connected across the secondary of the transformer. If the maximum and minimum values of collector-emitter voltage and currents are 20 V and 2.4 V and 225 mA and 25 mA respectively. Determine (*i*) appropriate value of zero signal collector current (*ii*) zero signal base current (*iii*) dc power input (*iv*) ac power output (*v*) transformer turn ratio (*vi*) collector efficiency and (*vii*) transistor power rating.

[**Ans.** (*i*) 125 mA (*ii*) 1.25 mA (*iii*) 1.4 W (*iv*) 0.44 W (*v*) 3.317 : 1 (*vi*) 31.43% (*vii*) 1.4 W]

6. An output signal displayed on an oscilloscope provided the maximum, minimum and quiescent values of collector current to be 240 mA, 24 mA and 125 mA respectively. Determine the second harmonic distortion. [**Ans.** 3.24%]

7. Using a 5-point method of determination of harmonic components gives the following results : D_2 = 0.1, D_3 = 0.03 and D_4 = 0.015, with I_1 = 5 A and R_C = 10 Ω. Determine the total distortion, fundamental power component, total power and percentage increase in power because of distortion.

[**Ans.** 10.55%; 125 W; 126.4 W; 1.12%]

8. A class B push-pull amplifier is supplied with V_{CC} = 40 V. The signal swings the collector voltage down to V_{min} = 8 V. The dissipation in both transistors total 38 W. Determine the (*i*) total power input (*ii*) total power developed across the load (*iii*) power rating of each transistor and (*iv*) overall efficiency.

[**Ans.** (*i*) 101.86 W (*ii*) 64 W (*iii*) 19 W (*iv*) 62.83%]

9. Determine the input power, output power and efficiency resulting in a class B push-pull amplifier providing a signal of 20 V peak to a 16 Ω load, using a single supply of V_{CC} = 30 V.

[U.P.S.C. I.E.S. Electronics & Telecommunication Engg.-II, 2006]

[**Ans.** 35.81 W; 28.125 W; 78.5%]

❑❑❑

Unit-IV

Feedback Amplifiers and Oscillators

Contents of This Unit

UNIT-IV : FEEDBACK AMPLIFIERS AND OSCILLATORS : Feedback Concept : Effect of Negative Feedback on Gain, Bandwidth, Stability, Distortion and Frequency Response.

Sinusoidal Oscillators : Criterion For Oscillation; Different Types of Oscillators—R-C Phase Shift, Wein Bridge, Hartley, Colpitts and Crystal Oscillators, Derivation of Expression For Frequency and Amplitude of These Oscillators.

CHAPTER

7

FEEDBACK AMPLIFIERS

7.1. INTRODUCTION

INSIDE THIS CHAPTER

The voltage gain, input impedance, output impedance, bandwidth etc., are few important characteristics of an amplifier. These parameters are more or less constant for a given amplifier. Quiet often, these parameters are required to be controlled. This can be easily achieved by employing a technique, called the *feedback.*

When a part or fraction of output is combined to the input, feedback is said to exist. Thus the process of combining a fraction of output energy (voltage or current) back to the input is called the *feedback*. When the feedback voltage (or current) is so applied as to increase the input signal (*i.e.,* when the feedback is in phase with input signal), it is called the *positive* or *direct* or *regenerative feedback.* Since positive feedback causes excessive distortion and instability, it is seldom used in amplifiers. However, because of its capability of increasing the power of the original signal it is used in oscillator circuits.

When the feedback voltage (or current) is so applied as to weaken the input signal (*i.e.,* when the feedback is in phase opposition to the input signal), it is called *negative*, *inverse* or *degenerative feedback.* Negative feedback reduces the amplifier gain but it has numerous advantages (such as gain stability, reduction in nonlinear distortion, reduction in noise, increase in bandwidth or improvement in frequency response, increase in input impedance and decrease in output impedance) and is frequently used in amplifier circuits. The reduction in gain due to negative feedback, can always be compensated by increasing the number of stages.

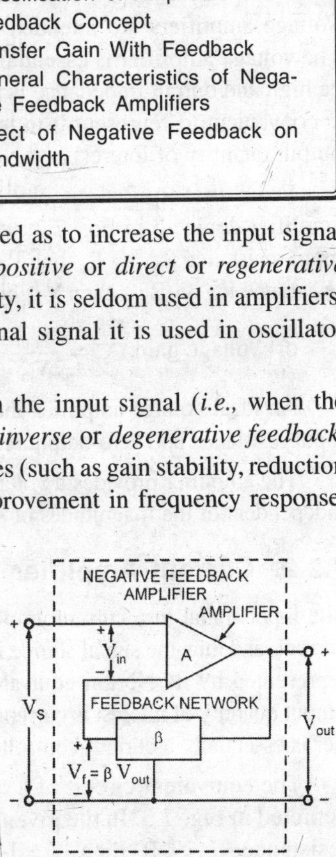

Fig. 7.1

An amplifier with two input terminals and one output terminal, in triangular representation, is shown in Fig. 7.1. The amplifier has a voltage gain A and output voltage V_{out} is applied to a feedback network which reduces V_{out} by a factor β to give a feedback voltage V_f. The feedback network also ensures that feedback voltage V_f is in phase opposition to input signal voltage V_s. The instantaneous voltage at the amplifier input terminals is now $v_{in} = v_s - v_f$. Thus, output voltage will be less than what could be obtained if negative feedback were not employed. This is due to the fact that input voltage has become less than the input signal voltage because of negative feedback. The overall voltage gain $\frac{V_{out}}{V_s}$ is, therefore, reduced. However, as will be demonstrated, the stability of the voltage gain is increased when negative feedback is employed.

A more and more negative feedback is used, the characteristics of a feedback amplifier become more and more independent of the characteristics of the *open-loop* amplifier (amplifier with zero feedback) and ultimately depends on the properties of the feedback network itself.

7.2. CLASSIFICATION OF AMPLIFIERS

The amplifiers without negative feedback are known as *basic* amplifiers. Their gain is unstable [due to (*i*) change in supply voltage, (*ii*) change in parameters, (*iii*) ageing of device and (*iv*) due to rise in leakage current with temperature], distortion is large, noise is more, bandwidth is not sufficiently large and their input and output impedances do not meet the requirements of general applications.

Amplifiers may be usefully classified into four broad categories viz. (*i*) Voltage amplifiers, (*ii*) Current amplifiers, (*iii*) Transconductance amplifiers and (*iv*) Transresistance amplifiers. Their relative magnitudes of input impedance in comparison with the source impedance and that of output impedance in comparison with the load impedance are the determining factors of amplifier category.

7.2.1. Voltage Amplifier

Voltage amplifiers are intended to amplify an input voltage signal and provide an output voltage signal. The voltage amplifier is essentially a voltage-controlled voltage source. The input impedance is required to be high and output impedance is required to be low. Since the signal source is essentially a voltage source, it is convenient to represent it in terms of a Thevenin equivalent circuit (Fig. 7.2). In a voltage amplifier the output quantity of interest is the output voltage.

From the equivalent circuit of a voltage amplifier depicted in Fig. 7.2, we find that if the input resistance $R_{in} >> R_s$ then $V_{in} \approx V_s$ and if load resistance $R_L >> R_{out}$, then $V_{out} \simeq A_V V_{in}$

or Voltage gain, $A_V = \frac{V_{out}}{V_{in}} \simeq \frac{V_{out}}{V_s}$ ...(7.1)

An ideal voltage amplifier should have infinite input resistance and zero output resistance.

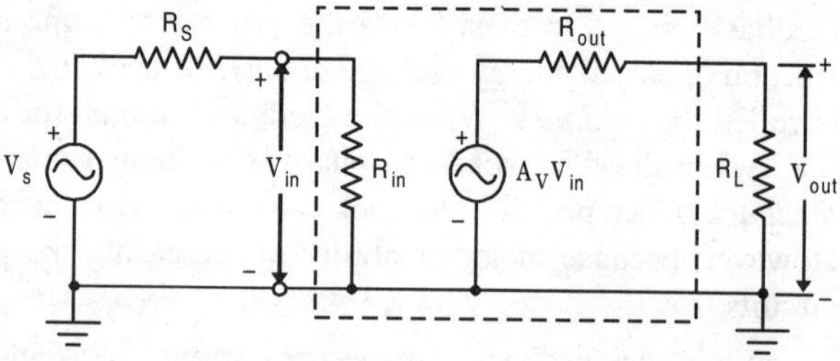

Fig. 7.2. ***Equivalent Circuit of Voltage Amplifier***

The amplifier provides a voltage output proportional to the voltage input and the proportionality constant is independent of the magnitudes of source and load resistances.

7.2.2. Current Amplifier

The input signal in a current amplifier is essentially a current, and thus the signal source is most conveniently represented by its Norton equivalent (Fig. 7.3). The output quantity of interest is current. The current amplifier is essentially a current-controlled current source.

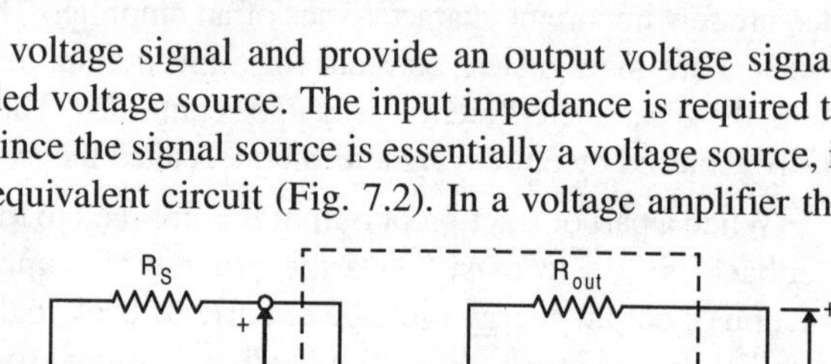

Fig. 7.3. ***Equivalent Circuit of Current Amplifier***

The equivalent circuit of a current amplifier is depicted in Fig. 7.3. In the given circuit if the input resistance $R_{in} << R_s$ then $I_{in} \simeq I_s$ and if $R_L << R_{out}$ then $I_L = I_{out} = A_i I_{in}$

Current gain, $A_i = \frac{I_{out}}{I_{in}} = \frac{I_L}{I_s}$ ...(7.2)

An ideal current amplifier should have zero input resistance and infinite output resistance.

An ideal current amplifier is defined as an amplifier which provides an output current proportional to the signal current, and proportionality constant is independent of magnitudes of source and load resistances (R_s and R_L respectively).

7.2.3. Transconductance Amplifier

In transconductance amplifiers the input signal is a voltage and its output signal is a current. The ideal transconductance amplifier provides an output current which is proportional to the signal voltage and the proportionality constant is independent of the magnitudes of R_s and R_L.

From the equivalent circuit of a transconductance amplifier depicted in Fig. 7.4, we find that if $R_s << R_{in}$, then $V_{in} \simeq V_s$ and if load resistance $R_L << R_{out}$,

then $I_L = I_{out} = G_m V_{in} = G_m V_s$

and $G_m = \frac{I_L}{V_s}$ ...(7.3)

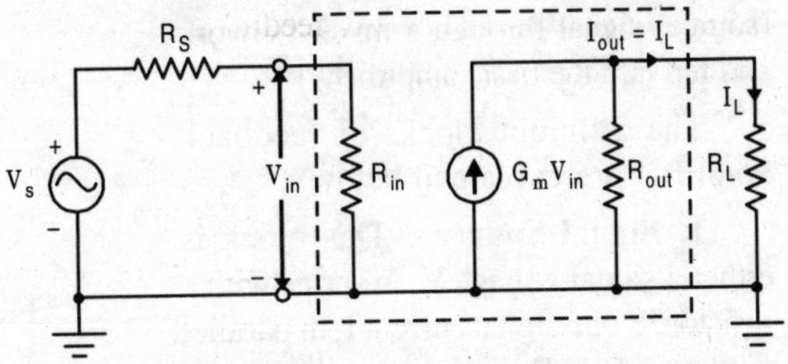

Fig. 7.4. *Equivalent Circuit of Transconductance Amplifier*

An ideal transconductance amplifier should have infinite input and output resistances.

7.2.4. Transresistance Amplifier

In transresistance amplifiers the input signal is current and the output signal is voltage. The ideal transresistance amplifier provides an output voltage in proportion to the signal current, and the proportionality constant is independent of the magnitudes of R_s and R_L. From the equivalent circuit of a transresistance amplifier shown in Fig. 7.5, we find that if $R_{in} << R_s$ then $I_{in} \simeq I_s$ and if $R_{out} << R_L$, then

$$V_{out} = R_m I_{in} = R_m I_s$$

and $R_m = \frac{V_{out}}{I_s}$ ...(7.4)

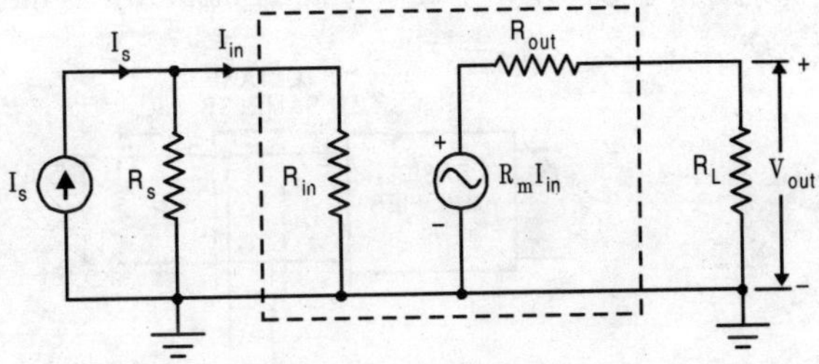

Fig. 7.5. *Equivalent Circuit of Transresistance Amplifier*

An ideal transresistance amplifier should have zero input and output resistances.

The characteristics of ideal amplifiers are summarized below in tabular form :

TABLE 7.1. *Characteristics of Ideal Amplifier*

Type of Amplifier	Input	Output	R_{in}	R_{out}	Transfer Characteristic
Voltage Amplifier	V	V	∞	0	$A_v = \frac{V_{out}}{V_s}$
Current amplifier	I	I	0	∞	$A_i = \frac{I_L}{I_s}$
Transconductance amplifier	V	I	∞	∞	$G_m = \frac{I_L}{V_s}$
Transresistance amplifier	I	V	0	0	$R_m = \frac{V_{out}}{I_s}$

7.3. FEEDBACK CONCEPT

Figure 7.6 represents the block diagram of an amplifier with feedback. The output quantity (either voltage or current) is sampled by means of a suitable sampling network (or sampler) and is fed to the *feedback network*. The output of feedback network, which has a fraction of output signal is combined with the external (source) signal through a mixer network and fed into the basic amplifier.

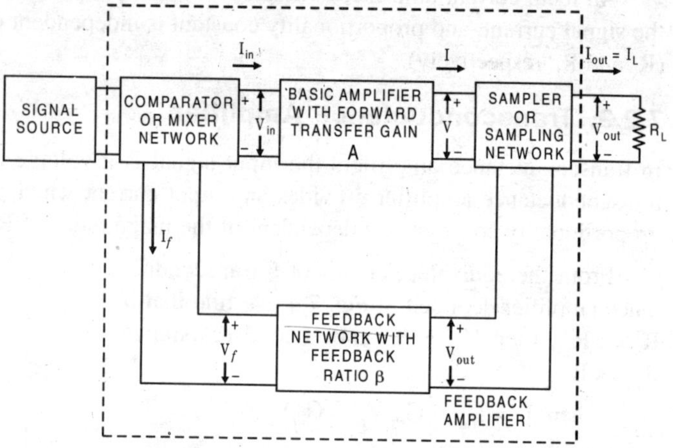

Fig. 7.6. ***Block Diagram of an Amplifier With Feedback***

The different blocks of feedback amplifier are explained below :

1. Signal Source : This block is either a signal voltage V_s in series with a resistor R_s or a signal current I_s in parallel with a resistor R_s.

2. Feedback Network : The feedback network is usually in the form of a passive two-port network and may be formed of resistors, inductor and capacitors (most often of resistors). Its function is to return a function of the output energy (voltage or current) to the input of the amplifier.

3. Sampling Network : Sampling networks are of two types, namely, voltage sampling network and current sampling network, as shown in Fig. 7.7. Output voltage is sampled by connecting the feedback network either shunt across the output or in series with the output. The former is known as *voltage* or *node sampling* [Fig. 7.7 (*a*)] and the latter is known as the current or loop sampling [Fig. 7.7 (*b*)].

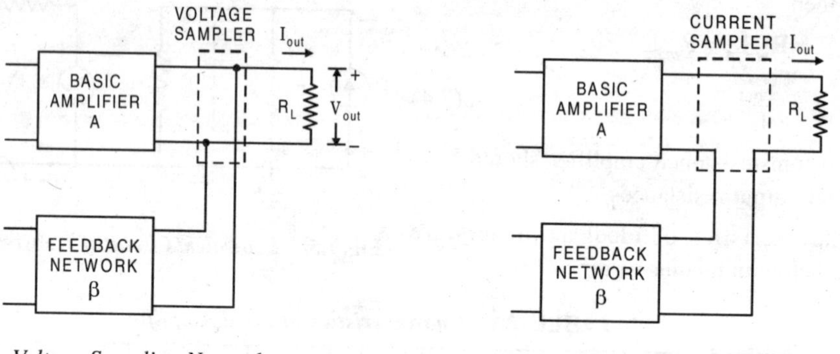

(*a*) *Voltage Sampling Network* (*b*) *Current Sampling Network*

Fig. 7.7

4. Comparator, or Mixer, Network : Mixer, also known as *comparator*, is of two types, namely, series mixer and shunt mixer, as shown in Fig. 7.8. A differential amplifier, which has two inputs and one output proportional to difference between the signals at the two inputs, is usually used as a mixer or comparator.

Types of Feedback Connections

Any of the output connections shown in Fig. 7.7 may be combined with any of the input connections shown in Fig. 7.8 to form the feedback amplifier shown in Fig. 7.6. The resulting feedback amplifier topologies are given below :

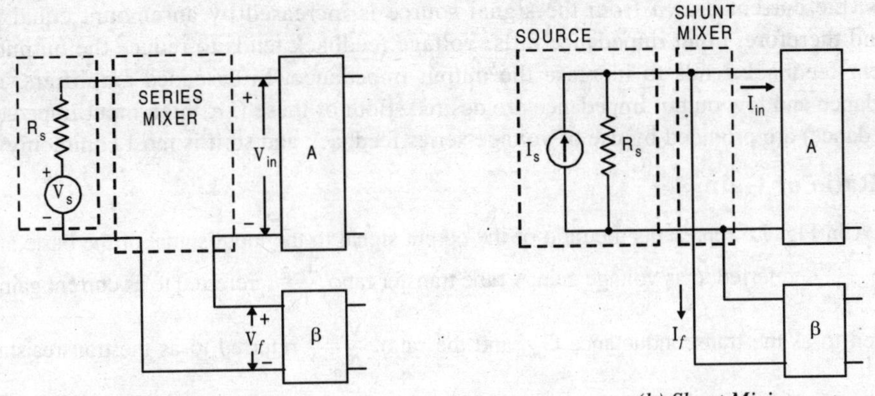

(a) Series Mixing (b) Shunt Mixing

Fig. 7.8

1. Voltage-series feedback, as shown in Fig. 7.9 (*a*)
2. Voltage-shunt feedback, as shown in Fig. 7.9 (*b*).
3. Current-series feedback, as shown in Fig. 7.9 (*c*).
4. Current-shunt feedback, as shown in Fig. 7.9 (*d*).

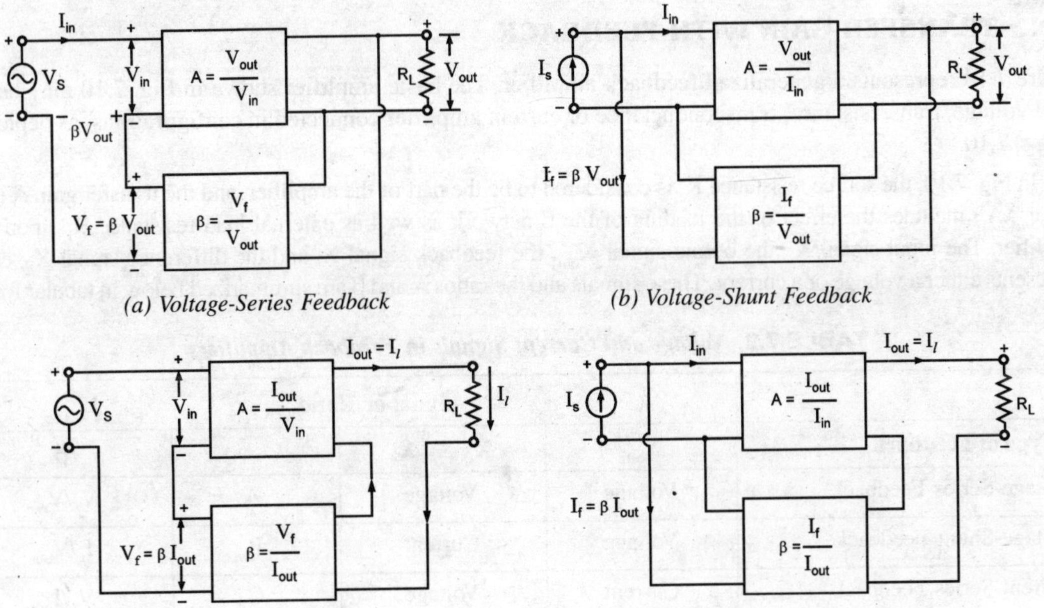

(a) Voltage-Series Feedback (b) Voltage-Shunt Feedback

(c) Current-Series Feedback (d) Current-Shunt Feedback

Fig. 7.9

In the list above, voltage refers to connecting the output voltage as input to the feedback network and current refers to tapping off some output current through the feedback network. Series refers to connecting the feedback signal in series with the input signal voltage; and shunt refers to connecting the feedback signal in parallel (shunt) with an input current source.

Series feedback connections tend to increase the input impedance of the amplifier. This is because in this case feedback signal is returned to the input in series to oppose the applied voltage causing input current to fall and consequently makes the input impedance to increase. In case of shunt or parallel connections, the current drawn from the signal source is increased by an amount equal to feedback current I_f and therefore, input impedance falls. Voltage feedback tends to reduce the output impedance while current feedback tends to increase the output impedance. In cascaded amplifiers, mostly high input impedance and low output impedance are desired. Both of these (*i.e.*, high input impedance and low output impedance) are provided by use of voltage-series feedback and so it is most commonly employed.

Transfer Ratio or Gain

The symbol A in Fig. 7.6 represents the ratio of the output signal to the input signal of the basic amplifier. The transfer ratio $\frac{V_{out}}{V_{in}}$, referred to as voltage gain A_v, the transfer ratio $\frac{I_{out}}{I_{in}}$, referred to as current gain A_i, the ratio $\frac{I_{out}}{V_{in}}$, referred to as the transconductance G_m, and the ratio $\frac{V_{out}}{I_{in}}$, referred to as the transresistance R_m, are referred to as a *transfer gain of the basic amplifier without feedback* and represented by symbol A.

The symbol A_f, defined as the ratio of the output signal to the input signal of the amplifier configuration shown in Fig. 7.6, is shown as the *transfer gain of the amplifier with feedback.* Thus, A_f is used to represent any one of the four ratios $\frac{V_{out}}{V_s} = A_{vf}$, $\frac{I_{out}}{I_s} = A_{if}$, $\frac{I_{out}}{V_s} = G_{mf}$ and $\frac{V_{out}}{I_s} = R_{mf}$. The relationship between the transfer gain A_f with feedback and the gain A of the ideal amplifier without feedback is derived below (Art. 7.4).

7.4. TRANSFER GAIN WITH FEEDBACK

Figure 7.10 represents a generalized feedback amplifier. The basic amplifier shown in Fig. 7.10 may be an ideal voltage, transresistance, transconductance or current amplifier connected in configurations, as depicted in Fig. 7.10.

In Fig. 7.10, the source resistance R_s is considered to be the part of the amplifier, and the transfer gain A (A_v, R_m, g_m, A_i) includes the effect of the loading of the β network as well as external load resistance R_L upon the amplifier. The input signal X_s, the output signal X_{out}, the feedback signal X_f and the difference signal X_d, each represents either a voltage or a current. These signals and the ratios A and β are summarized below in tabular form.

TABLE 7.2. ***Voltage and Current Signals in Feedback Amplifiers***

	Signal or Ratio			
Type of Feedback	X_{out}	X_s, X_f, X_d	**A**	**β**
Voltage-Series Feedback	Voltage	Voltage	A_v	V_f/V_{out}
Voltage-Shunt Feedback	Voltage	Current	R_m	I_f/V_{out}
Current-Series Feedback	Current	Voltage	G_m	V_f/I_{out}
Current-Shunt Feedback	Current	Current	A_i	I_f/I_{out}

The symbol indicated by the circle in Fig. 7.10 represents a mixing or comparison network, whose output is the sum of the inputs, considering the sign indicated at each input. Thus

$$\text{Difference signal, } X_d = X_s - X_f = X_{in} \quad ...(7.5)$$

X_d is also called the *difference, error,* or *comparison signal* because it represents the difference of applied signal and that fedback to the input

The *reverse transmission factor* β, defined by the ratio of feedback signal X_f and output signal, is usually a positive or a negative real number

$$i.e., \quad \beta = \frac{X_f}{X_{out}} \quad ...(7.6)$$

But in general, β is a complex function of the signal frequency. The symbol β should neither be confused with transistor current gain factor or symbol β used for CE short-circuit gain.

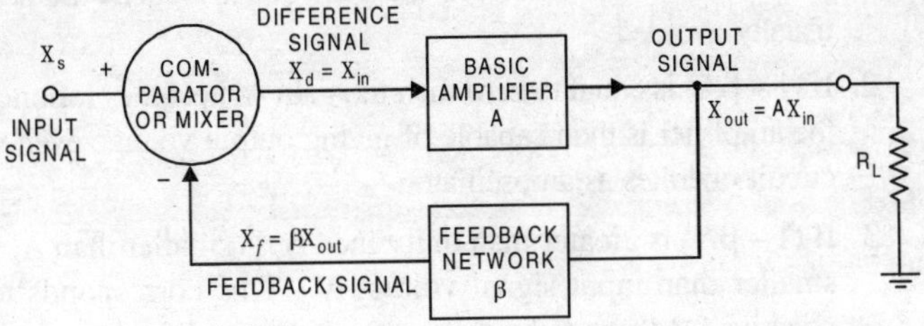

Fig. 7.10. ***Schematic Representation of a Single Loop Feedback Amplifier***

The transfer gain A is defined as the ratio of the output signal X_{out} and the input signal X_{in}, so

$$A = \frac{X_{out}}{X_{in}} \quad ...(7.7)$$

Substituting Eqs. (7.5) and (7.6) into Eq. (7.7), we have the expression for transfer gain with feedback

$$A_f = \frac{X_{out}}{X_s} = \frac{X_{out}}{X_{in} + X_f} = \frac{\dfrac{X_{out}}{X_{in}}}{1 + \dfrac{X_f}{X_{out}} \times \dfrac{X_{out}}{X_{in}}} = \frac{A}{1 + \beta A} \quad ...(7.8)$$

The quantity A in Eqs. (7.7) and (7.8) represents the transfer gain of the corresponding amplifier without feedback, but including the loading of β network, R_L and R_s.

A_f will be less than A if the feedback is *negative, inverse* or *degenerative.* A_f will be more than A if the feedback is *positive, direct* or *regenerative.*

Transfer gain with feedback is given by equation

$$A_f = \frac{X_{out}}{X_s} = \frac{A}{1 + \beta A} \quad \text{for negative feedback} \quad ...(7.9)$$

$$\text{and} \quad A_f = \frac{X_{out}}{X_s} = \frac{A}{1 - \beta A} \quad \text{for positive feedback} \quad ...(7.10)$$

Loop Gain : The difference signal X_d in Fig. 7.10 is multiplied by A in passing through the amplifier, is multiplied by β in transmission through the feedback network, and is multiplied by –1 in the mixing network. Such a path takes us from the input terminals around the loop consisting of the amplifier and feedback network back to the input; the product (–Aβ) is known as the *loop gain* or *return ratio.* The difference between unity and the loop gain is called the return difference D = 1 – (–Aβ) = 1 + Aβ. Also, the amount of feedback introduced into an amplifier is usually expressed in decibels by the definition

$$N = \text{dB of feedback} = 20 \log_{10} \left| \frac{A_f}{A} \right| = 20 \log_{10} \left| \frac{1}{1 + \beta A} \right| \quad ...(7.11)$$

N will be a negative number for negative feedback.

In general both A and β are phasor quantities (having magnitude as well as phase). Every amplifier stage introduces a phase shift of 180° (assuming CE configuration, as usual). Thus for a 3-stage amplifier, total phase shift between the output and input will be 3 × 180° = 540° or 180°. β is usually arranged to have a phase value of 0° or 180° (*i.e.,* either in phase with input signal voltage or in phase opposition to it).

Since $(1 - \beta A)$ is a complex quantity, it can be observed that

1. If $(1 - \beta A)$ is less than unity for Eq. (7.10), then A_f exceeds A. This condition corresponds to positive feedback because voltage feedback adds to input signal voltage and increases input voltage V_{in}. Positive feedback, though, increases the gain but it reduces the stability and increases the distortion and so it is usually avoided.
2. If $(1 - \beta A)$ is equal to zero then the gain A_f becomes infinite. This is only possible when input is zero. Thus the amplifier is then capable of giving output voltage even with zero input signal. Under such situation the circuit operates as an oscillator.
3. If $(1 - \beta A)$ is greater than unity then A_f is smaller than A. This means the feedback voltage V_{in} becomes smaller than input signal voltage V_s. This corresponds to negative feedback in an amplifier. Though negative feedback reduces the gain of the amplifier but improves its performance in several aspects, given in the succeeding articles. It may be noted that negative voltage feedback does not affect the current of the circuit.

Note : The assumptions made in investigating the effect of negative feedback on the amplifier characteristics are:

1. Basic amplifier is unilateral from input to the output (this is not valid if $h_{re} \neq 0$ for a CE transistor amplifier).
2. Passive feedback network is unilateral and transmits a signal from the output to the input but not in the opposite direction.
3. Feedback network presents no loading on the output of the basic amplifier.
4. There is no forward transmission through the β feedback network.

Example 7.1. The voltage gain of an amplifier without feedback is 3,000. Calculate the voltage gain of the amplifier if the negative feedback is introduced in the circuit. Given that feedback fraction $m = 0.01$.

[Pb. Technical Univ. Analog Electronics May 2006]

Solution : Open-loop voltage gain (voltage gain without feedback), A = 3000

Feedback fraction, $\beta = 0.01$

$$\text{Voltage gain with negative feedback, } A_f = \frac{A}{1+\beta A} = \frac{3{,}000}{1+0.01\times 3{,}000} = \frac{3{,}000}{31} = 96.774 \textbf{ Ans.}$$

Example 7.2. Calculate the gain of a negative feedback amplifier with an internal gain $A_v = 75$ and feedback fraction $m_v = \frac{1}{15}$. What will be the gain if A_v doubles ?

[Pb. Technical Univ. Analog Electronics December 2005]

Solution : Internal gain (open-loop voltage gain), A = 75

$$\text{Feedback fraction, } \beta = \frac{1}{15}$$

$$\text{Voltage gain with negative feedback, } A_f = \frac{A}{1+\beta A} = \frac{75}{1+\frac{1}{15}\times 75} = 12.5 \textbf{ Ans.}$$

When A_v doubles *i.e.*, when $A' = 2 \times 75 = 150$

$$A'_f = \frac{A}{1+\beta A'} = \frac{150}{1+\frac{1}{15}\times 150} = 13.64 \textbf{ Ans.}$$

Example 7.3. An amplifier has a voltage gain of 40. The amplifier is now modified to provide a 10% negative feedback in series with the input. Calculate (*i*) voltage gain with feedback, (*ii*) amount of feedback in dB, (*iii*) loop gain.

[Rajasthan Univ. 2003]

Solution : Open-loop voltage gain, A = 40

Feedback ratio, β = 10% or 0.1

(*i*) Voltage gain with feedback, $A_f = \frac{A}{1+\beta A} = \frac{40}{1+0.1 \times 40} = 8$ **Ans.**

(*ii*) Amount of feedback in dB $= 20 \log_{10} \left| \frac{1}{1+\beta A} \right| = 20 \log_{10} \left| \frac{1}{1+40 \times 0.1} \right|$

$= 20 \log_{10} 0.2 = -13.98$ **Ans.**

(*iii*) Loop gain $= A\beta = 40 \times 0.1 = 4$ **Ans.**

Example 7.4. An amplifier with voltage gain of 60 dB uses $\frac{1}{20}$ of its output in negative feedback. Calculate the gain with feedback in dB.

Solution : Open-loop voltage gain, A = 60 dB or antilog $\frac{60}{20} = 1{,}000$

Feedback ratio, $\beta = \frac{1}{20} = 0.05$

Gain with feedback, $A_f = \frac{A}{1+\beta A} = \frac{1{,}000}{1+0.05 \times 1{,}000} = 19.6$ or $20 \log_{10} 19.6$ dB = 25.85 dB **Ans.**

Example 7.5. The voltage gain of an amplifier without feedback is 2,500. If 40 dB of negative feedback is added :

(*i*) what is the gain with feedback ?

(*ii*) for obtaining same output, how much must be the input increased ?

[*M.D. Univ. Analog Electronic Circuits May 2009*]

Solution : Open-loop gain, A = 2,500

Desensitivity of transfer gain, $(1 + A\beta)$ = 40 dB or antilog $\frac{40}{20} = 100$

(*i*) Gain with feedback, $A_f = \frac{A}{1+\beta A} = \frac{2{,}500}{100} = 25$ **Ans.**

(*ii*) Input for same output will become $\frac{A}{A_f}$ times the input without feedback *i.e.,* $\frac{2500}{25}$ *i.e.,* 100 times the input without feedback. **Ans.**

Example 7.6. Voltage gain of an amplifier without feedback is 60 dB. It decreases to 40 dB with feedback. Calculate the feedback factor.

Solution : Voltage gain of amplifier without feedback,

A = 60 dB or 1,000

Voltage gain with feedback, A_f = 40 dB or 100 $\quad \because 20 \log A_f = 40$ dB or $A_f = 100$

$\therefore$ Gain with feedback, $A_f = \frac{A}{1+\beta A}$

or Feedback factor, $\beta A = \frac{A}{A_f} - 1 = \frac{1{,}000}{100} - 1 = 9$ **Ans.**

Example 7.7. A single stage transistor amplifier has a voltage gain of 600 without feedback, and 50 with feedback. Calculate the percentage of output which is fedback to the input.

Solution : Voltage gain without feedback, A = 600

Voltage gain with feedback, $A_f = 50$

$\because A_f = \frac{A}{1+\beta A}$

$$\therefore\ 50 = \frac{600}{1+600\beta}$$

$$\text{or } \beta = 0.01833$$

Percentage of output voltage that is fedback to the input

$$= \frac{V_f}{V_{out}} \times 100 = \beta \times 100 = 0.01833 \times 100 = 1.833\% \text{ Ans.}$$

Example 7.8. An amplifier with negative feedback gives an output of 12.5 volts with an input of 1.5 V. When feedback is removed, it requires 0.25 V input for the same output. Find (*i*) value of voltage gain without feedback, and (*ii*) value of β, if the input and output are in phase and β is real. [*Rajasthan Technical Univ. Analog Electronics 2009*]

Solution : (*i*) Voltage gain without feedback,

$$A_v = \frac{\text{Output voltage}}{\text{Input voltage}} = \frac{12.5}{0.25} = 50$$

∵ Output voltage is same with and without feedback

$$\text{Voltage gain with feedback, } A_{vf} = \frac{\text{Output voltage}}{\text{Input voltage}} = \frac{12.5}{1.5} = 8 \qquad \text{with feedback}$$

$$\text{Since, } A_{vf} = \frac{A_v}{1+\beta A_v},$$

$$\text{Feedback ratio, } \beta = \frac{\frac{A_v}{A_{fv}}-1}{A_v} = \frac{\frac{50}{8}-1}{50} = \frac{21}{200} = 0.105 \text{ or } 10.5\% \text{ Ans.}$$

Example 7.9. An amplifier with negative feedback has a gain of –100. It is found that without feedback an input signal of 50 mV is required to produce a given output whereas with feedback the input signal must be 0.06 V for the same output. Calculate the amount of feedback in dB and the values A_v and β. [*Rajasthan Univ. 2000*]

Solution : Voltage gain with feedback, $A_f = -100$

Output voltage with negative feedback = A_f × input voltage = –100 × 0.06 = – 6 V

$$\text{Voltage gain without feedback A} = \frac{\text{Output voltage without feedback}}{\text{Input voltage}} = \frac{-6}{50 \times 10^{-3}} = -120$$

∵ Output voltage is same with and without feedback

$$\text{Now since } A_f = \frac{A}{1+\beta A}$$

$$\text{so } 100 = \frac{120}{1+\beta \times 120}$$

$$\text{or } \beta = \frac{120-100}{100 \times 120} = \frac{20}{100 \times 120} = \frac{1}{600} \text{ Ans.}$$

$$\text{Amount of feedback in dB} = 20\log_{10}\left|\frac{1}{1+A\beta}\right| = 20\log_{10}\left|\frac{1}{1+100 \times \frac{1}{600}}\right| = -1.34$$

Example 7.10. A negative feedback of β = 0.002 is applied to an amplifier of gain 1,000. Calculate the change in overall gain of the feedback amplifier if the internal amplifier is subjected to a gain reduction of 15%.

[*M.D. Univ. Rohtak Analog Electronic Circuits Dec. 2006*]

Solution : Voltage gain without feedback, A = 1,000

$$\beta = 0.002$$

$$\text{Voltage gain with feedback, } A_f = \frac{A}{1+\beta A} = \frac{1{,}000}{1+0.002 \times 1{,}000} = 333.33$$

When open-loop gain is reduced by 15%, $A' = (1 - 0.15) \times 1{,}000 = 850$

$$\text{Voltage gain with feedback, } A'_f = \frac{850}{1 + 0.002 \times 850} = 314.8$$

$$\text{Percentage change in overall gain} = \frac{A_f - A'_f}{A_f} \times 100 = \frac{333.333 - 314.8}{333.333} \times 100 = 5.6\% \textbf{ Ans.}$$

Example 7.11. Find (*a*) feedback ratio (*b*) feedback factor (*c*) voltage gain without feedback (*d*) voltage gain with feedback for a circuit given below. Assume $\beta = 200$ and neglect V_{BE}. [*Pb. Technical Univ. Analog Electronics Dec. 2007*]

Solution : Collector load resistance, $R_C = 2.5\ k\Omega$

Emitter resistance, $R_E = 1\ k\Omega$

Assume $h_{ie} = 1.1\ k\Omega$

and $h_{fe} = \beta = 200$

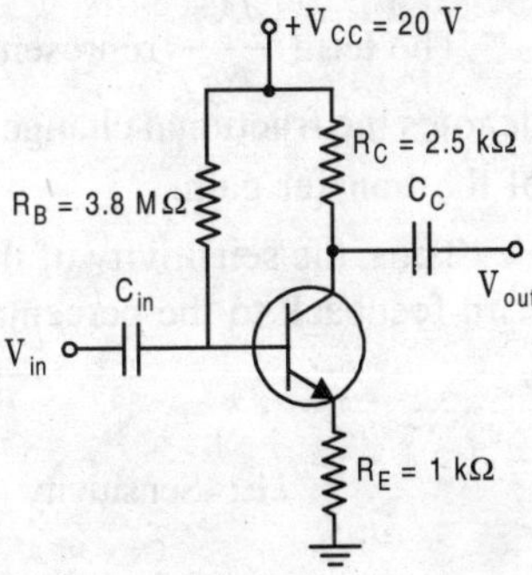

Fig. 7.11

(*c*) Voltage gain without feedback, $A = -\frac{h_{fe}}{h_{ie}} R_C$

$$= -\frac{200}{1.1\ k\Omega} \times 2.5\ k\Omega = -455 \textbf{ Ans.}$$

(*d*) Voltage gain with feedback, $A_f = -\frac{R_C}{R_E} = \frac{-2.5}{1} = -2.5$ **Ans.**

$$\text{Since } A_f = \frac{A}{1 + \beta A} \text{ where } \beta \text{ is the feedback ratio}$$

(*a*) So feedback ratio, $\beta = \frac{A - A_f}{AA_f} = \frac{455 - 2.5}{455 \times 2.5} = 0.398$ **Ans.**

(*b*) Feedback factor $= A\beta = 455 \times 0.398 = 181$ **Ans.**

7.5. GENERAL CHARACTERISTICS OF NEGATIVE FEEDBACK AMPLIFIERS

Inspite of the fact that negative feedback reduces the transfer gain, it is used. Why so ? The simple answer is that it improves many desirable characteristics. We will now examine the effects of negative feedback on different characteristics of an amplifier.

7.5.1. Stabilization of Gain With Negative Feedback

The variations in temperature, supply voltages, ageing of components or variations in transistor parameters with replacement are some of the factors that affect the gain of an amplifier and cause it to change. However, the overall gain of the amplifier can be made independent of these variations if negative feedback is used. This is an important advantage of negative feedback.

From Eq. (7.8), the voltage gain with negative feedback is given as

$$A_f = \frac{A}{1 + \beta A} \simeq \frac{A}{\beta A} \simeq \frac{1}{\beta} \quad \text{...(7.12)} \quad \text{if } A\beta \text{ is made much larger than unity}$$

The gain is thus independent of internal gain of the amplifier and depends on the passive elements such as resistors. The values of resistors remain fairly constant because they can be chosen very precisely with almost zero temperature coefficient of resistance. Thus the gain is stabilized.

Even if open-loop gain A is not very large, some improvement in gain stability can be achieved.

Differentiating Eq. (7.8) w.r.t. A we have

$$\frac{dA_f}{dA} = \frac{(1+\beta A) - A\times\beta}{(1+\beta A)^2} = \frac{1}{(1+\beta A)^2}$$

$$\text{or} \quad dA_f = \frac{dA}{(1+\beta A)^2}$$

Dividing above equation by Eq. (7.9) we have

$$\frac{dA_f}{A_f} = \frac{dA}{A}\times\frac{1+\beta A}{(1+\beta A)^2} = \frac{1}{(1+\beta A)}\frac{dA}{A} \quad \text{...(7.13)}$$

The term $\frac{dA_f}{A_f}$ represents the fractional change in amplifier transfer gain with feedback and $\frac{dA}{A}$ denotes the fractional change in transfer gain without feedback. The term $\frac{1}{1+A\beta}$ is called the *sensitivity* of the transfer gain.

Thus, the sensitivity of the transfer gain is defined as the ratio of percentage change in transfer gain with feedback to the percentage change in transfer gain without feedback.

$$\textit{i.e.,}\ \text{Sensitivity} = \frac{\dfrac{dA_f}{A_f}}{\dfrac{dA}{A}} = \frac{1}{1+A\beta} \quad \text{...(7.14)}$$

The reciprocal of the term sensitivity is called the *desensitivity,* which is given as

$$\text{Desensitivity} = (1+A\beta) \quad \text{...(7.15)}$$

In negative feedback (1 + βA) >> 1, the percentage change in gain with negative feedback is less than the percentage change in gain without feedback. Thus negative feedback improves the gain stability of the amplifier.

The topology determines which transfer ratio (Table 7.2) is stabilized. For instance, for voltage-series feedback Eq. (7.12) signifies that $A_{fv} \simeq \frac{1}{\beta}$ and it is the voltage gain which is stabilized. Similarly, in cases of voltage-shunt feedback, current-series feedback and current-shunt feedback transresistance gain R_m, transconductance gain G_m and current gain A_i respectively are stabilized.

7.5.2. Reduction in Frequency Distortion With Negative Feedback

For a negative feedback amplifier having βA >> 1 the gain with feedback is given as $A_f \cong 1/\beta$. It follows from this that if the feedback network is purely resistive, the gain with feedback becomes independent of frequency even though the basic amplifier is frequency dependent. Practically, the frequency distortion arising because of varying amplifier gain with frequency is considerably reduced with negative feedback.

7.5.3. Reduction in Nonlinear Distortion With Negative Feedback

Harmonic distortion occurs when a transistor or other device is driven beyond its range of linear operation. Now negative feedback is used and the strength of the input signal is increased by the same factor by which the gain is reduced. Thus the magnitude of output signal remains unaltered and thus there is a reduction in harmonic or nonlinear distortion. It may be proved as below.

Let the amplifier with open-loop gain A produce a distortion D in the output signal without feedback. Now when feedback is applied, output as well distortion is fed back to the input. Let the gain with feedback be A_f and distortion in the output D_f. A part βD_f of this distortion is fed back to the input. It gets amplified by factor A and becomes $\beta A D_f$.

Net distortion, $D_f = D - \beta A D_f$ $\quad\because$ distorted output is in phase opposition

or $D_f = \dfrac{D}{1+\beta A}$ ...(7.16)

Thus the distortion is reduced by the same factor as the gain.

However, it is to be noted that improvement in distortion is possible only when the distortion is introduced by the amplifier itself, not when it is already present in the input signal.

7.5.4. Reduction in Noise With Negative Feedback

Noise in an amplifier is reduced in the same way as distortion. However, distortion is usually introduced in the last stage of the amplifier and, therefore, is always within the feedback loop whereas noise in an amplifier may be generated or introduced from outside source at any stage that may not be within the feedback loop. The noise can be reduced by negative feedback only if its source lies within the feedback loop after the first stage.

As nonlinear distortion generated within the amplifier is reduced with negative feedback by a factor $(1 + \beta A)$, the noise introduced within amplifier also gets reduced by the same factor $(1 + \beta A)$ on using negative feedback and the performance of the amplifier is thus much improved.

Example 7.12. An amplifier with a closed loop gain of 100 is required and this gain should not vary more than 1% when the inherent gain of the amplifier without feedback varies by 20%. Find the values of A_v and β.

[Rajasthan Technical Univ. Analog Electronics 2009]

Solution : Variation in open-loop gain, $\dfrac{dA_v}{A_v} = 20\% = 0.2$

Variation in closed-loop gain, $\dfrac{dA_{vf}}{A_{vf}} = 1\% = 0.01$

From Eq. (7.14) Feedback factor $\beta A_v = \dfrac{\dfrac{dA_v}{A_v}}{\dfrac{dA_{vf}}{A_{vf}}} - 1 = \dfrac{0.2}{0.01} - 1 = 19$

Open-loop voltage gain, $A_v = A_{vf}(1 + \beta A_v) = 100(1 + 19) = 2{,}000$ **Ans.**

$$\beta = \frac{\dfrac{A_v}{A_{vf}} - 1}{A_v} = \frac{\dfrac{2{,}000}{100} - 1}{2{,}000} = 0.0095 \quad \textbf{Ans.}$$

Example 7.13. If the open-loop gain $A_v = 10{,}000$ and feedback ratio $\beta = \dfrac{1}{10}$, then for a negative feedback, find the percentage change in closed loop gain when open-loop gain varies 50% with temperature.

[Rajasthan Technical Univ. Analog Electronics 2008]

Solution : Open-loop gain, $A_v = 10{,}000$

Feedback ratio, $\beta = \dfrac{1}{10} = 0.1$

Closed-loop gain, $A_{vf} = \dfrac{A_v}{1 + A_v\beta} = \dfrac{10{,}000}{1 + 10{,}000 \times 0.1} = 9.99$

Change in open-loop gain, $\dfrac{dA_v}{A_v} = 50\%$ or 0.5

Change in closed-loop gain, $\dfrac{dA_{vf}}{A_{vf}} = \dfrac{1}{(1 + \beta A_v)} \dfrac{dA_v}{A_v} = \dfrac{1}{1 + 0.1 \times 10{,}000} \times 0.5$

$= 0.0004995$ or 0.04995% **Ans.**

Example 7.14. The gain and distortion of an amplifier are 150 and 5% respectively without feedback. If the stage has 10% of its output voltage applied as negative feedback, find the distortion of the amplifier with feedback.

[*Pb. Technial Univ. Analog Electronics May 2005*]

Solution : Open-loop voltage gain, A = 150

Feedback ratio, $\beta = 10\% = 0.1$

Distortion without feedback, D = 5% = 0.05

$$\text{Distortion with feedback, } D_f = \frac{D}{1+\beta A} = \frac{0.05}{1+0.1\times 150} = 0.003125 \text{ or } 0.3125\% \textbf{ Ans.}$$

Example 7.15. An amplifier has an input of 10 mV and a gain of 200 without feedback. The distortion produced at the output of the amplifier is 10%. It is desired to reduce the distortion to 1% by using negative feedback. Calculate the gain and output voltage with feedback.

Solution : Distortion without feedback, D = 10% = 0.1

Distortion with feedback, D_f = 1% = 0.01

Amplifier gain without feedback, A = 200

$$D_f = \frac{D}{1+\beta A}$$ ...Refer to Eq. (7.16)

$$\text{or } 0.01 = \frac{0.1}{1+200\beta}$$

$$\text{or } \beta = \left(\frac{0.1}{0.01}-1\right)\Big/200 = 0.045 \text{ or } 4.5\%$$

$$\text{Gain with feedback, } A_f = \frac{A}{1+\beta A} = \frac{200}{1+0.045\times 200} = 20 \textbf{ Ans.}$$

New output voltage, $V_{out} = A_f V_s = 20 \times 10 \text{ mV} = 200 \text{ mV or } 0.2 \text{ V}$ **Ans.**

Example 7.16. An amplifier with an open-loop voltage gain of 1,000 delivers 10 W of output power at 10% second harmonic distortion, when the input signal is 10 mV. If 40 dB negative voltage-series feedback is applied and the output power is to remain at 10 W, determine (*i*) the required input signal (*ii*) percentage second harmonic distortion and (*iii*) closed loop voltage gain.

Solution : Open-loop gain, A = 1,000

Distortion without feedback, D = 10% or 0.1

Input signal voltage, V_s = 10 mV

$$\beta A = 40 \text{ dB or antilog } \frac{40}{20} = 100$$

Since with feedback the output is reduced by $\frac{1}{1+A\beta}$, so the input must be increased by a factor $(1+\beta A)$ so as to deliver the same output. Thus

(*i*) Input signal voltage $V_s' = V_s \times (1+\beta A) = 10 \text{ mV} \times (1+100) = 1.01 \text{ V}$ **Ans.**

$$(ii) \text{ Percentage second harmonic distortion, } D_f = \frac{D}{1+\beta A} = \frac{10}{1+100}\% = 0.1\% \textbf{ Ans.}$$

$$(iii) \text{ Closed-loop voltage gain, } A_f = \frac{A}{1+\beta A} = \frac{1{,}000}{1+100} = 9.9 \textbf{ Ans.}$$

Example 7.17. An amplifier without feedback gives fundamental output 36 V with 7% second harmonic distortion; when the input is 0.028 V.

(*i*) If 1.2% of the output is fed back into the input in a negative voltage series feedback, what is the output voltage ?

***(ii)* If the fundamental output is maintained at 36 V but the second harmonic distortion is reduced to 1%, what is the input voltage ?**

[*M.D. Univ. Analog Electronics May 2009*]

Solution : Amplifier gain without feedback, $A = \dfrac{\text{Output voltage without feedback}}{\text{Input voltage}} = \dfrac{36}{0.028} = 1{,}285.7$

(*i*) Feedback ratio, $\beta = 1.2\% = 0.012$

$$\text{Amplifier gain with feedback, } A_f = \frac{A}{1+\beta A} = \frac{1{,}285.7}{1+0.012\times 1{,}285.7} = 78.26$$

Output voltage, $V_{out} = A_f \times V_{in} = 78.26 \times 0.028 = 2.19$ V **Ans.**

(*ii*) Distortion without feedback, $D = 7\% = 0.07$

Distortion with feedback, $D_f = 1\% = 0.01$

Amplifier gain without feedback, $A = 1{,}285.7$

$$\text{Since } D_f = \frac{D}{1+\beta A}$$

$$\text{So } \beta = \left(\frac{D}{D_f} - 1\right) \times \frac{1}{A} = \left(\frac{0.07}{0.01} - 1\right) \times \frac{1}{1{,}285.7} = 0.004667$$

$$\text{Gain with feedback, } A_f = \frac{A}{1+\beta A} = \frac{1{,}285.7}{1+0.004467\times 1{,}285.7} = 183.67$$

Input voltage required to provide output voltage of 36 V,

$$V_{in} = \frac{V_{out}}{A_f} = \frac{36}{183.67} = 0.196 \text{ V } \textbf{Ans.}$$

7.6. EFFECT OF NEGATIVE FEEDBACK ON BANDWIDTH

From Eq. (7.8) voltage gain with feedback is given as

$$A_{vf} = \frac{A_v}{1+\beta A_v} \simeq \frac{1}{\beta} \quad \text{if } \beta A_v >> 1 \qquad ...(7.17)$$

and from the above result it may be concluded that the voltage gain may be made to depend entirely on the feedback network. However, it is now important to consider the fact that even if β is constant, the voltage gain A_v is not, since it depends on frequency. This means that at certain low or high frequencies βA_v will not be much larger than unity, and, therefore, Eq. (7.17) will not be valid.

When negative feedback is applied in an amplifier, cutoff frequencies are also affected—lower cutoff frequency is lowered by a factor of (1 + βA) while upper cutoff frequency is raised by the same factor (1 + βA). This is explained below.

(*i*) *Lower Cutoff Frequency* f_1. The voltage gain at a frequency f in low frequency range of R-C coupled amplifier is given as

$$A_{vl} = \frac{A_{vm}}{1 - \dfrac{jf_1}{f}} \qquad ...(7.18)$$

where f_1 is lower cutoff frequency and $j = \sqrt{-1}$.

When negative feedback is applied

$$A_{vmf} = \frac{A_{vm}}{1+\beta A_{vm}} \qquad ...(7.19)$$

$$\text{and } A_{vlf} = \frac{A_{vl}}{1+\beta A_{vl}} \qquad ...(7.20)$$

Substituting the value of A_{vl} from Eq. (7.18) in Eq. (7.20) we have

$$A_{vlf} = \frac{\dfrac{A_{vm}}{1 - j f_1/f}}{1 + \beta \dfrac{A_{vm}}{1 - j f_1/f}} = \frac{A_{vm}}{1 - j f_1/f + \beta A_{vm}} = \frac{A_{vm}}{1 + \beta A_{vm} - j f_1/f}$$

By dividing numerator and denominator by $1 + \beta A_{vm}$, this equation may be rewritten as

$$A_{vlf} = \frac{A_{vmf}}{1 - j f_1'/f} \quad ...(7.21)$$

$$\text{where } A_{vmf} = \frac{A_{vm}}{1 + \beta A_{vm}} \text{ and } f_1' = \frac{f_1}{1 + \beta A_{vm}} \quad ...(7.22)$$

From above equations we see that the midband amplification with feedback A_{vmf} equals the midband amplification without feedback divided by $1 + \beta A_{vm}$. Also the lower cutoff frequency with feedback f_1' equals the lower cutoff frequency without feedback f_1 divided by the same factor $(1 + \beta A_{vm})$ *i.e.,* lower cutoff frequency is reduced when negative feedback is applied.

(*ii*) *Upper Cutoff Frequency*. The voltage gain at frequency f in the high frequency range of R-C coupled amplifier is given as

$$A_{vh} = \frac{A_{vm}}{1 + jf/f_2} \quad ...(7.23)$$

When negative feedback is applied

$$A_{vhf} = \frac{A_{vh}}{1 + \beta A_{vh}} \quad ...(7.24)$$

Substituting the value of A_{vh} from Eq. (7.23) in Eq. (7.24), we have

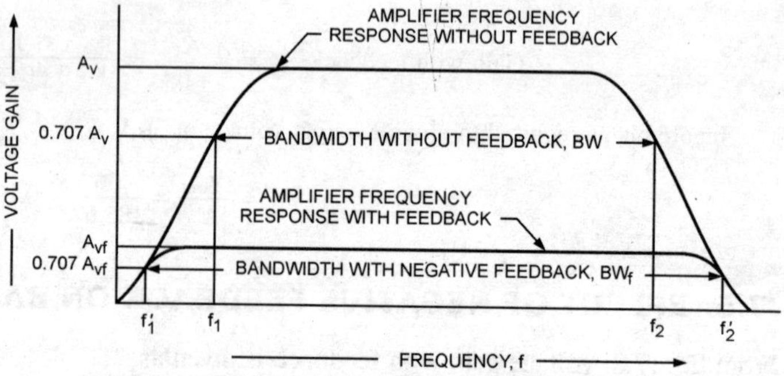

Fig. 7.12. ***Amplifier Frequency Response With and Without Negative Feedback***

$$A_{vhf} = \frac{\dfrac{A_{vm}}{1 + jf/f_2}}{1 + \beta \dfrac{A_{vm}}{1 + jf/f_2}} = \frac{A_{vm}}{1 + \beta A_{vm} + jf/f_2}$$

By dividing the numerator and denominator by $1 + \beta A_{vm}$, the above equation may be rewritten as

$$A_{vhf} = \frac{A_{vmf}}{1 + jf/f_2'} \quad ...(7.25)$$

$$\text{where } A_{vmf} = \frac{A_{vm}}{1 + \beta A_{vm}} \text{ and } f_2' = f_2(1 + \beta A_{vm}) \quad ...(7.26)$$

Thus we see that upper cutoff frequency with feedback equals the corresponding cutoff frequency without feedback f_2 multiplied by the factor $(1 + \beta A_{vm})$ *i.e.,* upper cutoff frequency is raised when negative feedback is applied in an amplifier.

The effect of negative feedback on frequency response curve is shown in Fig. 7.12.

Bandwidth. The bandwidth with negative feedback is given as

$$BW_f = f_2' - f_1' = f_2(1 + \beta A_{vm}) - \frac{f_1}{1 + \beta A_{vm}} \quad ...(7.27)$$

Since $f'_2 > f_2$ and $f'_1 < f_1$ hence the bandwidth with negative feedback is increased.

Assuming $f'_2 >> f'_1$ and $f_2 >> f_1$ bandwidth may be rewritten as

$$BW_f = f_2 (1 + \beta A_{vm}) = (1 + \beta A_{vm}) \text{ times bandwidth without feedback.}$$

Thus we see that bandwidth of an amplifier with negative feedback is increased by the same factor by which voltage gain is reduced and gain bandwidth product remains the same. However, since the amplifier with feedback has lower gain, the net operation is to trade bandwidth for gain.

Example 7.18. An R-C coupled amplifier has a voltage gain of 1,000, f_1 = 50 Hz, f_2 = 200 kHz and a distortion of 5% without feedback. Find the amplifier voltage gain, f'_1, f'_2 and distortion when negative feedback is applied with feedback ratio of 0.01. *[U.P. Technical Univ. Analog and Digital Electronics 2009-10]*

Solution : Open-loop voltage gain, A = 1,000

Feedback ratio, β = 0.01

$$\text{Amplifier voltage gain with feedback, } A_f = \frac{A}{1+\beta A} = \frac{1,000}{1+0.01\times 1,000} = 90.91 \textbf{ Ans.}$$

$$\text{Lower cutoff frequency with feedback, } f'_1 = \frac{f_1}{(1+\beta A)} = \frac{50}{1+0.01\times 1,000} = 4.55 \text{ Hz } \textbf{Ans.}$$

$$\text{Upper cutoff frequency with feedback, } f'_2 = f_2 (1+\beta A) = 200 \times 10^3 (1 + 0.01 \times 1,000) = 2.2 \text{ MHz } \textbf{Ans.}$$

$$\text{Distortion with feedback, } D_f = \frac{D}{1+\beta A} = \frac{5\%}{1+0.01\times 1,000} = 0.455\% \textbf{ Ans.}$$

Example 7.19. An amplifier has the midband gain of 1,500 and a bandwidth of 4 MHz. The midband gain reduces to 150 when a negative feedback is applied. Determine the value of feedback factor and the bandwidth. *[U.P.S.C. I.E.S. Electrical Engineering II, 1998; U.P. Technical Univ. Analog & Digital Electronics 2009-10]*

Solution : Midband gain A_{vm} = 1,500

$$\text{Midband gain with feedback, } A_{vmf} = \frac{A_{vm}}{1+\beta A_{vm}}$$

Substituting the value of A_{vmf} = 150 and A_{vm} = 1,500 in above equation, we have

$$150 = \frac{1,500}{1+\beta A_{vm}}$$

$$\text{or Feedback factor, } \beta A_{vm} = \frac{1,500}{150} - 1 = 9 \textbf{ Ans.}$$

$$\text{Bandwidth with feedback, } BW_f = (1+\beta A_{vm}) \times \text{Bandwidth without feedback} = (1+9)\times 4 = 40 \text{ MHz } \textbf{Ans.}$$

Example 7.20. An amplifier has a mid-frequency gain of 100 and a bandwidth of 200 kHz.

(i) What will be the new bandwidth and gain, if 5% negative feedback is introduced ?

(ii) What should be the amount of feedback, if the bandwidth is to be restricted to 1 MHz ?

[Rajasthan Technical Univ. Analog Electronics 2009]

Solution : Mid-frequency gain, A = 100

Bandwidth without feedback, BW = 200 kHz

Feedback ratio, β = 5% = 0.05

$$(i) \text{ Bandwidth with feedback, } BW_f = (1+\beta A)\times BW = (1+0.05\times 100)\times 200 \text{ kHz} = 1,200 \text{ kHz or } 1.2 \text{ MHz } \textbf{Ans.}$$

$$\text{Gain with feedback, } A_f = \frac{A}{1+\beta A} = \frac{100}{1+0.05\times 100} = 16.67 \textbf{ Ans.}$$

Feedback ratio to restrict the bandwidth to 1 MHz or 1,000 kHz

$$\beta = \frac{\frac{BW_f}{BW} - 1}{A} = \frac{\frac{1000}{200} - 1}{100} = 0.04 \text{ or } 4\% \textbf{ Ans.}$$

Highlights

1. When a part or fraction of output is combined to the input, feedback is said to exist. Thus the process of combining a fraction of output energy (voltage or current) back to the input is called the *feedback*. When the feedback voltage (or current) is so applied as to increase the input signal (*i.e.*, when the feedback is in phase with input signal), it is called the *positive* or *direct* or *regenerative feedback*. Since positive feedback causes excessive distortion and instability, it is seldom used in amplifiers. However, because of its capability of increasing the power of the original signal it is used in oscillator circuits.
2. When the feedback voltage (or current) is so applied as to weaken the input signal (*i.e.*, when the feedback is in phase opposition to the input signal), it is called *negative*, *inverse* or *degenerative feedback*. Negative feedback reduces the amplifier gain but it has numerous advantages (such as gain stability, reduction in nonlinear distortion, reduction in noise, increase in bandwidth or improvement in frequency response, increase in input impedance and decrease in output impedance) and is frequently used in amplifier circuits. The reduction in gain due to negative feedback, can always be compensated by increasing the number of stages.
3. If an amplifier having a voltage gain A has some of the output voltage fed back in series with the input, the voltage gain with feedback is

$$A_f = \frac{A}{1-\beta A}$$

where β is the feedback factor, negative for negative feedback and positive for positive feedback. The term βA is called the feedback factor.
4. Though negative feedback reduces the gain of the amplifier but improves its performance in several aspects such as gain stability, reduction in nonlinear distortion, reduction in noise, increase in bandwidth or improvement in frequency response, increase in input impedance and decrease in output impedance.
   (*a*) Gain is stabilized with negative feedback because the voltage gain with negative feedback is given as

$$A_f \simeq \frac{1}{\beta}$$

   (*b*) A large signal stage has nonlinear distortion which is reduced by a factor (1 + Aβ) when negative feedback is used.
   (*c*) With negative feedback the amplifier gain is reduced and since gain bandwidth has to remain constant, obviously the bandwidth will increase to compensate for reduction in gain.
   (*d*) There is always a noise voltage in the amplifier which is reduced by a factor (1 + β) when negative feedback is used.

Exercises

1. What are different types of feedback ? **[Pb. Technical Univ. Analog Electronics May 2006]**
2. Explain types of negative feedback in trasistor circuits. **[Pb. Technical Univ. Analog Electronics May 2007]**
3. Explain the concept of feedback. **[Pb. Technical Univ. Analog Electronics May 2007]**
4. With negative feedback voltage gain reduces, explain why ? **[Pb. Technical Univ. Analog Electronics December 2006]**
5. Derive an expression for the gain of negative voltage feedback amplifier. **[Pb. Technical Univ. Analog Electronics May 2005]**
6. State the three fundamental assumptions which are made in order that the expression $A_f = \frac{A}{1+\beta A}$ be satisfied exactly.
   List five characteristics of an amplifier which are modified by negative feedback.
   Draw a feedback amplifier in block diagram form. Identify each block, and state its function. **[Pb. Univ. Applied Electronics I, May 1994]**

7. Differentiate between positive and negative feedback. How does negative feedback modify the gain of an amplifier ? **[Pb. Univ. Applied Electronics I, Dec. 1995]**

8. Discuss with the help of a circuit example, the purpose of providing (*i*) negative feedback (*ii*) positive feedback in amplifiers. **[U.P.S.C. I.E.S. Elec. Engineering II, 1996]**

9. Derive an expression for the overall gain of a voltage series-feedback amplifier. **[U.P.S.C. I.E.S. Electrical Engineering II, 1998]**

10. What are the advantages of negative feedback in amplifiers ? Explain them. **[U.P.S.C. I.E.S. Elec. Engineering 1992]**

Short Answer Type Questions With Answers

Q. 1. What is feedback in amplifiers ?

Ans. The process of combining a fraction of output energy back to the input is called the feedback.

Q. 2. Why voltage-series feedback is most commonly used in cascaded amplifiers ?

Ans. In cascaded amplifiers, high input impedance and low output impedance are usually the main requirements and both of these requirements can be met by voltage series feedback, so it is most commonly used in cascaded amplifiers.

Q. 3. Define the term stability. **[P.T.U. Analog Electronics December 2004]**

Ans. The variations in temperature, supply voltages, ageing of components or variations in transistor parameters with replacement are some of the factors that affect the gain of an amplifier and cause it to change.

The stability of an amplifier is defined as the ability of the amplifier to maintain the voltage gain constant as possible over as long a time interval as possible.

The overall voltage gain of the amplifier can be stabilized with negative feedback.

Q. 4. Why is negative feedback employed in high gain amplifiers ? **[P.T.U. Analog Electronics May 2005]**

Ans. Negative feedback is employed in high gain amplifiers to stabilise the gain. With the negative feedback voltage gain of an amplifier is given as $A_f = \frac{A}{1 + A\beta}$. If $A\beta >> 1$, then the expression for voltage gain becomes $A_f = \frac{1}{\beta}$

i.e., overall voltage gain of feedback amplifier A_f becomes independent of internal gain of the amplifier and depends only on feedback ratio β, and β in turn depends on passive elements such as resistors. Resistors remain fairly constant and so the gain is stabilised. Other advantages of negative feedback are reduced nonlinear distortion, reduced noise, increased bandwidth, increased input impedance and reduced output impedance.

Q. 5. What is the effect of negative feedback on the bandwidth of an amplifier ? **[Pb. Univ. Dec. 1995]**

Ans. Bandwidth of an amplifier is increased, when negative feedback is introduced, by a factor $(1 + \beta A)$, by which its gain is reduced.

Q. 6. How does negative feedback increases bandwidth of an amplifier ? **[P.T.U. Analog Electronics December 2005]**

Ans. The bandwidth (BW) of an amplifier without feedback is equal to the separation between 3-dB frequencies f_1 and f_2. If A is the gain then gain-bandwidth product is A × BW. With the negative feedback the amplifier gain is reduced and since gain bandwidth product has to remain constant in both cases, so obviously the bandwidth will increase to compensate for the reduction in gain.

Problems

1. A feedback amplifier has a voltage gain of 500 without feedback. Determine the voltage gain with feedback if feedback ratio is 0.1. **[Ans.** 9.8**]**

2. The overall gain of an amplifier is 200. When negative feedback is applied, the gain decreases to 10. Determine the fraction of the output that is fedback to the input. **[Ans.** 0.095**]**

3. In a negative feedback amplifier A = 100; $\beta = 0.02$ and input signal voltage is 40 mV. Determine (*i*) voltage gain with feedback (*ii*) feedback factor, (*iii*) feedback voltage and (*iv*) output voltage.

[**Ans.** (*i*) 33.33 (*ii*) 2 (*iii*) 26.666 m V (*iv*) 1.333 V]

4. To an amplifier of 60 dB gain a feedback (negative) of $\beta = 0.006$ is applied. What would be the change in the overall gain of the feedback amplifier if the gain of the amplifier decreases by 15%. [**Ans.** 2.462%]

5. An amplifier has a gain of 200 and distortion 10% with an input signal voltage of 200 mV. Determine (*i*) output signal voltage (*ii*) distortion voltage and (*iii*) output voltage when an input signal voltage is 200 mV.

[**Ans.** (*i*) 40 V (*ii*) 4 V (*iii*) 44 V]

6. An amplifier has gain of 60 and distortion 10% without feedback. Determine (*i*) gain and (*ii*) distortion when negative feedback is applied, the feedback factor being 6. [**Ans.** (*i*) 8.57 (*ii*) 1.43%]

7. An amplifier with an open loop voltage gain of 1,000 delivers 10 W of output power at 10 per cent harmonic distortion when the input signal is 10 mV. If 40 dB negative voltage-series feedback is applied and the output power is to remain at 10 W, determine required input signal. [**Ans.** 1.01 V]

8. An amplifier has a midband gain of 200 without feedback. The 3-dB frequency width of it is 200 kHz. The amplifier is to be used as a video amplifier that requires 5 MHz bandwidth. What gain can be obtained and what feedback must be used ? What bandwidth could be obtained if the feedback were 100%.

[Rajasthan Univ. 2001]

[**Ans.** 0.12; 8; 40.2 MHz]

❑❑❑

CHAPTER

8

SINUSOIDAL OSCILLATORS

8.1. INTRODUCTION

An oscillator is the basic element of all ac signal sources and generates sinusoidal signals of known frequency and amplitude. It is one of the basic and useful instruments used in electrical and electronic measurements. For example, an oscillator finds wide applications in electronic communication equipment. In AM (amplitude modulation) and FM (frequency modulation) superheterodyne receivers, "local" oscillators are used to assist in the reduction of the incoming radio frequency (RF) to a lower intermediate frequency (IF). Oscillator circuits are also employed in the "exciter" section of a transmitter to generate the RF carrier. Other applications include their use as "clocks" in digital systems such as microcomputers, in the sweep circuits found in TV sets and oscilloscopes.

Since sinusoidal waveforms are encountered so frequently in electronic measurement work, the oscillator (sinewave generator) represents the largest single category of signal generators. This device covers the frequency range from a few Hz to many GHz.

Although we speak of an oscillator as "generating" a sinusoidal signal, it is to be noted that it does not create energy, but merely acts as an *energy converter*. It simply converts unidirectional current drawn from a dc source of supply into alternating current of desired frequency. The function of an oscillator is reverse of that of a rectifier and, therefore, sometimes called *inverter*. However, we generally think of oscillator circuits as providing an ac *voltage signal*.

It is noteworthy here that although an alternator (ac generator) generates sinusoidal ac power of 50 Hz, it cannot be called an oscillator. An alternator is a mechanical device that has rotating parts, converts mechanical energy into ac energy but cannot produce ac energy of high frequency* whereas an oscillator

* An alternator cannot be employed for generating frequencies exceeding 1,000 Hz because for generating higher frequencies, either the speed of rotation of the armature has to be made extremely high or the number of poles has to be made large and both options are impracticable.

is a non-rotating electronic device that converts dc energy into ac energy of frequency ranging from a few Hz to many GHz.

Though alternator generates large amount of ac power but for several applications such as radio transmitters and receivers, radars etc., an oscillator is preferred owing to its numerous advantages over alternators for such applications. These advantages are :

(*i*) Portable and cheap in cost.

(*ii*) An oscillator is a non-rotating device. Consequently, there is no wear and tear and hence longer life.

(*iii*) Frequency of oscillation may be conveniently varied.

(*iv*) Voltage or currents of any frequency (20 Hz to extremely high frequencies) adjustable over a wide range can be generated.

(*v*) Frequency once set remains constant for a considerable period of time.

(*vi*) Voltages free from harmonics as well as rich with harmonics can be generated by sinusoidal oscillators and relaxation oscillators respectively.

(*vii*) Operation of an oscillator is silent, as there is no moving part in it.

(*viii*) High operation efficiency—due to absence of moving part, there is no wastage of energy owing to friction.

For having a sinusoidal oscillator, we require an amplifier with positive feedback.* The idea is to use the feedback signal in place of an input signal. If the loop gain and phase are correct, there will be an output signal even though there is no external input signal. In other words, an oscillator is an amplifier that is modified by positive feedback to supply its own input signal.

The difference between an amplifier and an oscillator can be explained with the help of Fig. 8.1.

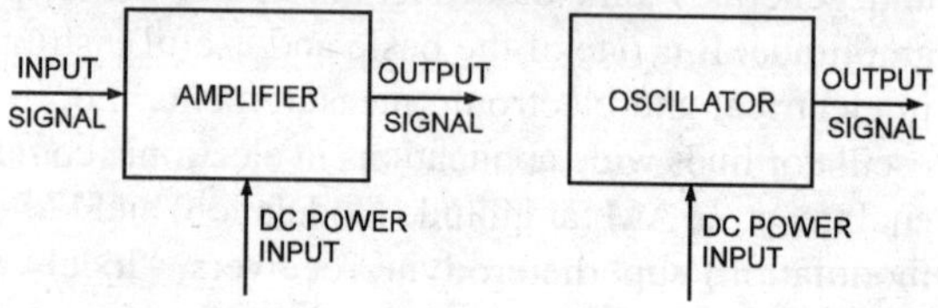

Fig. 8.1

As we know that, an amplifier strengthens the input signal without any change in its waveform and frequency. The additional power required comes from the external dc source. Thus an amplifier is essentially an *energy converter* that draws energy from a dc supply and converts it into ac energy at signal frequency, the energy conversion process being controlled by the input signal. On the other hand an oscillator does not require any external signal either to start or maintain the process of energy conversion and the energy conversion process is controlled by the oscillator circuit itself, as illustrated in Fig. 8.1. In an oscillator, the output signal frequency depends on the passive components employed in the circuit and can be varied as per needs. Oscillator may provide fixed or variable frequency.

When employed as an instrument, an oscillator may be called with different names such as *test oscillator* (or *signal generator*), standard signal generator (or function generator) depending on the function performed by the device, type of signal produced by it, order of sophistication etc.

Oscillators may be broadly divided into two categories viz. *harmonic* oscillators and *relaxation* oscillators. Both types can include active devices such as BJTs, FETs and op-amps, and passive components such as resistors, inductors and capacitors. In harmonic oscillators, the energy always flows in one direction—from the active to passive components. However, in relaxation oscillators, the energy is exchanged between the active and passive components. In harmonic oscillators, the frequency of oscillations is determined by the feedback path. However, in relaxation oscillators, the frequency is determined by time constants–specifically, the charge and discharge time constants during the exchange of energy. Harmonic oscillators can develop low–distortion sinusoidal output waveforms, but relaxation oscillators can only generate nonsinusoidal waveforms such as sawtooth, square or triangular. Harmonic oscillators will be discussed here.

* If the amplifier output signal increases in magnitude after applying the feedback, then it is called the positive feedback. The positive feedback is cumulative. The amplified signal when fed back and passed through the amplifier becomes even stronger. Theoretically signal approaches infinity but in practice a condition is reached when the amplifier gives an output signal without any need of input signal.

Oscillators can also be categorised on the basis of design principle used, and the frequency range over which they are employed.

The oscillators, according to the design principle used are feedback oscillators and negative resistance oscillators. *Feedback oscillators* are widely employed.

The oscillators, according to operating frequency, may be classified as follows :

Type of Oscillators	*Approximate Range*
1. Audio frequency (AF) oscillators	20 Hz–20 kHz
2. Radio frequency (RF) oscillators	20 kHz–30 MHz
3. Very low frequency (VLF) oscillators	15–100 kHz
4. Low frequency (LF) oscillators	100–500 kHz
5. Broadcast oscillators	500 kHz–1.5 MHz
6. Video frequency oscillators	0–5 MHz
7. High frequency (HF) oscillators	1.5–30 MHz
8. Very high frequency (VHF) oscillators	30–300 MHz
9. Ultra high frequency (UHF) oscillators	300–3,000 MHz
10. Microwave oscillators	Beyond 3 GHz (3,000 MHz)

8.2. OPERATION OF OSCILLATOR

The use of positive feedback that results in a feedback amplifier having closed-loop gain $|A_f|$ exceeding unity and satisfies the phase conditions results in operation as an oscillator circuit. An oscillator circuit then provides a constantly varying output signal. If the output signal varies sinusoidally, the circuit is referred to as a *sinusoidal oscillator* and on the other hand if the output voltage rises quickly to one voltage level and later drops quickly to another voltage level, the circuit is usually referred to as a pulse or *square-wave generator*.

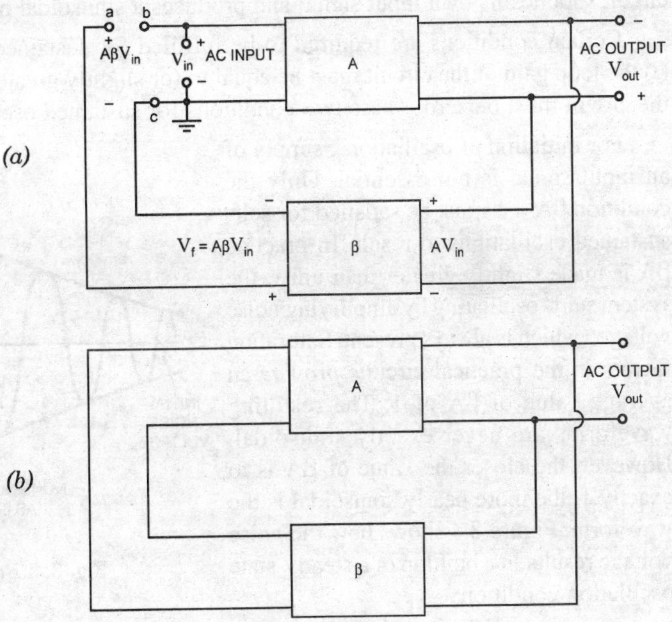

Fig. 8.2

To understand how an oscillator produces an output signal without an external input signal, let us consider the feedback circuit shown in Fig. 8.2 (*a*), where V_{in} is the voltage of ac input driving the input terminals *bc* of an amplifier having voltage gain A. The amplified output voltage is

$$V_{out} = AV_{in} \quad ...(8.1)$$

* If the amplifier output signal increases in magnitude after applying the feedback, then it is called the positive feedback. The positive feedback is cumulative. The amplified signal when fed back and passed through the amplifier becomes even stronger. Theoretically signal approaches infinity but in practice a condition is reached when the amplifier gives an output signal without any need of input signal.

This voltage drives a feedback circuit that is usually a *resonant circuit*, as we get maximum feedback at one frequency. The feedback voltage returning to point *a* is given by equation

$$V_f = A\beta V_{in} \quad ...(8.2)$$ where β is the gain of feedback network.

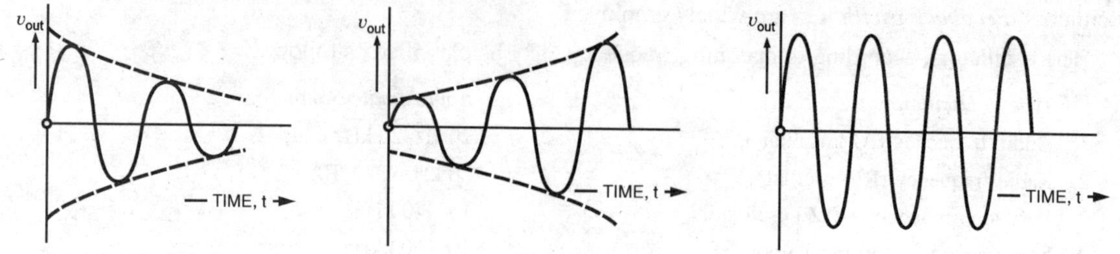

(*a*) *Oscillations Die out* (*b*) *Oscillations Buildup* (*c*) *Oscillations of Constant Amplitude*

Fig. 8.3

If the phase shift through the amplifier and feedback circuit is zero, then $A\beta V_{in}$ is in phase with the input signal V_{in} that drives the input terminals of the amplifier.

Now we connect point '*a*' to point '*b*' and simultaneously remove voltage source V_{in}, then feedback voltage $A\beta V_{in}$ drives the input terminals *bc* of the amplifier, as shown in Fig. 8.2 (*b*). In case Aβ is less than unity, $A\beta V_{in}$ is less than V_{in} and the output signal will die out, as illustrated in Fig. 8.3 (*a*). On the other hand if Aβ is greater than unity, the output signal will build up, as illustrated in Fig. 8.3 (*b*). If Aβ is equal to unity, $A\beta V_{in}$ equals V_{in} and the output signal is a steady sinewave, as illustrated in Fig. 8.3 (*c*). In this case the circuit supplies its own input signal and produces a sinusoidal output.

Certain conditions are required to be fulfilled for sustained oscillations and these conditions are that (*i*) the loop gain of the circuit must be equal to (or slightly greater than) unity and (*ii*) the phase shift around the circuit must be zero. These two conditions for sustained oscillations are called *Barkhausen criteria.*

For initiation of oscillations, supply of an input signal is not essential. Only the condition βA = 1 must be satisfied for self-sustained oscillations to result. In practice βA is made slightly greater than unity, the system starts oscillating by amplifying noise voltage* which is always present. Saturation factors in the practical circuits *provide* an average value of βA of 1. The resulting waveforms are never exactly sinusoidal. However, the closer the value of βA is to exactly 1, the more nearly sinusoidal is the waveform. Figure 8.4 shows how the noise voltage results in a buildup of a steady-state oscillation condition.

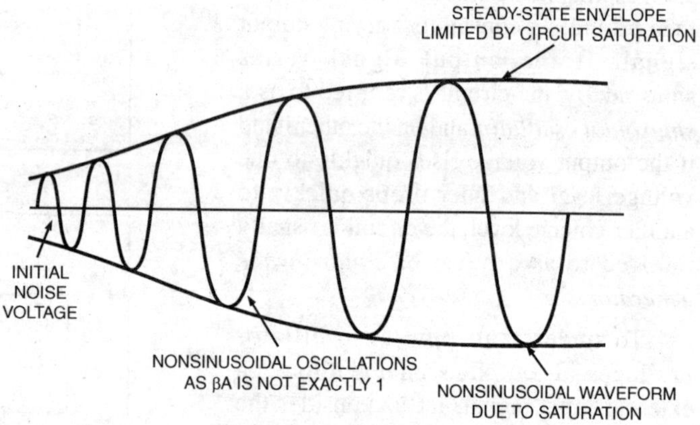

Fig. 8.4. ***Buildup of Steady-state Oscillations***

Another way of seeing how the feedback circuit provides operation as an oscillator is obtained by noting the denominator in the basic feedback equation, $A_f = \frac{A}{1+\beta A}$. When βA = – 1 or magnitude 1 at a phase angle of 180°, the denominator becomes zero and the gain with feedback, A_f, becomes infinite. Thus, an infinitesimal

* Because of the ambient temperature, the free electrons contained by resistors move randomly in different directions and generate a *noise voltage* across the resistor. The motion is so random that it contains frequencies to over 1,000 GHz. Each resistor may, thus be thought of as a small ac voltage source producing all frequencies.

signal (noise voltage) can provide a measurable output voltage, and the circuit acts as an oscillator even without an input signal.

By deliberate design the phase shift around the loop is made 0º at the resonant frequency. Above and below the resonant frequency, the phase shift is different from 0º. Thus, oscillations are obtained at only one frequency, the resonant frequency of the feedback circuit.

To understand and apply the Barkhausen criterion, we must consider both the gain and phase shift of $A\beta$ as a function of frequency. Reactive elements, capacitance in particular, contained in the amplifier and/or feedback causes the gain magnitude and phase shift to vary with the frequency. In general, there will be only one frequency at which the gain magnitude is unity and at which, simultaneously, the total phase shift is equivalent to 0° (in phase—a multiple of 360°). The *system will oscillate at the frequency that satisfies those conditions.* Designing an oscillator amounts to selection of reactive components and their incorporation into circuit in such a way that the conditions will be satisfied at a desired frequency.

To show the dependence of the loop gain $A\beta$ on frequency, we express $A\beta(j\omega)$, a complex phasor that can be expressed in both polar and rectangular form :

$$A\beta(j\omega) = |A\beta|\angle\theta = |A\beta| \cos\theta + j|A\beta| \sin\theta \qquad ...(8.3)$$

where $|A\beta|$ is the magnitude of the loop gain, a function of frequency, and θ is the phase shift, also a function of frequency. For satisfaction of Barkhausen criterion, we must have

$$|A\beta| = 1 \qquad ...(8.4)$$

$$\text{and } \theta = \pm 360° n \qquad ...(8.5)$$

where n is any integer, including 0. In polar and rectangular forms, the **Barkhausen criterion** is expressed as

$$A\beta(j\omega) = 1 \angle \pm 360° n = 1 + j0 \qquad ...(8.6)$$

8.3. ESSENTIALS OF AN OSCILLATOR CIRCUIT

From the above discussion it can be inferred that an oscillator circuit must have the following three elements :

1. Oscillatory circuit or frequency determining element.
2. Amplifier.
3. Positive feedback network.

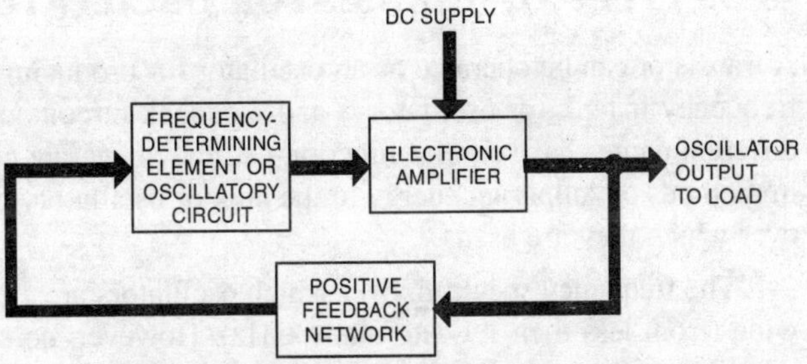

Fig. 8.5. *Block Diagram of An Oscillator*

The oscillatory circuit or a frequency-determining element can be an inductance-capacitance network (L-C tank), resistance-capacitance (R-C network), or a quartz crystal, depending on the frequency and waveshape desired.

The electronic amplifier receives dc power from the battery or dc power supply and converts it into ac power for supply to the tank circuit. The oscillations occurring in the tank circuit are applied to the input of the electronic amplifier. Because of the amplifying properties of the amplifier, we get increased output of these oscillations. This amplified output of oscillations is because of dc power supplied from the external source (a battery or power supply). The output of the amplifier can be supplied to the tank circuit to meet the losses.

The *feedback network* supplies a part of the output power to the frequency-determining element in correct phase to aid the oscillations. In other words, feedback circuit must be a *positive feedback network.* The block diagram of an oscillator is shown in Fig. 8.5.

8.4. FREQUENCY STABILITY OF OSCILLATOR

An oscillator having initially been set at a particular frequency does not maintain its initial frequency, but instead drift and wander about in frequency, sometimes uniformly in one direction, sometimes quite erratically. The *frequency stability* of an oscillator is a measure of its ability to maintain as nearly a fixed frequency as possible over as long a time interval as possible. These deviations in frequency are caused due to variations in the values of circuit features (circuit components, transistor parameters, supply voltage, stray capacitances, output load etc.) that determine the oscillator frequency.

As the circuit operates, it warms up. Consequently the values of resistors, inductors and capacitors, which are frequency determining factors in such circuits, are changed with temperature. Variations in temperature also cause variations in transistor parameters. However, the variation or drift in oscillator frequency is slow because of slow variations in R, L, C and transistor parameters with time.

The other major factor responsible for deviation in frequency is *variations in power supply* (operating voltage applied to the active device). However, this problem can be overcome by using regulated power supply.

Any variation in load coupled to the tank circuit may cause a change in effective resistance of the circuit by transformer action which, in turn, causes the drift in frequency.

Other factors responsible for drift in oscillator frequency are :

Inter-element capacitances, operating point of active device and mechanical vibrations. The effects of *variations in inter-element capacitances* can be neutralised by introducing a swamping capacitor across the offending elements (the introduced capacitor becomes the part of the tank circuit). The *operating point Q of the active device* is so chosen as to restrict the operation of the device on the linear portion of its characteristic because operation on nonlinear portion causes the variations in transistor parameters which, in turn, affect the oscillator frequency stability. The instability in oscillator frequency caused due to *mechanical vibrations* can be avoided by isolating the oscillator circuit from the source of mechanical vibrations.

8.5. TYPES OF TRANSISTOR OSCILLATORS

A transistor can be operated as an oscillator for producing continuous undamped oscillations of any desired frequency if tank (or oscillatory) and feedback circuits are properly connected to it. All oscillators under different names have similar function *i.e.,* they generate continuous undamped output. However, they differ in methods of supplying energy to the tank or oscillatory circuit to meet the losses and the frequency ranges over which they are used.

The frequency spectrum over which oscillators are employed to produce sinusoidal signals is extremely wide (from less than 1 Hz to many GHz). However, no single oscillator design is practical for generating signals over this entire range. Instead, a variety of designs are employed, each of which generates sinusoidal outputs most advantageously over various portions of the frequency spectrum.

Oscillators, which use *inductance-capacitance* (L-C) circuits as their tank or oscillatory circuits, are very popular for generating high frequency (*e.g.,* 10 kHz to 100 MHz) outputs. The most widely used LC oscillators are the *Hartley* and *Colpitt's oscillators*. Although they slightly differ from one another in their electronic circuitry but they have virtually identical frequency ranges and frequency-stability characteristics. However, such oscillators are not suitable for generating low- frequency sinusoidal outputs. This is due to the fact that the components required in construction of low-frequency L-C resonant circuits are too bulky and heavy. So *resistor-capacitor* (R-C) *oscillators* are generally employed for generating low-frequency (from 1 Hz to about 1 MHz) sinusoidal signals. Two most common R-C oscillators are the *Wien bridge* and *phase shift types.*

Other less frequently used oscillators are the *crystal oscillators* and the *negative resistance oscillators*. The operating frequency ranges of various types of most commonly used oscillators are given here :

Type of Oscillator	*Approximate Frequency Ranges*
Wien bridge oscillator	1 Hz — 1 MHz
Phase shift oscillator	1 Hz — 10 MHz
Hartley oscillator	10 kHz — 100 MHz
Colpitt's oscillator	10 kHz — 100 MHz
Negative resistance oscillator	> 100 MHz
Crystal oscillator	Fixed frequency

The oscillators of various types will be discussed in the following sections.

8.6. L-C OSCILLATORS

Oscillators, which use inductance-capacitance (L-C) circuits as their tank or oscillatory circuits are called L-C *oscillators*. L-C oscillators are very popular for generating high-frequency outputs (*e.g.*, 10 kHz to 100 MHz). There is a large variety of L-C oscillators such as tuned collector oscillators, tuned base oscillators, Colpitt's oscillators, Hartley oscillators, Clapp oscillators, crystal oscillators etc.

8.7. GENERAL FORM OF AN L-C OSCILLATOR

In the general form of an oscillator depicted in Fig. 8.6 (*a*), any of the active devices such as vacuum tube, BJT, FET or op-amp may be used in the amplifier section, Z_1, Z_2 and Z_3 are the reactive elements constituting the feedback tank circuit which determines the frequency of oscillation. Here, Z_1 and Z_2 serve as an ac voltage divider for the output voltage and feedback signal. Thus, the voltage across Z_2 is the feedback signal. The equivalent circuit is drawn in Fig. 8.6 (*b*) with the following two assumptions :

(*i*) h_{re} of transistor is negligibly small and, therefore, the feedback source $h_{re}V_{out}$ is negligible.

(*ii*) h_{oe} of the transistor is very small *i.e.*, the output resistance $\frac{1}{h_{oe}}$ is very large and, therefore, $\frac{1}{h_{oe}}$ is omitted from the equivalent circuit.

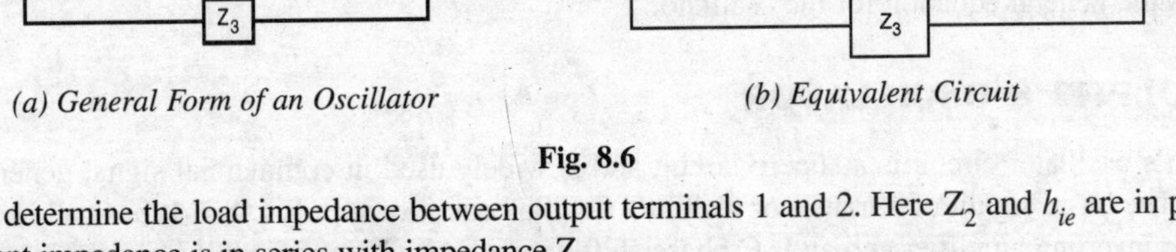

(*a*) *General Form of an Oscillator* (*b*) *Equivalent Circuit*

Fig. 8.6

Let us determine the load impedance between output terminals 1 and 2. Here Z_2 and h_{ie} are in parallel and their resultant impedance is in series with impedance Z_3.

The equivalent impedance is in parallel with impedance Z_1.

Thus load impedance between output terminals is given as

$$\mathbf{Z}_L = \mathbf{Z}_1 \parallel [\mathbf{Z}_3 + (\mathbf{Z}_2 \parallel h_{ie})]$$

$$= \mathbf{Z}_1 \parallel \left[\mathbf{Z}_3 + \frac{\mathbf{Z}_2 h_{ie}}{\mathbf{Z}_2 + h_{ie}}\right] = \mathbf{Z}_1 \parallel \left[\frac{\mathbf{Z}_3\,(\mathbf{Z}_2 + h_{ie}) + \mathbf{Z}_2 h_{ie}}{\mathbf{Z}_2 + h_{ie}}\right] = \mathbf{Z}_1 \parallel \left[\frac{h_{ie}\,(\mathbf{Z}_2 + \mathbf{Z}_3) + \mathbf{Z}_2\,\mathbf{Z}_3}{\mathbf{Z}_2 + h_{ie}}\right]$$

$$\text{or } \frac{1}{Z_L} = \frac{1}{Z_1} + \frac{Z_2 + h_{ie}}{h_{ie}(Z_2 + Z_3) + Z_2 Z_3}$$

$$= \frac{h_{ie}(Z_1 + Z_2 + Z_3) + Z_1 Z_2 + Z_2 Z_3}{Z_1 [h_{ie}(Z_2 + Z_3) + Z_2 Z_3]}$$

$$\text{or } Z_L = \frac{Z_1 [h_{ie}(Z_2 + Z_3) + Z_2 Z_3]}{h_{ie}(Z_1 + Z_2 + Z_3) + Z_1 Z_2 + Z_2 Z_3} \quad ...(8.7)$$

The voltage gain of a CE amplifier without feedback is given as

$$A = \frac{-h_{fe} Z_L}{h_{ie}} \quad ...(8.8)$$

The output voltage between terminals 1 and 2 is given as

$$V_{out} = \left[Z_3 + \frac{Z_2 h_{ie}}{Z_2 + h_{ie}} \right] I_1 = \left[\frac{h_{ie}(Z_2 + Z_3) + Z_2 Z_3}{Z_2 + h_{ie}} \right] I_1$$

The voltage feedback to the input terminals 2 and 3 is given as

$$V_f = \frac{Z_2 h_{ie}}{Z_2 + h_{ie}} I_1$$

$$\text{So feedback fraction, } \beta = \frac{V_f}{V_{out}} = \frac{Z_2 h_{ie}}{h_{ie}(Z_2 + Z_3) + Z_2 Z_3} \quad ...(8.9)$$

Applying the criterion of oscillation *i.e.*, $A\beta = 1$, we have

$$\frac{-h_{fe} Z_L}{h_{ie}} \cdot \frac{Z_2 h_{ie}}{h_{ie}(Z_2 + Z_3) + Z_2 Z_3} = 1$$

$$\text{or } \frac{h_{fe} Z_1 [h_{ie}(Z_2 + Z_3) + Z_2 Z_3]}{h_{ie}(Z_1 + Z_2 + Z_3) + Z_1 Z_2 + Z_2 Z_3} \cdot \left[\frac{Z_2}{h_{ie}(Z_2 + Z_3) + Z_2 Z_3} \right] = -1$$

$$\text{or } \frac{h_{fe} Z_1 Z_2}{h_{ie}(Z_1 + Z_2 + Z_3) + Z_1 Z_2 + Z_2 Z_3} = -1$$

$$\text{or } h_{ie}(Z_1 + Z_2 + Z_3) + Z_1 Z_2 + Z_2 Z_3 = -h_{fe} Z_1 Z_2$$

$$\text{or } h_{ie}(Z_1 + Z_2 + Z_3) + Z_1 Z_2 (1 + h_{fe}) + Z_2 Z_3 = 0 \quad ...(8.10)$$

This is the general equation for the oscillator.

8.8. COLPITT'S OSCILLATOR

The Colpitt's oscillator circuit is a superb circuit and is widely used in commercial signal generators up to 100 MHz. The basic circuit of a transistor Colpitt's Oscillator is shown in Fig. 8.7. It basically consists of a single stage inverting amplifier and an L-C phase shift network, as obvious from the circuit diagram shown. The two series capacitors C_1 and C_2 form the potential divider used for providing the feedback voltage—the voltage developed across capacitor C_2 provides the regenerative feedback required for sustained oscillations. Parallel combination of R_E and C_E along with resistors R_1 and R_2 provides the stabilized self bias. The collector supply voltage V_{CC} is applied to the collector through a radio frequency choke (RFC) which permits an easy flow of direct current but at the same time it offers very high impedance to the high frequency currents. The presence of coupling capacitor C_{C_2} in the output circuit does not permit the dc currents to go to the tank circuit (the flow of dc current in a tank circuit reduces its Q). The radio frequency energy developed across RFC is capacitively coupled to the tank circuit through the capacitor C_{C_2}. The output of the phase shift L-C network is coupled from the junction of L and C_2 to the amplifier input at base through coupling capacitor C_{C_1}, which blocks dc but provides path to ac. Transistor itself produces a phase shift of 180° and another phase shift of 180° is provided by the capacitive feedback. Thus a total phase shift of 360° is obtained which is an essential condition for developing oscillations. The output voltage is derived from a secondary winding L' coupled to the inductance L. The frequency is determined by the tank circuit and is varied by gang-tuning the two capacitors C_1 and C_2. It is to be noted that capacitors C_1 and C_2 are ganged. As the tuning is varied, values of both capacitors vary simultaneously, the ratio of the two capacitances remaining the same.

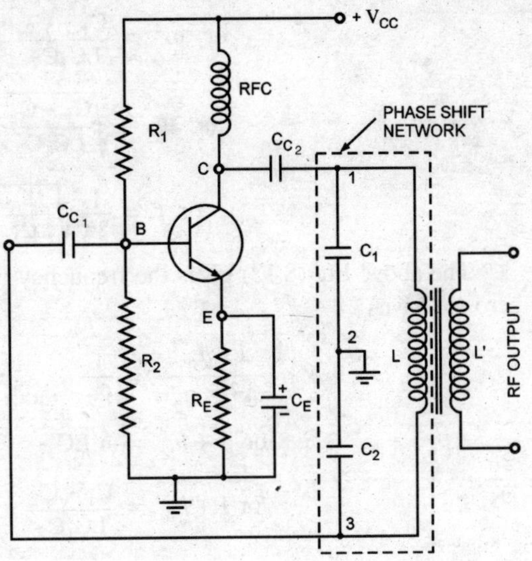

Fig. 8.7. ***Basic Circuit For Transistor Colpitt's Oscillator***

Working. When the collector supply voltage V_{CC} is switched on, the capacitors C_1 and C_2 are charged. These capacitors C_1 and C_2 discharge through the coil L, setting up oscillations of frequency $f = \dfrac{1}{2\pi}\sqrt{\dfrac{1}{LC_1}+\dfrac{1}{LC_2}}$. The oscillations across capacitor C_2 are applied to the base-emitter junction and appear in the amplified form in the collector circuit. Of course, the amplified output in the collector circuit is of the same frequency as that of the oscillatory circuit. This amplified output in the collector circuit is supplied to the tank circuit in order to meet the losses. Thus the tank circuit is getting energy continuously from the collector circuit to make up for the losses occurring in it and, therefore, ensures undamped oscillations. The energy supplied to the tank circuit is of correct phase, as already explained, and if Aβ exceeds unity, oscillations are sustained in the circuit.

Frequency of Oscillation :

Here referring to Fig. 8.6 $\quad \mathbf{Z}_1 = \dfrac{1}{j\omega C_1} = -\dfrac{j}{\omega C_1}$; $\mathbf{Z}_2 = \dfrac{1}{j\omega C_2} = \dfrac{-j}{\omega C_2}$ and $\mathbf{Z}_3 = j\omega L$

Substituting these values in Eq. (8.10) we have

$$h_{ie}\left[\frac{-j}{\omega C_1}-\frac{j}{\omega C_2}+j\omega L\right]+\left[\frac{-j}{\omega C_1}\cdot\frac{-j}{\omega C_2}\right](1+h_{fe})+\left[\frac{-j}{\omega C_2}\right]j\omega L = 0$$

$$\text{or } -jh_{ie}\left[\frac{1}{\omega C_1}+\frac{1}{\omega C_2}-\omega L\right]-\frac{1+h_{fe}}{\omega^2 C_1 C_2}+\frac{L}{C_2}=0 \qquad \text{...(8.11)}$$

Equating the imaginary component of the above Eq. (8.11) to zero we have

$$h_{ie}\left[\frac{1}{\omega C_1}+\frac{1}{\omega C_2}-\omega L\right]=0$$

$$\text{or } \quad \frac{1}{\omega C_1}+\frac{1}{\omega C_2}=\omega L \qquad \because h_{ie}\neq 0$$

$$\text{or } \quad \frac{C_1+C_2}{\omega C_1 C_2}=\omega L$$

$$\text{or } \omega^2 = \frac{C_1 + C_2}{LC_1C_2}$$

$$\text{or } \omega = \sqrt{\frac{C_1 + C_2}{LC_1C_2}} = \sqrt{\frac{1}{LC_1} + \frac{1}{LC_2}}$$

$$\text{or } f = \frac{1}{2\pi}\sqrt{\frac{1}{LC_1} + \frac{1}{LC_2}} \quad ...(8.12)$$

The above Eq. (8.12) gives the frequency of oscillation. Equating the real component of Eq. (8.11) to zero we have

$$\frac{1 + h_{fe}}{\omega^2 C_1 C_2} = \frac{L}{C_2}$$

$$\text{or } 1 + h_{fe} = \omega^2 LC_1$$

$$\text{or } 1 + h_{fe} = \frac{C_1 + C_2}{LC_1C_2} \times LC_1 = \frac{C_1 + C_2}{C_2} = 1 + \frac{C_1}{C_2}$$

$$\text{or } h_{fe} = \frac{C_1}{C_2} \quad ...(8.13)$$

As for other oscillator circuits, the loop gain must be greater than unity to ensure that the circuit oscillates.

$$\text{So } A\beta \geq 1 \text{ or } A \geq \frac{1}{\beta} \geq \frac{C_2}{C_1} \quad ...(8.14)$$

FET Colpitt's Oscillator. A practical version of an FET Colpitt's oscillator is shown in Fig. 8.8. Here, R_G is the gate biasing resistor. The RFC choke performs two functions. (*i*) It keeps ac current out of the dc drain supply and (*ii*) it provides drain load. As seen, C_1 is in the input circuit whereas C_2 is in the output circuit. The oscillator frequency can be determined by the same relation as in case of transistor Colpitt's oscillator *i.e.,* by relation

$$f_0 = \frac{1}{2\pi}\sqrt{\frac{1}{LC_1} + \frac{1}{LC_2}} \quad ...(8.15)$$

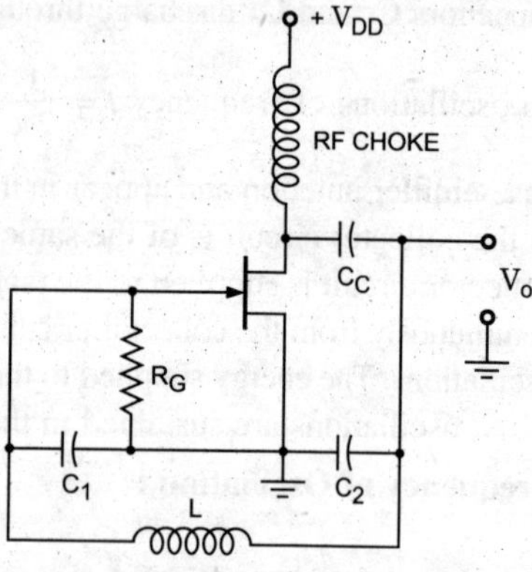

Fig. 8.8. ***FET Colpitt's Oscillator***

Example 8.1. In a transistor Colpitt's oscillator shown in Fig. 8.7, L = 100 μH, L_{RFC} = 0.6 mH, C_1 = 0.001 μF, C_2 = 0.01 μF and C_C = 10 μF. Determine (*i*) operating frequency (*ii*) feedback fraction (*iii*) minimum gain to sustain oscillations and emitter resistance if R_C = 2.5 kΩ.

Solution :

Inductance, $L = 100\ \mu H = 100 \times 10^{-6}$ or 1×10^{-4} H

Capacitance, $C_1 = 0.001\ \mu F = 0.001 \times 10^{-6}$ F

Capacitance, $C_2 = 0.01\ \mu F = 0.01 \times 10^{-6}$ F

(*i*) Operating frequency, $$f = \frac{1}{2\pi}\sqrt{\frac{1}{LC_1} + \frac{1}{LC_2}} = \frac{1}{2\pi}\sqrt{\frac{1}{10^{-4} \times 0.001 \times 10^{-6}} + \frac{1}{10^{-4} \times 0.01 \times 10^{-6}}}$$

$$= 528 \text{ kHz}$$

(*ii*) Feedback fraction, $$\beta = h_{fe} = \frac{C_1}{C_2} = \frac{0.001 \times 10^{-6}}{0.01 \times 10^{-6}} = 0.1 \textbf{ Ans.}$$

(*iii*) Minimum gain to sustain oscillations, $A_{min} = \frac{1}{\beta} = \frac{1}{0.1} = 10$ **Ans.**

(*iv*) Since $A \cong R_C/R_E$, $R_E \cong \frac{R_C}{A_{min}} = \frac{2.5\ k\Omega}{10} = 250\ \Omega$ **Ans.**

Example 8.2. A Colpitt's oscillator is designed with $C_1 = 100$ pF and $C_2 = 7{,}500$ pF. The inductance is variable. Determine the range of inductance values, if the frequency of oscillations is to vary between 950 kHz and 2,050 kHz.

[*U.P. Technical Univ. Analog and Digital Electronics 2009-10*]

Solution : Resonant frequency is given as

$$f_0 = \frac{1}{2\pi}\sqrt{\frac{1}{LC_1}+\frac{1}{LC_2}} \qquad \text{...Refer to Eq. (8.12)}$$

$$\text{or} \quad L = \frac{1}{4\pi^2 f_0^2}\left(\frac{1}{C_1}+\frac{1}{C_2}\right)$$

For frequency of oscillations of 950 kHz, $C_1 = 100$ pF and $C_2 = 7{,}500$ pF.

$$L_1 = \frac{1}{4\pi^2(950\times10^3)^2}\left(\frac{1}{100\times10^{-12}}+\frac{1}{7{,}500\times10^{-12}}\right)$$

$$= \frac{1}{4\pi^2(950\times10^3)^2}\times\frac{76\times10^{12}}{7{,}500} = 0.284 \text{ mH}$$

For frequency of oscillations of 2,050 kHz, $C_1 = 100$ pF and $C_2 = 7{,}500$ pF

$$L_2 = \frac{1}{4\pi^2(2{,}050\times10^3)^2}\times\left(\frac{1}{100\times10^{-12}}+\frac{1}{7{,}500\times10^{12}}\right)$$

$$= \frac{1}{4\pi^2(2{,}050)^2\times10^6}\times\frac{76\times10^{12}}{7{,}500} = 0.061 \text{ mH}$$

Thus, the range of required inductance is 0.061 mH to 0.284 mH **Ans.**

8.9. HARTLEY OSCILLATOR

The transistor Hartley oscillator is as popular as Colpitt's oscillator and is widely used as a local oscillator in radio receivers. The circuit arrangement is shown in Fig. 8.9. Hartley oscillator circuit is similar to Colpitt's oscillator circuit, except that phase shift network consists of two inductors L_1 and L_2 and a capacitor C instead of two capacitors and one inductor. The output of the amplifier is applied across inductor L_1 and the voltage across inductor L_2 forms the feedback voltage. The coil L_1 is inductively coupled to coil L_2, the combination functions as an auto-transformer. However, because of direct connection, the junction of L_1 and L_2 cannot be directly grounded. Instead, another capacitor C_L is used. The operation of the circuit is similar to that of the Colpitt's oscillator circuit.

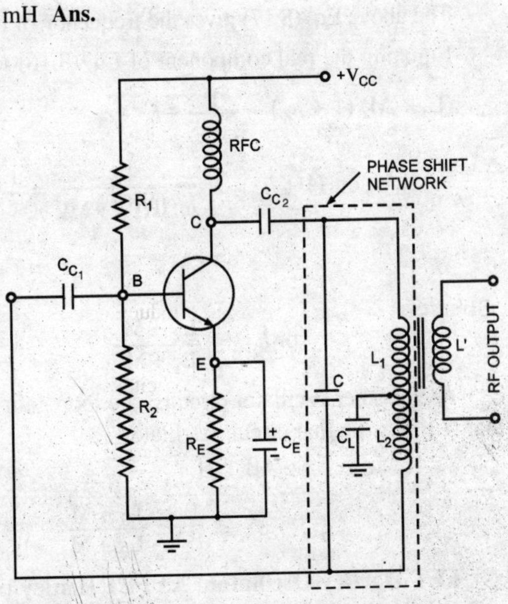

Fig. 8.9. ***Basic Circuit For Hartley Oscillator***

Considering the fact that there exists mutual inductance between coils L_1 and L_2 because the coils are wound on the same core, their net effective inductance is increased by mutual inductance M. So in this case effective inductance is given by the equation

$$L = L_1 + L_2 + 2M$$

and resonant or oscillation frequency is given by the equation

$$f = \frac{1}{2\pi\sqrt{[C(L_1 + L_2 + 2M)]}}$$, as derived below :

Frequency of Oscillation. The general Eq. (8.10) derived in Art. 8.7 is reproduced here

$$h_{ie}(\mathbf{Z}_1 + \mathbf{Z}_2 + \mathbf{Z}_3) + \mathbf{Z}_1 \mathbf{Z}_2 (1 + h_{fe}) + \mathbf{Z}_2 \mathbf{Z}_3 = 0$$

Here $\mathbf{Z}_1 = j\omega L_1 + j\omega M$; $\mathbf{Z}_2 = j\omega L_2 + j\omega M$ and $\mathbf{Z}_3 = \frac{1}{j\omega C} = -j/\omega C$

Substituting these values in general Eq. (8.10), we get

$$h_{ie}\left[(j\omega L_1 + j\omega M) + (j\omega L_2 + j\omega M) - \frac{j}{\omega C}\right] + (j\omega L_1 + j\omega M)(j\omega L_2 + j\omega M)(1 + h_{fe}) + (j\omega L_2 + j\omega M)\left(\frac{-j}{\omega C}\right) = 0$$

$$\text{or} \quad j\omega h_{ie}\left[L_1 + L_2 + 2M - \frac{1}{\omega^2 C}\right] - \omega^2 (L_2 + M)\left[(L_1 + M)(1 + h_{fe}) - \frac{1}{\omega^2 C}\right] = 0 \quad \text{...(8.16)}$$

Equating the imaginary part of above Eq. (8.16) to zero, we get

$$\omega h_{ie}\left[L_1 + L_2 + 2M - \frac{1}{\omega^2 C}\right] = 0$$

$$\text{or} \quad L_1 + L_2 + 2M - \frac{1}{\omega^2 C} = 0 \qquad \because \omega h_{ie} \neq 0$$

$$\text{or } \omega^2 C = \frac{1}{L_1 + L_2 + 2M}$$

$$\text{or } f = \frac{\omega}{2\pi} = \frac{1}{2\pi\sqrt{C(L_1 + L_2 + 2M)}} \quad \text{...(8.17)}$$

The above Eq. (8.17) gives the frequency of oscillation.

Equating the real component of Eq. (8.16) to zero, we get

$$(L_1 + M)(1 + h_{fe}) - \frac{1}{\omega^2 C} = 0$$

$$\text{or } (1 + h_{fe}) = \frac{1}{\omega^2 C(L_1 + M)} = \frac{L_1 + L_2 + 2M}{L_1 + M}$$

$$= 1 + \frac{L_2 + M}{L_1 + M}$$

$$\text{or } h_{fe} = \frac{L_2 + M}{L_1 + M} \quad \text{...(8.18)}$$

As for other oscillator circuits, the loop gain must be greater than 1 to ensure that circuit oscillates.

$$\text{So } A\beta \geq 1$$

$$\text{or } A \geq \frac{1}{\beta} \geq \frac{L_1 + M}{L_2 + M} \quad \text{...(8.19)}$$

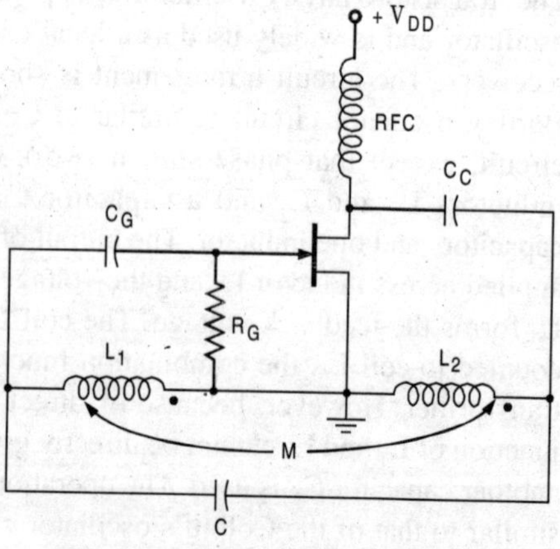

Fig. 8.10. ***FET Hartley Oscillator***

FET Hartley Oscillator. An FET Hartley oscillator circuit is shown in Fig. 8.10. The noteworthy point is that inductors L_1

and L_2 have a mutual coupling M, which is to be taken into account in determination of the equivalent inductance for the resonant circuit. This circuit operates at a frequency given by Eq. (8.17).

Example 8.3. In a transistorized Hartley oscillator, the tank circuit has the capacitance of 100 pF. The value of inductance between the collector and tapping point is 30 mH and the value of inductance between the tapping point and the transistor base is 1×10^{-8} H. Determine the frequency of oscillations. Neglect the mutual inductance.

[*Pb. Technical University Electronic Devices & Circuits 2003-04*]

Solution : Effective inductance, $L = L_1 + L_2 + 2M = 30 \times 10^{-3} + 1 \times 10^{-8} + 0 \simeq 0.03$ H

Capacitance, $C = 100 \text{ pF} = 100 \times 10^{-12} \text{ F} = 1 \times 10^{-10}$ F

Frequency of oscillation, $$f_o = \frac{1}{2\pi\sqrt{LC}} = \frac{1}{2\pi \times \sqrt{0.03 \times 10^{-10}}} = 91.9 \text{ kHz}$$ **Ans.**

Example 8.4. Determine the oscillation frequency of a transistor Hartley oscillator shown in Fig. 8.9 with circuit values $L_1 = 1$ mH, $L_2 = 100$ μH, M = 50 μH and C = 100 pF.

Solution : Effective inductance, $L = L_1 + L_2 + 2M = 1 \times 10^{-3} + 100 \times 10^{-6} + 2 \times 50 \times 10^{-6} = 1{,}200$ μH

Capacitance, $C = 100 \text{ pF} = 10^{-10}$ F

Frequency of oscillation, $$f_o = \frac{1}{2\pi\sqrt{LC}} = \frac{1}{2\pi\sqrt{1{,}200 \times 10^{-6} \times 10^{-10}}} = 459 \text{ kHz}$$ **Ans.**

8.10. CRYSTAL OSCILLATORS

In crystal oscillators, the usual electrical resonant circuit is replaced by a mechanically vibrating crystal. The crystal (usually quartz) has a high degree of stability in holding constant at whatever frequency the crystal is originally cut to operate. The crystal oscillators are, therefore, used whenever great stability is needed, for example, in communication transmitters, and receivers, digital clocks etc.

A quartz crystal exhibits a very important property known as piezoelectric effect. When a mechanical pressure is applied across the faces of the crystal, a voltage proportional to the applied mechanical pressure appears across the crystal. Conversely, when a voltage is applied across the crystal surfaces, the crystal is distorted by an amount proportional to the applied voltage. An alternating voltage applied to a crystal causes it to vibrate at its natural frequency.

Besides quartz, the other substances that exhibit the piezoelectric effect are Rochelle salt and tourmaline. Rochelle salt exhibits the greatest piezoelectric effect, but its applications are limited to manufacture of microphones, headsets and loudspeakers. It is because the Rochelle salt is mechanically the weakest and strongly affected by moisture and heat. Tourmaline is most rugged but shows the least piezoelectric effect. Quartz is a compromise between the piezoelectric effect of Rochelle salt and the mechanical strength of tourmaline. It is inexpensive and readily available in nature. It is mainly the quartz crystal that is used in radio frequency (RF) oscillators.

For use in electronic oscillators, the crystal is suitably cut and then mounted between two metal plates, as shown in Fig. 8.11 (*a*). Although the crystal has electromechanical resonance but the crystal action can be represented by an electrical resonant circuit, as shown in Fig. 8.11 (*b*). The crystal actually behaves as a series R-L-C circuit in parallel with C_M where C_M is the capacitance of the mounting electrodes. Because the crystal losses, represented by R, are small, the equivalent crystal Q is high—typically 20,000. Values of Q up to 10^6 can be obtained by making use of crystals.

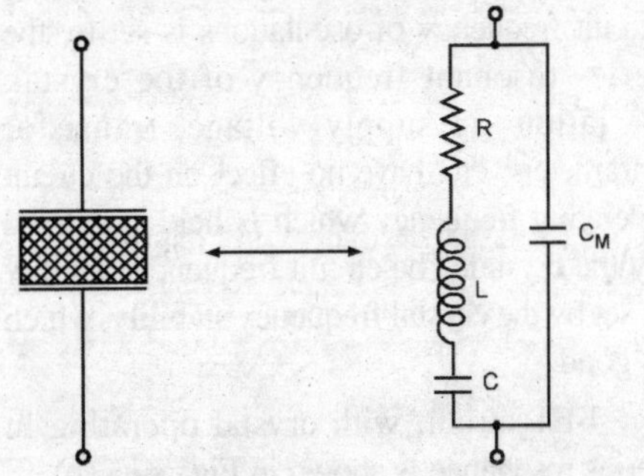

(*a*) Crystal Mounting (*b*) Electrical Equivalent Circuit of a Crystal

Fig. 8.11

Because of presence of C_M, the crystal has two resonant frequencies. One of these is the series resonant frequency f_s at which $2\pi fL = \frac{1}{2\pi fC}$ and in this case the crystal impedance is very low. The other is parallel resonance frequency f_p which is due to parallel resonance of capacitance C_M and the reactance of the series circuit. In this case crystal impedance is very high. The impedance versus frequency curve of the crystal is shown in Fig. 8.12. In order to use the crystal properly it must be connected in a circuit so that its low impedance in the series-resonant operating mode or high impedance in the antiresonant or parallel resonant operating mode is selected.

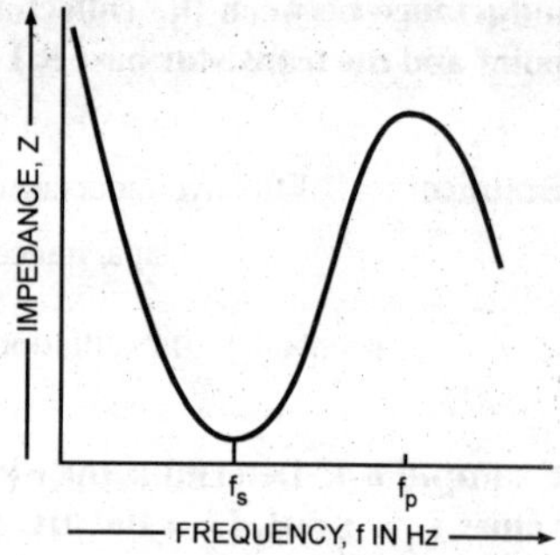

Fig. 8.12. ***Crystal Impedance Versus Frequency Curve***

Two resonant frequencies are given by the equations

$$\text{Series resonant frequency, } f_s = \frac{1}{2\pi\sqrt{LC}} \quad \text{...(8.20)}$$

$$\text{Parallel resonant frequency, } f_p = \frac{1}{2\pi}\sqrt{\frac{1 + C/C_M}{LC}} \quad \text{...(8.21)}$$

It appears that f_p is higher than f_s but the two frequencies are very close to each other. It is due to the fact that the ratio $\frac{C}{C_M}$ is very small.

To stabilize the frequency of an oscillator, a crystal may be operated at either its series or parallel resonant frequency.

To excite a crystal for operation in the series-resonant mode it may be connected as a series element in a feedback path, as shown in Fig. 8.13 (*a*). In this mode of operation the crystal impedance is the smallest and the amount of positive feedback is the largest. Resistors R_1, R_2 and R_E provide a voltage-divider stabilized dc bias circuit, the capacitor C_E provides ac bypass of the emitter resistor R_E and the radio frequency coil (RFC) provides for dc bias while decoupling any ac signal on the power lines from affecting the output signal. The voltage feedback signal from the collector to the base is maximum when the crystal impedance is minimum (*i.e.*, in series-resonant mode). The coupling capacitor C_C has negligible impedance at the circuit operating frequency but blocks any dc between collector and base. The circuit shown in Fig. 8.13 (*a*) is generally called the *Pierce crystal.* The resulting circuit frequency of oscillations is set by the series resonant frequency of the crystal. Variations in supply voltage, transistor parameters, etc. have no effect on the circuit operating frequency which is held stabilized by the crystal. The circuit frequency stability is set by the crystal frequency stability, which is good.

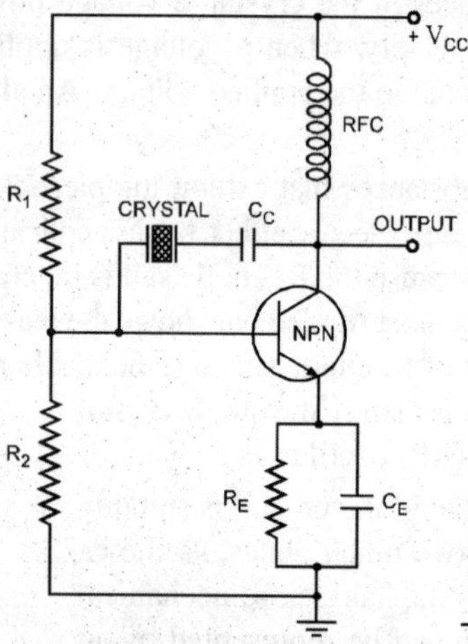

(a) Oscillator With Crystal Operating in Series Resonance

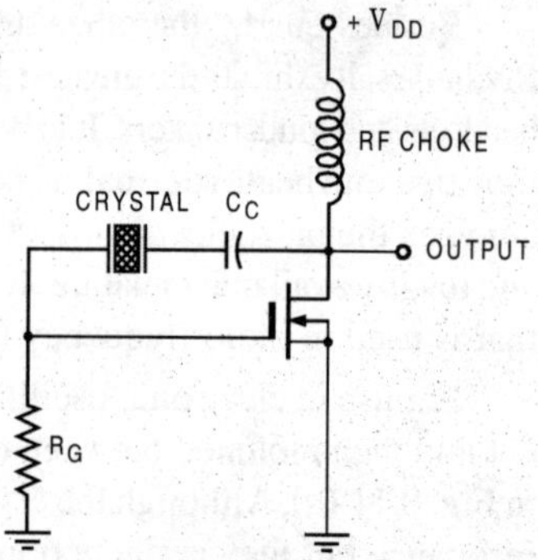

(b) FET Circuit

Fig. 8.13

FET circuit with crystal operating in series resonance is shown in Fig. 8.13 (*b*).

Use of FET is desirable because its high input impedance results in light loading of the crystal which (*i*) results in good stability and (*ii*) does not lower the Q value.

Oscillator circuit with crystal operating in parallel resonance (a modified Colpitt's oscillator circuit) is illustrated in Fig. 8.14. Since the parallel-resonant impedance of a crystal is of a maximum value, it is connected in parallel. C_1 and C_2 form a capacitive voltage divider which returns a portion of the output voltage to the transistor emitter. Transistor NPN combined with R_1, R_2, RFC, and R_E, constitutes a common base circuit. Capacitor C_3 provides an ac short circuit across R_2 to ensure that the transistor base remains at a fixed voltage level. As the output voltage increases positively, the emitter voltage also increases, and since the base voltage is fixed, the base-emitter voltage is reduced. The reduction in V_{BE} causes collector current I_C to diminish, and this in turn causes the collector voltage V_C to increase positively. Thus, the circuit is applying its own input, and a state of oscillation exists. The crystal in parallel with C_1 and C_2 permits maximum voltage feedback from the collector to emitter when its impedance is maximum, *i.e.*, at its parallel resonant frequency. At other frequencies, the crystal impedance is low, and so the resultant feedback voltage is too small to sustain oscillations. The oscillation frequency is stabilized at the parallel resonant frequency of the crystal.

A Miller crystal controlled oscillator circuit is shown in Fig. 8.15. A tuned L-C circuit in the drain section is adjusted near the crystal parallel-resonant frequency. The maximum gate-source signal occurs at the crystal antiresonant frequency, controlling the circuit operating frequency.

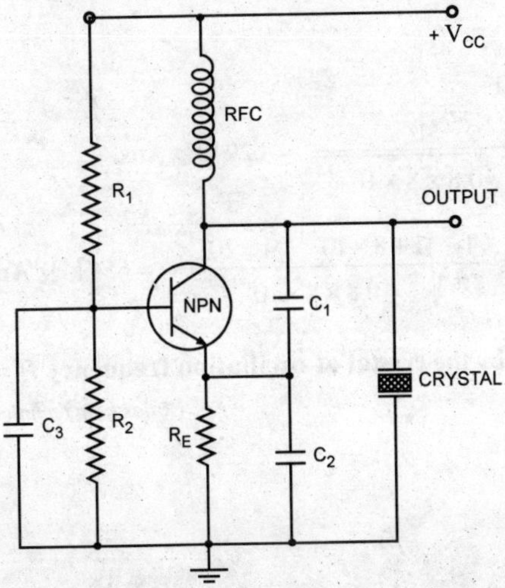

Fig. 8.14. ***Oscillator With Crystal Operating in Parallel Resonance***

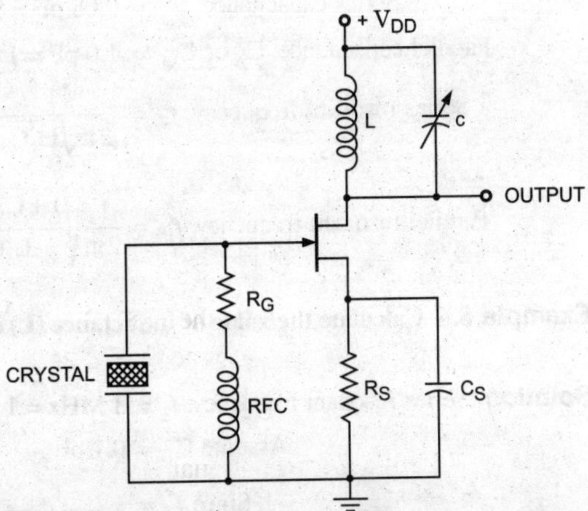

Fig. 8.15. ***Miller Crystal Controlled Oscillator***

The advantages and disadvantages of a crystal oscillator are given below :

Advantages. 1. It is very simple circuit as it does not need any tank circuit other than crystal itself.

2. Different oscillation frequencies can be had by simply replacing one crystal with another.
3. The Q factor, which is a measure of the quality of resonance circuit of a crystal, is very high. The Q factor of a crystal may range from 10^4 to 10^6 whereas the L-C circuit may have a Q factor only of the order of 100.
4. Most crystals will maintain frequency drift to within a few cycles at 25ºC. For greater frequency stability, the crystal is often contained in an insulated enclosure termed as crystal oven in which the temperature is thermostatically controlled. In this way it is possible to have frequency drifts less than 1 part in 10^{10}.

Disadvantages. 1. The crystal oscillators have a very limited tuning range (or not at all). They are used for frequencies exceeding 100 kHz.

2. The crystal oscillators are fragile and, therefore, can only be used in low power circuits.

Crystal oscillators must be designed to provide a load capacitance on the crystal as per specifications listed by the manufacturer. This requirement is essential for obtaining oscillations at the specified frequency. It is also important from the point of view of limiting the power supplied to the crystal to the specified maximum. Too much crystal power causes distortion in the oscillator waveform. It also causes overheating of the crystal, consequently rendering the resonant frequency unstable. More important is that the thin plated electrodes may be melted off an overdriven crystal, destroying the device. Typical maximum drive levels for plated crystals vary from 2 mW to 10 mW.

The maximum permissible drive power limits the ac voltages that may be applied across the crystal and consequently affects the design of oscillator circuits. Crystal manufacturers usually specify the resistance of individual crystal, as well as maximum drive power. From these two, the maximum crystal ac voltage may be determined by using the relation $P = V^2/R$.

Example 8.5. The parameters of a crystal oscillator equivalent circuit are $L_s = 0.8$ H; $C_s = 0.08$ pF, $R_s = 5$ kΩ and $C_p = 1.0$ pF. Determine the resonance frequencies f_s and f_p.

[U.P.S.C. I.E.S. Electrical Engineering II, 1999; Pb. Technical Univ. Analog Electronics, May 2007; Rajasthan Univ. 2003, 2005, 2006]

Solution : Series resistance, $R_s = 5\ \text{k}\Omega = 5{,}000\ \Omega$

Series inductance, $L_s = 0.8$ H

Series capacitance, $C_s = 0.08\ \text{pF} = 8 \times 10^{-14}$ F

Parallel capacitance, C_p or $C_M = 1.0\ \text{pF} = 1 \times 10^{-12}$ F

$$\text{Series resonant frequency, } f_s = \frac{1}{2\pi\sqrt{L_s C_s}} = \frac{1}{2\pi\sqrt{0.8 \times 8 \times 10^{-14}}} = 629 \text{ kHz } \textbf{Ans.}$$

$$\text{Parallel resonant frequency, } f_p = \frac{1}{2\pi}\sqrt{\frac{1 + C_s/C_p}{L_s C_s}} = \frac{1}{2\pi}\sqrt{\frac{1 + 8 \times 10^{-14}/1 \times 10^{-12}}{0.8 \times 8 \times 10^{-14}}} = 654 \text{ kHz } \textbf{Ans.}$$

Example 8.6. Calculate the value of inductance (L) offered by the crystal at oscillation frequency $f_s = 1$ MHz.

[Rajasthan Technical Univ. 2008]

Solution : Series resonant frequency, $f_s = 1\ \text{MHz} = 1 \times 10^6$ Hz

Assume $C_s = 0.1$ pF

$$\text{Since } f_s = \frac{1}{2\pi\sqrt{L_s C_s}}$$

$$\text{So } L_s = \frac{1}{4\pi^2 C_s f_s^2} = \frac{1}{4\pi^2 \times 0.1 \times 10^{-12} \times 1 \times 10^{12}} = 0.2533 \text{ H } \textbf{Ans.}$$

Example 8.7. Prove that ratio of the parallel to series resonant frequencies of crystal is given approximately by $\frac{f_p}{f_s} \cong \left(1 + \frac{1}{2}\frac{C}{C'}\right)$, where C′ = mounting capacitance. If C = 0.04 pF and C′ = 2.0 pF, by what percentage is the parallel resonant frequency greater than the series resonant frequency.

[Rajasthan Univ. 2004]

Solution :

$$f_p = \frac{1}{2\pi}\sqrt{\frac{1 + \frac{C}{C'}}{LC}} \qquad \text{...}(i)$$

$$f_s = \frac{1}{2\pi\sqrt{LC}} \qquad \text{...}(ii)$$

Dividing Eq. (*i*) by Eq. (*ii*), we have

$$\frac{f_p}{f_s} = \left(1 + \frac{C}{C'}\right)^{1/2} \cong 1 + \frac{1}{2}\frac{C}{C'} \quad \textbf{Proved.}$$

Parallel resonant frequency is greater than the series resonant frequency by percentage

$$= \frac{1}{2}\frac{C}{C'} \times 100 = \frac{1}{2} \times \frac{0.04}{2.0} \times 100 = 1\% \text{ Ans.}$$ by approximate expression

$$= \left[\left(1 + \frac{C}{C'}\right)^{1/2} - 1\right] \times 100 = \left(\sqrt{1 + \frac{0.04}{2.0}} - 1\right) \times 100 = 0.995\% \text{ Ans.}$$

8.11. AUDIO OSCILLATORS

So far, we have considered the oscillators which use L-C tuned circuit that causes a phase shift of 180° due to inductive or capacitive coupling in addition to a 180° phase shift produced by the transistor itself. The oscillators employing L-C elements, called the *L-C oscillators,* are very popular for generating high frequency oscillations but they cannot be employed for generation of low frequency oscillations as they become too bulky and expensive. *R-C oscillators* are commonly used for generating audio frequencies as they provide good frequency stability and proper waveform. Also, with the advent of IC technology, R-C network is the only feasible solution, as it is very difficult to make a too high value inductance in an integrated circuit. Two commonly used R-C oscillators are (*i*) R-C phase shift oscillator and (*ii*) Wien bridge oscillator. So these two audio oscillators will be discussed here.

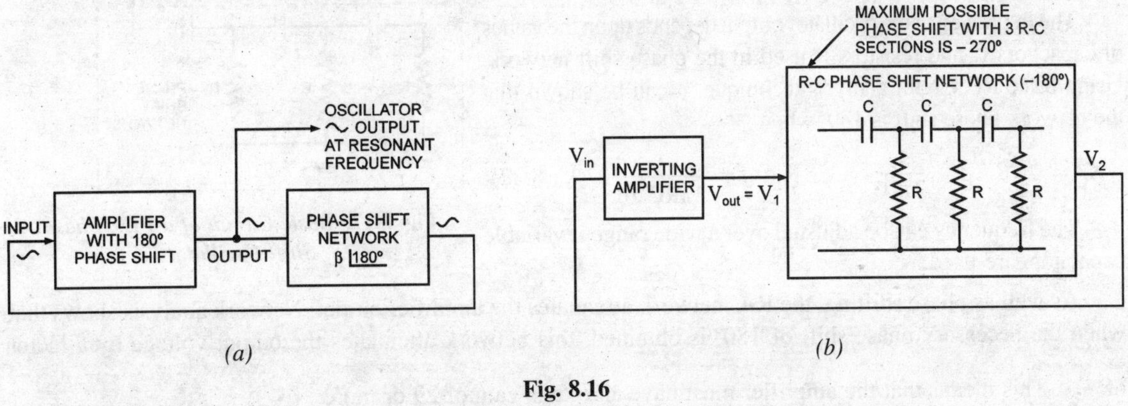

Fig. 8.16

8.11.1. Basic Principles of R-C Oscillators

For producing oscillations in an oscillator circuit we need positive feedback which means that the voltage signal feedback should be in phase with the input signal. For providing a positive feedback at one particular frequency, an inverting amplifier may be used with a feedback network that causes a phase shift of 180° at the desired frequency of oscillation, as shown in Fig. 8.16 (*a*). The 180° phase shift in the feedback signal can be obtained by a suitable network consisting of three R-C sections, as shown in Fig. 8.16 (*b*).

When a phase shift network such as that indicated in Fig. 8.16 is used in a phase shift oscillator, the R's and C's must be selected so as to produce a phase shift of 180° at the desired frequency of oscillation. The output of the voltage amplifier is fed to the input to the phase shift network. Thus $V_1 = V_{out}$. The output resistance of the amplifier is designed to be very small in comparison to the input impedance of the phase shift network. The output voltage of the phase shift network, V_2 is fed into the input of the amplifier *i.e.,* $V_2 = V_{in}$. The amplifier's input impedance must be much larger than the output impedance of the phase shift network.

Alternatively, a positive feedback can be obtained by using two stages of amplifiers each giving a phase shift of 180°. A part of this output is fed back to the input through a feedback network without causing any further phase shift. Wien bridge oscillator operates on this principle.

8.12. PHASE SHIFT OSCILLATOR

We select the so *called phase shift oscillator* (Fig. 8.17) as a first example as it exemplifies very simply the principles set forth above. The circuit is drawn to show clearly the amplifier and feedback network. The circuit consists of a common source FET amplifier followed by a three-section R-C phase shift network. The amplifier stage is self-biased with a capacitor bypassed source resistor R_S and a drain bias resistance R_D. The output of the last section is supplied back to the gate. If the loading of the phase shift network on the amplifier can be assumed to be negligible, a phase shift of 180° between the amplified output voltage V_{out} and the input voltage V_{in} at the gate is produced by the amplifier itself. The three-section R-C phase shift network produces an additional phase shift, which is a function of frequency and equals 180° at some frequency of operation. At this frequency the total phase shift from the gate around the circuit and back to gate will be exactly zero. This particular frequency will be the one at which the circuit will oscillate provided that the magnitude of the amplification is sufficiently large. In a FET phase shift oscillator voltage series feedback [*i.e.*, feedback voltage proportional to the output voltage V_{out} and supplied in series with the input signal at the gate] is used.

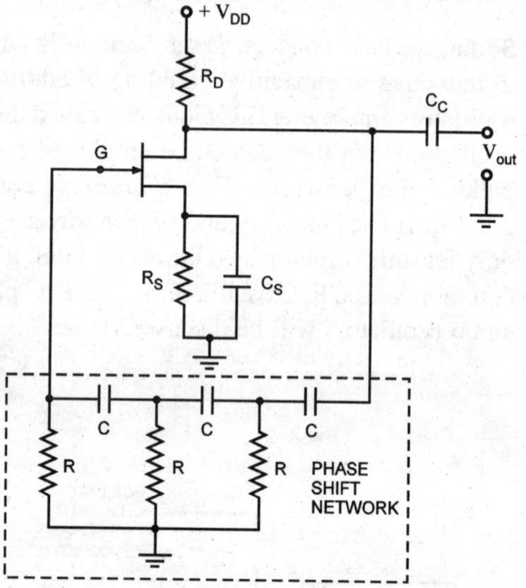

Fig. 8.17. ***Basic Circuit of a FET Phase Shift Oscillator***

The frequency of the oscillator output depends upon the values of capacitors C and resistors R used in the phase shift network. Using basic RC circuit analysis technique, it can be shown that the network phase shift is 180° when

$$X_C = \sqrt{6}R \quad \text{or} \quad \frac{1}{2\pi fC} = \sqrt{6}R \text{ or } f = \frac{1}{2\pi RC\sqrt{6}} \qquad ...(8.22)$$

The frequency can be adjusted over a wide range if variable capacitors are used.

As well as phase shifting, the R-C network attenuates the amplifier output. Network analysis shows that when the necessary phase shift of 180° is obtained, this network attenuates the output voltage by a factor of $\frac{1}{29}$. This means that the amplifier must have a voltage gain of 29 or more.

When the amplifier voltage gain is 29 and feedback factor of R-C network, $\beta = \frac{1}{29}$ then the loop gain is $\beta A = \frac{1}{29} \times 29 = 1$, the amplifier phase shift of –180° combined with the network phase shift of +180° gives a loop phase shift of zero. Both of these conditions are necessary to satisfy the Barkhausen criteria.

If the amplifier gain is much greater than 29, the oscillator output waveform is likely to be distorted. When the gain is slightly greater than 29, the output is usually a reasonably pure sinusoidal.

The advantages and disadvantages of phase shift oscillators are given below :

Advantages. 1. It is cheap and simple circuit as it contains resistors and capacitors (not bulky and expensive high value inductors).

2. It provides good frequency stability.
3. The phase shift oscillator circuit is much simpler than the Wien bridge oscillator circuit because it does not need negative feedback and the stabilization arrangements.
4. The output is sinusoidal that is quite distortion free.
5. They have a wide frequency range (from a few Hz to several hundred kHz).

6. They are particularly suitable for low frequencies, say of the order of 1 Hz, as these frequencies can be easily obtained by using R and C of large values.

Disadvantages. 1. The output is small. It is due to smaller feedback.

2. It is difficult for the circuit to start oscillations as the feedback is usually small.

3. The frequency stability is not as good as that of Wien bridge oscillator.

4. It needs high voltage (12 V) battery so as to develop sufficiently large feedback voltage.

Applications. FET phase shift oscillator is used for generating signals over a wide frequency range. The frequency may be varied from a few Hz to 200 Hz by employing one set of resistors with three capacitors ganged together to vary over a capacitance range in the 1 : 10 ratio. Similarly the frequency ranges of 200 Hz to 2 kHz, 2 kHz to 20 kHz and 20 kHz to 200 kHz can be obtained by using other sets of resistors.

Transistor Phase Shift Oscillator. The circuit arrangement of a phase shift oscillator using N-P-N transistor in CE configuration is shown in Fig. 8.18. As usual, the voltage divider $R_1 - R_2$ provides dc emitter base bias, R_E and C_E combination provides temperature stability and prevent ac signal degeneration and collector resistor R_C controls the collector voltage. The oscillator output voltage is capacitively coupled to the load by C_C.

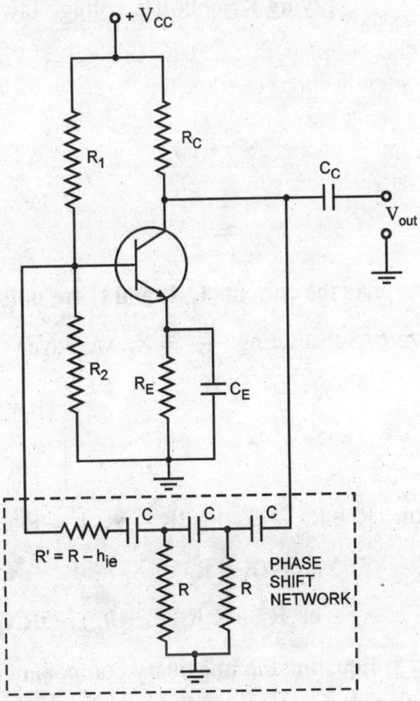

Fig. 8.18. *Basic Circuit of a Transistor Phase Shift Oscillator*

In case of a transistor phase shift oscillator, the output of the feedback network is loaded appreciably by the relatively small input resistance (h_{ie}) of the transistor. Hence, instead of employing voltage-series feedback (as used in case of FET phase shift oscillator), voltage-shunt feedback is used for a transistor phase shift oscillator, as shown in Fig. 8.18. In this circuit, the feedback signal is coupled through the feedback resistor R' in series with the amplifier stage input resistance h_{ie}. The value of R' should be such that when added with amplifier stage input resistance h_{ie}, it is equal to R *i.e.*, $R' + h_{ie} = R$.

Operation. The circuit is set into oscillations by any random or variation caused in the base current, that may be either due to noise inherent in the transistor or minor variation in voltage of dc power supply. This variation in base current is amplified in collector circuit. The output of the amplifier is supplied to an R-C feedback network. The R-C network produces a phase shift of 180° between output and input voltages. Since CE amplifier produces a phase reversal of the input signal, total phase shift becomes 360° or 0° which is essential for *regeneration* or for sustained oscillations. The output of this network is thus in the same phase as the originally assumed input to the amplifier and is applied to the base terminal of the transistor. Thus, sustained variation in collector current between saturation and cutoff values are obtained. R-C phase shift network is the frequency determining network, as already explained in case of FET phase shift oscillator.

Frequency of Oscillation. The equivalent circuit for the analysis of a transistor phase shift oscillator (circuit shown in Fig. 8.18) is shown in Fig. 8.19 (*a*). The equivalent circuit shown in Fig. 8.19 (*a*) is simplified if the following assumptions are made.

(*i*) h_{re} of the transistor is usually negligibly small and, therefore, $h_{re} V_{out}$ is omitted from the circuit.

(*ii*) h_{oe} of the transistor is very small *i.e.*, $\frac{1}{h_{oe}}$ is much larger than R_C. Thus the effect of h_{oe} can be neglected.

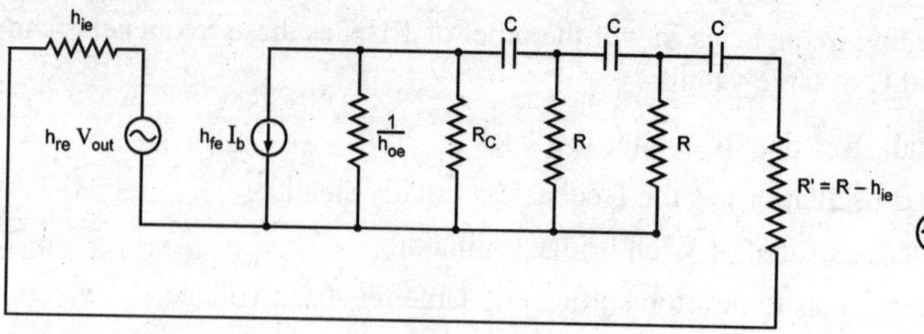

Fig. 8.19 (*a*). *Equivalent Circuit of Transistor Phase-Shift Oscillator Circuit Shown in Fig. 8.18*

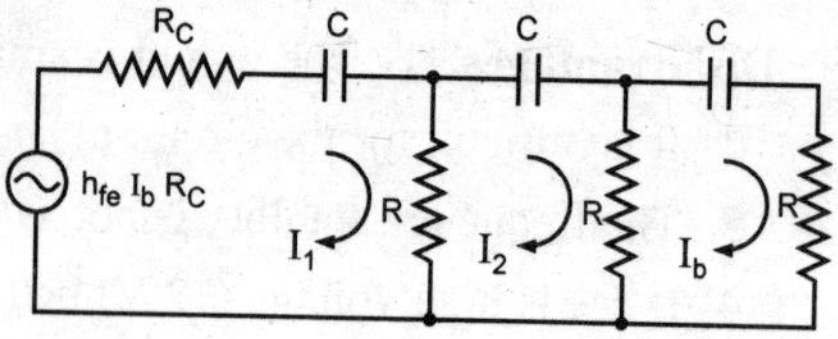

Fig. 8.19 (*b*)

Making above assumptions and replacing current source by equivalent voltage source, the simplified equivalent circuit is shown in Fig. 8.19 (*b*).

Applying Kirchhoff's voltage law to the three loops shown in Fig. 8.19 (*b*), we have

$$\left(R + R_C + \frac{1}{j\omega C}\right) I_1 - RI_2 + h_{fe}\, I_b R_C = 0 \qquad ....(8.23)$$

$$-RI_1 + \left(2R + \frac{1}{j\omega C}\right) I_2 - RI_b = 0 \qquad ...(8.24)$$

$$0 - RI_2 + \left(2R + \frac{1}{j\omega C}\right) I_b = 0 \qquad ...(8.25)$$

As the currents I_1, I_2 and I_b are non-vanishing, the determinant of the coefficients of I_1, I_2 and I_b must be zero. Substituting $\frac{1}{\omega C} = X_C$ we have

$$\begin{vmatrix} (R + R_C - jX_C) & (-R) & (h_{fe} R_C) \\ (-R) & (2R - jX_C) & (-R) \\ 0 & (-R) & (2R - jX_C) \end{vmatrix} = 0 \qquad ...(8.26)$$

or $(R + R_C - jX_C)\,[(2R - jX_C)^2 - R^2] + R\,[(-R)(2R - jX_C) - h_{fe}R_C(-R)] = 0$

or $(R + R_C - jX_C)(3R^2 - X_C^2 - j4RX_C) - R[2R^2 - jRX_C - h_{fe}R_CR] = 0$

or $R^3 + R^2R_C(3 + h_{fe}) - 5RX_C^2 - R_CX_C^2 - 6jR^2X_C - j4RR_CX_C + jX_C^3 = 0 \qquad ...(8.27)$

Equating the imaginary component of the above equation to zero, we have

$$6R^2X_C + 4RR_CX_C - X_C^3 = 0$$

$$\text{or } X_C = \sqrt{6R^2 + 4RR_C}$$

$$\text{or } 2\pi fC = \frac{1}{\sqrt{6R^2 + 4RR_C}} \qquad \because X_C = \frac{1}{\omega C} = \frac{1}{2\pi fC}$$

$$\text{or } f = \frac{1}{2\pi RC\sqrt{6 + \frac{4R_C}{R}}} \qquad ...(8.28)$$

If $R = R_C$, then

$$f = \frac{1}{2\pi RC\sqrt{10}} \qquad ...(8.29)$$

The above equation gives frequency of oscillation.

Equating the real component of Eq. (8.27) to zero we have

$$R^3 + R^2R_C(3 + h_{fe}) - X_C^2(5R + R_C) = 0$$

$$\text{or } R^3 + R^2R_C(3 + h_{fe}) - (6R^2 + 4RR_C)(5R + R_C) = 0$$

$$\text{or } -29R^3 - 23R^2R_C + h_{fe}R^2R_C - 4RR_C^2 = 0$$

$$\text{or } \frac{-29R}{R_C} - 23 + h_{fe} - 4\frac{R_C}{R} = 0$$

$$\text{or } h_{fe} = 23 + 29\frac{R}{R_C} + 4\frac{R_C}{R} \quad ...(8.30)$$

For the loop gain to be greater than unity, the requirement of the current gain of the transistor is found to be

$$h_{fe} > 23 + 29\frac{R}{R_C} + 4\frac{R_C}{R} \quad ...(8.31)$$

If $R = R_C$, then $h_{fe} > (23 + 29 + 4)$ *i.e.*, > 56 ...(8.32)

Applications. The phase shift oscillator is well suited to the range of frequencies from several hertz to several hundred kilohertz (20 Hz to 200 kHz), and so includes the audio frequency range (up to 20 kHz). For generating different audio frequencies, variable air capacitors are employed as circuit elements in the phase shift network. It is possible to vary the frequency in the range of about 1 : 10 because the range of capacitors can be varied in the ratio of 10 : 1 (typically from 40 pF to 450 pF). For variations of frequency over a large range, the three capacitors are usually ganged so as to vary the capacitance of the three capacitors simultaneously. Such a variation keeps the input impedance to the phase shift network constant and also keeps constant the magnitudes of β and $A\beta$. Thus the amplitude of oscillations will remain unaffected as the frequency is adjusted. The phase shift oscillator is operated in class A so as to keep distortion to the minimum. Frequency range from 20 Hz to 200 Hz, 200 Hz to 2 kHz, 2 kHz to 20 kHz and 20 kHz to 200 kHz can be obtained by using different set of resistors.

Phase shift oscillators are not suitable for higher frequency operation because at higher frequency, the internal phase shift of the transistor and reduction in h_{fe} cause difficulties in designing the circuit. The frequency of the oscillator cannot be changed easily.

Example 8.8. In a phase shift oscillator shown in Fig. 8.17, $R_1 = R_2 = R_3 = 800$ kΩ and $C_1 = C_2 = C_3 = 100$ pF. Determine the frequency of oscillation.

Solution : $R_1 = R_2 = R_3 = R = 800 \text{ k}\Omega = 8 \times 10^5\ \Omega$

$C_1 = C_2 = C_3 = C = 100 \text{ pF} = 10^{-10}$ F

$$\text{Frequency of oscillation, } f_o = \frac{1}{2\pi RC\sqrt{6}} = \frac{1}{2\pi \times 8 \times 10^5 \times 10^{-10} \times \sqrt{6}} = 812 \text{ Hz } \textbf{Ans.}$$

Example 8.9. An R-C phase shift oscillator has to be designed to provide a sine wave of frequency 1 kHz. If the value of capacitance is 0.016 μF, calculate the value of the resistances.

[M.D. Univ. Rohtak Analog Electronic Circuits May 2009]

Solution : Let the phase shift oscillator be BJT one.

$C_1 = C_2 = C_3 = C = 0.016 \times 10^{-6}$ F

$f_0 = 1 \text{ kHz} = 1{,}000$ Hz

$$\text{Since } f_0 = \frac{1}{2\pi RC\sqrt{10}} \quad \text{Refer to Eq. (8.29)}$$

$$R = \frac{1}{2\pi f_0 C\sqrt{10}} = \frac{1}{2\pi \times 1{,}000 \times 0.016 \times 10^{-6}\sqrt{10}} = 3{,}146\ \Omega = 3.3 \text{ k}\Omega \text{ (standard value) } \textbf{Ans.}$$

8.13. WIEN BRIDGE OSCILLATOR

It is one of the most popular type of oscillators used in audio and sub-audio frequency ranges (20 – 20 kHz). This type of oscillator is simple in design, compact in size, and remarkably stable in its frequency output. Furthermore, its output is relatively free from distortion and its frequency can be varied easily. However, the maximum frequency output of a typical Wien bridge oscillator is only about 1 MHz.

This is also, in fact, a phase shift oscillator. It employs two transistors, each producing a phase shift of 180°, and thus producing a total phase shift of 360° or 0°.

The circuit diagram of Wien bridge oscillator is shown in Fig. 8.20. It is essentially a two-stage amplifier with an R-C bridge circuit. R-C bridge circuit (Wien bridge) is a lead-lag network (R_1-C_1 and R_2-C_2). The phase shift across the network lags with increasing frequency and leads with decreasing frequency. By adding Wien bridge feedback network, the oscillator becomes sensitive to a signal of only one particular frequency. This particular frequency is that at which Wien bridge is balanced and for which the phase shift is 0°. If the Wien bridge feedback network is not employed and output of transistor Q_2 is fed back to transistor Q_1 for providing regeneration required for producing oscillations, the transistor Q_1 will amplify signals over a wide range of frequencies and thus direct coupling would result in poor frequency stability. Thus, by employing Wien bridge feedback network frequency stability is increased.

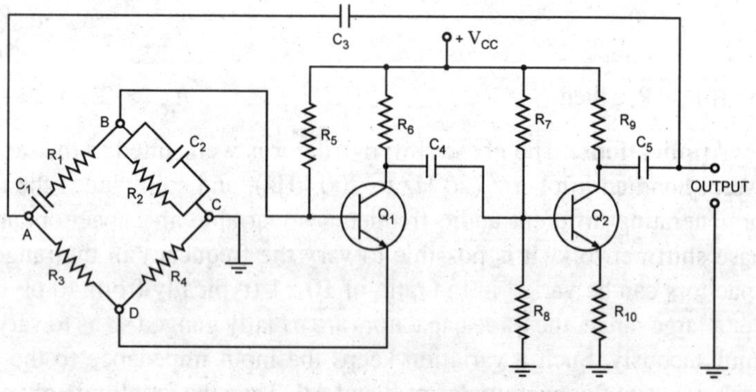

Fig. 8.20. ***Wien Bridge Oscillator Circuit***

In the bridge circuit R_1 in series with C_1, R_3, R_4 and R_2 in parallel with C_2 form the four arms.

From the analysis of the bridge circuit it is obvious that the bridge will be balanced only when

$$R_3\left[\frac{R_2}{1+j\omega C_2R_2}\right] = R_4\left(R_1 - \frac{j}{\omega C_1}\right)$$

$$\text{or } R_2R_3 = R_4(1 + j\omega C_2R_2)(R_1 - j/\omega C_1)$$

$$\text{or } R_2R_3 - R_4R_1 - \frac{C_2}{C_1}R_2R_4 + \frac{jR_4}{\omega C_1} - j\omega C_2R_2R_1R_4 = 0$$

Separating real and imaginary terms we have

$$R_2R_3 - R_4R_1 - \frac{C_2}{C_1}R_2R_4 = 0$$

$$\text{or } \frac{C_2}{C_1} = \frac{R_3}{R_4} - \frac{R_1}{R_2} \qquad ...(8.33)$$

$$\text{and } \frac{R_4}{\omega C_1} - \omega C_2R_2R_1R_4 = 0$$

$$\text{or } \omega^2 = \frac{1}{C_1C_2R_1R_2}$$

$$\text{or } \omega = \frac{1}{\sqrt{C_1 C_2 R_1 R_2}}$$

$$\text{and } f = \frac{1}{2\pi\sqrt{R_1 C_1 R_2 C_2}} \quad ...(8.34)$$

$$\text{if } C_1 = C_2 = C \text{ and } R_1 = R_2 = R, \text{ then } f = \frac{1}{2\pi CR} \quad ...(8.35)$$

$$\text{and } R_3 = 2R_4 \quad ...(8.36)$$

Thus, we see that in a bridge circuit the output will be in phase with the input only when the bridge is balanced *i.e.,* at resonant frequency given by Eq. (8.35). At all other frequencies the bridge is off-balance *i.e.,* the voltage fed back and output voltage do not have the correct phase relationship for sustained oscillations.

So this bridge circuit can be used as feedback network for an oscillator, provided that the phase shift through the amplifier is zero. This requisite condition is achieved by using a two-stage amplifier, as illustrated in Fig. 8.20. In this arrangement the output of the second stage is supplied back to the feedback network and the voltage across the parallel combination C_2 R_2 is fed to the input of the first stage. Transistor Q_1 serves as an oscillator and amplifier whereas the transistor Q_2 as an invertor to cause a phase shift of 180º. The circuit uses positive and negative feedbacks. The positive feedback is through R_1, C_1, R_2, C_2 to transistor Q_1 and negative feedback is through the voltage divider $R_3 - R_4$ to the emitter of transistor Q_1. Resistors R_3 and R_4 are used to stabilize the amplitude of the output.

The two transistors Q_1 and Q_2 thus cause a total phase shift of 360º and ensure proper positive feedback. The negative feedback is provided in the circuit to ensure constant output over a range of frequencies. This is achieved by taking resistor R_4 in the form of a temperature sensitive lamp, whose resistance increases with the increase in current. In case the amplitude of the output tends to increase, more current would provide more negative feedback. Thus the output would regain its original value. A reverse action would take place in case the output tends to fall.

$$\text{The amplifier voltage gain, } A = \frac{R_3 + R_4}{R_4} = \frac{R_3}{R_4} + 1 = 3 \quad ...(8.37) \quad \because \text{ from Eq. (8.36) } R_3 = 2R_4$$

The above corresponds with the feedback network attentuation of 1/3. Thus, in this case, voltage gain A must be equal to or greater than 3, to sustain oscillations.

To have a voltage gain of 3 is not difficult. On the other hand, to have a gain as low as 3 may be difficult. For this reason also negative feedback is essential.

Operation. The circuit is set in oscillation by any random change in base current of transistor Q_1, that may be due to noise inherent in the transistor or variation in voltage of dc supply. This variation in base current is amplified in collector circuit of transistor Q_1 but with a phase shift of 180º. The output of transistor Q_1 is fed to the base of second transistor Q_2 through capacitor C_4. Now a still further amplified and twice phase-reversed signal appears at the collector of the transistor Q_2. Having been inverted twice, the output signal will be in phase with the signal input to the base of transistor Q_1. A part of the output signal at transistor Q_2 is fed back to the input points of the bridge circuit (points A-C). A part of this feedback signal is applied to emitter resistor R_4 where it produces degenerative effect (or negative feedback). Similarly, a part of the feedback signal is applied across the base-bias resistor R_2 where it produces regenerative effect (or positive feedback). At the rated frequency, effect of regeneration is made slightly more than that of degeneration so as to obtain sustained oscillations.

The continuous frequency variation in this oscillator can be had by varying the two capacitors C_1 and C_2 simultaneously. These capacitors are variable air-gang capacitors. We can change the frequency range of the oscillator by switching into the circuit different values of resistors R_1 and R_2.

FET Wein bridge oscillator is shown in Fig. 8.21.

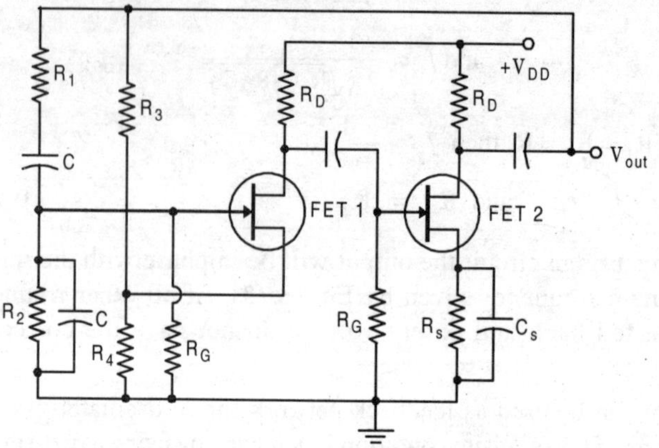

Fig. 8.21. ***FET Wein Bridge Oscillator***

The advantages and disadvantages of Wien bridge oscillators are given below :

Advantages. 1. It provides a stable low distortion sinusoidal output over a wide range of frequency.

2. The frequency range can be selected simply by using decade resistance boxes.
3. The frequency of oscillation can be easily varied by varying capacitances C_1 and C_2 simultaneously.
4. The overall gain is high because of two transistors.

Disadvantages. 1. The circuit needs two transistors and a large number of other components.

2. The maximum frequency output is limited because of amplitude and the phase shift characteristics of the amplifier.

Example 8.10. Design the R-C elements of a Wien bridge oscillator for operation at f_0 = 10 kHz.

[*M.D. Univ. Analog Electronic Circuits December 2006*]

Solution : Using values of R_1 equal to R_2 and C_1 equal to C_2 and taking each of R_1 and R_2 of 200 kΩ,

Required value of $C = \dfrac{1}{2\pi f R} = \dfrac{1}{2\pi \times 10 \times 10^3 \times 200 \times 10^3} = 79.6$ pF **Ans.** Refer to Eq. (8.35)

We can use R_3 = 400 kΩ and R_4 = 200 kΩ **Ans.**

Example 8.11. Design the R-C elements of a Wien bridge oscillator shown in Fig. 8.20 for operation at f = 15 kHz.

Solution : Using values of R_1 equal to R_2 and C_1 equal to C_2 and taking each of R_1 and R_2 of 200 kΩ,

Required value of $C = \dfrac{1}{2\pi f R} = \dfrac{1}{2\pi \times 15 \times 10^3 \times 200 \times 10^3} = 53$ pF **Ans.** Refer to Eq. (8.35)

We can use R_3 = 400 kΩ and R_4 = 200 kΩ **Ans.**

Example 8.12. Find the frequency of oscillations of a Wien bridge oscillator with R = 20 kΩ and C = 1,000 pF.

Solution : $R = R_1 = R_2 = 20\ k\Omega = 20{,}000\ \Omega$

$C = C_1 = C_2 = 1{,}000\ pF = 1 \times 10^{-9}\ F$

Frequency of oscillation, $f = \dfrac{1}{2\pi RC} = \dfrac{1}{2\pi \times 20{,}000 \times 1 \times 10^{-9}} = 7.96$ kHz **Ans.**

8.14. SELECTION OF OSCILLATOR

While selecting an oscillator for a particular application, the following particulars are to be considered.

1. ***Frequency Range.*** The oscillator selected for a particular application should be capable of supplying an output signal whose upper and lower frequency limits exceed those required by the application.
2. ***Power and/or Voltage Requirements.*** The oscillator selected for a particular application should be capable of generating the pertinent quantity with a magnitude large enough to meet the requirement.
3. ***Accuracy and Dial Resolution.*** The accuracy of an oscillator specifies how closely the output frequency corresponds to the frequency indicated on the dial of the instrument. Dial resolution indicates to what percentage of the output frequency value the dial setting can be read.
4. ***Amplitude and Frequency Stability.*** The amplitude stability is a measure of an oscillator's ability of maintaining a constant voltage amplitude with variations in the output signal frequency.

   Frequency stability determines how closely the oscillator maintains a constant frequency over a given time period. Sometimes the frequency stability is included in the accuracy specifications of the oscillator.
5. ***Waveform Distortion.*** This quantity is a measure of how closely the output waveform of the oscillator resembles a pure sinusoidal signal. Sometimes the oscillator is employed as a source in a test used for measuring the tendency of a circuit to distort a sinusoidal signal. In such tests, the distortion caused by the oscillator should be much less than the anticipated distortion because of the circuit under test.
6. ***Output Impedance.*** The output impedance of an oscillator specifies the impedance value of the load which must be connected to it for maximum power transfer. It is very important that the output impedance of the oscillator be equal to the characteristic impedance of the system to which it is to be connected.

Highlights

1. An oscillator is a device that generates a periodic, ac output signal without any form of input signal required. Since sinusoidal waveforms are encountered so frequently in electronic measurement work, sinusoidal oscillators represent the largest single category of signal generators. These devices cover the frequency range from a few Hz to many GHz.
2. The active devices used in oscillators may be BJTs, FETs or op-amps, and passive components used may be resistors, inductors and capacitors.
3. The introduction of positive feedback in an amplifier causes an increase in voltage gain. If the loop gain $A\beta \geq 1$, the Barkhausen criterion for oscillation is satisfied, and an output remains even without the input.

   For linear oscillation, in which the output is sinusoidal with little distortion, the loop gain, $A\beta$, should be only slightly greater than unity.

   The other condition that is required to be fulfilled for sustained oscillations is that the phase shift around the circuit must be zero.
4. The three elements that an oscillator must have are: 1. Oscillatory circuit or frequency determining element, 2. Amplifier, 3. Positive feedback network.
5. The oscillators may be categorised broadly into two classes viz. audio frequency oscillators and radio frequency oscillators. Phase shift and Wien bridge oscillators fall under the category of audio oscillators and are employed for generating low frequency sinusoidal signals. Colpitt's and Hartley oscillators fall under the category of radio frequency oscillators and are very popular for generating high frequency (*e.g.*, 10 kHz to 100 MHz) outputs.

   RF sinusoidal oscillators, also called the L-C oscillators, make use of L-C circuits as their tank or oscillatory circuits. L-C oscillators cannot be used for generation of AF signals because the components required in construction of low frequency L-C resonant circuits are too bulky and heavy.

6. Oscillators, which use inductance-capacitance (L-C) circuits as their tank or oscillatory circuits are called L-C *oscillators*. L-C oscillators are very popular for generating high frequency outputs (*e.g.*, 10 kHz to 100 MHz). There is a large variety of L-C oscillators such as tuned collector oscillators, tuned base oscillators, Colpitt's oscillators, Hartley oscillators, Clapp oscillators, crystal oscillators etc.
7. The Colpitt's oscillator is an example of a resonant circuit oscillator and is characterised by a split capacitor in its tank circuit: The frequency of oscillation is given by

$$f = \frac{1}{2\pi}\sqrt{\frac{1}{LC_1} + \frac{1}{LC_2}}$$

8. The Hartley oscillator has a tapped coil in the tank circuit. Like the Colpitt's, the Hartley oscillator is used for generation of radio frequency (RF) sinusoidal signals. The frequency of oscillation is given by

$$f = \frac{1}{2\pi\sqrt{C(L_1 + L_2 + 2M)}}$$

9. In crystal oscillators, the usual electrical resonant circuit is replaced by a mechanically vibrating crystal. The crystal (usually quartz) has a high degree of stability in holding constant at whatever frequency the crystal is originally cut to operate. The crystal oscillators are, therefore, used whenever great stability is needed, for example, in communication transmitters, and receivers, digital clocks etc.
10. A phase shift oscillator relies upon an R-C phase shift network to provide the necessary phase relationship between output and input to an amplifier. This network attenuates the output voltage by a factor $\beta = \frac{1}{29}$. So the amplifier must have a voltage gain of 29 or more. The frequency of oscillation is given by $f = \frac{1}{2\pi RC\sqrt{6}}$ where R is in ohms and C in farads.
11. A Wien bridge oscillator makes use of a frequency-selective bridge circuit to provide low distortion sinusoidal oscillations at a frequency given by

$$f = \frac{1}{2\pi\sqrt{R_1R_2C_1C_2}} = \frac{1}{2\pi RC}$$

if $R_1 = R_2 = R$ and $C_1 = C_2 = C$.
12. The (*i*) frequency range, (*ii*) power and/or voltage requirements, (*iii*) accuracy and dial resolution, (*iv*) amplitude and frequency stability, (*v*) waveform distortion and (*vi*) output impedance of an oscillator determines its suitability for a particular application.

Exercises

1. What is an oscillator ? How does it differ from an amplifier ?
[Pb. Technical Univ. Analog Electronics December 2007]
2. What are the essential parts of an oscillator circuit ?
What are the Barkhausen conditions of oscillations in electronic systems ? What are their significance ? What are the factors which affect the frequency stability of an oscillator ?
3. How an amplifier can be converted into an oscillator ? Write the oscillation criteria.
4. Explain the working of a Colpitt's oscillator with a neat diagram.
[Pb. Technical Univ. Analog Electronics May 2005]
5. Sketch the topology for a generalised resonant circuit oscillator, using impedances Z_1, Z_2, Z_3. At what frequency will the circuit oscillate ? Under what conditions does the configuration reduces to Hartley oscillator ? **[Pb. Technical Univ. Electronic Devices and Circuits 2005-06]**
6. Draw the circuit diagrams of Hartley and Colpitts oscillators and obtain the expression for the frequency of oscillation. **[U.P. Technical Univ. Solid State Devices and Circuits 2005-06]**
7. Draw the circuit diagram of Hartley oscillator and briefly explain, how the oscillations are maintained in this oscillator. **[Pb. Technical Univ. Electronic Devices and Circuits 2003-04]**

8. In detail discuss Hartley oscillator. **[Pb. Technical Univ. Analog Electronics May 2006]**

9. Explain the circuit of Hartley oscillator in detail. **[Pb. Technical Univ. Analog Electronics December 2005]**

10. Discuss the working of crystal oscillators. **[Pb. Technical Univ. Analog Electronics December 2004]**

11. What are the advantages of a crystal oscillator ? Draw the equivalent circuit of a piezoelectric crystal and show how its impedance varies with frequency.
[U.P. Technical Univ. Solid State Devices and Circuits 2006-07]

12. Why do we use a crystal in an oscillator circuit ? Obtain the expression of the impedance and show its plot w.r.t. the frequency. **[U.P. Technical Univ. Solid State Devices and Circuits 2005-06]**

13. Draw and explain the working of R-C phase shift oscillator and also derive the expression for its frequency of oscillations. **[Pb. Technical Univ. Analog Electronics December 2007]**

14. Explain with the aid of the circuit diagram the working of a transistor RC phase shift oscillator and derive the condition for sustained oscillations.
[U.P. Technical Univ. Solid State Devices and Circuits 2004-05]

15. Explain, with neat circuit diagram, the working of a phase shift oscillator using three sections of RC network. State the expression for the frequency of oscillations and minimum gain of the amplifier for sustained oscillations.

16. Point out the names of various sinusoidal oscillators in actual use. Draw the circuit diagram for a RC phase shift oscillator using JFET as the active device. Deduce that the frequency of oscillation (*f*) is equal to $\frac{1}{2\pi RC\sqrt{6}}$ in the tank circuit following the Barkhausen criterion.

17. Describe the construction, working and applications of Wien-bridge oscillator.
[Pb. Technical Univ. Analog Electronics December 2004]

18. What is Barkhausen criterion ? What do you mean by frequency stability and stability criterion in an oscillator ? Explain the Wien-bridge oscillator in detail. **[Pb. Technical Univ. Analog Electronics May 2006]**

19. Draw and explain the circuit of Wien-bridge oscillator. Derive the expression for frequency of oscillation. Will oscillation take place if the bridge is balanced ? Explain.
[Pb. Technical Univ. Analog Electronics May 2007]

20. Draw the circuit for Wien-bridge oscillator and hence obtain its condition of sustained oscillations.
[Pb. Technical Univ. Electronic Devices and Circuits 2005-06]

21. Draw a Wien-bridge oscillator circuit. Why is it mandatory that the minimum gain of the amplifier be equal to 3 ?

22. Compare R-C phase shift and crystal oscillators. **[Pb. Technical Univ. Analog Electronics May 2006]**

23. What are the main considerations which are to be kept in view while selecting an oscillator for a particular application ?

24. Write short notes on the following :
(*i*) R-C phase shift oscillator.
(*ii*) Wien-bridge oscillator.
(*iii*) Colpitt's oscillator.
(*iv*) Crystal oscillator.
(*v*) Signal generators.

Short Answer Type Questions With Answers

Q. 1. What is an oscillator ? **[P.T.U. Analog Electronics December 2006]**

Ans. Oscillator may be defined as a circuit which generates an ac output signal of very high frequency without requiring any externally applied input signal or it may be defined as an electronic source of alternating current (or voltage) having sinusoidal or non-sinusoidal (square, sawtooth or pulse) waveshape. It can also be defined as a circuit that converts dc energy into very high frequency ac energy.

Q. 2. What is an oscillator ? How does it differ from an amplifier ? **[P.T.U. Analog Electronics December 2007]**

Ans. Refer to Answer of Q. 1.

The difference between an amplifier and an oscillator can be explained with the help of Fig. 1.

As we know that, an amplifier strengthens the input signal without any change in its waveform and frequency. The additional power required comes from the external dc source. Thus an amplifier is essentially an *energy converter* that draws energy from a dc supply and converts it into ac energy at signal frequency, the energy conversion process being controlled by the input signal. On the other hand an oscillator does not require any external signal either to start or maintain the process of energy conversion and the energy conversion process is controlled by the oscillator circuit itself, as illustrated in Fig. 1. In an oscillator, the output signal frequency depends on the passive components employed in the circuit and can be varied as per needs. Oscillator may provide fixed or variable frequency.

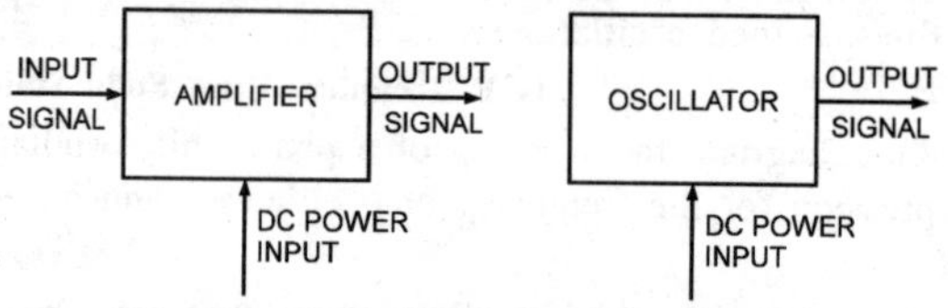

Fig. 1

Q. 3. What is Barkhausen criterion for the feedback oscillator. **[U.P. Technical Univ. 2009-10]**

Ans. Barkhausen conditions for sustained oscillations are : (*i*) the loop gain of the circuit must be ≥ 1 and (*ii*) the phase shift around the circuit must be zero.

Q. 4. Explain why positive feedback and not negative feedback is necessary to produce oscillations. **[U.P. Technical Univ. 2005-06]**

Ans. With negative feedback, feedback factor Aβ becomes less than unity, $A\beta V_{in}$ less than V_{in} and therefore output signal will die out while with positive feedback Aβ exceeds unity, $A\beta V_{in}$ exceeds V_{in} and oscillations are produced. That is why positive feedback is necessary to produce oscillations.

Q. 5. Every electronic oscillator can be considered to be an amplifier with infinite gain. Explain where you get the starting signal voltage from. **[U.P. Technical Univ. 2005-06]**

Ans. Due to the thermal energy, electrons are at random vibrations. These vibrations give rise to small potentials. All oscillators use positive feedback. The small voltage due to movement of electrons due to thermal energy is repeatedly amplified in the oscillator circuit and potential buildup occurs. This results in oscillations. To make these oscillations, dc biasing voltage is essential.

Q. 6. Why are L-C resonant circuits impractical at audio frequencies ?

Ans. L-C resonant circuits are impractical at audio frequencies because components required in their construction for low frequency operation are too bulky, heavy and expensive.

Q. 7. How does Hartley oscillator differ from Colpitt's oscillator in construction ?

Ans. Hartley oscillator circuit is similar to Colpitt's oscillator except that phase shift network consists of two inductors and one capacitor instead of two capacitors and one inductor.

Q. 8. Why is crystal oscillator used in radio transmitter ? **[P.T.U. Analog Electronics May 2005, Dec. 2005]**

Ans. In crystal oscillators, the usual electrical resonant circuit is replaced by a mechanically vibrating crystal. The crystal (usually quartz) has a high degree of stability in holding constant at whatever frequency the crystal is originally cut to operate. The crystal oscillators are, therefore, used whenever great stability is needed, for example, in communication transmitters, and receivers, digital clocks etc.

Q. 9. Why do you use three R-C sections in R-C phase shift oscillator ? **[P.T.U. Analog Electronics December 2005]**

Ans. In a phase shift oscillator, if the loading of the phase-shift network on the amplifier can be assumed to be negligible, a phase shift of 180° between the amplified output voltage v_{out} and the input voltage v_{in} is produced by the amplifier. The three section R-C phase-shift network is provided to produce an additional phase shift, which is a function of frequency and equals 180° at some frequency of operation. Thus, at the frequency of operation, the total shift from the gate around the circuit and back to the gate will be exactly zero.

Q. 10. Why we use two transistors in a Wien bridge oscillator ?

Ans. Wien bridge oscillator employs two transistors so as to obtain 360° or 0° phase shift between output and input voltages.

Q. 11. Sometimes a lamp is used in one of the resistance arms of a Wien bridge oscillator. Why ?

Ans. A lamp is used in one of the resistance arms of a Wien bridge oscillator so as to provide a negative feedback, in proportion to the output current, in the circuit and thus ensure constant output over a range of frequencies.

Q. 12. Why is –ve feedback provided in Wien-bridge oscillator ? **[P.T.U. Analog Electronics May 2005]**

Ans. Negative feedback is provided in Wien-bridge oscillator to stabilize the amplitude of the output.

Problems

1. Determine the operating frequency of a Colpitt's oscillator shown in Fig. 8.7 if L = 50 μH, L_{RFC} = 0.8 mH, C_1 = 0.01 μF, C_2 = 0.02 μF, C_C = 20 μF. **[Ans.** 276 kHz]
2. Determine the oscillation frequency of a transistor Hartley oscillator shown in Fig. 8.9 with circuit values L_1 = 150 μH, L_2 = 1.5 mH, M = 75 μH and C = 150 pF. **[Ans.** 306 kHz]
3. In a phase shift oscillator shown in Fig. 8.17 if $R_1 = R_2 = R_3$ = 400 kΩ and $C_1 = C_2 = C_3$ = 50 pF. Determine the frequency of oscillation. **[Ans.** 3.25 kHz]
4. Design the phase shifting network for oscillating frequency of 1 kHz.
   [Ans. $R_1 = R_2 = R_4$ = 200 kΩ, R_3 = 400 kΩ, C = 796 pF]
5. In a Wien bridge oscillator shown in Fig. 8.10 if $R_1 = R_2$ = 110 kΩ and $C_1 = C_2$ = 1,600 pF. Determine the frequency of oscillation. **[Ans.** 904 Hz]

❑❑❑

Unit-V

Low and High Frequency Transistor Model

CONTENTS OF THIS UNIT

CHAPTER

9

LOW AND HIGH FREQUENCY TRANSISTOR MODEL

9.1. INTRODUCTION

One of the most important characteristics of a signal is its amplitude. Sometimes, a signal may have the desired waveform and frequency, and yet cannot be used because of its small amplitude. This is because the circuits or devices, that are to be operated with such signals as their input, need definite levels of amplitude of signal for proper operation. Thus the amplification or strengthening of the input signal becomes imperative. The device that amplifies the input signal is called the *amplifier*.

Almost no electronic system can work without an amplifier. For instance radio receivers amplify very weak signals until they are strong enough to make the sound audible. The transducers employed in medical and scientific investigations generate signals in microvolt or millivolt range which are incapable to operate indicating instruments. So, these signals need amplification thousands and million times so as to be strong enough to operate indicating instruments.

We have studied that a properly biased transistor increases the amplitude of a weak signal and so acts as an amplifier. So, an amplifier may be defined as a device that increases the current, voltage or power of an input signal with the help of a transistor by furnishing the additional power from a separate source of supply.

INSIDE THIS CHAPTER

When only one transistor with associated circuit is used for increasing the strength of a weak signal, the circuit is known as *single stage transistor amplifier*. It may be emphasised here that a practical amplifier is always a multistage amplifier *i.e.,* an amplifier using a number of stages of amplification.

We know that a transistor alone cannot perform the function of amplification and some passive components, such as resistors and capacitors, and a biasing supply is to be connected. When an ac signal is applied between the base and emitter terminals of a properly biased transistor, a small base current starts flowing. Because of transistor action, a much larger ac current (β times the base current) flows through the collector resistor R_C. Since the value of collector resistance is quite high (usually 5–10 kΩ), a large voltage appears across R_C. Thus, a weak signal applied between base and emitter circuit appears in the amplified form between collector and emitter. This explains the *amplifying action of a transistor*. When the input signal is quite weak and produces *small fluctuations* in the collector current in comparison to its quiescent value, the amplifier is called the *small signal amplifier* (or *voltage amplifier*). On the other hand, when the fluctuations produced in collector current are quite large (beyond the linear portion of the output characteristics), the amplifier is called the *large signal* amplifier (or *power amplifier*). Small signal or voltage amplifier is used as the first stage of the amplifier used in radio and TV receivers, tape recorders, stereos and measuring instruments.

A dc voltage source connected to an amplifier is shown in Fig. 9.1. The amplifier consists of transistors that must be biased in the forward active region so that the transistors can act as amplifying devices. It is desirable that the output signal is linearly proportional to the input signal so that the output of the speakers is an exact (as much as possible) reproduction of the signal generated from the CD player. Thus, it is desired that the amplifier be linear one.

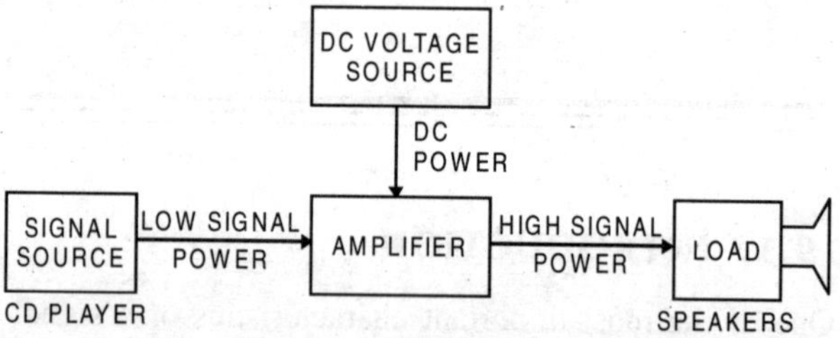

Fig. 9.1. ***Block Diagram of a Compact Disc Player System***

Figure 9.1 suggests that there are two types of analysis of the amplifier that must be considered. The first is a dc analysis because of the applied dc voltage source, and the second is a time-varying or ac analysis because of the time-varying signal source. A linear amplifier means that the superposition principle applies. The principle of superposition states : *The response of a linear circuit excited by multiple input signals is the sum of the responses of the circuit to each of the input signals alone.*

For the linear amplifier, then, the dc analysis can be performed with the ac source set to zero. This analysis, called a *large signal* analysis, establishes the operating or Q-point of the transistors in the amplifier. This analysis and design was the primary objective of the chapter 4. The ac analysis, called a *small-signal analysis,* can be performed with the dc source set to zero. The total response of the amplifier circuit is the sum of the two individual responses.

9.2. TRANSISTOR AMPLIFIER PRACTICAL CIRCUIT

A practical circuit of a single stage transistor amplifier in CE configuration and using self-biasing is shown in Fig. 9.2. The resistors R_1, R_2 and R_E form the *biasing* and stabilization circuit. The biasing circuit must establish a proper operating point otherwise a part of the –ve half cycle of the signal may be cut off in the output. The resistor R_L connected across the output terminals is called the load. When a number of stages are employed then R_L represents the input resistance for the next stage.

An electrolytic capacitor, called the *input capacitor*, C_{in} of capacity of about 10 μF is used to couple the signal to the transistor base. In the absence of this capacitor, the signal source resistance will come across R_2 and thus change the bias. This capacitor allows only ac signal to flow but isolates the signal source from R_2. Another capacitor, called the emitter bypass capacitor C_E, of capacity of about 100 μF is used in parallel with emitter resistance R_E in order to provide a low reactance path to the amplified ac signal. In the absence of this capacitor, amplified ac signal flowing through R_E will cause a voltage drop across it which in turn will feedback the input side and reduce the output voltage.

For coupling of one stage of amplifier to the next stage another capacitor C_C, called the coupling or blocking capacitor, of capacity of about 10 μF is used. Because of its presence, the output across the load resistance R_L is free from the collector dc voltage. In its absence R_C will come in parallel with the resistor R_1 of the biasing network of next stage and thereby change the biasing conditions of the next stage.

Various currents flowing in different branches of the amplifier circuit are indicated in Fig. 9.2.

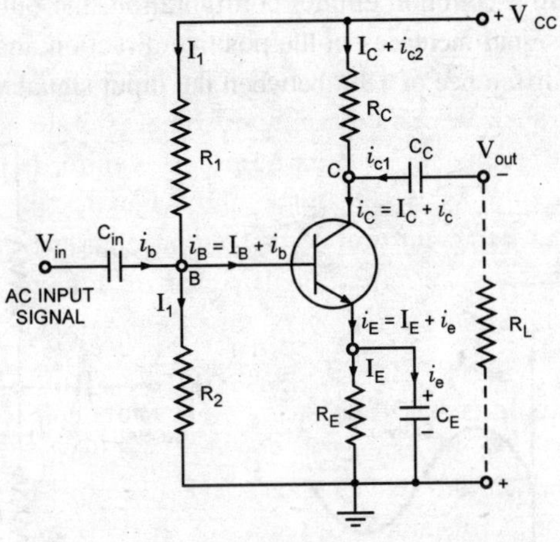

Fig. 9.2. ***Practical Circuit of a Transistor Amplifier in CE Configuration With Self-Biasing***

With the application of input signal

Base current, i_B = DC base current due to biasing circuit + ac base current due to ac input signal

$$= I_B + i_b \quad ...(9.1)$$

Collector current, i_C = DC collector current due to biasing circuit + ac collector current due to ac input signal

$$= I_C + i_c = \beta (I_B + i_b) \quad ...(9.2)$$

Emitter current i_E = DC emitter current + ac emitter current due to ac input signal

$$= I_E + i_e \quad ...(9.3)$$

9.2.1. Emitter Bypass Capacitor

As mentioned above, it is included in the circuit to provide a low reactance path to the amplified ac signal. As shown in Fig. 9.2 it is not connected in series like a coupling capacitor, instead, it is connected across the emitter resistor R_E. The reason for doing so is to bypass or shunt ac current away from the emitter resistor R_E. When the frequency is high enough it looks like a short. Thus emitter bypass capacitor C_E is used to prevent *negative feedback* in the emitter circuit.

Very simply, negative feedback involves feeding back a portion of the amplifier *output* signal to its *input* circuit. Further, the signal is fed back in such a manner that it *opposes* the input signal. Also, the **removal of bypass capacitor** results in an increase in the amplifier's *input resistance,* a reduction in its *voltage gain,* and an *increase* in the *output resistance* "looking into the collector". Obviously, an increase in input resistance is a step in the right direction for voltage amplifier. However, a reduction in the voltage gain and increase in the output resistance are generally undesirable effects. Yet there are some advantages offered by these sacrifices. First, even though the voltage gain is reduced, but *stability* is increased. Specifically, it tends to be less dependent on the transistor's parameters (*e.g.*, its g_m). Second, the increase in the collector-to-ground output resistance makes our approximation of R_{out} equal to R_C even *more realistic.*

9.2.2. AC Ground

You may never have heard of an *ac ground* before, but it really does exist as a separate kind of ground that is different from a *mechanical ground.* A mechanical ground is achieved by connecting a wire between the point to be grounded and ground.

An ac ground is different because it is produced by connecting a *bypass capacitor* between the point to be shorted to ground, as far as ac signals are concerned, and the ground, this kind of ground (*i.e.,* ac ground) exists only at high frequencies. In Fig. 9.2 emitter terminal is *ac grounded.* A designer uses an ac ground when any point needs to be frequency sensitive. Such a point remains normal at low frequencies but grounded at high frequencies.

9.3. PHASE REVERSAL

In a common emitter configuration, the output voltage increases in the negative direction when the input signal increases in the positive direction and vice versa. This is called phase reversal* and causes a phase difference of 180° between the input signal voltage V_{in} and output voltage V_{out}.

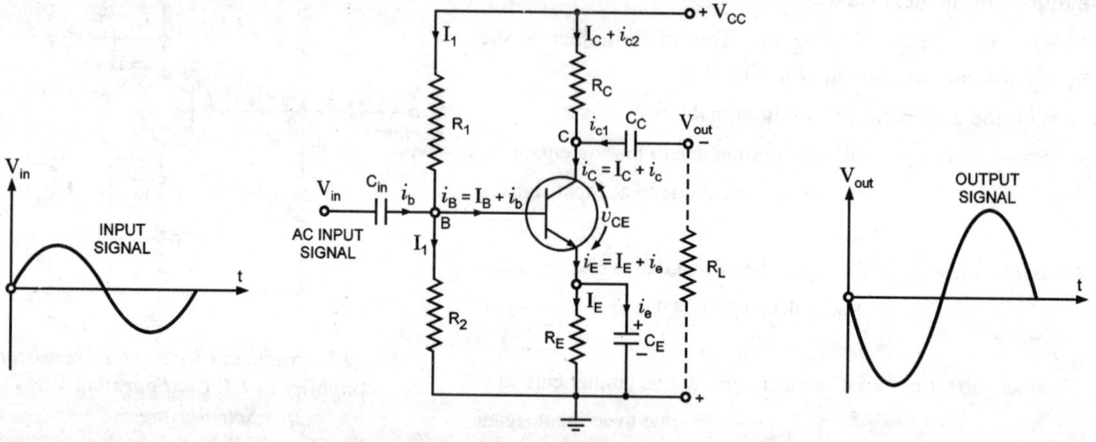

Fig. 9.3. ***Practical Circuit of a Transistor Amplifier in CE Configuration With Self-Biasing***

Consider a CE amplifier circuit shown in Fig. 9.3, the signal being fed at the input terminals (*i.e.,* between base and emitter) and the output is taken from the collector and emitter ends.

The instantaneous output voltage v_{CE} is given as

$$v_{CE} = V_{CC} - i_C R_C \qquad ...(9.4)$$

Reactance of coupling capacitor C_C is negligible at ordinary signal frequencies and so it may be considered a short for the output signal.

With the increase in the signal voltage in the positive half cycle, the base current increases causing increase in collector current and so the voltage drop $i_C R_C$. As V_{CC} is constant, therefore, output voltage v_{CE} decreases. Thus we can say that as the signal voltage increases in positive direction, the output voltage increases in the negative direction *i.e.,* there is a phase difference of 180° between output and input. Conversely on negative half cycle of the input signal voltage, lesser current flows through collector resistance R_C and so voltage drop across it decreases. For this reason output voltage v_{CE} increases.

The above phenomenon can be proved mathematically as below :

Differentiating Eq. (9.4), we have

$$dv_{CE} = 0 - di_C R_C$$

or $$dv_{CE} = - di_C R_C$$

The negative sign indicates that the output voltage is 180° out of phase with the input signal voltage.

The above fact *i.e.,* phase reversal in CE configuration can be illustrated graphically with the aid of output characteristics and ac load line, as shown in Fig. 9.4. The base current varies between I_{B_1} and I_{B_2} with I_B as the zero signal base current. From figure we see that when the base current is maximum in the positive direction, v_{CE} becomes maximum in the negative direction and vice versa. Thus, the input and output voltages are in phase opposition *i.e.,* the transistor has produced a phase reversal of output voltage w.r.t. the input signal voltage.

* This will be so when output is taken from collector and emitter ends, as is always done. However, if the output is taken across collector resistance R_C, it will be in phase with the input.

The noteworthy points are that common base and common collector amplifiers do not produce reversal, and input and output currents are in phase in all the three configurations.

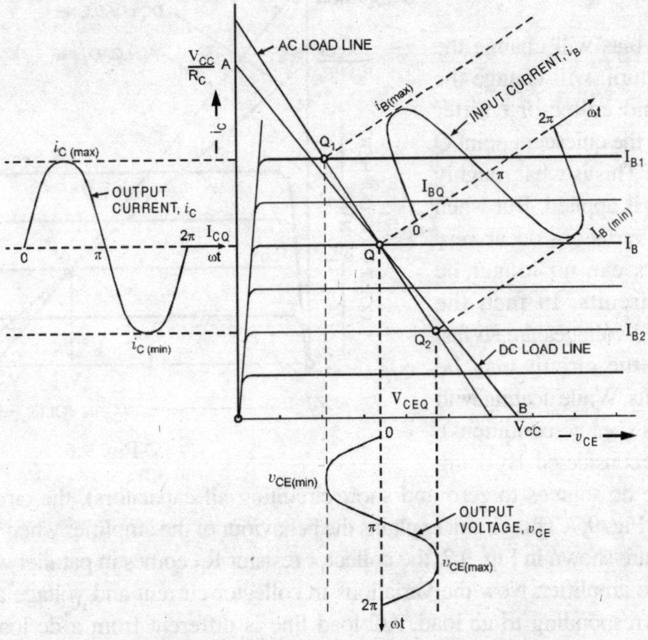

Fig. 9.4. ***Graphical Illustration of Phase Reversal***

9.4. LOAD LINE ANALYSIS

As already discussed, the relationship between collector-emitter output voltage V_{CE} and collector current I_C is linear and, therefore, it can be represented by a straight line, on the output characteristics giving relation between V_{CE} and I_C. This is known as load line. There are two types of load lines, namely dc load line and ac load line. The former determines the values of collector current I_C and collector-emitter voltage V_{CE} corresponding to zero signal conditions whereas the latter gives their (i_C and v_{CE}) values when an ac signal is applied.

Consider the transistor amplifier circuit shown in Fig. 9.2. As regards dc conditions, the amplifier circuit shown in Fig. 9.2 is reduced to that shown in Fig. 9.5. The dc equivalent circuit shown in Fig. 9.5 is so because the capacitors do not allow the flow of the direct current through them and, therefore, act as open circuits. While drawing dc equivalent circuit it is presumed that no signal is applied.

DC equivalent circuit of an amplifier circuit is drawn by reducing all the ac sources to zero and opening all the capacitors.

Applying Kirchhoff 's voltage law to the output section of the basic amplifier circuit shown in Fig. 9.5 we have

$$V_{CC} = I_C R_C + V_{CE} + I_E R_E$$

$$= V_{CE} + I_C (R_C + R_E) \qquad \because I_C = I_E$$

$$\text{or } I_C = \frac{-V_{CE}}{R_C + R_E} + \frac{V_{CC}}{R_C + R_E} \qquad ...(9.5)$$

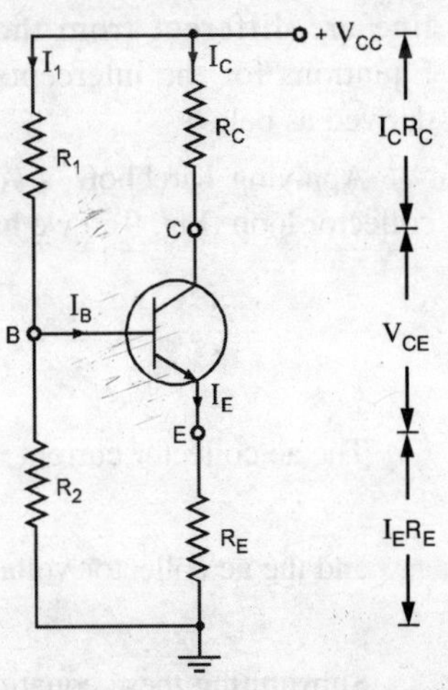

Fig. 9.5. ***DC Equivalent Circuit of Circuit Given in Fig. 9.2***

Since V_{CC} and $(R_C + R_E)$ are constant, therefore, it represents a straight line known as dc load line. By plotting this line on the output characteristics, collector-emitter voltage V_{CE} and collector current I_C can be determined for the given value of base current.

Any change in dc bias will change the base current which, in turn, will change the collector current I_C and collector-emitter voltage V_{CE}. As a result, the quiescent point Q shift on the dc load line. This is what roughly happens when ac signal is applied. But when ac signal is applied the variations occur very fast and the capacitors can no longer be considered as open circuits. In fact, the variations in currents and voltages are so fast that the capacitors in the circuit may be considered as short circuits. While dealing with ac currents and voltages (*i.e.,* ac conditions), dc supplies need not to be considered. By doing so (*i.e.,* reducing all the dc sources to zero and short circuiting all capacitors), the circuit given in Fig. 9.2 is reduced to that shown in Fig. 9.7. This circuit explains the behaviour of the amplifier when viewed in ac conditions. As obvious from the figure shown in Fig. 9.7, the collector resistor R_C comes in parallel with load resistor R_L and forms the ac load for the amplifier. Now the variations in collector current and voltage are determined with the help of ac load line corresponding to ac load. AC load line is different from a dc load line because ac load resistance is different from dc load resistance.

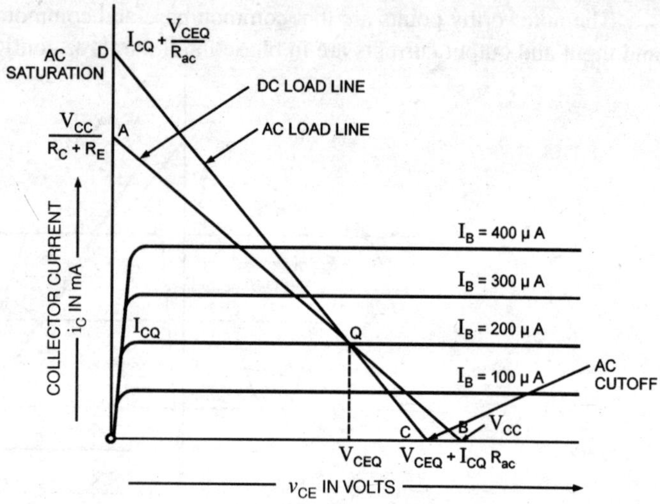

Fig. 9.6

Effective load for ac signal is given as

$$R_{ac} = R_C \parallel R_L = \frac{R_C R_L}{R_C + R_L} \quad ...(9.6)$$

When no signal is applied, the operating point is Q, as shown in Fig. 9.6. When signal is applied Q point swings along ac load line.

The saturation and cutoff points on the ac load line are different from those of dc load line. Equations for the intercepts of ac load line are derived as below.

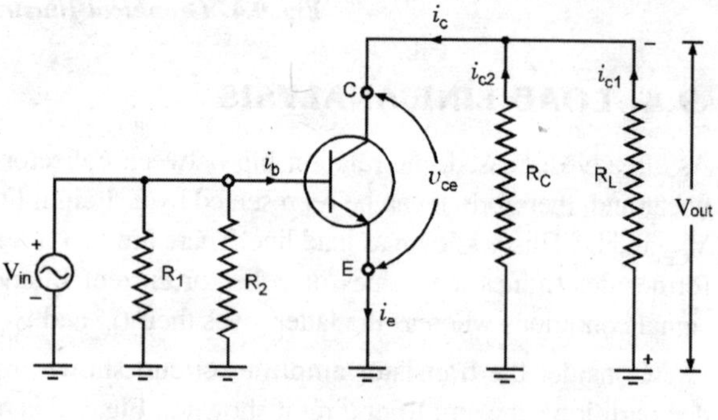

Fig. 9.7. ***AC Equivalent Circuit of Circuit Shown in Fig. 9.2***

Applying Kirchhoff 's voltage law around the collector loop (Fig. 9.7) we have

$$v_{ce} + i_c R_{ac} = 0$$

$$\text{or } i_c = \frac{-v_{ce}}{R_{ac}} \quad ...(9.7)$$

The ac collector current is given by

$$i_c = \Delta I_C = i_C - I_{CQ}$$

and the ac collector voltage is given as

$$v_{ce} = \Delta V_{CE} = v_{CE} - V_{CEQ}$$

Substituting these equations in Eq. (9.7) we have

$$v_{CE} - V_{CEQ} + (i_C - I_{CQ}) R_{ac} = 0$$

$$\text{or } i_C = I_{CQ} + \frac{V_{CEQ}}{R_{ac}} - \frac{v_{CE}}{R_{ac}} \quad ...(9.8)$$

This is the equation of the ac load line. We can find the intercepts in the usual way. When the transistor goes into saturation, v_{CE} is zero and Eq. (9.8) gives

$$i_{C\,(SAT)} = I_{CQ} + \frac{V_{CEQ}}{R_{ac}} \text{ (upper end)} \quad ...(9.9)$$

where $i_{C\,(SAT)}$ is the ac saturation current, I_{CQ} is dc collector current, V_{CEQ} is dc collector-emitter voltage and R_{ac} is the ac resistance seen by the collector.

When the transistor goes into cutoff, i_C equals zero and we get the ac cutoff voltage of

$$v_{CE\,(OFF)} = V_{CEQ} + I_{CQ} R_{ac} \text{ (lower end)} \quad ...(9.10)$$

By joining these two points line CQD can be drawn as shown in Fig. 9.6. This is the ac load line because it represents all possible ac operating points. At any instant during the ac cycle, the operating point of the transistor is somewhere along the ac load line, the exact point is determined by the amount of change that takes place from the Q point.

The slope of the ac load line is given as

$$-\frac{OD}{OC} = -\frac{I_{CQ} + \dfrac{V_{CEQ}}{R_{ac}}}{V_{CEQ} + I_{CQ} R_{ac}} = -\frac{I_{CQ} + (V_{CEQ}/R_{ac})}{R_{ac}\left[I_{CQ} + \dfrac{V_{CEQ}}{R_{ac}}\right]} = -\frac{1}{R_{ac}} \quad ...(9.11)$$

It is seen from Fig. 9.6 that maximum possible positive signal swing is $I_{CQ}R_{ac}$ and maximum possible negative signal swing is V_{CEQ}. It means peak signal handling capacity is limited to $I_{CQ}R_{ac}$ or V_{CEQ} whichever is smaller.

Example 9.1. Draw the dc and ac load lines for the CE circuit shown in Fig. 9.8. What is the maximum peak-to-peak signal that can be obtained ?

Solution : $V_{CE\,(cutoff)} = V_{CC} = 20$ volts representing point B

$$I_C \text{ (saturation)} = \frac{V_{CC}}{R_C + R_E} = \frac{20}{(3+2)\text{ k}\Omega} = 4 \text{ mA representing point A}$$

Hence line AB gives the dc load line for the given circuit.

Approximate bias conditions can be quickly determined by assuming that base current is too small to effect the base bias.

Voltage across emitter resistance R_E,

V_E = Voltage across resistance R_2, if V_{BE} is neglected

$$\text{So } I_E R_E = V_{CC} \times \frac{R_2}{R_1 + R_2} = 20 \times \frac{4}{16+4} = 4 \text{ V}$$

$$\text{or } I_E = \frac{V_E}{R_E} = \frac{4}{2\text{ k}\Omega} = 2 \text{ mA}$$

Collector current $I_C \simeq I_E = 2$ mA

Hence $I_{CQ} = 2$ mA

Collector-emitter voltage,

$$V_{CEQ} = V_{CC} - I_{CQ}(R_C + R_E) = 20 - 2\text{ mA} \times (3+2)\text{ k}\Omega = 10 \text{ V}$$

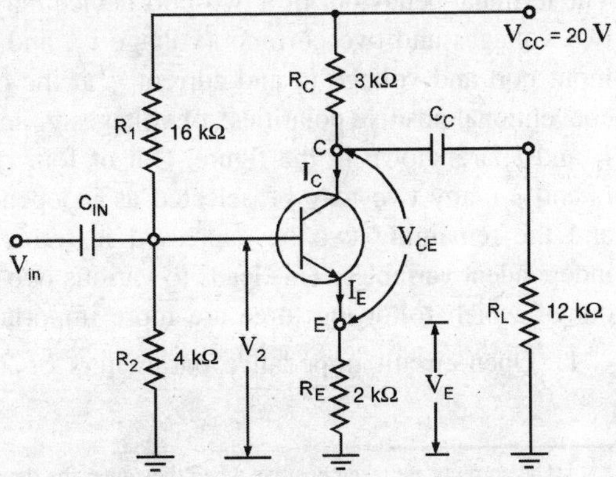

Fig. 9.8

$$R_{ac} = R_C \parallel R_L = \frac{R_C R_L}{R_C + R_L} = \frac{3 \times 12}{3 + 12} = 2.4 \text{ k}\Omega$$

$$i_{C\,(SAT)} = I_{CQ} + \frac{V_{CEQ}}{R_{ac}} = 2 + \frac{10}{2.4} = 6.17 \text{ mA}$$

giving saturation or upper end of ac load line

$$v_{CE\,(OFF)} = V_{CEQ} + I_{CQ} R_{ac} = 10 + 2 \text{ mA} \times 2.4 \text{ k}\Omega = 14.8 \text{ V}$$

giving cutoff point or lower end of ac load line

Hence line joining 14.8 V on V_{CE} – axis and 6.17 mA on I_C – axis gives the ac load line, as illustrated in Fig. 9.9. As expected this line passes through quiescent point Q.

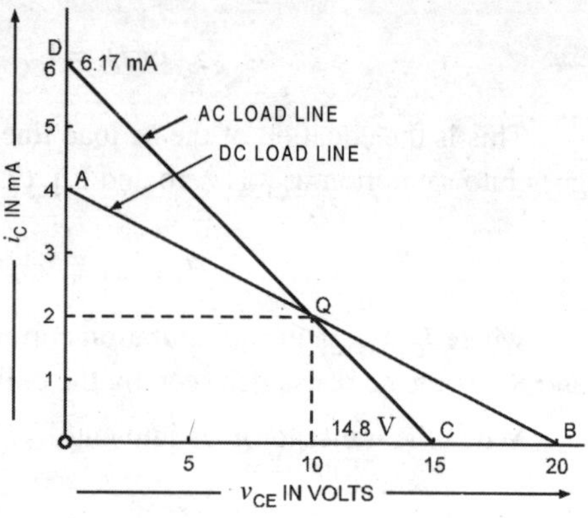

Fig. 9.9

Now, $I_{CQ} R_{ac} = 2 \text{ mA} \times 2.4 \text{ k}\Omega = 4.8 \text{ V}$ and $V_{CEQ} = 10 \text{ V}$

Taking the small quantity,

Maximum peak output signal = 4.8 V

Hence peak-to-peak value of output signal

= 2 × 4.8 = 9.6 V **Ans.**

9.5. HYBRID PARAMETERS

One of our first concerns in the sinusoidal ac analysis of transistor networks is the magnitude of the input signal (an amplifier may use either small-signal operation or large-signal operation). Study of small-signal operation may be done either graphically or by using small-signal equivalent circuit for the BJT operating in the active region, however, the second method is more convenient. The large-signal operation may be best studied graphically because of involvement of certain nonlinear operation in it. For study of small-signal operation, the transistor may be replaced by its equivalent circuit (or model) and then the usual method of network analysis may be used to obtain expressions for its operating characteristics *e.g.*, input impedance, output impedance, voltage gain, current gain etc.

We here take up the study of hybrid (h) parameter equivalent circuit (or model) for the transistor and analysis of low- frequency small-signal common-emitter (CE) amplifier using this model.

9.6. TWO-PORT DEVICES AND THE HYBRID MODEL

A box representing a two-port network is illustrated in Fig. 9.10. The terminal behaviour of a two-port device may be specified by two voltages and two currents (voltage v_1, and current i_1 at the input port and voltage v_2 and current i_2 at the output port). The conventional positive polarities* of voltages v_1 and v_2 and currents i_1 and i_2 are shown in the figure. Out of four quantities (v_1, v_2, i_1 and i_2), any two may be selected as independent variables** and the remaining two be expressed in terms of the selected independent variables. This leads to various two-port parameters, out of which following three are more important.

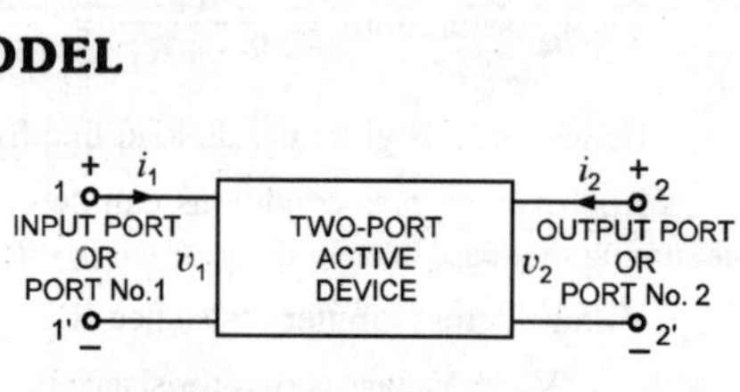

Fig. 9.10. Conventional Positive Polarities of Voltages and Currents in a Two-Port Network

1. Open-circuit impedance parameters or Z-parameters Z_{11}, Z_{12}, Z_{21} and Z_{22}

---

* The currents are taken positive when they enter the device and negative when they leave it.

** In general, we are not free to select the independent variables arbitrarily—for example in case of an ideal transformer, two voltages v_1 and v_2 cannot be selected as the independent variables because their ratio is a constant and is equal to the turn-ratio of the transformer.

2. Short-circuit admittance parameters or Y-parameters Y_{11}, Y_{12}, Y_{21} and Y_{22}
3. Hybrid parameters or the h parameters.

In transistor amplifier analysis, Z- and Y-parameters were used earlier. But now hybrid or the h parameters alone are used in a transistor circuit analysis and, therefore, only the h parameters will be taken here for discussion.

9.6.1. Hybrid Parameters or h Parameters

For the two-port network illustrated in Fig. 9.10, if input current i_1 and the output voltage v_2 are taken as independent variables and the two-port shown in the figure is linear, we may write

$$v_1 = h_{11}i_1 + h_{12}v_2 \quad ...(9.12)$$

$$i_2 = h_{21}i_1 + h_{22}v_2 \quad ...(9.13)$$

In the above equations, the *hs* are fixed for a given circuit and are called the hybrid or h parameters. Because these four parameters have mixed dimensions (h_{11} has dimension of ohm, h_{12} and h_{21} are dimensionless, and h_{22} has dimension of mho or siemen) so they are called hybrid or h parameters.

1. *Meaning of h Parameters.* By assuming that the given two-port network has no reactive element and by applying open circuit ($i_1 = 0$) or short circuit ($v_2 = 0$) conditions to Eqs. (9.12) and (9.13), the h parameters can be defined as below :

If the output terminals are short circuited, (Fig. 9.11) output voltage v_2 becomes zero and Eqs. (9.12) and (9.13) become

$$v_1 = h_{11}i_1 + h_{12} \times 0 = h_{11}i_1$$

and $$i_2 = h_{21}i_1 + h_{22} \times 0 = h_{21}i_1$$

or $$h_{11} = \left.\frac{v_1}{i_1}\right|_{v_2=0}$$

and $$h_{21} = \left.\frac{i_2}{i_1}\right|_{v_2=0}$$

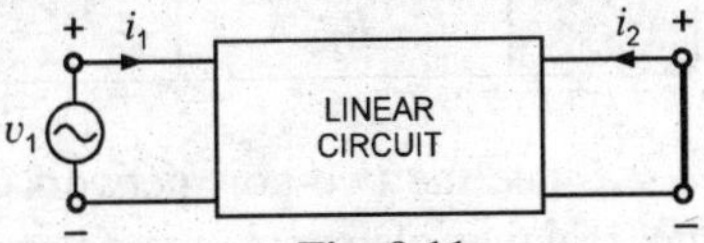

Fig. 9.11

Since h_{11} is the ratio of input voltage and input current with output terminals short circuited, it is called the *input impedance with output short circuited.* The subscript 11 of h_{11} defines the fact that the parameter is determined by the ratio of quantities measured at the input terminals. Its unit is ohm.

Similarly h_{21} is the ratio of output and input currents (*i.e.*, i_2/i_1) with output terminals short circuited, so it is called the *forward transfer current gain with output short circuited.* Obviously it is dimensionless quantity.

If the input terminals are open circuited and we drive the output terminals with voltage v_2, as shown in Fig. 9.12, input current i_1 becomes zero and Eqs. (9.12) and (9.13) become

$$v_1 = h_{11} \times 0 + h_{12}\, v_2 = h_{12}\, v_2$$

and $$i_2 = h_{21} \times 0 + h_{22}\, v_2 = h_{22}\, v_2$$

or $$h_{12} = \left.\frac{v_1}{v_2}\right|_{i_1=0}$$

and $$h_{22} = \left.\frac{i_2}{v_2}\right|_{i_1=0}$$

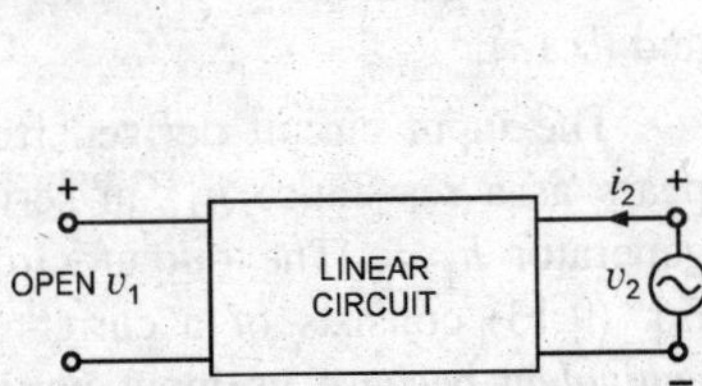

Fig. 9.12

Thus the parameter h_{12} is the ratio of input voltage to the output voltage with zero input current (*i.e.*, $i_1 = 0$). It is dimensionless quantity and is called the open-circuit *reverse transfer voltage ratio*, the subscript 12 of h_{12} reveals that the parameter is a transfer quantity determined by the ratio of input to output measurements.

Similarly h_{22} is the ratio of output current to the output voltage with zero input current (*i.e.*, $i_1 = 0$). It is called the *open-circuit output admittance* and is measured in siemens. The subscript 22 in h_{22} indicates that it is determined by a ratio of output quantities.

2. *Notations.* The convenient alternative subscript notations recommended by the IEEE Standards are given below :

i = 11 = input $\qquad$ o = 22 = output

f = 21 = forward transfer $\qquad$ r = 12 = reverse transfer

In case of transistors, another subscript (*b, e* or *c*) is added to designate the type of transistor configuration. For example $h_{ie} = h_{11e}$ = input resistance in common-emitter configuration.

Notations used in transistor amplifier for the three configurations are tabulated in Table 9.1.

TABLE 9.1

S. No.	*h parameter*	*Common Base Configuration*	*Common Emitter Configuration*	*Common Collector Configuration*
1.	h_{11}	h_{ib}	h_{ie}	h_{ic}
2.	h_{12}	h_{rb}	h_{re}	h_{rc}
3.	h_{21}	h_{fb}	h_{fe}	h_{fc}
4.	h_{22}	h_{ob}	h_{oe}	h_{oc}

Since the two-port network (or the device) described by Eqs. (9.12) and (9.13) is assumed to have no reactive elements, the four parameters h_{11}, h_{12}, h_{21} and h_{22} are real numbers and voltages and currents v_1, v_2, and i_1, i_2 are functions of time. However, if the reactive elements had been included in the device, the excitation would be considered to be sinusoidal, the h parameters would in general be functions of frequency, and the voltages and currents would be represented by phasors V_1, V_2, and I_1, I_2.

9.6.2. Hybrid Model

The hybrid circuit for any two-port network characterized by Eqs. (9.12) and (9.13) is shown in Fig. 9.13. If Kirchhoff's voltage law and Kirchhoff's current law are applied to the input and output ports, Eqs. (9.12) and (9.13) respectively will be obtained. Thus model given in Fig. 9.13 truely satisfies Eqs. (9.12) and (9.13).

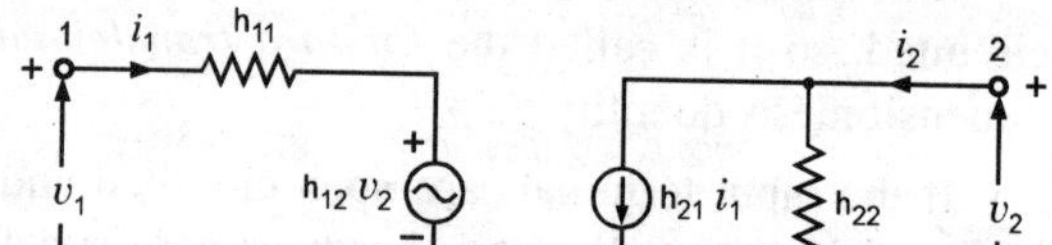

Fig. 9.13. ***Hybrid Model For The Two Port Network Shown in Fig. 9.10***

The input circuit derived from Eq. (9.12) appears as a resistance h_{11} in series with a voltage generator $h_{12}v_2$. The output circuit, derived from Eq. (9.13) consists of a current generator $h_{21}i_1$ and shunt resistance h_{22}. This circuit is called *hybrid equivalent* because its input portion is a Thevenin's equivalent (or a voltage generator in series with a resistance) while the output portion is a Norton's equivalent (or a current generator with shunt resistance). Thus it is a mixture or hybrid. The symbol '*h*' is simply the abbreviation of the word hybrid (hybrid means "mixed").

The hybrid equivalent circuit (or model) given in Fig. 9.13 is an extremely important one in the area of electronics today. It will appear over and over again in the analysis to follow. There are two main reasons of popularity of hybrid model. First, it isolates the input and output circuits, their interaction being accounted for by the two controlled voltage and current sources—the effect of output upon input is represented by the equivalent voltage generator $h_{12}v_2$ and the effect of input upon output is represented by the current generator $h_{21}i_1$. The value of the former depends upon the output voltage v_2 while the value for the latter depends upon the input current i_1. Secondly, the two portions of the circuit are in a form which makes it simple to take into account the source and the load circuits.

9.7. TRANSISTOR HYBRID MODEL

The basic assumption in arriving at a transistor linear model or equivalent circuit is that the variations about the operating or quiescent point are small and, therefore, the transistor parameters can be considered constant over the small range of operation.

Many transistor models have been proposed, each one having its particular merits and demerits. The transistor model presented here, is given in terms of the h parameters, which are real numbers at audio frequencies, are easy to measure, can also be obtained from the static characteristics of a transistor, and are particularly convenient to use in analysis and design of circuit. Furthermore, a set of h parameters is specified for many transistors by the manufacturers.

To derive a hybrid model for a transistor, let us consider the basic CE amplifier circuit given in Fig. 9.14. The variables i_B, i_C, v_B and v_C represent the total instantaneous values of currents and voltages. We may select the input current i_B and output voltage v_C as independent variables. Since input voltage v_B is some function f_1 of i_B and v_C and output current i_C is another function f_2 of i_B and v_C, we may write

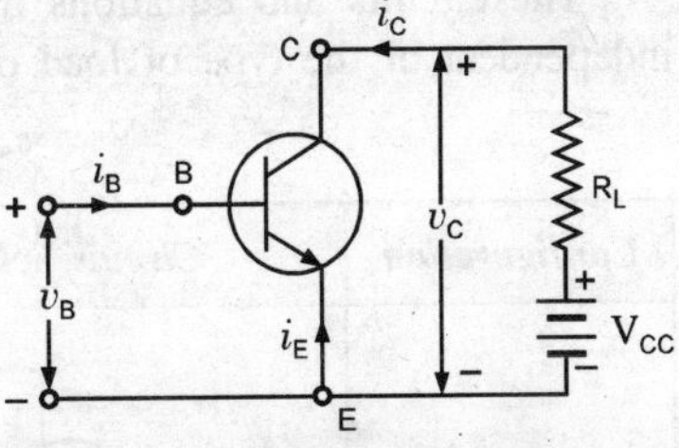

Fig. 9.14. *Basic CE Configuration*

$$v_B = f_1\,(i_B, v_C) \quad \text{...(9.14)}$$

$$i_C = f_2\,(i_B, v_C) \quad \text{...(9.15)}$$

Making a Taylor's series expansion of Eqs. (9.14) and (9.15) about the zero signal operating point (I_B, V_C) and neglecting higher-order terms we have

$$\Delta v_B = \left.\frac{\delta f_1}{\delta i_B}\right|_{V_C} \Delta i_B + \left.\frac{\delta f_1}{\delta v_C}\right|_{I_B} \Delta v_C \quad \text{...(9.16)}$$

$$\Delta i_C = \left.\frac{\delta f_2}{\delta i_B}\right|_{V_C} \Delta i_B + \left.\frac{\delta f_2}{\delta v_C}\right|_{I_B} \Delta v_C \quad \text{...(9.17)}$$

where partial derivatives $\frac{\delta f_1}{\delta i_B}$ and $\frac{\delta f_2}{\delta i_B}$ are taken, keeping collector voltage V_C constant while partial derivates $\frac{\delta f_1}{\delta v_C}$ and $\frac{\delta f_2}{\delta v_C}$ are taken, keeping base current I_B constant.

The quantities Δv_B, Δv_C, Δi_B and Δi_C represent the small-signal (incremental) base and collector voltages and currents and may be represented as v_b, v_c. i_b, and i_c respectively as per standard notations. We may now write Eqs. (9.16) and (9.17) as

$$v_b = h_{ie} i_b + h_{re} v_c \qquad ...(9.18)$$

$$i_c = h_{fe} i_b + h_{oe} v_c \qquad ...(9.19)$$

$$\text{where } h_{ie} \equiv \left.\frac{\delta f_1}{\delta i_B}\right|_{V_C} = \left.\frac{\delta v_B}{\delta i_B}\right|_{V_C} = \left.\frac{v_b}{i_b}\right|_{V_C=0} \qquad ...(9.20a)$$

$$h_{fe} \equiv \left.\frac{\delta f_2}{\delta i_B}\right|_{V_C} = \left.\frac{\delta i_C}{\delta i_B}\right|_{V_C} = \left.\frac{i_c}{i_b}\right|_{V_C=0} \qquad ...(9.20b)$$

$$h_{re} \equiv \left.\frac{\delta f_1}{\delta v_C}\right|_{I_B} = \left.\frac{\delta v_B}{\delta v_C}\right|_{I_B} = \left.\frac{v_b}{v_c}\right|_{I_B=0} \qquad ...(9.20c)$$

$$h_{oe} \equiv \left.\frac{\delta f_2}{\delta v_C}\right|_{I_B} = \left.\frac{\delta i_C}{\delta v_C}\right|_{I_B} = \left.\frac{i_c}{v_c}\right|_{I_B=0} \qquad ...(9.20d)$$

The partial derivatives of Eq. (9.20) define the h parameters for the transistor in common-emitter (CE) configuration.

Equations (9.18) and (9.19) are found to be of exactly the same form as Eqs. (9.12) and (9.13) hence, the model shown in Fig. 9.13 can be used to represent a transistor.

The common-emitter (CE), common-base (CB), and common-collector (CC) configurations, their hybrid models and their terminal volt-ampere equations are summarized in Table 9.2.

The circuits and equations in Table 9.2 are valid for either an N-P-N or P-N-P transistor and are independent of the type of load or biasing method.

TABLE 9.2

Configuration	***Circuit Schematic***	***Hybrid Model***	***v-i Equations***
Common Emitter, CE	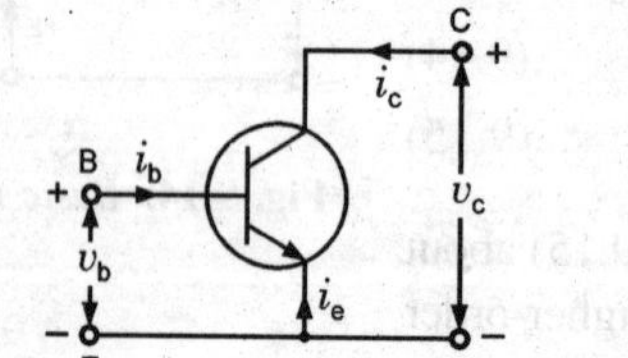	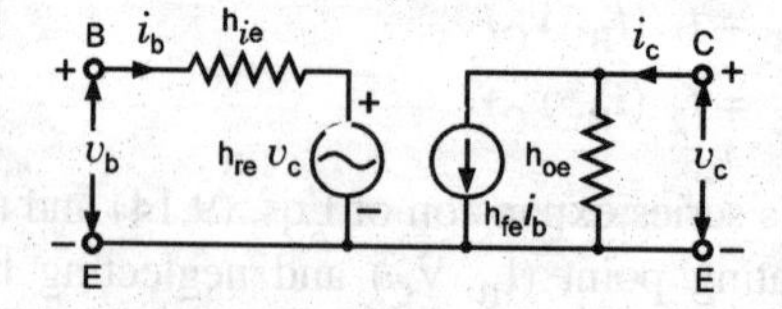	$v_b = h_{ie}\, i_b + h_{re}\, v_c$ $i_c = h_{fe}\, i_b + h_{oe}\, v_c$
Common Base, CB	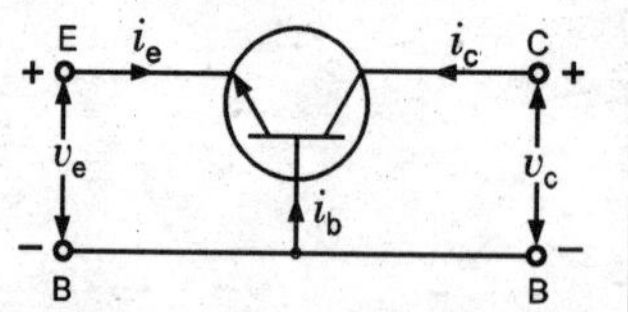	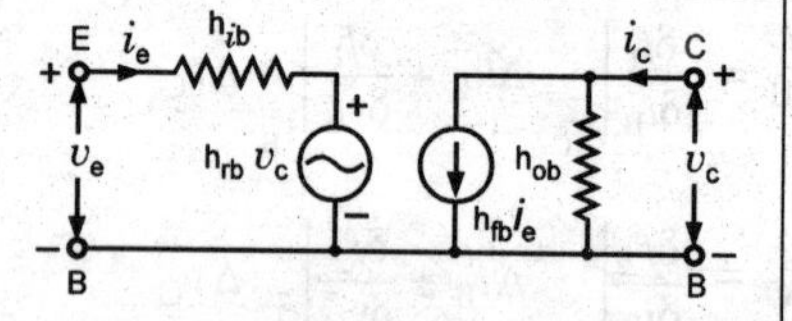	$v_e = h_{ib}\, i_e + h_{rb}\, v_c$ $i_c = h_{fb}\, i_e + h_{ob}\, v_c$
Common Collector CC	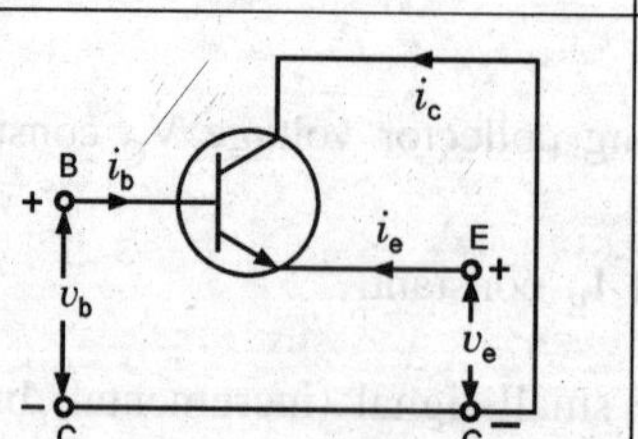	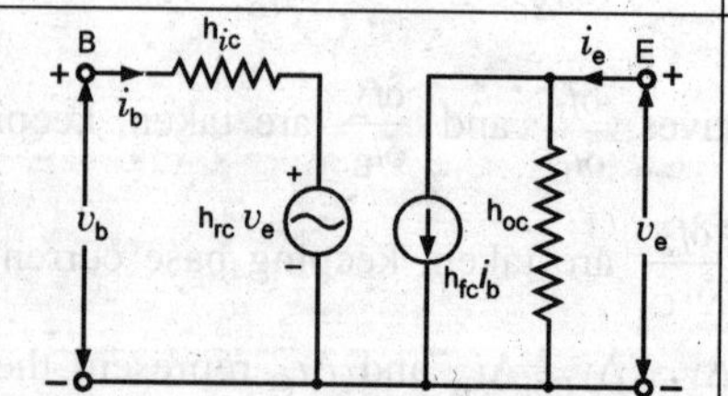	$v_b = h_{ic}\, i_b + h_{rc}\, v_e$ $i_e = h_{fc}\, i_b + h_{oc}\, v_e$

9.8. TYPICAL VALUES OF h PARAMETERS FOR A TRANSISTOR

Typical values of h parameters of a transistor for CE, CB and CC configurations at I_E = 1.3 mA are given in Table 9.3.

TABLE 9.3

Configuration / *Parameter*	*Common Emitter CE*	*Common Base CB*	*Common Collector CC*
h_i	1.1 kΩ	21.6 Ω	1.1 kΩ
h_r	2.5×10^{-4}	2.9×10^{-4}	$\simeq 1$
h_f	50	– 0.98	– 51
h_o	25 µS (µA/V)	0.49 µS (µA/V)	25 µS (µA/V)
$1/h_o$	40 kΩ	2.04 MΩ	40 kΩ

9.9. CONVERSION OF HYBRID PARAMETERS IN TRANSISTOR THREE CONFIGURATIONS

Some transistor manufacturers provide only the four CE hybrid parameters while others provide CB hybrid parameters. Sometimes, for a specific purpose, it becomes necessary to convert one set of h parameters in one configuration to another set in another configuration. The conversion formulas can be obtained using the definitions of the parameters involved and Kirchhoff's laws.

For example, let us convert common-emitter h parameters to the common-base h parameters. For this purpose first the CB hybrid model is drawn as shown in Fig. 9.15 (*a*) and then it is redrawn in CE configuration, as shown in Fig. 9.15 (*b*). The latter corresponds in every detail to the former except that the emitter terminal E is made common to the input and output ports.

By definition
$$h_{re} = \left.\frac{V_{be}}{V_{ce}}\right|_{I_b=0} = \left.\frac{V_{bc}+V_{ce}}{V_{ce}}\right|_{I_b=0} = 1+\left.\frac{V_{bc}}{V_{ce}}\right|_{I_b=0}$$

If $I_b = 0$, then $I_c = -I_e$, and the current I in h_{ob} in Fig. 9.15 (*b*) is $I = (1 + h_{fb})\, I_e = h_{ob} V_{bc}$

Applying Kirchhoff's voltage law to the output mesh of Fig. 9.15 (*b*) we have

$$h_{ib} I_e + h_{rb} V_{cb} + V_{bc} + V_{ce} = 0$$

Combining the above two equations we have

$$\frac{h_{ib}h_{ob}}{1+h_{fb}} V_{bc} - h_{rb} V_{bc} + V_{bc} + V_{ce} = 0$$

$$\text{or} \quad \frac{V_{bc}}{V_{ce}} = \frac{-(1+h_{fb})}{h_{ib}\, h_{ob} + (1-h_{rb})(1+h_{fb})}$$

$$\text{Thus} \quad h_{re} = 1+\frac{V_{bc}}{V_{ce}} = \frac{h_{ib}h_{ob} - (1+h_{fb})h_{rb}}{h_{ib}h_{ob} + (1-h_{rb})(1+h_{fb})} \qquad ...(9.21)$$

This is an exact expression. The simpler approximate formula is obtained by noting that, for typical values given in Table 9.3 $h_{rb} \ll 1$ and $h_{ob}\, h_{ib} \ll (1 + h_{fb})$

$$\text{So } h_{re} \simeq \frac{h_{ib}\, h_{ob}}{1 + h_{fb}} - h_{rb} \quad \text{...(9.22)}$$

$$\text{By definition } h_{ie} = \left.\frac{V_{be}}{I_b}\right|_{V_{ce}=0}$$

If terminals C and E are connected together in Fig. 9.15 (*b*), we have circuit shown in Fig. 9.16. From Fig. 9.16, we see that $V_{cb} = -V_{bc} = -V_{be}$

Applying Kirchhoff's voltage law to the left-hand mesh, we have

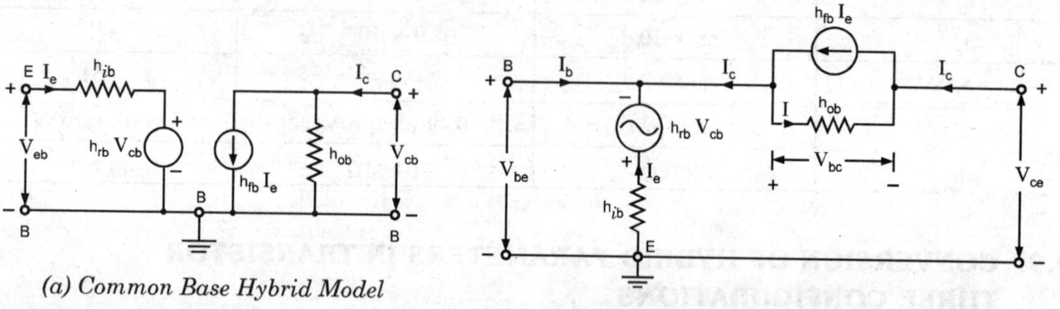

(a) Common Base Hybrid Model

(b) Circuit of Fig (a) Redrawn in CE Configuration

Fig. 9.15

$$V_{be} + h_{ib} I_e + h_{rb} V_{cb} = 0$$

Combining the above two equations we have

$$I_e = -\frac{1 - h_{rb}}{h_{ib}} V_{be}$$

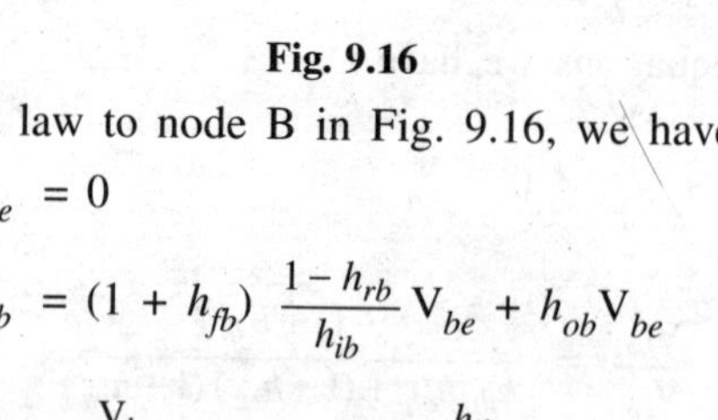

Fig. 9.16

Applying Kirchhoff's current law to node B in Fig. 9.16, we have

$$I_b + I_e + h_{fb} I_e - h_{ob} V_{be} = 0$$

$$\text{or } I_b = (1 + h_{fb}) \frac{1 - h_{rb}}{h_{ib}} V_{be} + h_{ob} V_{be}$$

$$\text{Thus } h_{ie} = \frac{V_{be}}{I_b} = \frac{h_{ib}}{h_{ib}\, h_{ob} + (1 - h_{rb})(1 + h_{fb})} \quad \text{...(9.23)}$$

This is an exact expression. Since $h_{rb} << 1$ and $h_{ob}\ h_{ib} << (1 + h_{fb})$, so the above equation is reduced to

$$h_{ie} \simeq \frac{h_{ib}}{1 + h_{fb}} \quad \text{...(9.24)}$$

Approximate conversion formulas for transistor parameters are given in Table 9.4.

TABLE 9.4

Configuration / Parameter	Common Emitter, CE	Common Base, CB	Common Collector, CC	Equivalent T-Circuit
h_{ie}	1.1 kΩ	$\frac{h_{ib}}{1+h_{fb}}$	h_{ic} *	$r_b + \frac{r_e}{1-a}$
h_{re}	2.5×10^{-4}	$\frac{h_{ib}\, h_{ob}}{1+h_{fb}} - h_{rb}$	$1 - h_{rc}$ *	$\frac{r_e}{r_c\,(1-a)}$
h_{fe}	50	$-\frac{h_{fb}}{1+h_{fb}}$	$-(1+h_{fc})$*	$\frac{a}{1-a}$
h_{oe}	25 μS (μA/V)	$\frac{h_{ob}}{1+h_{fb}}$	h_{oc} *	$\frac{1}{r_c\,(1-a)}$
h_{ib}	$\frac{h_{ie}}{1+h_{fe}}$	21.6 Ω	$-h_{ic}/h_{fc}$	$r_e + (1-a)\, r_b$
h_{rb}	$\frac{h_{ie}\, h_{oe}}{1+h_{fe}} - h_{re}$	$2.9. \times 10^{-4}$	$h_{rc} - \frac{h_{ic} h_{oc}}{h_{fc}} - 1$	r_b/r_c
h_{fb}	$\frac{-h_{fe}}{1+h_{fe}}$	– 0.98	$-\frac{1+h_{fc}}{h_{fc}}$	$-a$
h_{ob}	$\frac{h_{oe}}{1+h_{fe}}$	0.49 μS (μA/V)	$\frac{-h_{oc}}{h_{fc}}$	$1/r_c$
h_{ic}	h_{ie} *	$\frac{h_{ib}}{1+h_{fb}}$	1.1 kΩ	$r_b + \frac{r_e}{1-a}$
h_{rc}	$1 - h_{re} \simeq 1$*	1	1	$1 - \frac{r_e}{r_c\,(1-a)}$
h_{fc}	$-(1+h_{fe})$*	$\frac{-1}{1+h_{fb}}$	– 51	$-\frac{1}{1-a}$
h_{oc}	h_{oe} *	$\frac{h_{ob}}{1+h_{fb}}$	25 μS (μA/V)	$\frac{1}{r_c\,(1-a)}$
a	$\frac{h_{fe}}{1+h_{fe}}$	$-h_{fb}$	$\frac{1+h_{fc}}{h_{fc}}$	0.98
r_c	$\frac{1+h_{fe}}{h_{oe}}$ *	$\frac{1}{h_{ob}}$	$\frac{-h_{fc}}{h_{oc}}$ *	2.04 MΩ
r_e	$\frac{h_{re}}{h_{oe}}$ *	$h_{ib} - \frac{h_{rb}}{h_{ob}}\,(1+h_{fb})$*	$\frac{1-h_{rc}}{h_{oc}}$ *	10 Ω
r_b	$h_{ie} - \frac{h_{re}}{h_{oe}}\,(1+h_{fe})$*	$\frac{h_{rb}}{h_{ob}}$ *	$h_{ic} + \frac{h_{fc}}{h_{oc}}\,(1-h_{rc})$*	590 Ω

* Stands for exact.

Example 9.2. $h_{fe} = 50$, $h_{ie} = 0.83$ kΩ, find out the current gain (h_{fb}) and input impedance (h_{ib}) for a transistor in CB configuration.

Solution : $h_{fb} = \frac{-h_{fe}}{1+h_{fe}} = \frac{-50}{1+50} = -0.98$ **Ans.**

$$h_{ib} = \frac{h_{ie}}{1+h_{fe}} = \frac{0.83\times10^3}{1+50} = 16.27\ \Omega \text{ Ans.}$$

Example 9.3. The h parameters for a CE configuration are h_{ie} = 2,600 Ω, h_{fe} = 100, $h_{re} = 0.02 \times 10^{-2}$ and $h_{oe} = 5 \times 10^{-6}$ S. Find h parameters for CC configuration.

Solution :

$h_{ic} = h_{ie}$ = 2,600 Ω **Ans.**

$h_{fc} = -(1 + h_{fe}) = -(1 + 100) = -101$ **Ans.**

$h_{rc} = 1 - h_{re} \simeq 1$ **Ans.**

$h_{oc} = h_{oe} = 5 \times 10^{-6}$ S **Ans.**

9.10. TRANSISTOR AMPLIFIER CIRCUIT PERFORMANCE IN h PARAMETER

A transistor amplifier can be formed simply by connecting a signal source to the input and an external load to the output terminals of a transistor, as shown in Fig. 9.17 and giving proper bias to it. All transistor amplifiers, connected in any one of the three possible configurations, are basically two-port devices, as shown in Fig. 9.18, that is, there are a pair of input terminals and a pair of output terminals. In Fig. 9.18, the transistor has been replaced with its small-signal hybrid model without specifying the configuration. The circuit shown in Fig. 9.18 is valid for any type of load whether it be a pure resistance, an impedance or another transistor. This is true because the transistor hybrid model was derived without any regard to the external circuit in which the transistor is incorporated. The only limitation is the requirement that the h parameters remain substantially constant over the working range. For an amplifier there are six quantities of great interest (input impedance, output impedance, current gain, voltage gain, power gain and phase relationship), each of which will be discussed in detail here.

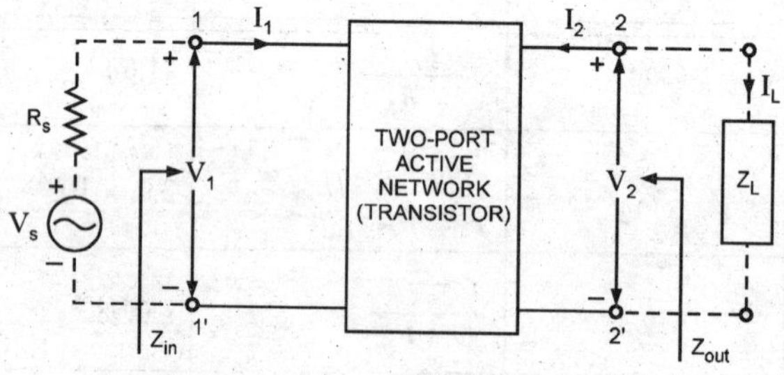

Fig. 9.17. *Basic Amplifier Circuit*

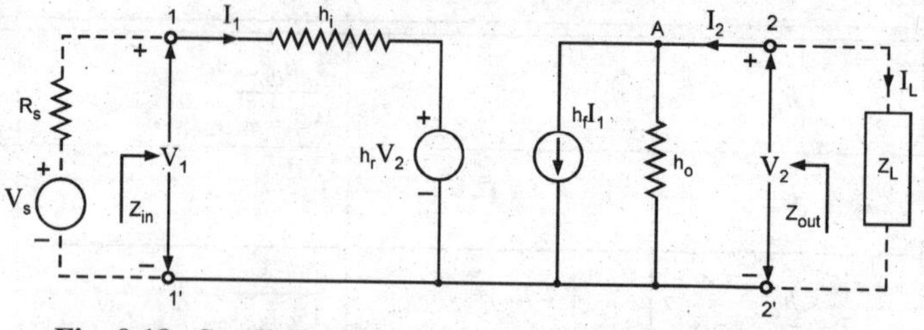

Fig. 9.18. *Small Signal Hybrid Model of Transistor Amplifier Circuit Shown in Fig. 9.17*

For the analysis of circuit, let us assume sinusoidally varying voltages and currents :

1. ***Input Impedance.*** The input impedance Z_{in} is defined as the ratio of input voltage to the input current *i.e.*,

$$Z_{in} = \frac{V_1}{I_1} \quad ...(9.25)$$

From the input circuit shown in Fig. 9.18 we have

$$V_1 = h_i I_1 + h_r V_2 \quad ...(9.26)$$

Substituting $V_1 = h_i I_1 + h_r V_2$ in Eq. (9.25) we have

$$Z_{in} = \frac{h_i I_1 + h_r V_2}{I_1} = h_i + h_r \frac{V_2}{I_1} \quad ...(9.27)$$

From the output circuit shown in Fig. 9.18 we have

$$I_2 = h_f I_1 + h_o V_2 = \frac{-V_2}{Z_L} \qquad ...(9.28)$$

The minus sign is used here because the actual current flows in the direction opposite to that of I_2.

$$\text{So} \quad \frac{V_2}{I_1} = \frac{-h_f}{h_o + \frac{1}{Z_L}} \qquad ...(9.29)$$

Substituting value of $\frac{V_2}{I_1}$ from Eq. (9.29) in Eq. (9.27), we have

$$\text{Input impedance } Z_{in} = h_i - \frac{h_r h_f}{h_o + \frac{1}{Z_L}} \qquad ...(9.30)$$

$$\simeq h_i \qquad \text{...if } h_r \text{ or } Z_L \text{ is very small.}$$

It is seen that Z_{in} depends on Z_L *i.e.,* ac resistance of the load across the output terminals of the transistors.

2. *Current Gain.* The current gain is denoted by A_i and is defined as the ratio of output current to input current.

$$\text{i.e., } A_i = \frac{I_L}{I_1} = \frac{-I_2}{I_1} \qquad ...(9.31)$$

Applying Kirchhoff's current law to node A in the output circuit

$$I_2 = h_f I_1 + h_o V_2$$

Substituting $V_2 = -I_2 Z_L$ from Eq. (9.28), we have

$$I_2 = h_f I_1 - h_o I_2 Z_L$$

$$\text{or Current gain } A_i = \frac{-I_2}{I_1} = \frac{-h_f}{1 + h_o Z_L} \simeq -h_f \text{ if } Z_L \text{ is zero or } h_o Z_L << 1 \qquad ...(9.32)$$

Current Gain Taking R_s into Account. The source current is not the transistor input current because it partly flows through R_s and partly through Z_{in}. So the voltage source V_s with series source resistance R_s is replaced by the Norton's equivalent source, shown in Fig. 9.19, consisting of current source I_s with source resistance R_s in shunt. The overall current gain A_{is} is given as

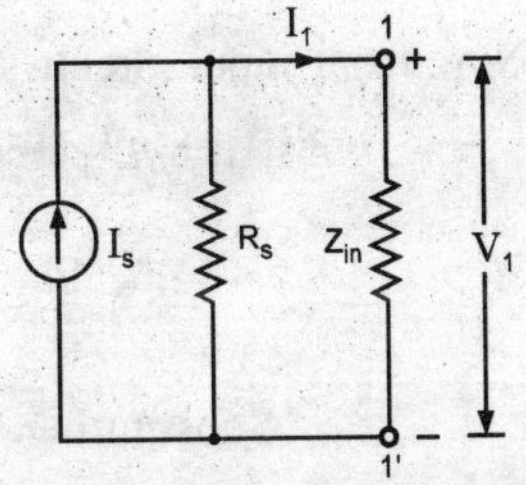

Fig. 9.19. ***Norton's Equivalent For The Source***

$$A_{is} \equiv \frac{-I_2}{I_s} = \frac{-I_2}{I_1} \cdot \frac{I_1}{I_s} = A_i \frac{I_1}{I_s} \qquad ...(9.33)$$

$$\text{From Fig. 9.19,} \quad \frac{I_1}{I_s} = \frac{R_s}{Z_{in} + R_s}$$

$$\text{So } A_{is} = \frac{A_i R_s}{Z_{in} + R_s} \qquad ...(9.34)$$

Note that if $R_s = \infty$ then $A_{is} = A_i$. Hence A_i is the *current gain for an ideal current source* (one with infinite source resistance).

3. *Voltage Gain.* The voltage gain is denoted by A_v and is defined as the ratio of output voltage to the input voltage *i.e.,*

$$A_v = \frac{V_2}{V_1} = \frac{V_2}{I_1 Z_{in}} = \frac{-h_f}{\left(h_o + \frac{1}{Z_L}\right) Z_{in}} \qquad ...(9.35) \quad \because \text{ from Eq. (9.29) } \frac{V_2}{I_1} = \frac{-h_f}{h_o + \frac{1}{Z_L}}$$

Voltage Gain Taking R_s into Account. Overall voltage gain A_{vs} is given as

$$A_{vs} = \frac{V_2}{V_s} = \frac{V_2}{V_1}\cdot\frac{V_1}{V_s} = A_v \frac{V_1}{V_s} \qquad ...(9.36)$$

From the equivalent input circuit of the amplifier given in Fig. 9.20, we have

$$\frac{V_1}{V_s} = \frac{Z_{in}}{Z_{in} + R_s},$$

$$\text{So } A_{vs} = A_v \frac{Z_{in}}{Z_{in} + R_s} \qquad ...(9.37)$$

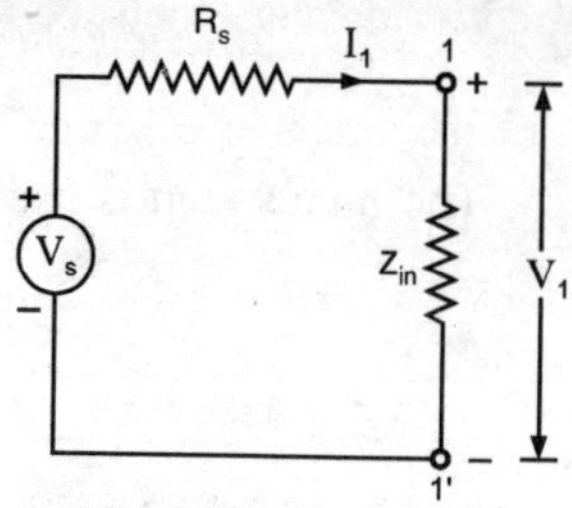

Fig. 9.20. ***Thevenin's Equivalent For The Source***

Note that if $R_s = 0$ the $A_{vs} = A_v$. Hence A_v is the voltage gain for an ideal voltage source (one with zero internal resistance). In practice, the quantity A_{vs} is more meaningful than A_v since usually, the source resistance has significant effect on the overall voltage amplification.

Independent of the transistor characteristics, the voltage and current gains taking source impedance into account, are related as

$$A_{vs} = A_{is} \frac{Z_L}{R_s} \qquad ...(9.38)$$

provided that the current and voltage generators have the same source resistance R_s.

4. *Output Admittance.* The output admittance Y_{out} is defined as the ratio of the output current to output voltage with $V_s = 0$ *i.e.,*

$$Y_{out} \equiv \frac{I_2}{V_2} \text{ with } V_s = 0 \qquad ...(9.39)$$

$$= h_f \frac{I_1}{V_2} + h_o \qquad ...(9.40)$$

$\because$ from Eq. (9.28), $I_2 = h_f I_1 + h_o V_2$

Now from input circuit shown in Fig. 9.18 we have

$$R_s I_1 + h_i I_1 + h_r V_2 = 0 \qquad ...(9.41) \text{ taking } V_s = 0$$

$$\text{or } \frac{I_1}{V_2} = -\frac{h_r}{h_i + R_s} \qquad ...(9.42)$$

Substituting $\frac{I_1}{V_2} = \frac{-h_r}{h_i + R_s}$ from Eq. (9.42) in Eq. (9.40) we have

$$Y_{out} = h_o - \frac{h_f h_r}{h_i + R_s} \qquad ...(9.43)$$

It is to be noted that *output admittance is a function of source resistance* R_s. If the source impedance is purely resistive, as it has been assumed, then Y_{out} is real (purely conductance).

In the above definition of Y_{out}, the load Z_L has been considered external to the amplifier. If the output impedance of the amplifier stage with Z_L included is required, the load impedance can be determined as the parallel combination of Z_L and Z_{out} (*i.e.,* $Z_L \parallel Z_{out}$).

5. *Power Gain.* The average power delivered to the load shown in Fig. 9.17, P_2 equals $|V_2|\, I_L|$ cos ϕ where ϕ is the phase angle between V_2 and I_L. For situation under discussion power delivered, $P_2 = -\, V_2 I_2 \cos\phi$. The minus sign arises for the same reason discussed in the derivation of input impedance. It indicates that the load is absorbing power and not supplying to the amplifier circuit. For purely resistive load, cos ϕ = 1 and $P_2 = -V_2 I_2$. The input power is $V_1 I_1$. So power gain

$$A_P = \frac{P_2}{P_1} = \frac{-V_2}{V_1} \cdot \frac{I_2}{I_1} = -A_v A_i \quad ...(9.44)$$

As $V_2 = -I_2 R_L$ and $I_2 = A_i I_1$

$$\therefore A_v = \frac{V_2}{V_1} = \frac{-A_i I_1 R_L}{V_1} = \frac{-A_i R_L}{V_1/I_1} = \frac{-A_i R_L}{Z_{in}}$$

$$\text{So power gain, } A_p = -A_v A_i = -\left(\frac{-A_i R_L}{Z_{in}}\right) A_i = \frac{A_i^2 R_L}{Z_{in}} \quad ...(9.45)$$

6. *Phase Relationship*. The phase relationship between the output current or voltage and the input current and voltage can be determined simply examining the Eqs. (9.31), (9.32) and (9.35).

All the h parameters are positive except h_f for the common-base and common-collector configurations (Table 9.3). For the common-emitter configuration it is obvious from Eqs. (9.31), (9.32) and (9.35) that for the purely resistive loads, the transistor output current I_2 is in phase with the input current I_1 while the output voltage, due to the negative sign, has a polarity opposite to that of the input voltage *i.e.*, output voltage attains its negative maximum when the input signal attains its positive maximum. For the common-collector and common-base configurations the opposite is true because of negative value of h_f.

The important relations derived above are summarized below :

$$\text{Current gain, } A_i = \frac{-h_f}{1 + h_0 Z_L} \simeq -h_f \text{ if } h_0 Z_L << 1$$

$$A_{is} = \frac{A_i R_s}{Z_{in} + R_s}$$

$$\text{Voltage gain, } A_v = \frac{-h_f}{\left(h_0 + \frac{1}{Z_L}\right) Z_{in}} = \frac{A_i Z_L}{Z_{in}}$$

$$A_{vs} = A_V \frac{Z_{in}}{Z_{in} + R_s}$$

$$\text{Input impedance, } Z_{in} = h_i - \frac{h_r h_f}{h_0 + \frac{1}{Z_L}} \simeq h_i \quad \text{if } h_r \text{ or } Z_L \text{ is very small}$$

$$\text{Output admittance, } Y_{out} = h_o - \frac{h_f h_r}{h_i + R_s} = \frac{1}{Z_{out}}$$

Example 9.4. A CE transistor amplifier circuit has the following characteristics :

h_{ie} = 1,000 Ω; h_{fe} = 50; $h_{re} = 2.5 \times 10^{-4}$; $h_{oe} = 25 \times 10^{-6}$ A/V. If the load resistance R_L = 10 kΩ, and source resistance is 100 Ω, find the input resistance, output resistance, voltage, current and power gains.

[M.D. Univ. Rohtak Analog Electronics December 2008]

Solution : Current gain, $A_i = \frac{-h_{fe}}{1 + h_{oe} R_L} = \frac{-50}{1 + 25 \times 10^{-6} \times 10 \times 10^3} = -40$ **Ans.**

$$\text{Input resistance, } R_{in} = h_{ie} - \frac{h_{re}h_{fe}}{h_{oe} + \frac{1}{R_L}} = 1{,}000 - \frac{2.5\times10^{-4}\times50}{25\times10^{-6} + \frac{1}{10\times10^3}} = 900\ \Omega \textbf{ Ans.}$$

$$\text{Voltage gain, } A_v = \frac{A_i R_L}{R_{in}} = -40\times\frac{10{,}000}{900} = -444.44 \textbf{ Ans.}$$

$$\text{Overall current gain, } A_{is} = A_i \times \frac{R_s}{R_{in}+R_s} = -40\times\frac{100}{900+100} = -4 \textbf{ Ans.}$$

$$\text{Overall voltage gain, } A_{vs} = A_v \times \frac{R_{in}}{R_{in}+R_s} = -444.44\times\frac{900}{900+100} = -400$$

$$\text{Output conductance, } G_{out} = h_{oe} - \frac{h_{fe}h_{re}}{h_{ie}+R_s} = 25\times10^{-6} - \frac{50\times2.5\times10^{-4}}{1000+100} = 13.636\times10^{-6}\text{ S}$$

$$\text{Output resistance, } R_{out} = \frac{1}{G_{out}} = \frac{1}{13.636\times10^{-6}} = 73.3\text{ k}\Omega \textbf{ Ans.}$$

$$\text{Power gain, } A_p = A_{vs}A_{is} = -400\times-4 = 1{,}600 \textbf{ Ans.}$$

Example 9.5. A CB transistor amplifier uses a voltage source of internal resistance R_S = 200 Ω and the load resistance is R_L = 1,200 Ω. The h parameters are h_{ib} = 24 Ω, h_{rb} = 4 × 10^{-4}, h_{fb} = –0.98 and h_{ob} = 0.6 μA/V. Calculate the following :

(1) Current gain (2) Input impedance (3) Voltage gain
(4) Overall voltage gain (5) Overall current gain. [*Pb. Technical Univ. Electronic Devices & Circuits Dec. 2009*]

Solution :

$$\text{(1) Current gain, } A_i = \frac{-h_{fb}}{1+h_{ob}\cdot R_L} = \frac{-(-0.98)}{1+0.6\times10^{-6}\times1{,}200} = 0.979 \textbf{ Ans.}$$

$$\text{(2) Input impedance, } R_{in} = h_{ib} - \frac{h_{rb}h_{fb}}{h_{ob}+\frac{1}{Z_L}} = 24 - \frac{4\times10^{-4}\times(-0.98)}{0.6\times10^{-6}+\frac{1}{1{,}200}} = 24.47\ \Omega \textbf{ Ans.}$$

$$\text{(3) Voltage gain, } A_v = \frac{A_i R_L}{R_{in}} = \frac{0.979\times1{,}200}{24.47} = 48 \textbf{ Ans.}$$

$$\text{(4) Overall voltage gain, } A_{VS} = \frac{A_V\cdot R_{in}}{R_{in}+R_S} = \frac{48\times24.47}{24.47+200} = 5.23 \textbf{ Ans.}$$

$$\text{(5) Overall current gain, } A_{is} = \frac{A_i R_S}{R_{in}+R_S} = \frac{0.979\times200}{24.47+200} = 0.872 \textbf{ Ans.}$$

Example 9.6. For the emitter follower, with R_S = 0.1 kΩ, R_L = 5 kΩ, h_{fe} = 50, h_{ie} = 1000 Ω, h_{oe} = 25 μAV. Calculate A_i, A_V, A_{VS}, R_{in} and R_{out}. [*M.D. Univ. Rohtak Analog Electronics May 2006*]

Solution : Transistor parameters for emitter follower (common-collector) amplifier

$$h_{ic} = h_{ie} = 1{,}000\ \Omega$$

$$h_{fc} = -(1+h_{fe}) = -(1+50) = -51$$

$$h_{rc} = 1 - h_{re} = 1$$

$$h_{oc} = h_{oe} = 25\times10^{-6}\text{ S}$$

$$\text{Current gain, } A_i = \frac{-h_{fc}}{1+h_{oc}R_L} = \frac{51}{1+25\times10^{-6}\times5\times10^3} = 45.333 \textbf{ Ans.}$$

$$\text{Input resistance, } R_{in} = h_{ic} - \frac{h_{rc}h_{fc}}{h_{oc} + \frac{1}{R_L}} = 1{,}000 - \frac{1 \times (-51)}{25 \times 10^{-6} + \frac{1}{5{,}000}}$$

$$= 1{,}000 + 226{,}667\ \Omega = 227{,}667\ \Omega \text{ or } 227.667\ k\Omega \textbf{ Ans.}$$

$$\text{Voltage gain, } A_v = \frac{-h_{fc}}{\left(h_{oc} + \frac{1}{R_L}\right)R_{in}} = \frac{51}{\left(25 \times 10^{-6} + \frac{1}{5{,}000}\right) \times 227{,}667} = 0.9956 \textbf{ Ans.}$$

$$\text{Overall voltage gain, } A_{vs} = A_v \frac{R_{in}}{R_{in} + R_s} = 0.9956 \times \frac{227{,}667}{227{,}667 + 100} = 0.995 \textbf{ Ans.}$$

$$\text{Output conductance, } G_{out} = h_{oc} - \frac{h_{fc}h_{rc}}{h_{ic} + R_s} = 25 \times 10^{-6} - \frac{-51 \times 1}{1{,}000 + 100} = 0.046388\ S$$

$$\text{Output resistance, } R_{out} = \frac{1}{G_{out}} = \frac{1}{0.046388} = 21.56\ \Omega \textbf{ Ans.}$$

9.11. LIMITATIONS OF h PARAMETERS

Input impedance, output impedance, current gain, voltage gain and power gain of a transistor amplifier can be determined accurately by using h parameters of the circuit, as discussed in Art. 9.9. However, there are two major limitations in the use of these parameters, as mentioned below.

1. By using h parameters correct answers can be had for small ac signals only because a transistor behaves as a linear device for small signals only.
2. It is very difficult to get the exact values of h parameters for a particular transistor. It is due to the fact that h parameters are subject to considerable variations—unit to unit variation, variation due to change in junction temperature and variation due to change in operating point.

While predicting performance of an amplifier, care is required to be taken that h parameter values used are correct for the operating point under consideration.

9.12. PHYSICAL MODEL OF A CB TRANSISTOR

The small-signal model of the transistor described by hybrid parameters is found very convenient for circuit analysis. As already discussed in Art. 9.6, these h parameters characterize a general two-port network. While applying this model to a specific transistor, the values of hybrid parameters are measured experimentally by the user or by the manufacturer. The device designer, on the other hand, prefers to use a model containing circuit parameters whose values can be determined directly from the physical properties of the transistor. Such a model is called the *physical model* and takes into consideration the physical mechanism occurring within the device.

A simplified physical model of a CB transistor is shown in Fig. 9.21. Looking into the emitter, we observe a forward-biased diode. Hence, between the input terminals E and B', there is a dynamic resistance r'_e, which is the slope of the forward-biased emitter junction volt-ampere characteristic. Looking back into the output terminals C and B', we observe a reverse-biased diode. Hence, between these terminals, there is a dynamic

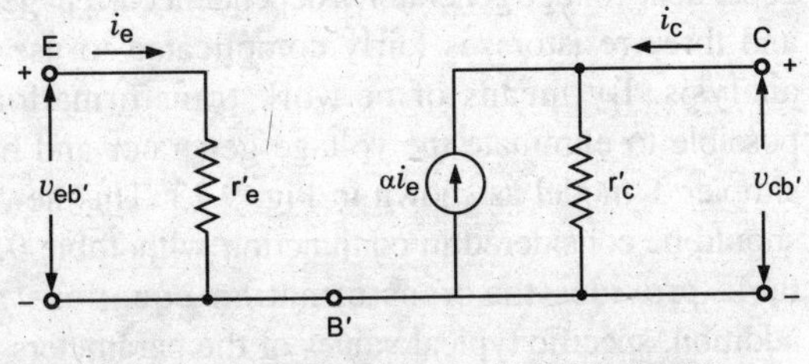

Fig. 9.21. *Simplified Physical Model of a CB Transistor*

resistance r'_c, which is a slope of the reverse-biased collector-junction volt-ampere characteristic. We know that collector current is proportional to emitter current. Hence a current generator αI_e has been added across r'_c resulting in equivalent circuit shown in Fig. 9.21.

The physical model, depicted in Fig. 9.21, is not realistic because of its following shortcomings.

1. It does not indicate the dependence of emitter current i_e on the collector voltage v_C.
2. It does not take into account the ohmic resistances of the three regions of the transistor.

By removing the above shortcomings we have a more complete physical model of the CB transistor, as illustrated in Fig. 9.22.

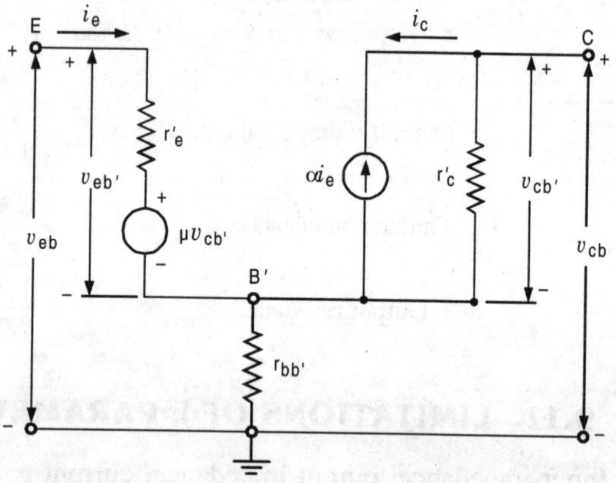

Fig. 9.22. ***Approximate Physical Model***

Early Feedback Generator. An increase in the magnitude of collector voltage effectively reduces the base width W, a phenomenon known as the *Early effect*. The minority-carrier current in the base in the active region is proportional to the slope of the injected minority-carrier density curve. This slope increases with the decrease in base width W. Thus the emitter current injected into the base increases with the increase in reverse collector voltage. This effect of collector-base voltage $v_{cb'}$ on emitter current may be taken into account by including a voltage source $\mu v_{cb'}$ in series with r'_e as shown in Fig. 9.22.

Base Spreading Resistance. For completing the equivalent circuit of Fig. 9.21, the ohmic resistances of the three regions of transistor are to be taken into account. Since base section is very thin, the base current passes through a region of extremely small cross section. Hence this resistance, called the *base spreading resistance* and denoted by $r_{bb'}$, is very large, and may be of the order of several hundred ohms. On the other hand, the collector and emitter ohmic resistances are only a few ohms, and may usually be neglected. If the external connection to the base is designated by B, then between the fictitious internal base node B' and B we must include a resistance $r_{bb'}$, as shown in Fig. 9.22.

If the base spreading resistance could be neglected so that B and B' coincided, the circuit depicted in Fig. 9.22 would be identical with the hybrid model depicted in Fig. 9.13, with

$$r'_e = h_{ib},\ \mu = h_{rb},\ \alpha = -h_{fb} \text{ and } r'_c = \frac{1}{h_{ob}}$$

T-Model. The circuit given in Fig. 9.22 contains elements each of which has been identified with the transistor physics. However, this circuit, which includes a dependent voltage generator, a dependent current generator, and three resistors, is fairly complicated to use for the analysis. By means of network transformations it is possible to eliminate the voltage generator and have the simpler T model as shown in Fig. 9.23. This new circuit should be considered in conjunction with Table 9.5. This table provides the transformation equations and, in addition, specific typical values of the parameters in each of the circuits.

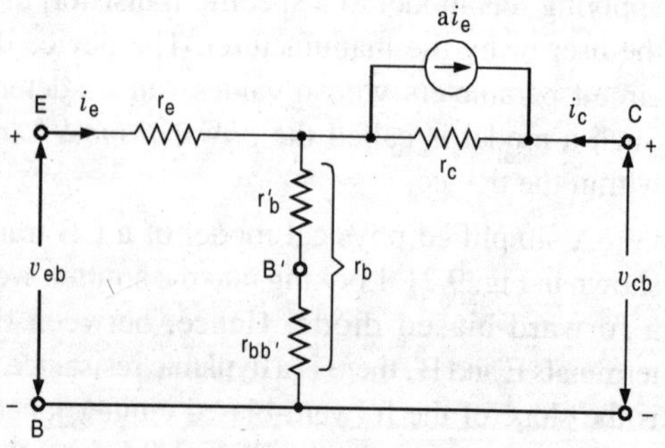

Fig. 9.23. ***T-Model of a CB Transistor***

TABLE 9.5. *Typical Parameter Values and The Equations of Transformation Between The Circuits Shown in Figs. 9.22 and 9.23*

Parameter in Physical Model Shown in Fig. 9.22	*Transformation Equations*	*Parameter in T Model Shown in Fig. 9.23*
$r'_e = 40\ \Omega$ $\mu = 5 \times 10^{-4}$ $r'_c = 2\ M\Omega$ $\alpha = 0.98$	$r_e = r'_e - (1 - \alpha)\ \mu r'_c$ $r'_b = \mu r'_c$ $r_c = (1 - \mu) r'_c$ $a = \dfrac{(\alpha - \mu)}{1 - \mu}$	$r_e = 20\ \Omega$ $r'_b = 1\ k\Omega$ $r_c = 2\ M\Omega$ $a = 0.98$

We observe that the transformed circuit accounts for the effect of the collector circuit on emitter circuit essentially through resistor r'_b rather than through the generator $\mu v_{cb'}$.

From Table 9.5 we note that

$$r_c \approx r'_c;\ a \approx \alpha \text{ and } r_e \simeq r'_e/2$$

The resistor r_b in the base leg is given by

$$r_b = r'_b + r_{bb'},$$ where r'_b and $r_{bb'}$ are resistances of comparable magnitudes.

T model have some shortcomings which are given below :

1. The circuit components in the T model cannot be interpreted directly in terms of the physical mechanisms in the transistor.
2. The values of the circuit components are difficult to obtain experimentally.
3. The analysis of a circuit is somewhat simpler in terms of the h parameters than through the use of the T equivalent circuit.

For the above reasons the T model is not used for analysis of small-signal low-frequency circuits.

The relationships between the hybrid parameters and those in the T equivalent circuit are given in Table 9.4.

9.13. ANALYSIS OF A CE TRANSISTOR AMPLIFIER USING h PARAMETERS

A common-emitter transistor amplifier circuit with self biasing is shown in Fig. 9.24. The circuit arrangement has already been explained in Art. 9.2. Various currents flowing in different branches, and various voltages across different components and branches are indicated in the figure.

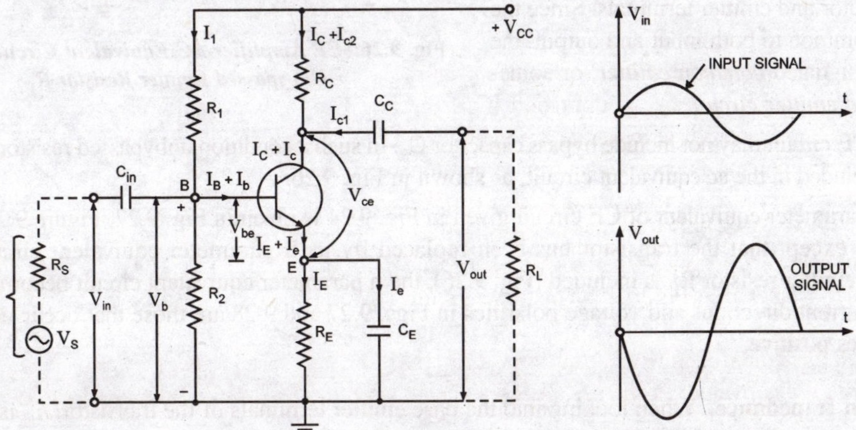

Fig. 9.24. *Common Emitter Transistor Amplifier Circuit With Self Biasing*

In the circuit NPN transistor is used. Input signal is applied to the base-emitter circuit while output is taken from the collector-emitter circuit. The emitter-base junction is forward biased and collector-emitter junction is reverse biased by supply V_{CC}. When the input ac signal increases in the positive direction, the voltage between base and emitter V_{be} increases because it is already positive with respect to ground. Thus forward bias is increased and causes increase in base current. Due to transistor action, collector current is increased β times. Due to flow of this current through collector resistance R_C, the voltage drop across it I_cR_C increases considerably. As a result the voltage between collector and emitter V_{ce} decreases. Thus amplified voltage appears across R_L. As explained in Art. 9.3, the positive going input signal appears as a negative going output signal *i.e.,* there is a phase shift of 180° between input and output signals. This is illustrated in Fig. 9.24.

Circuit analysis can be done rigorously using h parameters derived from a manufacturer's data sheet for a given transistor. However, the parameters are specified only for a particular set of bias conditions and, therefore, need correction when the circuit bias conditions differ from those mentioned on the data sheet. Correction involves use of graphs showing variation of each parameter with variations in bias conditions (I_C and V_{CE}). This process makes the analysis of even the simplest circuits quite complicated. The value of lengthy, rigorous analysis is questionable, especially when it is remembered that each h parameter can vary significantly from one transistor to another one of the same type. For these reasons, rigorous h parameter analysis is not taken here. Instead, practical, approximate methods of determining circuit performance are treated.

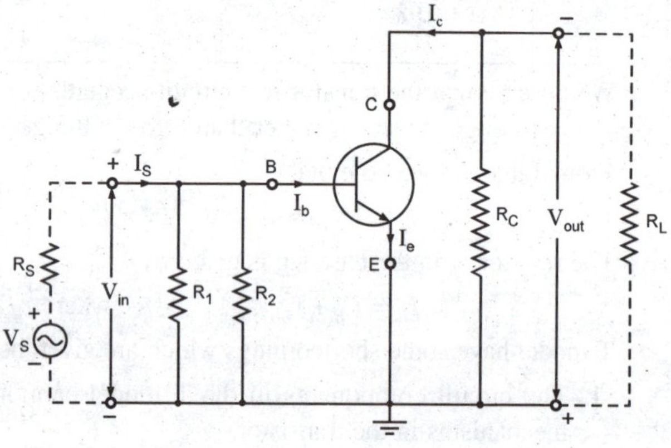

Fig. 9.25. *CE Amplifier AC Equivalent Circuit*

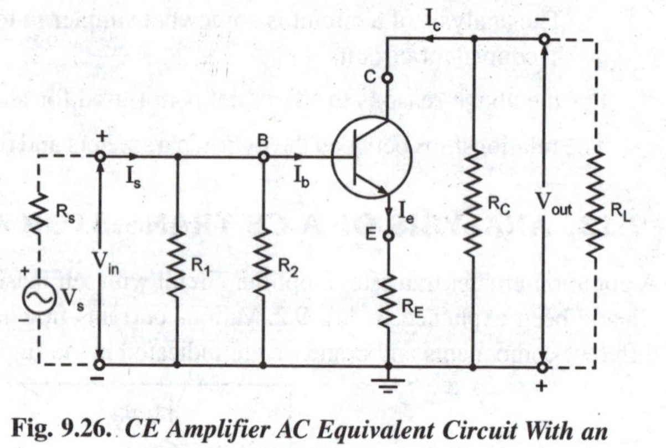

Fig. 9.26. *CE Amplifier AC Equivalent Circuit With an Unbypassed Emitter Resistor R_E*

AC equivalent circuit of a CE transistor amplifier shown in Fig. 9.24 is given in Fig. 9.25. From Fig. 9.25 it is obvious that the circuit input voltage V_{in} is developed across the transistor base and emitter terminals. The output is available across collector and emitter terminals. Since the emitter is common to both input and output, the circuit is designated *common emitter*, or sometimes *ground emitter circuit*.

Some CE circuit may not include bypass capacitor C_E. In such a condition unbypassed resistor R_E exists and it is to be included in the ac equivalent circuit, as shown in Fig. 9.26.

The h parameter equivalent of CE circuit given in Fig. 9.24 is given in Fig. 9.27. Figure 9.27 is the same as Fig. 9.25 except that the transistor has been replaced by its h parameter equivalent circuit. When an unbypassed emitter resistor R_E is included (Fig. 9.26), the h parameter equivalent circuit becomes as given in Fig. 9.28. Current directions and voltage polarities in Figs. 9.27 and 9.28 are those that occur when the input ac signal goes positive.

1. Input Impedance. When looking into the base emitter terminals of the transistor, h_{ie} is seen to be in series with h_{re} V_{out}. For a CE circuit h_{re} is normally very small quantity, so h_{re} V_{out} fed back from the output to

9.13.1. CE Circuit Using Fixed Bias

Common emitter transistor amplifier circuit using fixed bias is given in Fig. 4.5. The ac performance equations for the circuit given differ only slightly from those derived above. The presence of only one bias resistor makes $R_B = R_1$. This gives

$$Z_{in} = R_{in} \parallel h_{ie} \quad ...(9.58)$$

Note : For circuits using PNP transistors, the ac performance equations are exactly the same as those derived for NPN transistor circuits.

Example 9.7. A transistor amplifier is shown as under [Fig. 9.29 (*a*)]. The parameters of the transistor are h_{ie} = 2 kΩ, h_{oe} = 25 μS and h_{fe} = 55 and the output load resistor dissipates a signal power of 10 mW. Determine the power gain of the amplifier and the input signal emf E. The reactances of the capacitors may be neglected. Draw the equivalent circuit of the amplifier.

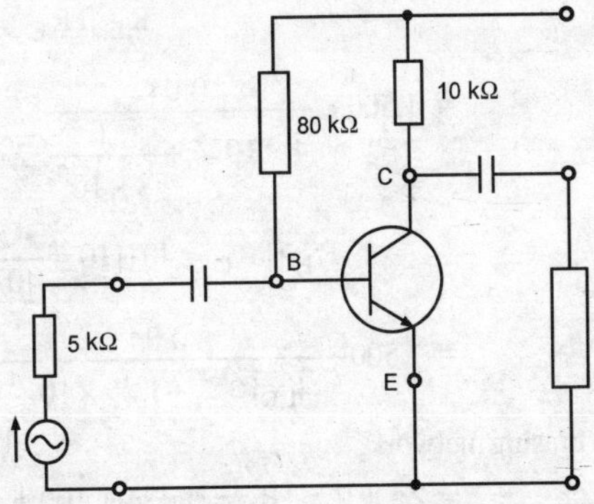

(*a*)

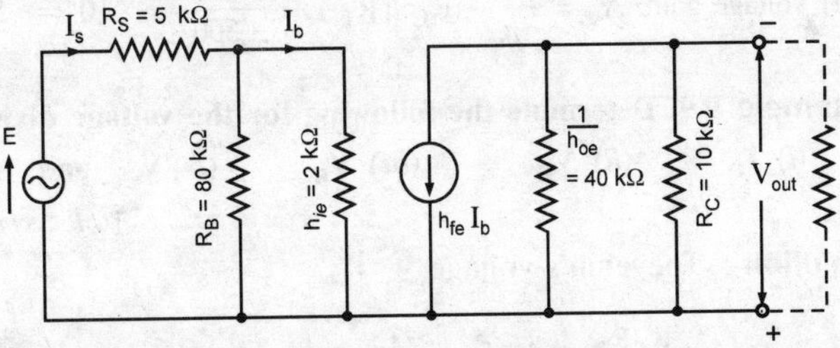

(*b*) *AC Equivalent Circuit of Circuit Shown in Fig. 9.29 (a)*

Fig. 9.29

Solution : $Z_b = h_{ie} = 2$ kΩ

$Z_{in} = R_B \parallel h_{ie} = 80 \parallel 2 = 1.95$ kΩ

Input impedance to E,

$Z_s = R_s + Z_{in} = 5 + 1.95 = 6.95$ kΩ

Power drawn from the source,

$$P_{in} = \frac{E^2}{Z_s} = \frac{E^2}{6.95} \text{ mW}$$

$$\text{Base current, } I_b = I_s \times \frac{R_B}{R_B + h_{ie}} = \frac{E}{6.95} \times \frac{80}{80+2}$$

$$= 0.14 \text{ E (mA)}$$

$$\text{Output impedance, } Z_{out} = \frac{1}{h_{oe}} \parallel R_C$$

$$= \frac{1}{25 \times 10^{-6}} \parallel 10 = 8 \text{ k}\Omega$$

AC load resistance, $R_{ac} = Z_{out} \parallel R_L = 8 \parallel 10 = 4.44$ kΩ

Output voltage, $V_{out} = -h_{fe} I_b R_{ac} = -55 \times 0.14 \text{ E} \times 4.44 = -34.3$ E volts

$$\text{Power dissipated, } P_{out} = \frac{(V_{out})^2}{R_L} = \frac{(-34.3\text{E})^2}{10 \times 10^3} = 10 \text{ mW or } 10 \times 10^{-3} \text{ W}$$

$$\text{So E} = \frac{\sqrt{10 \times 10^{-3} \times 10 \times 10^3}}{34.3} = 0.29 \text{ V } \textbf{Ans.}$$

$$\text{Power gain, } A_p = \frac{\text{Power dissipated}}{\text{Power input}} = \frac{10 \times 10^{-3}}{\frac{E^2}{6.95} \times 10^{-3}} = \frac{10 \times 6.95}{(0.29)^2} = 826 \textbf{ Ans.}$$

Example 9.8. A BJT amplifier is given in Fig. 9.30. The h parameters are : h_{ie} = 1,500 Ω; h_{re} = 4 × 10^{-4}; h_{fe} = 100; h_{oe} = 4 × 10^{-4} S. Calculate (*i*) Z_{in} (*ii*) Z_{in} with biasing network and (*iii*) voltage gain.

[*Rajasthan University 2005*]

Solution : (*i*) $Z_{in\ (base)}$,

$$Z_b = h_{ie} - \frac{h_{re}h_{fe}}{h_{oe} + \frac{1}{Z_L}}$$

$$= 1{,}500 - \frac{4\times10^{-4}\times100}{4\times10^{-4} + \frac{1}{R_L \,||\, R_C}}$$

$$= 1{,}500 - \frac{0.04}{4\times10^{-4} + \frac{1}{8\times10^3}}$$

$$\because R_L \,||\, R_C = 40\,||\,10 = \frac{4\times10}{40+10} = 8\ k\Omega$$

$$= 1{,}500 - \frac{0.04}{4\times10^{-4} + 1.25\times10^{-4}} = 1{,}500 - 76.2 = 1{,}423.8\ \Omega \text{ or } 1.424\ k\Omega \textbf{ Ans.}$$

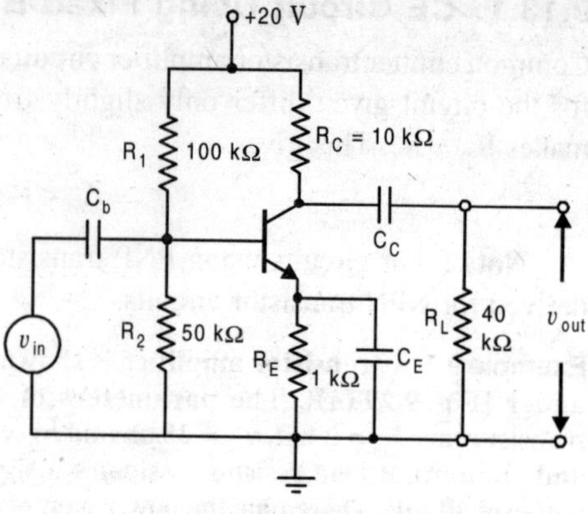

Fig. 9.30

(*ii*) Z_{in} with biasing network,

Z_{in} or $Z_{in\ (stage)} = Z_b \,||\, R_1 \,||\, R_2 = 1.424 \,||\, 100 \,||\, 50 = 1.366\ k\Omega$ **Ans.**

(*iii*) Voltage gain, $A_V = \frac{-h_{fe}}{h_{ie}}(R_C \,||\, R_L) = \frac{-100}{1{,}500}\times8\times10^3 = -533.33$ **Ans.**

Example 9.9. Determine the following for the voltage divider bias circuit shown in Fig. 9.31.
(*i*) I_C (*ii*) V_E (*iii*) V_B (*iv*) V_{CE} and (*v*) R_{in}

[U.P. Technical University Electronics Engg. Second Semester 2009-10]

Solution : Thevenin's voltage,

$$V_{Th} = \frac{R_2}{R_1+R_2}V_{CC} = \frac{12}{33+12}\times18 = 4.8\ V$$

Applying KVL to the base circuit, we have

$$V_B = V_{BE} + V_E = V_{BE} + I_ER_E$$

$$\text{or } I_E = \frac{V_B - V_{BE}}{R_E} = \frac{4.8-0.7}{1\times10^3} = 4.1\ mA$$

Assuming $V_{BE} = 0.7$ V

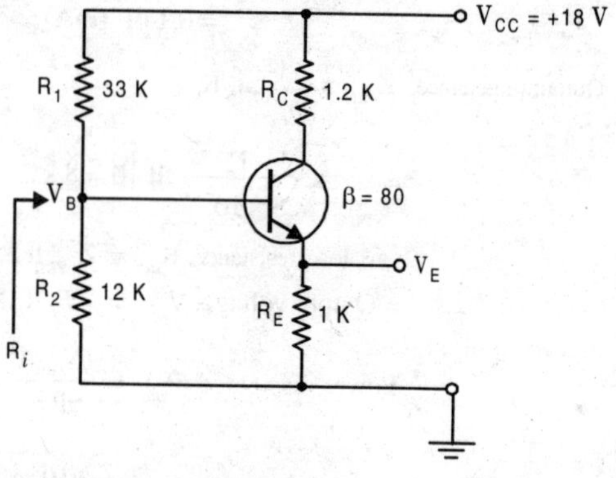

Fig. 9.31

(*i*) Collector current,

$$I_C = \frac{\beta}{\beta+1}I_E = \frac{80}{80+1}\times4.1 = 4.05\ mA \textbf{ Ans.}$$

(*ii*) Emitter voltage, $V_E = I_ER_E = 4.1\times10^{-3}\times1\times10^3 = 4.1$ V **Ans.**

(*iii*) Base voltage, $V_B = V_{Th} = 4.8$ V **Ans.**

(*iv*) $V_{CE} = V_{CC} - I_CR_C - I_ER_E = 18 - 4.05\times10^{-3}\times1.2\times10^3 - 4.1\times10^{-3}\times1\times10^3$
$= 9.04$ V **Ans.**

$R_b = h_{ie}(1+h_{fe})R_E = 1.1 + (1+80)\times1 = 82.1\ k\Omega$ Assuming $h_{ie} = 1.1\ k\Omega$

(*v*) Input resistance, $R_{in} = R_b \,||\, R_1 \,||\, R_2 = (82.1 \,||\, 33 \,||\, 12)\ k\Omega = 7.95\ k\Omega$ **Ans.**

Example 9.10. Calculate the following using hybrid equivalent model for the following circuit where h_{fe} = 110 and h_{oe} = 20 μA/V.

(i) Z_i, (ii) Z_0, (iii) A_V, (iv) A_i. [*U.P. Technical Univ. Electronics Engineering First Semester 2006-07*]

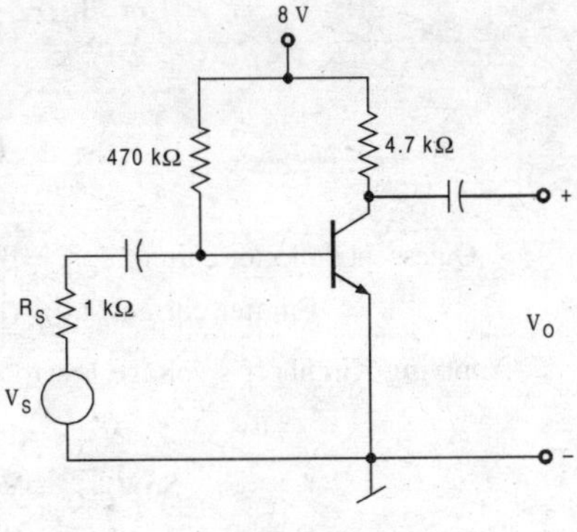

Fig. 9.32

Solution : Given that

R_B = 470 kΩ, R_C = 4.7 kΩ

R_s = 1 kΩ, V_{CC} = 8 V, h_{fe} = 110

h_{oe} = 20 μA/V

and h_{ie} = 1.1 kΩ (standard value)

(i) Input impedance $Z_i = R_B \parallel h_{ie}$ = 470 kΩ ∥ 1.1 kΩ ≃ 1.1 kΩ

$Z_s = R_s + Z_i$ = 1.1 kΩ + 1 kΩ = 2.1 kΩ **Ans.**

(ii) Output impedance $$Z_0 = \frac{1}{h_{oe}} \parallel R_C = \frac{10^6}{20} \parallel 4.7 \text{ k}\Omega$$

$$= 50 \text{ k}\Omega \parallel 4.7 \text{ k}\Omega = 4.3 \text{ k}\Omega \textbf{ Ans.}$$

(iii) Voltage gain $$A_V = \frac{-R_C \cdot h_{fe}}{h_{ie}} = \frac{-4.7 \text{ k}\Omega \times 110}{1.1 \text{ k}\Omega} = -470 \textbf{ Ans.}$$

(iv) Current gain, $$A_i = \frac{-h_{fe} \times R_B}{R_B + h_{ie}} = \frac{-110 \times 470}{470 + 1.1} = -109.75 \textbf{ Ans.}$$

Example 9.11. In a small signal amplifier shown below [Fig. 9.33], h_{fe} = 100, h_{ie} = 560 Ω, R_C = 2 kΩ, R_E = 1 kΩ, R_B = 600 kΩ, h_{re} and h_{oe} are negligible.

(a) Draw the h parameter equivalent for the parameter. Calculate the input and output impedances and the voltage gain.

(b) Give the dc load line of the circuit and find Q-point. [*Pb. Technical Univ. Electronic Devices and Circuits May 2007*]

Solution : CE amplifier h parameter equivalent circuit (with R_E included in the circuit) is shown in Fig. 9.28.

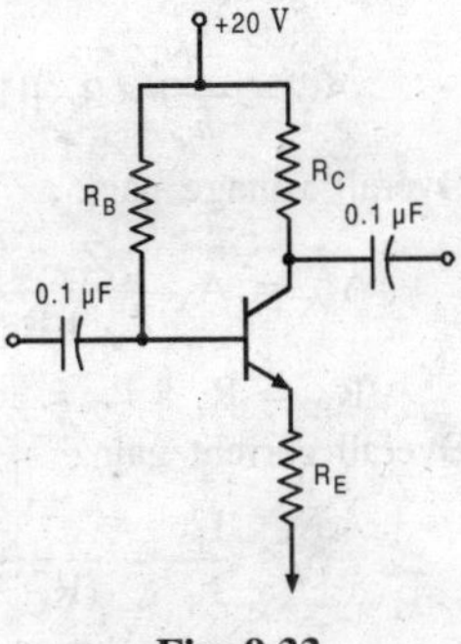

Fig. 9.33

Since h_{re} is negligible, $h_{re}V_{out}$ is short and since h_{oe} is negligible, *i.e.*, $\frac{1}{h_{oe}}$ is close to infinity and, therefore, open. Thus the h parameter equivalent circuit is reduced to that shown in Fig. 9.34.

Input impedance to the transistor base,

$Z_b = h_{ie} + R_E(1 + h_{fe})$ Refer to Eq. (9.47)

$= 560 + 1 \times 10^3 (1 + 100) = 101{,}560\ \Omega = 101.560$ kΩ

Circuit input impedance,

$Z_{in} = Z_b \parallel R_B = 101.560 \parallel 600$ kΩ Refer to Eq. (9.48)

$$= \frac{101.56 \times 600}{101.56 + 600} = 86.86 \text{ k}\Omega \textbf{ Ans.}$$

Output impedance,

$$Z_{out} = \frac{1}{h_{oe}} \parallel R_C$$ Refer to Eq. (9.50)

$= R_C = 2$ kΩ **Ans.** $\because \frac{1}{h_{oe}} \simeq \infty$, h_{oe} being negligible

Voltage gain,

$$A_V = -\frac{R_C \parallel R_L}{R_E} = -\frac{2 \text{ k}\Omega}{1 \text{ k}\Omega} = -2 \textbf{ Ans.}$$ Refer to Eq. (9.53)

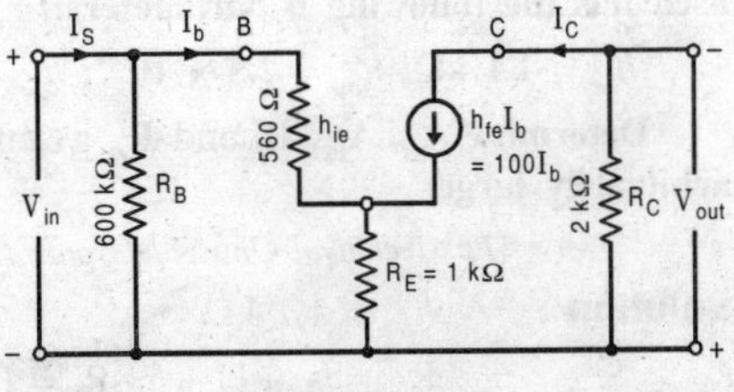

Fig. 9.34

Applying Kirchhoff's voltage law to base-emitter loop (Fig. 9.33), we have

$$V_{CC} = I_BR_B + V_{BE} + I_ER_E = I_BR_B + V_{BE} + (\beta + 1)I_BR_B \qquad \because I_E = (\beta + 1)I_B$$

$$\text{or } I_B = \frac{V_{CC} - V_{BE}}{R_B + (\beta+1)R_E}$$

$$= \frac{20 - 0.7}{600 \times 10^3 + (100+1) \times 1 \times 10^3} \qquad \because \beta = h_{fe} = 100 \text{ and assume } V_{BE} = 0.7 \text{ V}$$

$$= 0.0275 \text{ mA}$$

Quiescent collector current, $I_{CQ} = \beta I_B = 100 \times 0.0275 \text{ mA} = 2.75 \text{ mA}$

Emitter current, $I_E = (\beta + 1)I_B = (100 + 1) \times 0.0275 = 2.78 \text{ mA}$

Applying Kirchhoff's voltage law to the output loop (Fig. 9.33), we have

$$V_{CC} = I_{CQ}R_C + V_{CEQ} + I_ER_E$$

$$\text{So } V_{CEQ} = V_{CC} - I_{CQ}R_C - I_ER_E = 20 - 2.75 \times 10^{-3} \times 2 \times 10^3 - 2.78 \times 10^{-3} \times 1 \times 10^3 = 11.72 \text{ V}$$

Thus operating point (V_{CEQ}, I_{CQ}) = (11.72 V, 2.75 mA) **Ans.**

Example 9.12. A transistor amplifier is shown in Fig. 9.35. Compute $A_V = V_0/V_S$, $A_i = I_L/I_S$, R_i and R'_0. Assume $h_{ie} = 1$ kΩ, $h_{fe} = 100$, $h_{re} = h_{oe} = 0$.

[G.G.S.I.P. Univ. Analog Electronics December 2009]

Solution : Input impedance to transistor base,

$$Z_b = h_{ie} + R_E(1 + h_{fe}) = 1{,}000 + 100(1 + 100)$$

$$= 11{,}100 \ \Omega \text{ or } 11.1 \text{ k}\Omega$$

Actual circuit input impedance,

$$Z_{in} = R_1 \parallel R_2 \parallel Z_b = 20 \parallel 10 \parallel 11.1 = 4.165 \text{ k}\Omega \textbf{ Ans.}$$

Actual circuit output impedance,

$$R'_{out} = \frac{1}{h_{oe}} \parallel R_C = R_C = 2 \text{ k}\Omega \textbf{ Ans.}$$

Voltage gain,

$$A_V = \frac{-h_{fe}}{h_{ie}} \times R_C \parallel R_L = \frac{-100}{1 \text{ k}\Omega} \times 2 \text{ k}\Omega \parallel 2 \text{ k}\Omega = -100 \textbf{ Ans.}$$

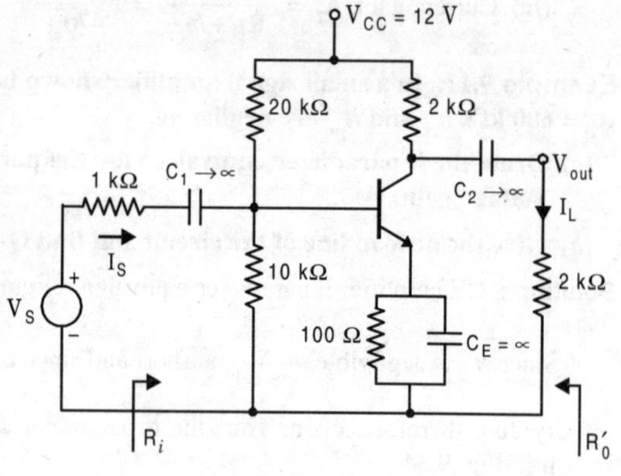

Fig. 9.35

Overall voltage gain,

$$A_{Vs} = A_V \frac{Z_{in}}{Z_{in} + R_S} = -100 \times \frac{4.165}{4.165 + 1} = -80.64 \textbf{ Ans.}$$

$$R_B = R_1 \parallel R_2 = 20 \text{ k}\Omega \parallel 10 \text{ k}\Omega = 6.667 \text{ k}\Omega$$

Overall current gain,

$$A_i = \frac{I_L}{I_S} = \frac{-h_{fe}R_BR_C}{(R_C + R_L)(R_B + Z_b)} = \frac{-100 \times 6.667 \times 2}{(2+2)(6.667 + 11.1)} = -18.76 \textbf{ Ans.}$$

Example 9.13. For the circuit shown in Fig. 9.36, the transistor used has the following h parameters:

$$\boldsymbol{h_{ie} = 1.1 \text{ k}\Omega;\ h_{oe} = 2.5 \times 10^{-4};\ h_{fe} = 50;\ h_{re} = 25 \ \mu\text{A/V}.}$$

Determine A_{is}, A_{vs}, R_i and R_o assuming all capacitors to be arbitrarily large.

[Pb. Technical Univ. Electronic Devices and Circuits May 2010]

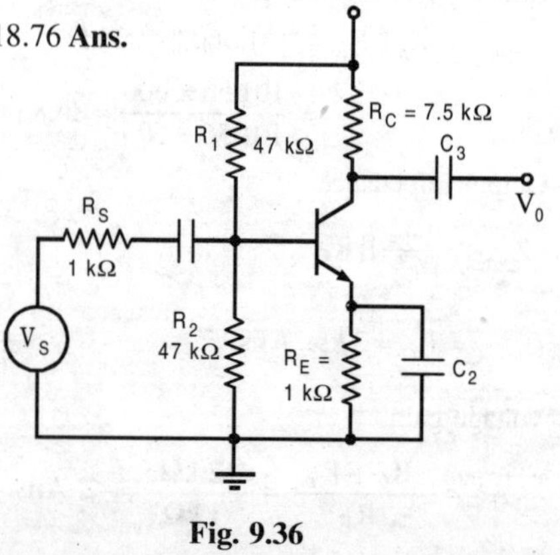

Fig. 9.36

Solution : $h_{ie} = 1.1$ kΩ

$$R_b = h_{ie} + (1 + h_{fe})R_E = 1.1 + (1 + 50) \times 1 = 52.1 \text{ k}\Omega$$

$$R_{in} = R_b \parallel R_1 \parallel R_2 = 52.1 \parallel 47 \parallel 47 = 16.195 \text{ k}\Omega \textbf{ Ans.}$$

Current gain, $A_i = h_{fe} = 50$ **Ans.**

$$A_{is} = A_i \times \frac{R_S}{R_{in}+R_S} = 50 \times \frac{1}{16.195+1} = 2.91 \textbf{ Ans.}$$

$$\text{Voltage gain, } A_v = \frac{-h_{fe}R_L}{h_{ie}+(1+h_{fe})R_E} = \frac{-50\times 7.5}{1.1+(1+50)\times 1} = -7.198$$

$$A_{vs} = A_v \times \frac{R_{in}}{R_{in}+R_S} = \frac{-7.198\times 16.195}{16.195+1} = -6.8 \textbf{ Ans.}$$

$$\text{Output impedance, } R_{out} = \frac{1}{h_{oe}} \| R_C = \frac{1}{2.5\times 10^{-4}} \| 7.5\times 10^3 = 4 \| 7.5 \text{ k}\Omega = 2.6 \text{ k}\Omega \textbf{ Ans.}$$

Example 9.14. **(*a*) Find I_{CQ} and V_{CEQ} for the following figure [Fig. 9.37 (*a*)]**

$\beta_{dc} = h_{FE} = 130$

(*b*) What will happen to V_{CE} if R_2 is open circuited in the circuit of part (*a*)

(*c*) Find A_v and R_{in} for the circuit of part (*a*) if $h_{fe} = 50$, $h_{ie} = 1$ kΩ, $h_{re} = 0$ and $h_{oe} = 0$.

[*U.P. Technical Univ. Basic Electronics First Semester 2000-01*]

Solution : Resistance seen into the base and ground terminals with V_{CC} short circuited or

$$R_{Th} = R_1 \| R_2 = 510 \| 510 = 255 \text{ k}\Omega$$

Open-circuit voltage across base and ground terminals, or

$$V_{Th} = V_{CC} \times \frac{R_2}{R_1+R_2} = 18 \times \frac{510}{510+510} = 9 \text{ V}$$

$$\text{Base current, } I_B = \frac{V_{Th}-V_{BE}}{R_{Th}+(\beta+1)R_E} = \frac{9-0}{[255+(130+1)\times 7.5]\times 10^3} = 7.27\ \mu\text{A}$$

$$\text{Collector current, } I_C = \beta I_B = 130 \times 7.27 \times 10^{-6} = 0.946 \text{ mA}$$

$$I_{CQ} = I_C = 0.946 \text{ mA } \textbf{Ans.}$$

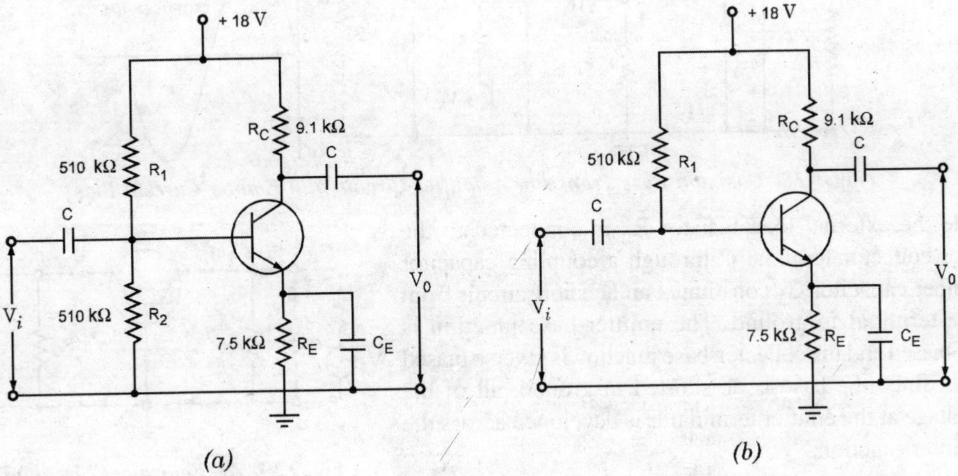

Fig. 9.37

$$V_{CEQ} = V_{CC} - I_{CQ}R_C - I_ER_E = 18 - 0.946 \times 10^{-3} \times 9.1 \times 10^3 - (\beta+1) \times I_B \times R_E$$

$$= 18 - 0.946 \times 9.1 - 131 \times 7.27 \times 10^{-6} \times 7.5 \times 10^3 = 2.246 \text{ V } \textbf{Ans.}$$

(*b*) When R_2 is open circuited, the circuit becomes as shown in Fig. 9.37 (*b*).

$$\text{Base current, } I_B = \frac{V_{CC}-V_{BE}}{R_B+(\beta+1)R_E} = \frac{18}{510\times 10^3+(130+1)\times 7.5\times 10^3} = 12.06\ \mu\text{A}$$

Collector Current, $I_C = \beta I_B = 130 \times 12.06 \times 10^{-6} = 1.568$ mA

$$V_{CE} = V_{CC} - I_C R_C - I_E R_E$$
$$= V_{CC} - I_C R_C - (\beta + 1)\, I_B R_E$$
$$= 18 - 1.568 \times 10^{-3} \times 9.1 \times 10^3 - 131 \times 12.06 \times 10^{-6} \times 7.5 \times 10^3 = -8.116 \text{ V}$$

But V_{CEQ} can't be less than 0 V. It means that the transistor is fully saturated and saturation current is given by

$$I_{C(sat)} = \frac{V_{CC}}{R_C + R_E} = \frac{18}{(9.1 + 7.5) \times 10^3} = 1.084 \text{ mA } \textbf{Ans.}$$

(*c*) Input resistance, $R_{in} = R_1 \parallel R_2 \parallel h_{ie} = 510 \parallel 510 \parallel 1 = 0.996$ kΩ **Ans.**

$$\text{Voltage gain, } A_v = \frac{-h_{fe}}{h_{ie}} \times R_C = \frac{-50}{1 \times 10^3} \times 9.1 \times 10^3 = -455 \textbf{ Ans.}$$

9.14. ANALYSIS OF A CB TRANSISTOR AMPLIFIER USING h PARAMETERS

The circuit arrangement of a practical common base amplifier using an emitter current bias is given in Fig. 9.38. The ac signal source is connected to the transistor emitter terminal E through a coupling capacitor

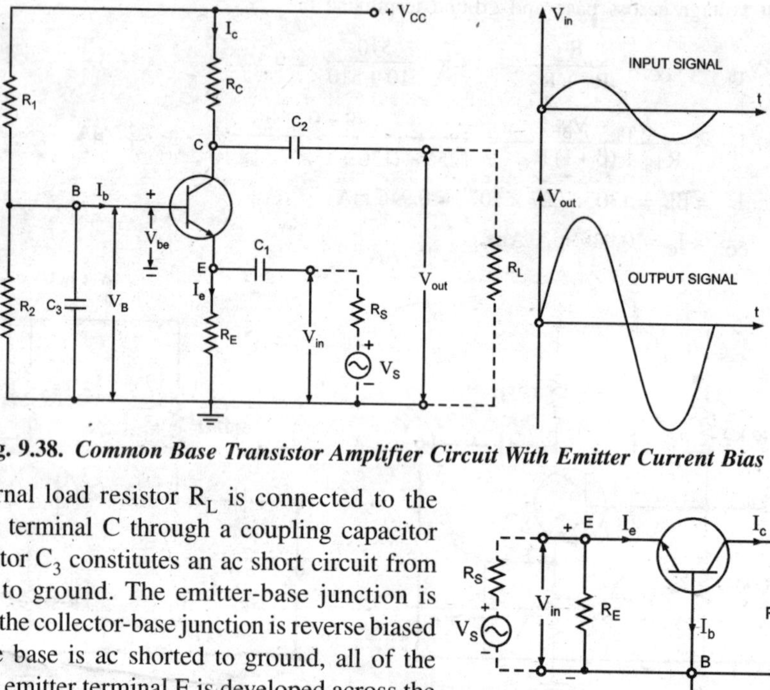

Fig. 9.38. ***Common Base Transistor Amplifier Circuit With Emitter Current Bias***

C_1 while the external load resistor R_L is connected to the transistor collector terminal C through a coupling capacitor C_2. Another capacitor C_3 constitutes an ac short circuit from the base terminal to ground. The emitter-base junction is forward biased and the collector-base junction is reverse biased as usual. Since the base is ac shorted to ground, all of the input voltage at the emitter terminal E is developed across the base-emitter junction.

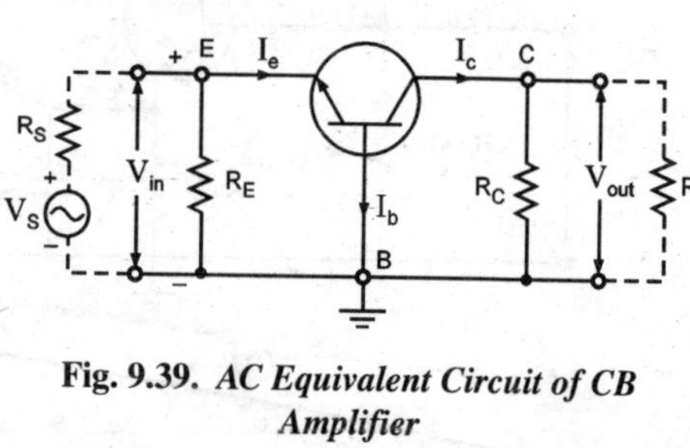

Fig. 9.39. ***AC Equivalent Circuit of CB Amplifier***

Now consider that the signal is applied. For positive half of the applied signal, the base-emitter voltage V_{be} is reduced, because V_B remains constant w.r.t. ground. As a result the base current I_b is reduced. Consequently both collector current I_c and emitter current I_e are reduced, as they are approximately β times of the base current I_b. Now the voltage drop across collector resistance R_C *i.e.,* $I_c R_C$ is reduced. Due to decrease in voltage drop across collector resistance (*i.e.,* $I_c R_C$), V_C is increased because $V_C = V_{CC} - I_c R_C$. Thus output voltage V_{out} is increased. Thus a positive-going input produces a positive going output and hence the input and output are in phase or there is *no phase reversal between input and output.*

Circuit Analysis. The ac equivalent circuit is drawn, as usual, by replacing the dc supply voltage and capacitors by short circuits. From the common base ac equivalent circuit given in Fig. 9.39 it is seen that V_{in} acts across the transistor emitter and base terminals, and output voltage V_{out} is available from the collector and base terminals. Because the base is common to both input and output, the circuit is termed *common base* or sometimes *grounded base circuit.*

The h parameter equivalent of the practical common base circuit given in Fig. 9.38 is shown in Fig. 9.40. Figure 9.40 showing the h parameter equivalent circuit is the same as Fig. 9.39 except that the transistor has been replaced by its h parameter equivalent circuit. Once again, the current directions and voltage polarities in Fig. 9.40 are indicated for a positive-going input signal. The noteworthy point is that it is assumed, while drawing Fig. 9.40, that the base of the transistor is ac shorted to ground through capacitor C_3 (refer to Fig. 9.38). In the absence of capacitor C_3, the parallel combination of R_1 and R_2 appears in series with the transistor base terminal, and the h parameter circuit becomes that given in Fig. 9.41, where $R_B = R_1 \parallel R_2$. The circuit performance is significantly affected in the absence of capacitor C_3.

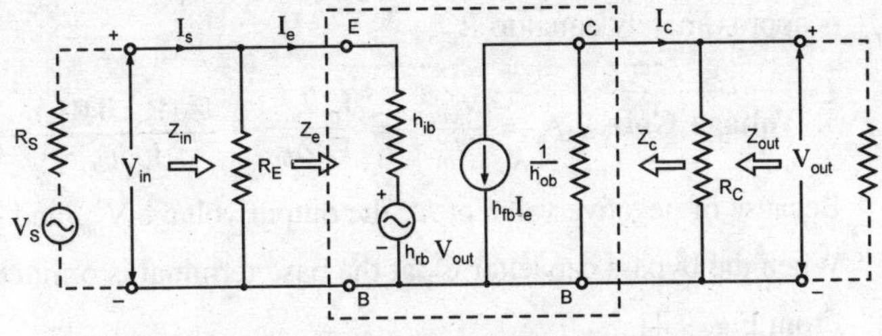

Fig. 9.40. *CB Amplifier h Parameter Equivalent Circuit*

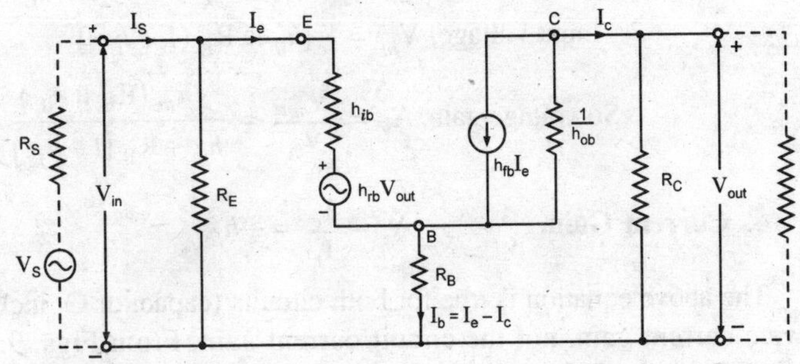

Fig. 9.41. *CB Amplifier h Parameter Equivalent Circuit With Bypass Capacitor At Transistor Base Terminal Omitted*

In the common base circuit, only a fraction of the voltage is fed back to the input *i.e.*, h_{rb} is very small. So $h_{rb} V_{out}$ can be neglected when deriving approximate equations for impedances and gains for a CB circuit.

1. Input Impedance. The input impedance to the transistor emitter is given as

$$Z_e \approx h_{ib} \qquad \text{...(9.59)}$$

When the bypass capacitor C_3 at the base terminal is omitted.

From circuit given in Fig. 9.41

$$V_{in} = h_{ib} I_e + I_b \times R_B = h_{ib} I_e + I_e R_B - I_c R_B \qquad \because I_b = I_e - I_c$$

$$= h_{ib} I_e + I_e R_B + h_{fb} I_e R_B \qquad \because I_c = -h_{fb} I_e$$

$$= I_e [h_{ib} + R_B (1 + h_{fb})] \qquad \text{...(9.60)}$$

and input impedance to the transistor emitter,

$$Z_e = \frac{V_{in}}{I_e} = h_{ib} + R_B (1 + h_{fb}) \qquad \text{...(9.61)}$$

The actual circuit input impedance is parallel combination of Z_e and R_E, so we have

$$Z_{in} = Z_e \parallel R_E \qquad \text{...(9.62)}$$

2. Output Impedance. Since the output voltage variations have little effect upon the input of a CB circuit, only the output of the circuit need be considered in determining the output impedance. Looking back into the collector and base terminals of the transistor in Fig. 9.40, a large resistance $\frac{1}{h_{ob}}$ is seen. Thus the device output impedance

$$Z_c \approx \frac{1}{h_{ob}} \qquad ...(9.63)$$

The actual output impedance is parallel combination of Z_c and R_C so

$$Z_{out} = \frac{1}{h_{ob}} \parallel R_C \qquad ...(9.64)$$

Since $\frac{1}{h_{ob}}$ is of the order of MΩ, and R_C is quite small as compared to $\frac{1}{h_{ob}}$, the circuit output impedance Z_{out} is approximately equal to R_C.

3. Voltage Gain. $A_v = \frac{V_{out}}{V_{in}} = \frac{I_c Z_{ac}}{I_e Z_{in}} \approx \frac{I_c (R_C \parallel R_L)}{I_e h_{ib}} \approx \frac{-h_{fb}(R_C \parallel R_L)}{h_{ib}}$ ...(9.65) $\because I_c = -h_{fb} I_e$

Because of negative value of h_{fb} the output voltage V_{out} and input voltage V_{in} are in phase.

When the bypass capacitor C_3 at the base terminal is omitted.

From Fig. 9.41

Output voltage, $V_{out} = I_c (R_C \parallel R_L) = -h_{fb} I_e (R_C \parallel R_L)$ $\qquad \because I_c = -h_{fb} I_e$

Input voltage, $V_{in} = I_e [h_{ib} + R_B (1 + h_{fb})]$ ...Refer to Eq. (9.60)

So voltage gain, $A_v = \frac{V_{out}}{V_{in}} = \frac{-h_{fb}(R_C \parallel R_L)}{h_{ib} + R_B(1 + h_{fb})}$ ...(9.66)

4. Current Gain. $A_i = \frac{I_c}{I_e} = -h_{fb}$ ...(9.67)

The above equation is true for both circuits (capacitor C_3 included and without C_3). However, this is the device current gain, not the circuit current gain. From Figs. 9.40 and 9.41 it is obvious that the signal current I_s is divided between R_E and Z_e, and the collector current is divided between R_C and R_L. Thus overall current gain,

$$A_i = \frac{I_L}{I_S} = \frac{-h_{fb} R_E R_C}{(R_E + Z_e)(R_C + R_L)} \qquad ...(9.68)$$

5. Power Gain. $A_p = A_v A_i$, as for all other circuits.

Example 9.15. For the common-base circuit given in Fig. 9.38

$R_1 = 20$ kΩ, $R_2 = 5$ kΩ, $R_C = 8$ kΩ, $R_E = 4$ kΩ and $R_L = 6$ kΩ. Determine the input and output impedances and the voltage gain for the circuit if the transistor parameters are as follows :

$h_{ib} = 25$ Ω, $h_{ob} = 0.5 \times 10^{-6}$ S and $h_{fb} = -0.98$.

Solution : Input impedance, $Z_{in} = Z_e \parallel R_E = h_{ib} \parallel R_E = 25 \parallel 4{,}000$ Ω $= 24.85$ Ω **Ans.**

Output impedance, $Z_{out} = \frac{1}{h_{ob}} \parallel R_C = \frac{1}{0.5 \times 10^{-6}} \parallel 8{,}000 = 7.97$ kΩ **Ans.**

Voltage gain, $A_v = \frac{-h_{fb}(R_C \parallel R_L)}{h_{ib}} = \frac{-(-0.98)(8 \parallel 6)}{25 \times 10^{-3}} = 134.4$ **Ans.**

9.15. ANALYSIS OF A CC TRANSISTOR AMPLIFIER USING h PARAMETERS

The circuit arrangement of a practical common collector amplifier using an emitter current bias is given in Fig. 9.42. The ac signal source is connected to the base terminal through coupling capacitor C_1, while the external load

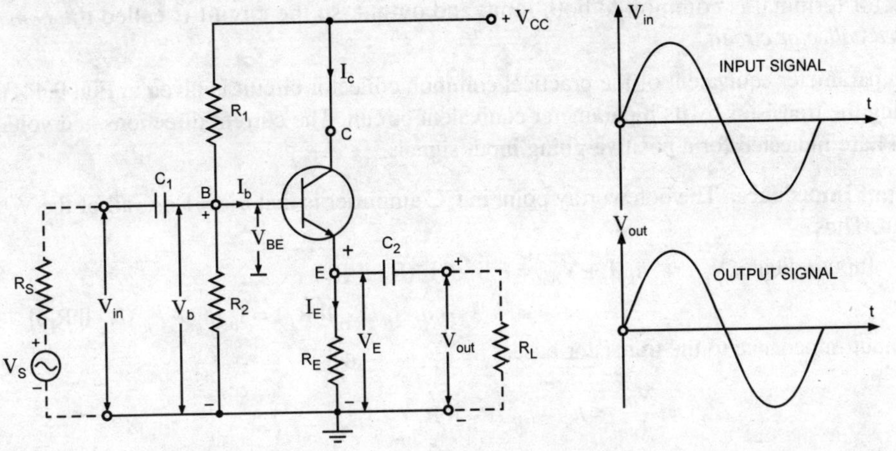

Fig. 9.42. ***Common Collector Transistor Amplifier Circuit With Emitter Current Biasing***

resistor R_L is connected to the emitter terminal of the transistor through coupling capacitor C_2. Base bias voltage is derived from dc supply V_{CC} by means of potential divider R_1 and R_2 (*i.e.*, by emitter current bias). No resistor is connected in series with the transistor collector terminal and no emitter bypass capacitor is used.

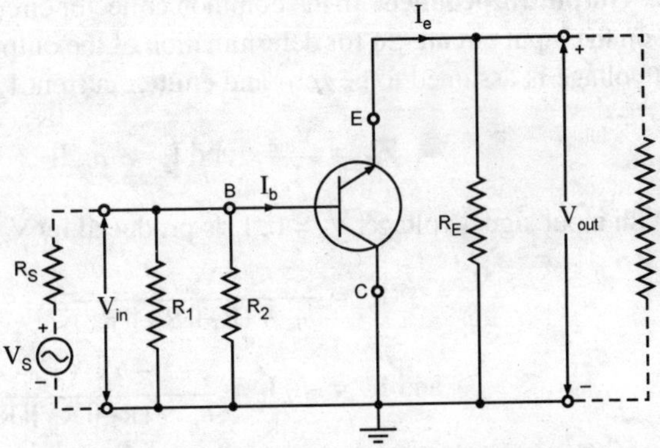

Fig. 9.43. ***Common Collector AC Equivalent Circuit***

When the signal applied to the base terminal of the transistor goes positive, the forward bias V_b is increased causing increase in emitter voltage. This is because emitter voltage V_e is equal to base voltage V_b less base-emitter voltage V_{BE} and V_{BE} remains substantially constant (*i.e.*, $V_e = V_b - V_{BE}$). Since emitter is the output terminal, it is seen that the output voltage from a CC circuit is almost* the same as its input voltage.

Because the output voltage at the emitter terminal follows the input signal applied to base terminal, the common collector circuit is also known as an *emitter follower*.

Circuit Analysis. The common collector ac equivalent circuit is shown in Fig. 9.43, that is obtained by replacing the dc supply voltage and capacitors by short circuits. The input terminals of the ac equivalent circuit are the base and the collector terminals while the output terminals are the emitter and the collector.

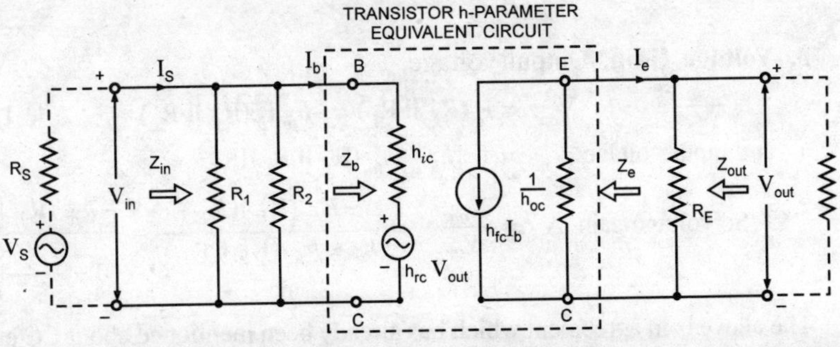

Fig. 9.44. ***CC Amplifier h Parameter Equivalent Circuit***

---

* The output voltage V_{out} can be shown slightly smaller than the input voltage V_{in} because of a very small change in V_{BE}.

The collector terminal is common to both input and output, so the circuit is called the *common collector* or *grounded collector circuit*.

The h parameter equivalent of the practical common collector circuit is given in Fig. 9.44, that is drawn by replacing the transistor by its h parameter equivalent circuit. The current directions and voltage polarities in Fig. 9.44 are indicated for a positive-going input signal.

1. Input Impedance. The noteworthy point in CC amplifier is that $h_{rc} = 1$ *i.e.*, all of the V_{out} is fed back to the input. Thus

$$\text{Input voltage, } V_{in} = h_{ic} I_b + V_{out} = h_{ic} I_b + I_e (R_E \parallel R_L)$$
$$= h_{ic} I_b - h_{fc} I_b (R_E \parallel R_L) = I_b [h_{ic} - h_{fc} (R_E \parallel R_L)] \quad ...(9.69)$$

The input impedance to the transistor base,

$$Z_b = \frac{V_{in}}{I_b} = h_{ic} - h_{fc} (R_E \parallel R_L) \quad ...(9.70)$$

As seen from Fig. 9.44, $Z_{in} = Z_b \parallel R_1 \parallel R_2$ ...(9.71)

2. Output Impedance. In the common collector circuit, any variation in output voltage has a significant effect on the input circuit. So for determination of the output impedance at the transistor emitter (*i.e.*, Z_e), the signal voltage is assumed to be zero and emitter current I_e is determined in terms of output voltage V_{out}

$$Z_e = \frac{V_{out}}{I_e} \text{ and } I_e = -h_{fc} I_b$$

With input signal voltage, $V_s = 0$, I_b is produced by V_{out}

$$I_b = \frac{V_{out}}{h_{ic} + (R_1 \parallel R_2 \parallel R_s)}$$

$$\text{and } I_e = -h_{fc} I_b = \frac{-h_{fc} V_{out}}{h_{ic} + (R_1 \parallel R_2 \parallel R_s)}$$

$$\text{So } Z_e = \frac{V_{out}}{I_e} = -\frac{h_{ic} + (R_1 \parallel R_2 \parallel R_s)}{h_{fc}} \quad ...(9.72)$$

The above equation gives the output impedance of the device only. For the circuit output impedance, R_E is in parallel with Z_e so

$$Z_{out} = Z_e \parallel R_E \quad ...(9.73)$$

Since R_E is usually much larger than Z_e,

$$Z_{out} = Z_e$$

3. Voltage Gain. Output voltage,

$$V_{out} = I_e (R_E \parallel R_L) = -h_{fc} I_b (R_E \parallel R_L) \qquad \because I_e = -h_{fc} I_b$$

$$\text{and input voltage } V_{in} = I_b [h_{ic} - h_{fc} (R_E \parallel R_L)] \quad \text{...refer to Eq. (9.69)}$$

$$\text{So voltage gain } A_v = \frac{V_{out}}{V_{in}} = \frac{-h_{fc} (R_E \parallel R_L)}{h_{ic} - h_{fc} (R_E \parallel R_L)} = \frac{-h_{fc} (R_E \parallel R_L)/h_{ic}}{1 - \dfrac{h_{fc} (R_E \parallel R_L)}{h_{ic}}} \simeq 1 \quad ...(9.74)$$

The above is in agreement which has already been mentioned about CC amplifier *i.e.*, it has a voltage gain of approximately unity and there is no phase difference between output and input voltages.

4. Current Gain. Current gain, $A_i = \dfrac{I_e}{I_b} = -h_{fc}$ ...(9.75) $\because I_e = -h_{fc} I_b$

The above Eq. (9.75) gives current gain of the device. The signal current divides between R_B and Z_b, where $R_B = R_1 \parallel R_2$. Using the same reasoning employed to arrive at the current gain for the CE circuit [Eq. (9.56)], the common collector circuit.

$$\text{Current gain, } A_i = \frac{-h_{fc} R_B R_E}{(R_B + Z_b)(R_L + R_E)} \qquad ...(9.76)$$

For the common collector circuit, R_B (*i.e.*, $R_1 \parallel R_2$) is usually very much smaller than Z_b, so that R_B has a significant lesser effect upon the circuit current gain.

5. Power Gain. $A_p = A_v A_i = A_i$ because $A_v \cong 1$

From the above discussions it is obvious that the CC configuration provides current gain and power gain but no voltage gain. The input impedance is also very high, and the output impedance is quite low. This high input impedance—low output impedance characteristic makes the amplifier suitable for supplying a low impedance load, which is to be supplied with a signal from a high impedance source.

Example 9.16. For an emitter follower, derive expressions for A_i, A_V, R_i and R_0. Compute these for $R_E = 5$ kΩ, $R_S = 1$ kΩ. Assume transistor parameters as $h_{ie} = 1$ kΩ, $h_{fe} = 100$, $h_{re} = 2 \times 10^{-4}$ and $h_{oe} = 20$ μA/V.

[*G.G.S.I.P. Univ. Analog Electronics December 2009*]

Solution :

$$h_{ic} = h_{ie} = 1 \text{ k}\Omega$$

$$h_{fc} = -(1 + h_{fe}) = -(1 + 100) = -101$$

$$Z_{in} = Z_b = h_{ic} - h_{fc}(R_E \parallel R_L) = 1 - (-101)(5) = 506 \text{ k}\Omega \quad \textbf{Ans.}$$

$$Z_e = \frac{h_{ic} + R_S}{-h_{fe}} = \frac{1+1}{101} = 0.0198 \text{ k}\Omega \text{ or } 19.8\ \Omega$$

$$\text{Output impedance, } Z_{out} = Z_e \parallel R_E = 19.8 \parallel 5{,}000 = 19.7\ \Omega \quad \textbf{Ans.}$$

$$\text{Voltage gain, } A_V = 1 \quad \textbf{Ans.}$$

$$\text{Current gain, } A_i = -h_{fc} = 101 \quad \textbf{Ans.}$$

Example 9.17. In a CC amplifier circuit shown in Fig. 9.42, $R_1 = 10$ kΩ, $R_2 = 10$ kΩ, $R_E = 5$ kΩ, $R_L = 5$ kΩ, $R_S = 1$ kΩ.

Determine the circuit input and output impedances and voltage, current and power gains if the transistor parameters are $h_{ie} = 2$ kΩ and $h_{fe} = 100$.

Solution :

$$h_{ic} = h_{ie} = 2 \text{ k}\Omega$$

$$h_{fc} = -(1 + h_{fe}) = -(1 + 100) = -101$$

$$Z_b = h_{ic} - h_{fc}(R_E \parallel R_L) = 2 - (-101)(5 \parallel 5) = 254.5 \text{ k}\Omega$$

$$\text{Input impedance, } Z_{in} = Z_b \parallel R_1 \parallel R_2 = 254.5 \parallel 10 \parallel 10 = 4.9 \text{ k}\Omega \textbf{ Ans.}$$

$$Z_e = \frac{h_{ic} + (R_1 \parallel R_2 \parallel R_s)}{-h_{fc}} = \frac{2 + (1 \parallel 10 \parallel 10)}{-(-101)} = 0.028 \text{ k}\Omega \text{ or } 28\ \Omega$$

$$\text{Output impedance, } Z_{out} = Z_e \parallel R_E = 28 \parallel 5{,}000 \simeq 28\ \Omega \textbf{ Ans.}$$

$$\text{Voltage gain } A_v \simeq 1 \textbf{ Ans.}$$

$$R_B = R_1 \parallel R_2 = 10 \parallel 10 = 5 \text{ k}\Omega$$

$$\text{Current gain, } A_i = \frac{-h_{fc} R_B R_E}{(R_B + Z_B)(R_L + R_E)} = \frac{-(-101) \times 5 \times 5}{(5 + 254.5)(5+5)} = 0.973$$

$$\text{Power gain, } A_p = A_i = 0.973 \textbf{ Ans.}$$

Example 9.18. For the emitter follower with $R_S = 0.5$ kΩ and $R_L = 5$ kΩ calculate A_i, R_i, A_V. Assume $h_{fe} = 50$, $h_{ie} = 1$ kΩ, $h_{oe} = 25$ μA/V.

Solution :

$$h_{ic} = h_{ie} = 1 \text{ k}\Omega$$

$h_{rc} \simeq 1$

$h_{fc} = -(1 + h_{fe}) = -(1 + 50) = -51$

$h_{oc} = h_{oe} = 25\ \mu A/V$

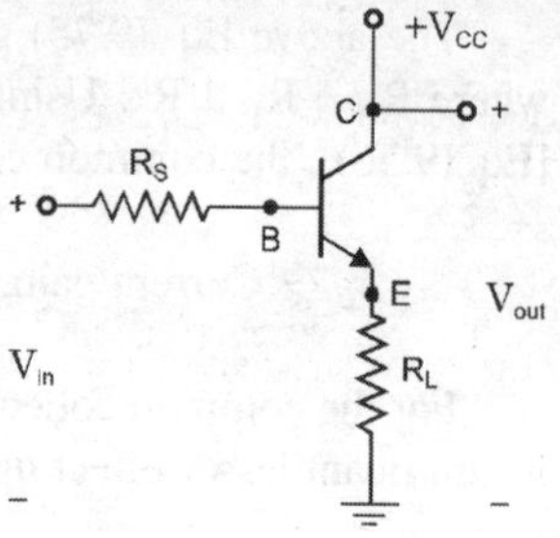

Fig. 9.45

$$\text{Current gain, } A_i = \frac{-h_{fc}}{1 + h_{oc} R_L}$$

$$= \frac{-(-51)}{1 + 25 \times 10^{-6} \times 5 \times 10^3}$$

$= 45.33$ **Ans.** ...Refer to Eq. (9.32)

$$\text{Input resistance, } R_{in} = h_{ic} - \frac{h_{rc} h_{fc}}{h_{oc} + \frac{1}{R_L}} = 1 \times 10^3 - \frac{1 \times (-51)}{25 \times 10^{-6} + \frac{1}{5 \times 10^3}}$$ ...Refer to Eq. (9.30)

$= 227.667\ k\Omega$ **Ans.**

$$\text{Voltage gain, } A_V = \frac{A_i R_L}{R_{in}} = \frac{45.33 \times 5}{227.667} = 0.9955 \text{ Ans.}$$

Example 9.19. For the circuit shown in Fig. 9.46, find A_V, A_i, Z_i and Z_o

$V_{CC} = 15$ V; $R_B = 100\ k\Omega$; $R_E = 1\ k\Omega$; $h_{ie} = 1{,}100\ \Omega$; $h_{fe} = 50$

[U.P. Tech. Univ. Basic Electronics Second Semester 2001-02]

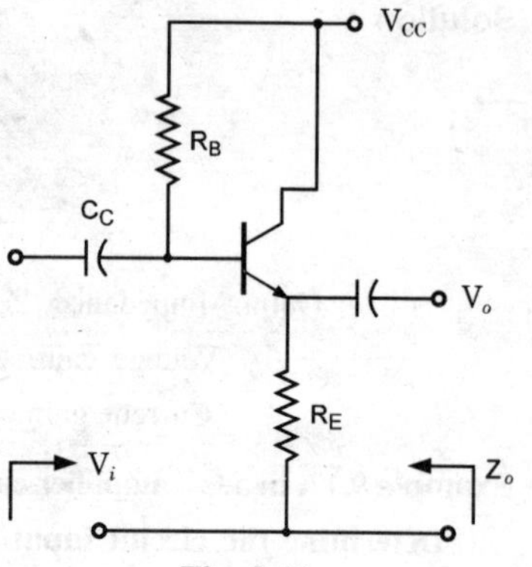

Fig. 9.46

Solution : Since in the given circuit, we get output at the emitter, it is a common collector amplifier circuit and we have

$h_{ic} = h_{ie} = 1{,}100\ \Omega$

$h_{fc} = -(1 + h_{fe}) = -(1 + 50) = -51$

Input impedance, $Z_{in} = [h_{ic} - h_{fc} R_E] \parallel R_B$

$= [1{,}100 - (-51) \times 1{,}000] \parallel 100{,}000$

$= 34.254\ k\Omega$ **Ans.**

$$\text{Output impedance, } Z_{out} = R_E \parallel \left(-\frac{h_{ic}}{h_{fe}}\right) = 1{,}000 \parallel \left(\frac{-1{,}100}{-51}\right)$$

$$= 1{,}000 \parallel \frac{1{,}100}{51} = 21\ \Omega \text{ Ans.}$$

$$\text{Voltage gain, } A_V = \frac{-h_{fc} R_E}{h_{ic} - h_{fc} R_E} = -\frac{-51 \times 1{,}000}{1{,}100 - (-51) \times 1{,}000} = 0.9789 \text{ Ans.}$$

$$\text{Current gain, } A_i = A_V \times \frac{Z_{in}}{R_E} = 0.9789 \times \frac{34.254\ k\Omega}{1\ k\Omega} = 33.53 \text{ Ans.}$$

Example 9.20. Input is given to Fig. 9.47 through the base and output is taken from the emitter. Draw the small signal equivalent circuit, using common emitter hybrid parameters. Hence compute the small signal voltage gain and input impedance. Given that $h_{ie} = 1\ k\Omega$, $h_{fe} = 100$, $h_{re} = 0$ and $h_{oe} = 0$. What is the configuration of the amplifier ?

[U.P. Technical Univ. Basic Electronics First Semester 2003-04]

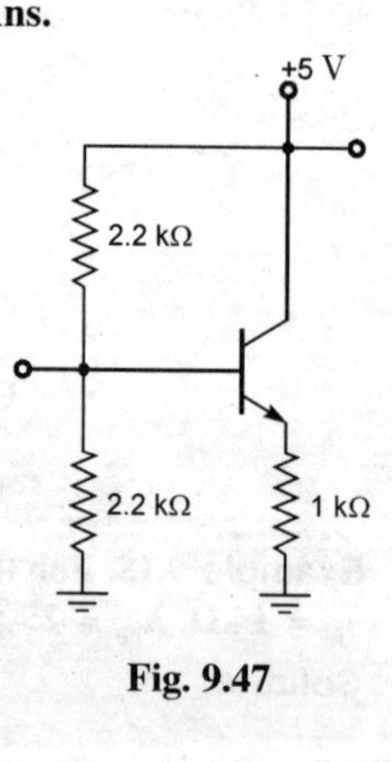

Fig. 9.47

Solution : Since the output is taken from the emitter, it is a common collector amplifier. **Ans.**

$h_{ic} = h_{ie} = 1\ k\Omega$

$h_{fc} = -(1 + h_{fe}) = -(1 + 100) = -101$

$h_{rc} = 1 - h_{re} = 1 - 0 = 1$

$h_{oc} = h_{oe} = 0$

Small signal equivalent circuit using common hybrid parameters is shown in Fig. 9.48.

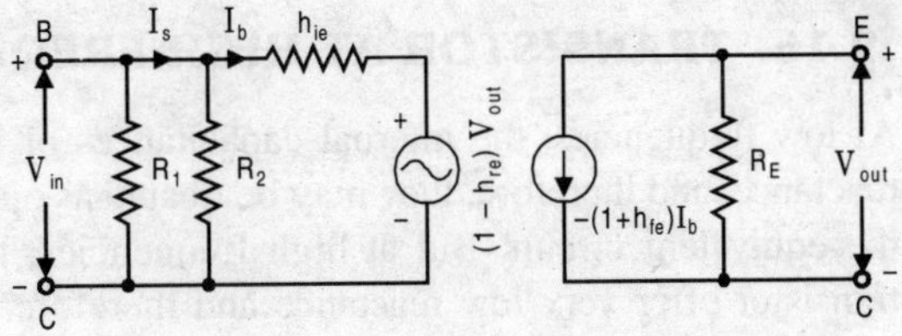

Fig. 9.48. ***Small Signal AC Equivalent Circuit of Emitter Follower Given in Fig. 9.47***

Input impedance, $Z_{in} = [h_{ic} - h_{fc}\ R_E] \parallel R_1 \parallel R_2$

$= [1 - (-101) \times 1] \parallel 2.2 \parallel 2.2$

$= 102 \parallel 1.1\ k\Omega = 1.0883\ k\Omega$ **Ans.**

Voltage gain, $A_V = \dfrac{-h_{fc}\ R_E}{h_{ic} - h_{fc}\ R_E} = \dfrac{-(-101)\times 1}{1-(-101)\times 1} = \dfrac{101}{102} = 0.99$ **Ans.**

Example 9.21. An emitter follower using transistor with $\beta = 150$ is biased at $I_C = 0.25$ mA. The voltage signal source has $R_s = 3\ k\Omega$.

(*a*) In order to make overall $R_{out} = 110\ \Omega$, determine, R_E.

(*b*) For this value of R_E obtain A_V and input resistance.

[Pb. Technical Univ. Electronic Devices and Circuits May 2005]

Solution :

$h_{fe} = \beta = 150$

$R_s = 3\ k\Omega$

$I_C = 0.25$ mA

$I_E \simeq I_C \simeq 0.25$ mA

$r'_e = \dfrac{25\ mV}{0.25\ mA} = 100\ \Omega$

$h_{ie} = \beta r'_e = 150 \times 100 = 15\ k\Omega$

$h_{ic} = h_{ie} = 15\ k\Omega$

Output impedance at the transistor emitter,

$Z_e = -\dfrac{h_{ic} + R_s}{h_{fc}}$ Refer to Eq. (9.72)

$= -\dfrac{15\times 10^3 + 3\times 10^3}{-(1+150)} = 119.2\ \Omega$ $\because h_{fc} = -(1+h_{fe})$

Output impedance of the circuit, $Z_{out} = R_E \parallel Z_e$

or $110 = \dfrac{R_E Z_e}{R_E + Z_e}$

or $R_E = \dfrac{110 Z_e}{Z_e - 110} = \dfrac{110\times 119.2}{119.2 - 110} = 1.425\ k\Omega$ **Ans.**

Input impedance to the transistor base,

$Z_b = h_{ie} - h_{fc} R_E = 15 - [-(1 + 150)] \times 1.425 = 230\ k\Omega$ Refer to Eq. (9.70)

Input impedance to the circuit, $Z_{in} = Z_b = 230\ k\Omega$ **Ans.**

Voltage gain $A_V = \dfrac{\dfrac{-h_{fc}R_E}{h_{ie}}}{1 - \dfrac{h_{fc}R_E}{h_{ic}}} = \dfrac{-h_{fc}R_E}{h_{ic} - h_{fc}R_E} = \dfrac{151\times 1.425}{15 + 151\times 1.425} = 0.9325$ **Ans.**

9.16. TRANSISTOR AT HIGH FREQUENCY

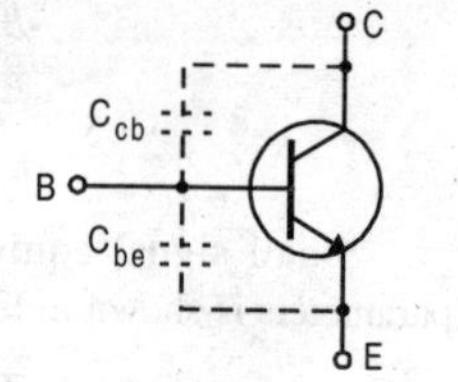

Fig. 9.49. *Internal Capacitnace of Transistor*

At low frequencies, the internal capacitances of the transistor offer a very high reactance, and therefore, they may be treated as open circuit and can be ignored in the equivalent circuit. But at high frequencies, the internal capacitance of the transistor offer very low reactance and therefore, they are to be included in the equivalent circuit.

It has been assumed that at low frequency, the response of the transistor is instantaneous to the variations of the input voltage or current, but actually it is not true. This is because of mechanism of the transport of charge carriers from emitter to collector being of diffusion. So, for knowing the behaviour of transistor at high frequency, diffusion mechanism needs detailed examination. Such an analysis is complicated, and needs a model that is based on the lossy transmission line equations.

Also the diffusion process result in time delay in response. At low frequencies, this time delay is negligibly small as compared to the periodic time T (or $1/f$) and may, therefore, be ignored. But at high frequencies, this time delay becomes significant part of the period time and therefore cannot be ignored in the analysis of the amplifier. Thus, high-frequency model for transistor can be arrived at by taking diffusion and consequent time delay in consideration. Such a model is known as high-frequency hybrid π-model.

9.17. HIGH FREQUENCY T MODEL

The diffusion phenomenon can be taken into account by modifying the common-base T model shown in Fig. 9.22. The high-frequency transistor T model is shown in Fig. 9.50, which is obtained by modifying the basic CB model as follows.

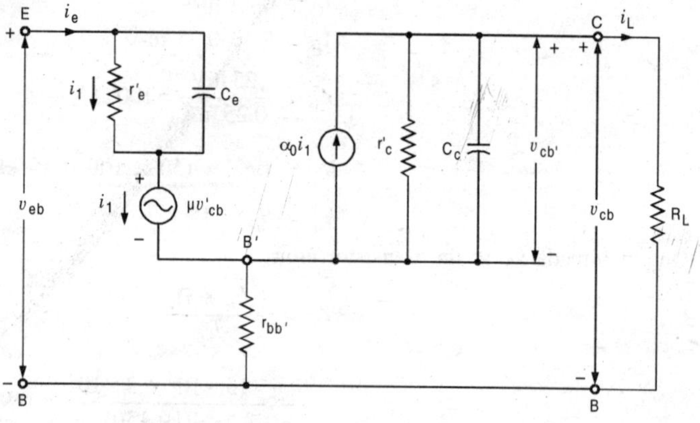

Fig. 9.50. *High-Frequency Transistor T Model*

The collector resistor r'_c is shunted by a capacitor C_c and the emitter resistance r'_e is shunted by a capacitor C_e. The dependent current source is made proportional to current i_1 in emitter resistance r'_e and not to the emitter current i_e. The low-frequency α is denoted by α_0. The capacitance C_e represents the sum of diffusion capacitance C_{De} and the transition capacitance C_{Te} across the emitter junction. Since $C_{De} > C_{Te}$, except for very small values of emitter current, C_e is roughly equal to diffusion capacitance C_{De} which is proportional to the quiescent emitter current. The collector junction is reverse biased, therefore, the collector diffusion capacitance C_{DC} is negligible, so that C_c is equal to collector transition capacitance C_{Te}. Usually, C_e is at least 30 times as large as C_c.

9.18. HYBRID-pi (π) CE TRANSISTOR MODEL

As we know that the CE configuration is the most significant configuration, so CE model is being considered most suitable for high frequencies. Hybrid-π or Giacoletto CE transistor model is given in Fig. 9.51. This circuit is quite simple and analysis of circuits using the hybrid π model is not very difficult and provides accurate results at all the frequencies for which the transistor provides adequate amplification. Secondly, the resistive components in this circuit may be had from the low-frequency h parameters. All the components (*i.e.*, resistors and capacitors) are assumed to be independent of frequency. The values of these components may vary with the quiescent operating point, but under given bias conditions, they are reasonably constant for small-signal variations.

Components in The Hybrid π Model. In the hybrid π model given in Fig. 9.51 C, B, E and B′ are the four terminals, out of which the internal node B′ is not physically accessible. It is known as the "virtual base" terminal.

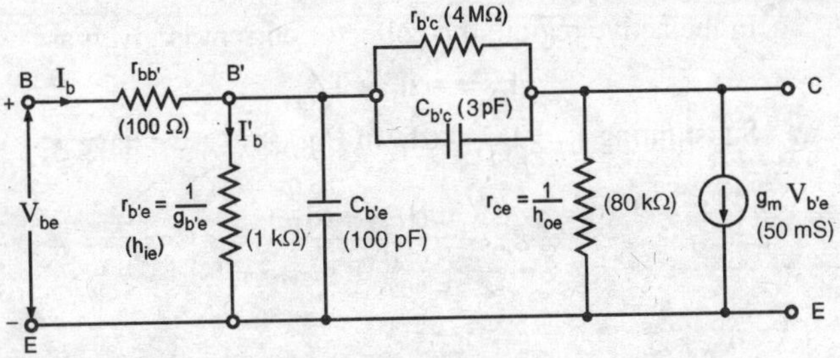

Fig. 9.51. *Giacoletto or Hybrid-π Model of a Transistor in CE Configuration*

The various parameter are explained below:

1. Base Spreading Resistance ($r_{bb'}$). This is represented as a lumped parameter between the external base terminal B and B′. This resistance $r_{bb'}$ includes the base contact, base bulk and base spreading resistances.

2. Resistance $r_{b'e}$. The resistance $r_{b'e}$ is the resistance that can be assumed to exist between the virtual base and emitter terminals. This resistance may be assumed to be existing in series with the collector junction.

3. Resistance $r_{b'c}$. Because of Early effect, the variations in the voltage across the collector-to emitter junction gives rise to base-width modulation. The change in effective base width will make the emitter and collector currents to change. This feedback effect between the input and output is taken into account by connecting resistance $r_{b'c}$ (or conductance $g_{b'c}$) between terminals B′ and C.

4. Resistance r_{ce}. This is the resistance existing between terminals C and E.

5. Diffusion Capacitance $C_{b'e}$ or C_e. This is the diffusion capacitance offered by the forward biased B-E junction and represents the excess-minority carrier storage in the base.

6. Transition Capacitance $C_{b'c}$ or C_c. This is the transition capacitance offered by the reverse biased collector-base junction and is known as the *collector-junction barrier capacitance.*

7. Current Source $g_m V_{b'e}$. The small variations in the voltage $V_{b'e}$ across the emitter-base junction will cause variations in the excess minority-carriers concentration injected into the base since the latter is proportional to the former. So the corresponding small signal collector current will vary in proportion to the voltage $V_{b'e}$. This effect is taken into consideration for the current generator $g_m V_{b'e}$ in the circuit.

For the typical h parameters given in Table 9.3, at I_C = 1.3 mA and room temperature, component values for a germanium transistor are

$$g_m = 50 \text{ mA/V} \; ; r_{b'e} = 1 \text{ k}\Omega; \; r_{bb'} = 100 \; \Omega, \; r_{b'c} = 4 \text{ M}\Omega; \; r_{ce} = 80 \text{ k}\Omega; \; C_c = 3 \text{ pF}, \; C_e = 100 \text{ pF}.$$

9.19. HYBRID-pi CONDUCTANCES IN TERMS OF LOW-FREQUENCY h PARAMETERS

All the resistive components in the hybrid π model can be had from the h parameters in the CE configuration. These h parameters are discussed briefly below :

1. Transistor Conductance g_m. P-N-P transistor connected in the CE configuration, with the collector shorted to emitter for time-varying signals, is shown in Fig. 9.52.

The transconductance g_m is given by

$$g_m = \left. \frac{\partial I_C}{\partial V_{B'E}} \right|_{V_{CE} = \text{constant}} \qquad ...(9.77)$$

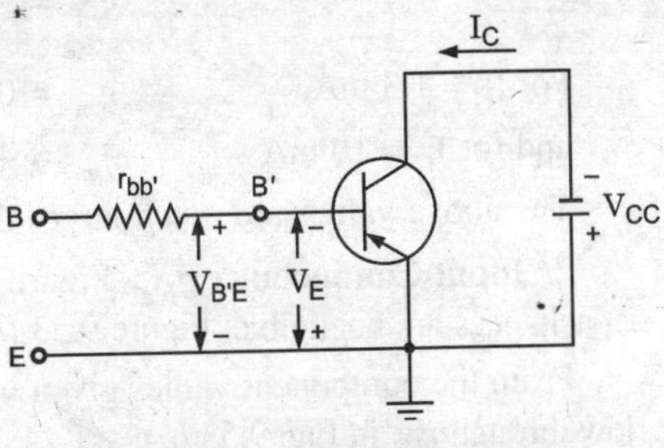

Fig. 9.52. *P-N-P Transistor in CE Configuration For The Derivation of Transconductance (g_m)*

In the active region, the collector current is given as

$$I_C = -\alpha I_E + I_{CO} \quad ...(9.78)$$

Substituting $I_C = I_{CO} - \alpha I_E$ in Eq. (9.77), we have

$$g_m = \frac{\partial(I_{CO} - \alpha I_E)}{\partial V_{B'E}} = -\alpha \frac{\partial I_E}{\partial V_{B'E}} = \alpha \frac{\partial I_E}{\partial V_E} \quad ...(9.79)$$

$\because$ For a PNP transistor $V_{B'E} = -V_E$

In the above expression α_N has been assumed independent of V_E

But the emitter diode resistance is defined as

$$r'_e = \frac{\partial V_E}{\partial I_E} \quad ...(9.80)$$

Substituting $\frac{\partial I_E}{\partial V_E} = \frac{1}{r'_e}$ from above equation in Eq. (9.79), we have

$$g_m = \frac{\alpha}{r'_e} = \frac{\alpha I_E}{V_T} = \frac{I_{CO} - I_C}{V_T} \quad ...(9.81) \qquad \because r'_e = \frac{V_T}{I_E} \text{ and } \alpha I_E = I_{CO} - I_C$$

For a P-N-P transistor I_C is negative and for an N-P-N transistor I_C is positive. But for the foregoing analysis with $V_E = +V_{B'E}$ we have

$$g_m = \frac{I_C - I_{CO}}{V_T}$$

Hence for both the types of transistors g_m will be positive.

Since $|I_C| >> |I_{CO}|$, g_m is given by

$$g_m \simeq \frac{|I_C|}{V_T} \quad ...(9.82)$$

We know that $V_T = \frac{T}{11,600}$

So $g_m = \frac{11,600|I_C|}{T}$

At room temperature of 27°C (300 K), the value of V_T is 26 mV, so the value of g_m is given by

$$g_m = \frac{|I_C| \text{ in mA}}{26} \quad ...(9.83)$$

For $I_C = 1.3$ mA, $\quad g_m = 0.05$ S (mho) = 50 mA/V

and for $I_C = 10$ mA, $\quad g_m = 400$ mS or 400 mA/V

The above values are much larger than the transconductance obtained with tubes and FETs.

2. Input Conductance $g_{b'e}$. Figure 9.53 (*a*) depicts the hybrid-π model valid at low frequencies, where all capacitances are negligible. Figure 9.53 (*b*) depicts the same transistor, using the *h* parameter equivalent circuit.

From the component values given in Art. 9.18, we note that $r_{b'c} >> r_{b'e}$. Hence all the base current I_{be} will flow through $r_{b'e}$ in Fig. 9.53 (*a*).

$$\text{Thus } V_{b'e} \approx I_b r_{b'e} \quad ...(9.84)$$

and the short-circuit collector current I_c is given by

$$I_c = g_m V_{b'e} \approx g_m I_b r_{b'e}$$

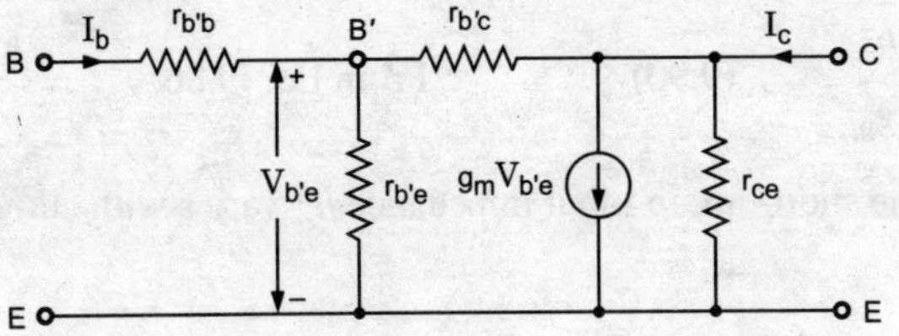

(a) The Hybrid π Model at Low Frequencies

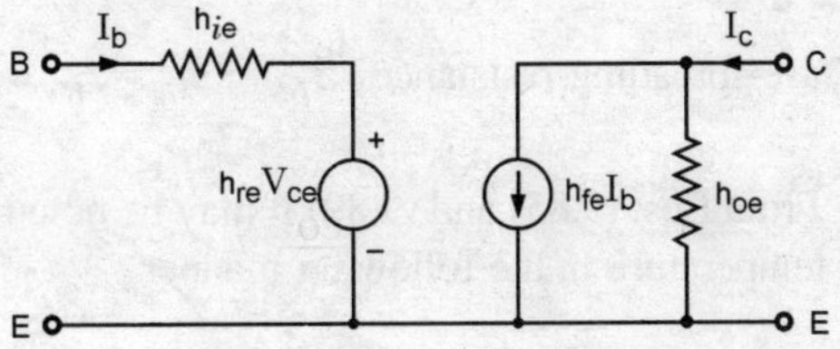

(b) The h Parameter Model at Low Frequencies

Fig. 9.53

The short-circuit gain h_{fe} is defined as

$$h_{fe} = \left.\frac{I_c}{I_b}\right|_{V_{CE} = \text{Constant}} = g_m r_{b'e}$$

$$\text{or } r_{b'e} = \frac{h_{fe}}{g_m} = \frac{h_{fe} V_T}{|I_c|}$$

$$\text{or } g_{b'e} = \frac{g_m}{h_{fe}} \quad \ldots(9.85)$$

The above equation is required expression for the input conductance.

Thus, over the range of currents for which h_{fe} remains fairly constant, $r_{b'e}$ is directly proportional to temperature and inversely proportional to current.

Since $g_m = \frac{\alpha_0}{r'_e}$, and $h_{fe} \simeq \frac{\alpha_0}{1-\alpha_0}$, $r_{b'e}$ may be expressed in terms of the T-model emitter resistance r'_e as

$$r_{b'e} = \frac{h_{fe}}{g_m} = \frac{r'_e}{1-\alpha_0} \quad \ldots(9.86)$$

3. Feedback Conductance $g_{b'c}$. With the input open circuited, h_{re} is defined as the reverse voltage ratio, or from Fig. 9.53 (*a*) with $I_b = 0$,

$$h_{re} = \left.\frac{V_{b'e}}{V_{ce}}\right|_{\text{For } I_b = 0} = \frac{r_{b'e}}{r_{b'e} + r_{b'c}} \quad \ldots(9.87)$$

$$\text{or } r_{b'e}(1 - h_{re}) = h_{re}\, r_{b'c}$$

Since $h_{re} << 1$, then to a good approximation

$$r_{b'e} = h_{re}\, r_{b'c} \quad \text{or} \quad g_{b'c} = \frac{h_{re}}{r_{b'e}} \quad \ldots(9.88)$$

Since $h_{re} \simeq 1 \times 10^{-4}$, Eq. (9.88) verifies that $r_{b'c} >> r_{b'e}$

It is found that h_{re} is quite insensitive to current and temperature. Therefore, $r_{b'c}$ has the same dependence on $|I_C|$ and T as does $r_{b'e}$.

4. Base-Spreading Resistance $r_{bb'}$. Parameter h_{ie} forms the input resistance of the transistor with the output shorted. With reference to the circuit given in Fig. 9.53 (*a*), with output shorted, $r_{b'c}$ is in parallel with $r_{b'e}$. Hence

$$h_{ie} = r_{bb'} + r_{b'e} \parallel r_{b'c} = r_{bb'} + r_{b'e} \quad \ldots(9.89)$$

$\because r_{b'c} >> r_{b'e}$ and, therefore, $r_{b'e} \parallel r_{b'c} = r_{b'e}$

or Base-spreading resistance $r_{bb'} = h_{ie} - r_{b'e} = h_{ie} - \frac{h_{fe}}{g_m}$ ...(9.90) $\because$ From Eq. (9.86) $r_{b'e} = \frac{h_{fe}}{g_m}$

From Eqs. (9.85) and (9.89) it may be noted that the short-circuit input impedance h_{ie} varies with current and temperature in the following manner :

$$h_{ie} = r_{bb'} + \frac{h_{fe} V_T}{|I_C|} \quad ...(9.91)$$

5. Output Conductance g_{ce}. Parameter h_{oe} constitutes the output conductance with open-circuited input

$$i.e., \; h_{oe} = \left.\frac{I_C}{V_{ce}}\right|_{\text{For } I_b = 0}$$

From the circuit shown in Fig. 9.53(*a*) with $I_b = 0$, we have

$$I_c = \frac{V_{ce}}{r_{ce}} + \frac{V_{ce}}{r_{b'c} + r_{b'e}} + g_m V_{b'e} \quad ...(9.92)$$

With $I_b = 0$, we have, from Eq. (9.87), $V_{b'e} = h_{re} V_{ce}$ and from Eq. (9.92), we have

$$h_{oe} = \frac{I_c}{V_{ce}} = \frac{1}{r_{ce}} + \frac{1}{r_{b'c}} + g_m h_{re} \quad ...(9.93)$$

Substituting the values of g_m and h_{re} from Eqs. (9.85) and (9.88) in Eq. (9.93) and assuming $r_{b'c} >> r_{b'e}$ we have

$$h_{oe} = g_{ce} + g_{b'c} + g_{b'e} h_{fe} \frac{g_{b'c}}{g_{b'e}} = g_{ce} + (1 + h_{fe}) g_{b'c} \quad ...(9.94)$$

but if $h_{fe} >> 1$, Eq. (9.94) may be written as

$$g_{ce} \simeq h_{oe} - g_m h_{re} \quad ...(9.95)$$

6. Feedback Conductance $g_{b'c}$:

$$g_{b'c} = h_{re} g_{b'e} \quad ...(9.96)$$

If the CE h parameters at low frequencies are known at a given collector current I_c, the conductances or resistances in the hybrid-π model can be calculated from the following equations :

$$g_m = \frac{|I_C|}{V_T} \quad ...(9.97)$$

$$r_{b'e} = \frac{h_{fe}}{g_m} \quad ...(9.98)$$

$$r_{bb'} = h_{ie} - r_{b'e} \quad ...(9.99)$$

$$r_{b'c} = \frac{r_{b'e}}{h_{re}} \quad ...(9.100)$$

$$g_{ce} = h_{oe} - (1 + h_{fe}) g_{b'c} = \frac{1}{r_{ce}} \quad ...(9.101)$$

For the typical h-parameters given in Table 9.3, at $I_C = 1.3$ mA and room temperature, the component values obtained are listed on pages 325.

Hybrid-π Capacitances. The basic high-frequency hybrid π model, discussed in Art. 9.18, includes two capacitances (*i*) $C_{b'c}$ or C_c and (*ii*) C_e.

1. Collector-Junction Capacitance C_c. The collector-junction capacitance C_c or $C_{b'c}$ is the measured CB output capacitance with the input open ($I_E = 0$), and is usually specified by the manufacturers as C_{ob}. Since in active region the collector junction is reverse biased, then C_c is a transition capacitance, and therefore, varies as V_{CE}^{-n}, when $n = \frac{1}{2}$ or $\frac{1}{3}$ for an abrupt or gradual junction respectively.

Since $C_e = C_{b'e}$ represents, mainly the diffusion capacitance across the emitter junction, it is directly proportional to the current and is almost independent of temperature. Experimentally, C_e is determined from a measurement of frequency f_T at which the CE short-circuit current gain falls to unity. C_e is given approximately by the following relation

$$C_e = \frac{g_m}{2\pi f_T} \quad \text{...(9.102)}$$

Reasonable values for $C_{b'c}$ and $C_{b'e}$ are 3 pF and 100 pF respectively.

Example 9.22. A BJT has $h_{ie} = 6$ kΩ and $h_{fe} = 224$ at $I_C = 1$ mA with $f_T = 80$ MHz and $C_{b'c} = 12$ pF. Determine (*i*) g_m (*ii*) $r_{b'e}$ (*iii*) $r_{bb'}$ and (*iv*) $C_{b'e}$ at room temperature and a collector current of 1 mA.

[U.P. Technical Univ. Solid State Devices and Circuits 2006-07]

Solution : (*i*)From Eq. (9.97);

$$g_m = \frac{I_C}{V_T} = \frac{1\text{ mA}}{26\text{ mV}} = \frac{1}{26}\text{S or } 38.46\text{ mS} \quad \textbf{Ans.}$$

(*ii*) From Eq. (9.98); $r_{b'e} = \frac{h_{fe}}{g_m} = \frac{224}{1/26} = 224 \times 26\ \Omega = 5.824\text{ k}\Omega$ **Ans.**

(*iii*) From Eq. (9.99); $r_{b'b} = h_{ie} - r_{b'e} = 6 \times 10^3 - 5.824 \times 10^3 = 0.176\text{ k}\Omega$ or $176\ \Omega$ **Ans.**

(*iv*) $C_{b'e} = \frac{g_m}{2\pi f_T} - C_{b'c} = \frac{38.46 \times 10^{-3}}{2\pi \times 80 \times 10^6} - 12 \times 10^{-12}$ F

$= 76.5 \times 10^{-12} - 12 \times 10^{-12}$ F $= 64.5$ pF **Ans.**

Example 9.23. The following low-frequency parameters are known for a given transistor at $I_C = 10$ mA and $V_{CE} = 10$ V and at room temperature $h_{ie} = 500\ \Omega$; $h_{oe} = 4 \times 10^{-5}$ A/V; $h_{fe} = 100$ and $h_{re} = 10^{-4}$.

At the same operating point $f_T = 50$ MHz and $C_{ob} = 3$ pF. Calculate the values of all the hybrid parameters.

[U.P. Technical Univ. Solid State Devices & Circuits 2004-05]

Solution : From Eq. (9.97); $g_m = \frac{I_C}{V_T} = \frac{10\text{ mA}}{26\text{ mV}} = 384.6$ mS **Ans.**

(*ii*) From Eq. (9.98); $r_{b'e} = \frac{h_{fe}}{g_m} = \frac{100}{384.6 \times 10^{-3}} = 260\ \Omega$ **Ans.**

(*iii*) From Eq. (9.99); $r_{b'b} = h_{ie} - r_{b'e} = 500 - 260 = 240\ \Omega$ **Ans.**

(*iv*) From Eq. (9.100); $r_{b'c} = \frac{r_{b'e}}{h_{re}} = \frac{260}{10^{-4}} = 2.6 \times 10^6\ \Omega$ or 2.6 MΩ **Ans.**

(*v*) From Eq. (9.101); $g_{ce} = h_{oe} - (1 + h_{fe}) g_{b'c} = 4 \times 10^{-5} - (1 + 100) \times \frac{1}{2.6 \times 10^6}$

$= 4 \times 10^{-5} - 3.885 \times 10^{-5} = 0.115 \times 10^{-5}$ S **Ans.**

$$C_{b'c} = C_{bo} = 3 \text{ pF } \textbf{Ans.}$$

$$C_{b'e} = \frac{g_m}{2\pi f_T} - C_{b'c} = \frac{384.6 \times 10^{-3}}{2\pi \times 50 \times 10^6} - 3 \times 10^{-12} = 1221 \times 10^{-12} \text{ F or } 1{,}221 \text{ pF } \textbf{Ans.}$$

Highlights

1. The device that amplifies the amplitude of the input signal is called the amplifier. An amplifier may be defined as a device that increases the current, voltage or power of an input signal with the help of a transistor by furnishing the additional power from a separate source of supply.
2. Amplification in the ac domain cannot be had without applying the dc biasing level.
3. For most applications, the BJT amplifier may be considered linear, permitting the use of the superposition theorem for separating the dc and ac analyses and designs.
4. The hybrid equivalent circuit for a transistor consists of a Thevenin's equivalent circuit at the input and a Norton's equivalent circuit at the output.
5. The equations relating the input and output voltages and currents in a transistor are given by

$$v_1 = h_i i_1 + h_r v_2$$
$$i_2 = h_f i_1 + h_0 v_2$$

where h_i, h_r, h_f and h_0 are the h parameters and have a second subscript of *b*, *e* or *c* according to the configuration under consideration.

6. Typical values of h parameters for CB, CE and CC configurations are given in Table 9.3.
7. If the h parameters are given for one type of amplifier configuration but are required for another type, the approximate conversion formulas in Table 9.4 may be used.
8. Knowing the h parameters for a transistor, the values of current gain A_i and A_{is}, voltage gain A_v and A_{vs}, R_{in} and Y_{out} can be had by using the exact equations given at the end of Art. 9.9 (page 303) by using a second subscript appropriate to the amplifier under consideration.
9. When introducing the ac model for a BJT, all dc sources are set to zero and replaced by a short circuit connection to ground, all capacitors are replaced by a short circuit equivalent and all elements in parallel with an introduced short circuit equivalent should be removed from the circuit.
   Elements that were isolated by capacitors for the dc analysis will appear in the ac analysis due to the short circuit equivalent for the capacitive elements.
10. The amplification factor (β or h_{fe}) is the least sensitive to variations in collector current, whereas the output impedance parameter is the most sensitive. The output impedance is also quite sensitive to variations in V_{CE}, whereas the amplification factor is the least sensitive. However, the output impedance is the least sensitive to variations in temperature, whereas the amplification factor is somewhat sensitive.
11. One of the main advantages of a CE amplifier is the ability to operate from a single supply voltage, using a resistor from V_{CC} to the base to provide bias current.
    A CE amplifier enjoys both a voltage and a current gain, with the result that its power gain is usually larger than in either a CB or CC amplifier.
    The input resistance to a CE amplifier is of the order of 1 kΩ and is thus much larger than that in a CB amplifier, but lower than that in a CC amplifier.
12. The CB amplifier has a very low input impedance but it can have significant voltage gain. The current gain is slightly less than unity, and the output impedance is simply R_C.
    The CB amplifier is used primarily to match a very low source impedance (roughly 20 Ω) to a high load impedance (100 kΩ and more).
13. The CC amplifier provides a current gain and power gain but no voltage gain. The input impedance is also very high, and the output impedance is quite low. The high input impedance-low output impedance characteristic makes the amplifier suitable for supplying a load impedance, which is to be supplied with a signal from a high impedance source. Thus, the CC amplifier finds wide application as a buffer amplifier between a high impedance source and a low impedance load.

14. The hybrid-π model of a transistor at high frequencies includes the capacitive effects of the P-N junctions and involves a base-spreading resistance that creates a virtual base.
15. In the hybrid-π model, there are four terminals namely C, B, E and B' out of which C, B and E have their normal meaning and they are user accessible terminals but B' is not a user terminal. Terminal B' is called the 'virtual base' terminal.
16. The resistances and capacitances involved in hybrid-π model are : (*i*) Base spreading resistance $r_{bb'}$ (*ii*) Input resistance $r_{b'e}$ (*iii*) Output resistance r_{ce} (*iv*) Feedback resistance $r_{b'c}$ (*v*) Diffusion capacitance $C_{b'e}$ or C_e (*vi*) Transition capacitance $C_{b'c}$ or Cc.
If the CE *h* parameters at low frequencies are known at a given collector current I_C, the conductance or resistances in the hybrid-π model can be calculated from Eqs. (9.97), (9.98), (9.99), (9.100) and (9.101)
Diffusion capacitance is determined from the relation:

$$C_e = \frac{g_m}{2\pi f_T}$$

Exercises

1. Draw the circuit diagram of a single stage transistor amplifier, stating the function of each component used in the circuit.
2. What do you understand by coupling capacitor in amplifiers ? Explain how it affects the operation of an amplifier. Draw required diagram also. **[Pb. Technical Univ. Electronic Devices and Circuits May 2008]**
3. Explain the working of a common emitter amplifier with the help of its circuit diagram. Explain how dc load line is drawn for a common emitter amplifier.
4. With the help of a circuit diagram, explain the operation of a simple common emitter amplifier and find the voltage gain expression. **[M.D. Univ. Analog Electronic Circuits May 2007]**
5. Explain how will you determine the voltage gain of the CE amplifier by plotting the dc load line on the output characteristics of the transistor.
6. Define h parameters for a BJT transistor. **[Pb. Technical Univ. Electronic Devices & Circuits 2003-04]**
7. Draw the hybrid model of a BJT. Define (*i*) h_{fe} (*ii*) h_{ie} (*iii*) h_{re} (*iv*) h_{oe}. **[U.P. Technical Univ. Solid State Devices & Circuits 2008-09]**
8. Draw the small signal equivalent circuit of a BJT and explain each component. **[U.P. Technical Univ. Electronics Engg. First Semester 2009-10]**
9. Draw the hybrid model of a low-frequency small-signal voltage amplifier. **[Pb. Technical Univ. Electronic Devices & Circuits Dec. 2006]**
10. Draw the equivalent circuit of a transistor with CE configuration. **[Pb. Technical Univ. Electronic Devices & Circuits Dec. 2006]**
11. Given h parameters of transistor in common-emitter configuration, determine the corresponding h parameters in common-collector configuration. **[Pb. Univ. Applied Electronics-I, May 1994]**
12. Determine the input impedance, current gain, voltage gain, output impedance of a transistor in terms of its parameters. **[M.D. Univ. Rohtak Analog Electronics Dec. 2007]**
13. Explain the h parameter model of a transistor at low frequencies and derive the formulas for (*i*) current gain (*ii*) input impedance (*iii*) voltage gain (*iv*) output impedance. **[Rajasthan Technical Univ. 2007]**
14. Draw h parameter equivalent circuit of a loaded amplifier in common emitter configuration and derive the expressions for current gain, voltage gain, input impedance, output impedance, overall voltage and power gains. **[Pb. Technical Univ. Analog Electronics December 2006]**
15. Using the approximate h parameter model, obtain the expression for a CE circuit for (*i*) A_i, (*ii*) R_i, (*iii*) A_V, (*iv*) R_0. **[U.P. Technical Univ. Electronics Engg. First Semester 2007-08]**
16. Explain the common emitter configuration of a BJT. Draw its equivalent circuit using h parameters. Also discuss its important features. **[Pb. Technical Univ. Electronic Devices & Circuits Dec. 2006]**
17. Define h parameters. Derive equations for current gain and voltage gain of a CE amplifier using h parameters. **[Pb. Technical Univ. Electronic Devices & Circuits May 2008]**

18. Draw h parameter equivalent circuit of a loaded amplifier in common emitter configuration and derive the expressions for current gain, voltage gain, input impedance, output impedance, overall voltage gain and current gain. **[U.P.S.C. I.E.S. Elec. Engineering II, 1997]**

19. Discuss analysis of transistor amplifier using h parameters in CB configuration. **[Pb. Technical Univ. Electronic Devices and Circuits May 2009]**

20. Draw the hybrid equivalent circuit for common base configuration and find the expression of current gain, voltage gain and input impedance. **[U.P. Technical Univ. Electronics Engg. Second Semester 2005-06]**

21. Draw hybrid π-model for a transistor in CE configuration. Define various parameters.

22. What are the physical origins of resistance in hybrid pi model of CE transistor amplifier at high frequencies ? **[Pb. Technical Univ. Analog Electronics December 2007]**

23. Discuss the hybrid pi CE transistor model. **[Pb. Technical Univ. Analog Electronics December 2004]**

24. Explain hybrid pi CE transistor model in detail. **[Pb. Technical Univ. Analog Electronics May 2005; December 2005]**

25. Draw the small signal hybrid π model at high frequency. Draw each component and prove that $h_{fe} = g_m r_{b'e}$. **[Pb. Technical Univ. Electronic Devices and Circuits 2006-07]**

26. Prove that $h_{ie} = r_{bb'} + r_{b'e}$ assuming $r_{bb'} << r_{b'e}$. How does h_{ie} vary with I_C ? **[Pb. Technical Univ. Electrical Devices and Circuits 2005-06]**

Short Answer Type Questions With Answers

Q. 1. What is an amplifier ? **[P.T.U. Analog Electronics May 2007]**

Ans. The device that amplifies the amplitude of the input signal is called the amplifier. An amplifier may be defined as a device that increases the current, voltage or power of an input signal with the help of a transistor by furnishing the additional power from a separate source of supply.

Q. 2. What is meant by small signal amplifier ?

Ans. When the input signal is quite weak and produces small fluctuations in the output current in comparison to its quiescent value, the amplifier is called the small signal or voltage amplifier.

Q. 3. Write the characteristics of an ideal voltage amplifier. **[Pb. Technical Univ. May 2006]**

Ans. Main characteristics of an ideal voltage amplifier is to present a high impedance to the input source so as to minimize the loading effects and to provide a large enough voltage signal to the output device like loudspeaker.

Q. 4. Explain how phase reversal of the signal takes place when it is amplified by a single stage CE voltage amplifier. **[P.T.U. Analog Electronics December 2006]**

Ans. In a CE configuration, the output voltage increases in the negative direction when the input signal voltage increases in the positive direction and vice versa. This is called the phase reversal and causes a phase difference of 180° between the input signal voltage and output voltage.

Q. 5. What is an ac emitter resistance ?

Ans. The dynamic resistance of the emitter-base junction diode is called the ac emitter resistance. It is given as $\frac{25 \text{ mV}}{I_E}$ where I_E is the dc emitter current at Q-point.

Q. 6. What is the effect of source resistance on voltage gain of a common base transistor amplifier ?

Ans. The voltage gain of a CB transistor amplifier will decrease if source resistance is considered because in such a case there will be a voltage drop across the source resistance and output voltage will decrease.

Q. 7. Why is coupling capacitor used to connect a signal source to an amplifier ? **[Pb. Technical Univ. December 2005]**

Ans. An electrolytic capacitor, called the input capacitor, C_{in} of capacity of about 10 μF is used to couple the signal source to the transistor base. In the absence of this capacitor, the signal source resistance will come across resistor R_2 of the biasing network and thus change the bias. This capacitor allows only ac signal to pass but isolates the signal source from R_2.

Q. 8. What is the function of bypass capacitor ? **[P.T.U. Analog Electronics December 2004]**

Ans. The capacitor C_E, called the emitter bypass capacitor, is connected in parallel with the emitter resistance R_E in order to provide a low reactance path to the amplified ac signal. In the absence of this capacitor, amplified ac signal will cause a voltage drop across it which in turn will feedback the input side and reduce the output voltage.

Q. 9. State what will happen to the voltage gain of an amplifier if the bypass capacitor is open circuited. **[P.T.U. Analog Electronics May 2007]**

Ans. Removal of bypass capacitor in a CE amplifier circuit causes excessive degeneration in the amplifier circuit and, therefore, reduction in voltage gain.

Q. 10. Why common collector amplifier is called an emitter follower ? **[P.T.U. Electronic Devices & Circuits May 2009]**

Ans. The CC circuit amplifier is called an emitter follower because in this circuit the output voltage at the emitter terminal follows the input signal applied to the base terminal.

Q. 11. What configuration of bipolar transistor is called as emitter follower ? For what purpose it is used. **[P.T.U. Analog Electronics December 2007]**

Ans. The common collector configuration of bipolar transistor is called as emitter follower because in this circuit the output voltage at the emitter terminal follows the input signal applied to the base terminal. This configuration finds wide application as a buffer amplifier between a high impedance source and a low impedance load.

Q. 12. What are the main purposes for which a common collector amplifier may be used ? **[U.P.S.C. I.E.S. Elec. Engineering II, 1997]**

Ans. For a common collector amplifier, current gain is as high as for CE amplifier, voltage gain is less than unity, input resistance is the highest and the output resistance is the lowest of all the three (CE, CC and CB) configurations. This circuit finds wide application as a buffer amplifier between a high impedance source and a low impedance load.

Q. 13. What are small signal amplifiers ? How are they different from large signal amplifiers ? **[R.G. Technical Univ. June 2009]**

Ans. When the input signal is so weak as to produce small fluctuations in the collector current compared to its quiescent value, the amplifier is known as *small signal amplifier.* On the other hand, when the fluctuations in collector current are large *i.e.,* beyond the linear portion of its characteristics, the amplifier is called the large signal amplifier. In case of a small signal amplifier, the transistor can be represented by a linear circuit model, and a simple analytical method can be used for determination of various performance characteristics of the amplifier. But in case of a large signal power amplifier, the equivalent linear circuit model of the transistor and analytical method cannot be applied and graphical method is preferred because of involvement of certain nonlinear operation in it. Voltage amplifiers and audio-power amplifiers are the small-signal power amplifiers while radio-power amplifiers are the large-signal power amplifiers.

Q. 14. Why is it not possible to use the h parameters at high frequencies and large signals.

Ans. h parameters cannot be used at high frequencies because at high frequencies the internal capacitances offer very low reactance and they are required to be considered in the analysis.

h parameters cannot be used for large signals because the transistor behaves as a linear device for small signals only.

Q. 15. What are hybrid capacitances ?

Ans. Forward biased P-N junction exhibits a capacitive effect, called the diffusion capacitance. This capacitive effect of the normally forward biased base-emitter junction of transistor is represented as $C_{b'e}$ or C_e. Similarly a reverse biased P-N junction has a transition capacitance and it is shown as $C_{b'c}$ or C_c.

Q. 16. What do you understand by the term 'virtual base' ?

Ans. The internal capacitances of transistor C_e and C_c have a common termination at terminal B', not the input base terminal B (Fig. 9.51). Resistance $R_{bb'}$, referred to as the base-spreading resistance, represents the built resistance of the base and is the portion of base through which the base current flows. Resistance $r_{b'e}$ is that portion of base and emitter which may be thought of as being "in series with" with the collector junction. This establishes a *virtual base* B' for the junction capacitances to be connected to instead of B.

Q. 17. How does h_{ie} vary with I_C if $r_{bb'} << r_{b'e}$.

Ans. If $r_{bb'} << r_{b'e}$, h_{ie} varies as reciprocal of collector current I_C.

Problems

1. If h_{ie} = 1,500 Ω, h_{fe} = 99, what is h_{ib} ? **[Ans.** 15 Ω]
2. If h_{ie} = 1,500 Ω, h_{fe} = 99, what is h_{ic} ? **[Ans.** 1,500 Ω]
3. Given h_{ie} = 2.4 kΩ, h_{fe} = 100, $h_{re} = 4 \times 10^{-4}$ and h_{oe} = 25 μS. Sketch the common-emitter equivalent model.

[U.P. Technical Univ. Basic Electronics Second Semester 2000-01]

Ans.

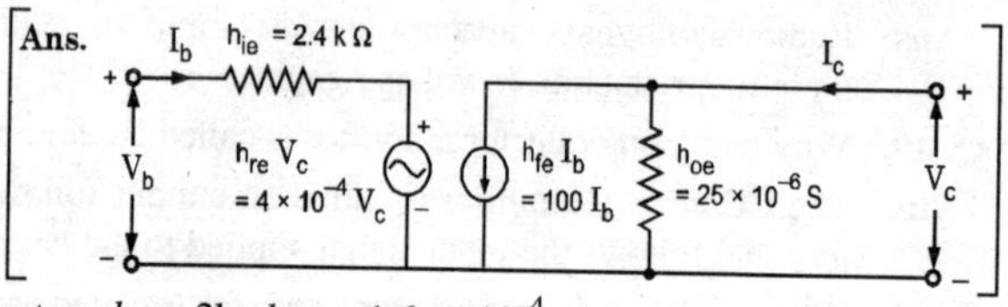

4. A junction transistor has the following h parameters h_{ie} = 2k, $h_{re} = 1.6 \times 10^{-4}$, h_{fe} = 50, h_{oe} = 50 μA/V. Determine the current gain, voltage gain, input resistance and output resistance of CE amplifier if the load resistance is 12 kΩ and source resistance is 500 Ω.

[Ans. A_i = –31.25; R_{in} = 1,940 Ω, A_V = –193.3, A_{is} = –6.4, A_{Vs} = –153.7; R_{out} = 21.37 kΩ]

5. For a CE amplifier circuit with h parameters : h_{ie} = 2 kΩ; $h_{re} = 6 \times 10^{-4}$, h_{fe} = 50, h_{oe} = 25 μA/V and load resistance R_L = 4 kΩ and source resistance R_s = 10 kΩ. Compute A_V, A_i, R_i and R_0.

[U.P. Technical Univ. Electronics Engg. First Semester 2008-09]

[Ans. A_i = –45.45; A_{is} = –38.226; R_{in} = 1,891 Ω; A_V = –96.15; A_{Vs} = –15.29, R_{out} = 44.45 kΩ]

6. A BJT has the following parameters :

$$h_{ie} = 2000\ \Omega,\ h_{re} = 16 \times 10^{-5},\ h_{fe} = 49 \text{ and } h_{oe} = 50\ \mu A/V.$$

Determine the current gain, voltage gain, input resistance and output resistance of the CE amplifier if the load resistance is 30 kΩ. Neglect source resistance. **[M.D. Univ. Rohtak Analog Electronics May 2007]**

[Ans. A_i = –19.6; R_{in} = 1.906 kΩ; A_V = –308.5; R_{out} = 21.7 kΩ]

7. A BJT has h_{ie} = 2 kΩ, h_{fe} = 100, $h_{re} = 2.5 \times 10^{-4}$ and h_{oe} = 25 μA/V as parameters in CE configuration. It is used as an emitter follower amplifier with R_s = 1 kΩ and R_L = 500 Ω. Determine for the amplifier the voltage gain $A_{Vs} = V_o/V_s$, the current gain $A_{is} = I_o/I_s$, the input resistance R_i and output resistance R_0.

[Ans. 0.9432; 1.886; 51.876 kΩ; 29.68 Ω]

8. Given h_{ie} = 1 kΩ, h_{fe} = 100, draw the h parameter equivalent circuit of the amplifier shown in Fig. 1. Calculate the input impedance, output impedance, current gain and voltage gain.

[Pb. Univ. Applied Electronics-I, Dec. 1995]

[Ans. 0.87 kΩ; 1.9 kΩ, –43.5, –100]

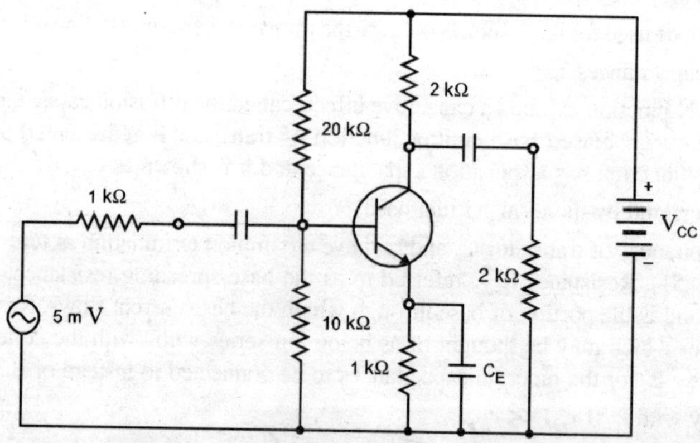

Fig. 1

9. For the circuit shown in Fig. 2, find A_V, A_i, Z_0 and Z_i. h parameters are $h_{ie} = 1\ k\Omega$; $h_{re} = 0$; $h_{fe} = 50$, $h_{oe} = 0$.

[Ans. –110, –4.4, 2.2 kΩ, 4.56 kΩ]

10. Compute $A_i = I_0/I_i$, A_V, A_{VS} and R_i for amplifier circuit shown in the following Fig. 3 for which the h-parameters are $h_{ie} = 1{,}100\ \Omega$, $h_{re} = 0$, $h_{fe} = 50$ and $\dfrac{1}{h_{oe}} = \infty$.

[U.P. Technical Univ. Basic Electronics Second Semester 2002-03]

[Ans. –50, –4.8, –2.1, 7.74 kΩ]

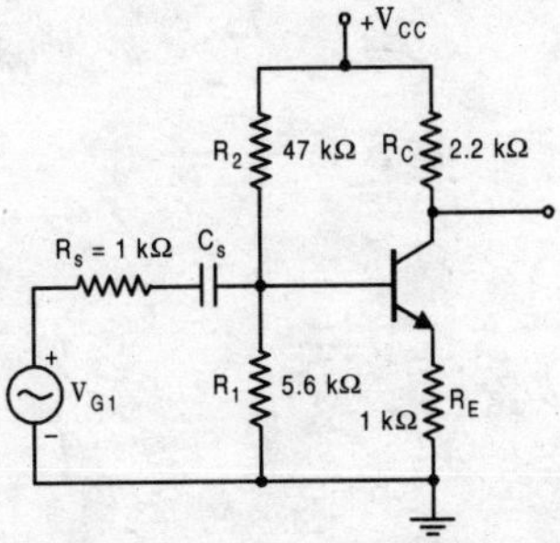

Fig. 2

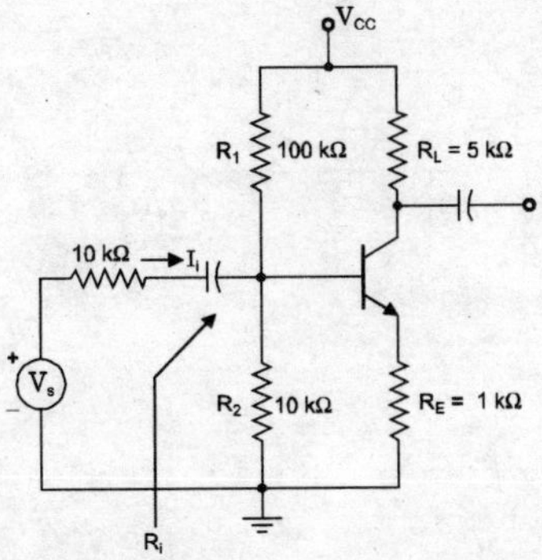

Fig. 3

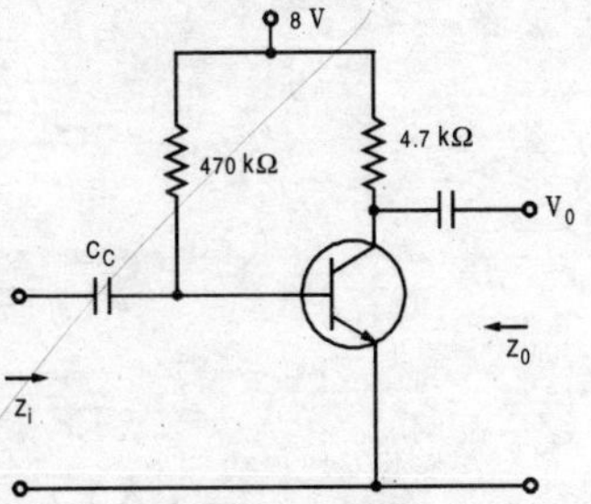

Fig. 4

11. Determine Z_i, Z_0 and A_v for the following network (Fig. 4)
[Given $h_{fe} = 110$, $h_{ie} = 1.6\ k\Omega$, $h_{re} = 2 \times 10^{-4}$ and $h_{oe} = 20\ \mu A / V$] **[Ans.** 1.595 kΩ, 4.296 kΩ, –323.125]

12. A transistor with $h_{ie} = 1.5\ k\Omega$ and $h_{fe} = 75$ is used as an emitter follower circuit where resistances R_1 and R_2 are used for normal biasing. Calculate (*i*) A_i, (*ii*) r_{in} (*iii*) r_o and (*iv*) A_v if $R_E = 860\ \Omega$, $R_1 \parallel R_2 = 20\ k\Omega$ and $R_s = 1\ k\Omega$.

[Indore Univ. Electronic Engineering 1983]

[Ans. $A_i = 17.5$, $r_{in} = 66.86\ k\Omega$, $r_o = 31.1\ \Omega$, $A_v \simeq 1$]

13. Given that $h_{fe} = 50$, $h_{oe} = 24\ \mu A$. Find out g_m and $g_{b'e}$ at $I_C = 1.3$ mA and room temperature.

[Rajasthan Univ. 2002]

[Ans. 0.05 S, 0.001 S]

14. Given the following parameters for a given transistor at $I_C = 1.3$ mA and at room temperature : $h_{oe} = 25\ \mu S$; $h_{fe} = 50$; $h_{ie} = 1.1\ k\Omega$ and $h_{re} = 2.5 \times 10^{-4}$. Determine g_m; $r_{b'e}$; $r_{bb'}$; $r_{b'c}$ and r_{ce}.

[Ans. $g_m = 50$ mS; $r_{b'e} = 1\ k\Omega$; $r_{bb'} = 100\ \Omega$; $r_{b'c} = 4\ M\Omega$; $r_{ce} = 81.6\ k\Omega$]

❑❑❑

APPENDICES

APPENDICES

APPENDIX–1

1N4001 THROUGH 1N4007 RECTIFIERS

SURMETIC RECTIFIERS

Subminiature size, axial lead-mounted rectifiers for general purpose, low power applications

Designer Data for "Worst Case" Conditions

The Designers Data Sheets permit the design of most circuits entirely from the information presented. Limit curves-representing boundaries on device characteristics — are given to facilitate "worst case" design.

MAXIMUM RATINGS*

Rating	Symbol	1N 4001	1N 4002	1N 4003	1N 4004	1N 4005	1N 4006	1N 4007	Unit
Peak Repetitive Reverse Voltage Working Peak Reverse Voltage DC Blocking Voltage	V_{RRM} V_{RWM} V_R	50	100	200	400	600	800	1000	Volts
Non-Repetitive Peak Reverse Voltage (Half Wave Single Phase 60 Hz)	V_{RSM}	60	120	240	480	720	1000	1200	Volts
RMS Reverse Voltage	$V_{R\,(RMS)}$	35	70	140	280	420	560	700	Volts
Average Rectified Forward Current (Single Phase Resistive Load, 60 Hz, T_A = 75ºC)	I_0	1.0							Amp
Non-Repetitive Peak Surge Current (Surge applied at rated load conditions	I_{FSM} or $I_{FM\,(surge)}$	30 (for 1 cycle)							Amp
Peak Repetitive Forward Current	$I_{FM\,(rep)}$	10							Amp
Operating and Storage Junction Temperature Range	T_j, T_{stg}	– 65 to + 175							ºC

LEAD-MOUNTED
SILICON RECTIFIERS
50-100 VOLTS
DIFFUSED JUNCTION

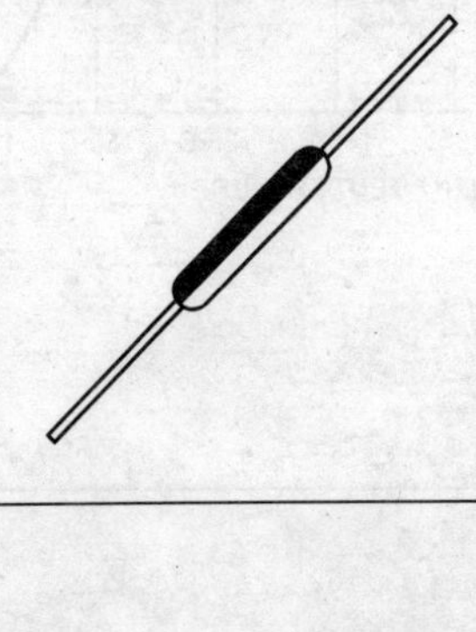

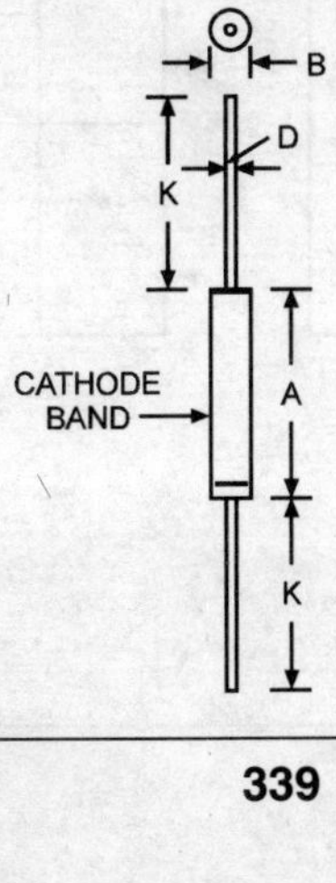

DIM	MILLIMETERS MIN	MILLIMETERS MAX	INCHES MIN	INCHES MAX
A	5.97	6.60	0.235	0.260
B	2.79	3.05	0.110	0.120
D	0.76	0.86	0.030	0.034
K	27.94	—	1.100	—

CASE 59-04
Does Not Conform to 00-41 outline

ELECTRICAL CHARACTERISTICS*

Characteristics and Conditions	Symbol	Typical	Max	Unit
Maximum Instantaneous Forward Voltage Drop (1.0 Amp Continuous DC, 25 ºC)	v_F	0.93	1.1	Volts
Maximum Full-Cycle Average Forward Voltage Drop (I_0 = 1.0 Amp, T_L = 75ºC 1 inch leads)	$V_{F(AV)}$	—	0.8	Volts
Maximum Reverse Current (rated dc voltage) T_j = 25 ºC T_j = 100 ºC	I_R	0.05 1.0	10 50	μA
Maximum Full-Cycle Average Reverse Current (I_0 = 1.0 Amp, T_L = 75 ºC, 1 inch lead)	$I_{R(AV)}$	—	30	μA

* Indicates JEDEC Registered Data

MECHANICAL CHARACTERISTICS

CASE : Void free, transfer Molded

MAXIMUM LEAD TEMPERATURE FOR SOLDERING PURPOSES :

350º C, $\frac{3}{8}$" from case for 10 seconds at 5 lbs tension

FINISH : All external surfaces are corrosion-resistant, leads are readily solderable

POLARITY : Cathode indicated by color band

WEIGHT : 0.40 grams (approximately)

TYPICAL FORWARD CHARACTERISTICS

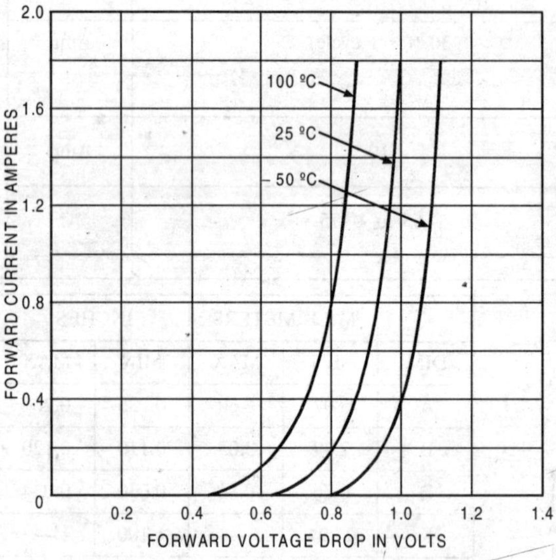

MAXIMUM DC OUTPUT

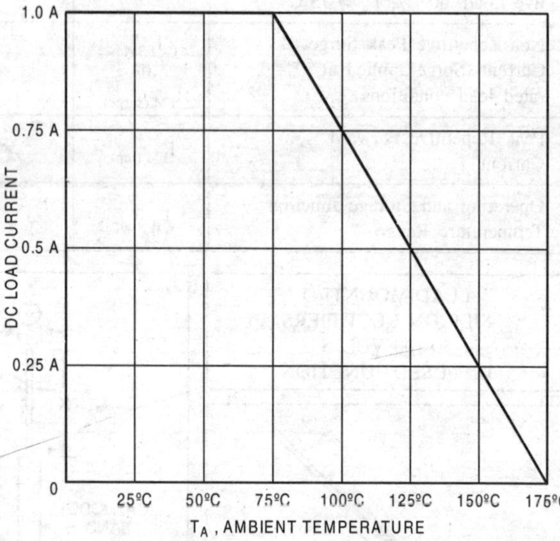

Courtesy of Motorola, Inc.

APPENDIX–2

1N746 through 1N759 Silicon

Voltage Regulator Diodes

3.3 To 12 V; 400 mW

Guaranteed Dynamic Zener Impedance

Available in 5 % and 10 % tolerances

– 65 ºC to 175 ºC operation and storage

* Electrical Characteristics at 25 ºC free-air temperature (unless otherwise stated)

Parameter	Zener Breakdown Voltage, V_Z					Temperature Coefficient of Breakdown Voltage, α_Z	Small Signal Breakdown Impedance r_Z	Static Reverse Current, I_R	
Test Conditions	I_{ZT} = 20 mA					I_{ZT} = 20 mA	I_{ZT} = 20 mA I_{ZT} = 1 mA	V_R = 1 V	V_R = 1 V T_A = 150ºC
Limit	NOM	1 N 746 – 1 N 759 MIN	1 N 746 – 1 N 759 MAX	1 N 746A – 1 N 759A MIN	1 N 746A – 1 N 759A MAX	TYP	MAX	MAX	MAX
UNIT	V	V	V	V	V	% / ºC	Ω	μ A	μ A
1 N 746	3.3	2.97	3.63	3.135	3.465	– 0.62	28	10	30
1 N 747	3.6	3.24	3.96	3.420	3.780	– 0.055	24	10	30
1 N 748	3.9	3.51	4.29	3.705	4.095	– 0.049	23	10	30
1 N 749	4.3	3.87	4.73	4.085	4.515	– 0.036	22	2	30
1 N 750	4.7	4.23	5.17	4.465	4.935	– 0.018	19	2	30
1 N 751	5.1	4.59	5.61	4.845	5.355	– 0.008	17	1	20
1 N 752	5.6	5.04	6.16	5.320	5.880	+ 0.006	11	1	20
1 N 753	6.2	5.58	6.82	5.890	6.510	+ 0.022	7	0.1	20
1 N 754	6.8	6.12	7.48	6.460	7.140	+ 0.035	5	0.1	20
1 N 755	7.5	6.75	8.25	7.125	7.875	+ 0.045	6	0.1	20
1 N 756	8.2	7.38	9.02	7.790	8.610	+ 0.052	8	0.1	20
1 N 757	9.1	8.19	10.01	8.645	9.555	+ 0.056	10	0.1	20
1 N 758	10.0	9.00	11.00	9.50	10.500	+ 0.060	17	0.1	20
1 N 759	12.0	10.80	13.20	11.400	12.000	+ 0.060	30	0.1	20

* **Absolute Maximum Ratings**

Average Rectified Forward Current at (or below) 25 ºC Free-Air Temperature .... 230 mA
Average Rectified Forward Current at 150 ºC Free-Air Temperature .... 85 mA
Continuous Power Dissipation at (or below) 50 ºC Free-Air Temperature .... 400 mW
Continuous Power Dissipation at 150 ºC Free-Air Temperature .... 100 mW
Operating Free-Air Temperature Range .... – 65 ºC to 175 ºC
Storage Temperature Range .... – 65 ºC to 175 ºC

* Indicates JEDEC Registered Data

Courtesy of Texas Instruments, Incorporated

APPENDIX–3

CASE 29
(TO-92)

2N3903, 2N3904 (Silicon) NPN Transistors

NPN silicon annular transistors, designed for general purpose switching and amplifier applications, features one-piece, injection-molded plastic package for high reliability. The 2N3903 and 2N3904 are complementary with types 2N3905 and 2N3906, respectively.

MAXIMUM RATINGS (T_A = 25 °C unless otherwise stated)

Characteristic	Symbol	Rating	Unit
Collector-Base Voltage	V_{CB}	60	Vdc
Collector-Emitter Voltage	V_{CEO}	40	Vdc
Emitter-Base Voltage	V_{EB}	6	Vdc
Collector Current	I_C	200	mAdc
Total Device Dissipation @ T_A = 60 °C	P_D	210	mW
Total Device Dissipation @ T_A = 25 °C Derate above 25 °C	P_D	310 2.81	mW mW/°C
Thermal Resistance, Junction to Ambient	θ_{JA}	0.357	°C/mW
Junction Operating Temperature	T_J	135	°C
Storage Temperature Range	T_{stg}	– 55 to + 135	°C

ELECTRICAL CHARACTERISTICS (T_A = 25 °C unless otherwise noted)

Characteristic	Symbol	Min	Max	Unit
OFF CHARACTERISTICS				
Collector-Base Breakdown Voltage (I_C = 10 μA dc, I_E = 0)	BV_{CBO}	60	—	Vdc
Collector-Emitter Breakdown Voltage* (I_C = 1 m A dc)	BV_{CEO}*	40	—	Vdc
Emitter-Base Breakdown Voltage (I_E = 10 μA dc, I_C = 0)	BV_{EBO}	6	—	Vdc
Collector Cut-off Current (V_{CE} = 40 V dc, V_{OB} = 3 Vdc)	I_{CEX}	—	50	nAdc
Base Cut-off Current (V_{CE} = 40 V dc, V_{OB} = 3 Vdc)	I_{BL}	—	50	nAdc

* Pulse Test. Pulse Width = 300 μ sec, Duty Cycle = 2 % V_{OB} = Base-Emitter Reverse Bias

* Courtesy of Motorola, Inc.

Appendix 1.3 (Continued)

Characteristic		Symbol	Min	Max	Unit
ON CHARACTERISTICS					
DC Current Gain*		h_{FE}*			
(I_C = 0.1 m A dc, V_{CE} = 1 Vdc)	2N3903		20	—	
	2N3904		40	—	
(I_C = 1.0 m A dc, V_{CE} = 1 Vdc)	2N3903		35	—	
	2N3904		70	—	
(I_C = 10 m A dc, V_{CE} = 1 Vdc)	2N3903		50	150	
	2N3904		100	300	
(I_C = 50 m A dc, V_{CE} = 1 Vdc)	2N3903		30	—	
	2N3904		60	—	
(I_C = 100 m A dc, V_{CE} = 1 Vdc)	2N3903		15	—	
	2N3904		30	—	
Collector-Emitter Saturation Voltage*		$V_{CE\ (sat)}$*			Vdc
(I_C = 10 m A dc, I_B = 1 m A dc)			—	0.2	
(I_C = 50 m A dc, I_B = 5 m A dc)			—	0.3	
Base-Emitter Saturation Voltage*		$V_{BE\ (sat)}$*			Vdc
(I_C = 10 m A dc, I_B = 1 m A dc)			0.65	0.85	
(I_C = 50 m A dc, I_B = 5 m A dc)			—	0.95	
SMALL SIGNAL CHARACTERISTICS					
High Frequency Current Gain	2N3903	$\|h_{fe}\|$	2.5	—	—
(I_C = 10 m A, V_{CE} = 20 V, f = 100 mc)	2N3904		3.0	—	
Current Gain– Bandwidth Product	2N3903	f_T	250	—	mc
(I_C = 10 m A, V_{CE} = 20 V, f = 100 mc)	2N3904		300	—	
Output Capacitance		C_{ob}			pf
(V_{CB} = 5 Vdc, I_E = 0, f = 100 kc)			—	4	
Input Capacitance		C_{ib}			pf
(V_{OB} = 0.5 V dc, I_C = 0, f = 100 kc)			—	8	
Small Signal Current Gain	2N3903	h_{fe}	50	200	—
(I_C = 1.0 m A, V_{CE} = 10 V, f = 1 kc)	2N3904		100	400	
Voltage Feedback Ratio	2N3903	h_{re}	0.1	5.0	$\times 10^{-4}$
(I_C = 1.0 m A, V_{CE} = 10 V, f = 1 kc)	2N3904		0.5	8.0	
Input Impedance	2N3903	h_{ie}	0.5	8	Kohms
(I_C = 1.0 m A, V_{CE} = 10 V, f = 1 kc)	2N3904		1.0	10	
Output Admittance		h_{oe}			μ mhos
(I_C = 1.0 m A, V_{CE} = 10 V, f = 1 kc)	Both Types		1.0	40	
Noise Figure		NF			db
(I_C = 100 μ A, V_{CE} = 5 V, R_g = 1 Kohms),	2N3903		—	6	
Noise Bandwidth = 10 cps to 15.7 kc)	2N3904		—	5	
SWITCHING CHARACTERISTICS					
Delay Time — V_{CC} = 3 V dc, V_{OB} = 0.5 V dc,		t_d	—	35	nsec
Rise Time — I_C = 10 m A dc, I_{B1} = 1 m A		t_T	—	35	nsec
Storage Time — V_{CC} = 3 V dc, I_C = 10 m A dc,	2N3903	t_E	—	175	nsec
I_{B1} = I_{B2} = 1 m A dc	2N3904		—	200	
Fall Time		t_f	—	50	nsec

* Pulse Test : Pulse Width = 300 μ sec, Duty Cycle = 2 % V_{OB} = Base-Emitter Reverse Bias

APPENDIX–4

Type 2N3055 NPN Single-Diffused Mesa Silicon Power Transistor FOR POWER-AMPLIFIER APPLICATIONS

Mechanical Data

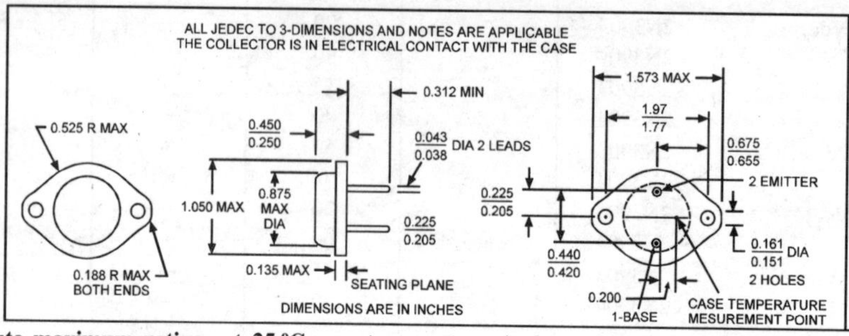

Absolute maximum ratings at 25 °C case temperature (unless otherwise noted)

- Collector-Base Voltage .... 100 V
- Collector-Emitter Voltage (See Note 1) .... 70 V
- Emitter-Base Voltage .... 7 V
- Continuous Collector Current .... 15 A
- Continuous Base Current .... 7 A
- Continuous Device Dissipation at (or below) 25 °C Case Temperature (See Note 2) .... 115 W
- Operating Case Temperature Range .... – 65 °C to 200 °C
- Storage Temperature Range .... – 65 °C to 200 °C
- Lead Temperature 1/32 Inch from Case for 10 Seconds .... 235 °C

NOTES 1. This value applies when the base-emitter resistance R_{BE} = 100 Ω

2. Derate linearly to 200 °C case temperature at the rate of 0.66 W/deg.

*** Electrical characteristics at 25 °C case temperature (unless otherwise noted)**

PARAMETER		TEST CONDITIONS	MIN	MAX	UNIT
$V_{(BR)CEO}$	Collector-Emitter Breakdown Voltage	I_C = 200 m A, I_B = 0, See Note 4 60		V	
$V_{(BR)CER}$	Collector-Emitter Breakdown Voltage	I_C = 200 m A, R_{BE} = 100 Ω	70		V
I_{CEO}	Collector Cutoff Current	V_{CE} = 30 V, I_B = 0		0.7	m A
I_{CEV}	Collector Cutoff Current	V_{CE} = 100 V, V_{BE} = – 1.5 V		5	m A
		V_{CE} = 100 V, V_{BE} = – 1.5 V, T_C = 150 °C		30	
I_{EBO}	Emitter Cutoff Current	V_{EB} = 7 V, I_C = 0		5	m A
h_{FE}	Static Forward Current Transfer Ratio	V_{CE} = 4 V, I_C = 4 A, See Notes 3 and 4	20	70	
		V_{CE} = 4 V, I_C = 10 A, See Notes 3 and 4	5		
V_{BE}	Base-Emitter Voltage	V_{CE} = 4 V, I_C = 4 A, See Notes 3 and 4		1.8	V
$V_{CE(sat)}$	Collector-Emitter Saturation Voltage	I_B = 400 m A, I_C = 4 A See Notes 3 and 4		1.1	V
		I_B = 3.3 A, I_C = 10 A See Notes 3 and 4		8	
h_{fe}	Small-Signal Common-Emitter Forward Current Transfer Ratio	V_{CE} = 4 V, I_C = 1 A, f = 1 kHz 15	120		
f_{hfe}	Small-Signal Common-Emitter Forward Current Transfer Ratio Cutoff Frequency	V_{CE} = 4 V, I_C = 1 A, See Notes 5	20		kHz

Thermal Characteristics

PARAMETER		MAX	UNIT
θ_{J-C}	Junction-to-Case Thermal Resistance	1.52	deg/W

NOTES 3. These parameters must be measured using pulse technique t_p = 300 μs, duty cycle ≤ 2 %.

4. These parameters are measured with voltage-sensing contacts separate from the current-carrying contacts.

5. f_{hfe} is the frequency at which the magnitude of the small-signal forward current transfer ratio is 0.707 of its low-frequency value. For this device, the reference measurement is made at 1 kHz.

* Courtesy of Texas Instruments, Incorporated

APPENDIX–5

2N5457, 2N5458, 2N5459 (Silicon) N-Channel FET

Silicon N-channel junction field-effect transistors depletion mode (Type A) designed for general-purpose audio and switching applications.

CASE 29 (5)
(TO-92)

Drain and source may be interchanged

MAXIMUM RATINGS

Rating	Symbol	Value	Unit
Drain-Source Voltage	V_{DS}	25	V dc
Drain-Gate Voltage	V_{DG}	25	V dc
Reverse Gate Source Voltage	$V_{GS\ (r)}$	25	V dc
Gate Current	I_G	10	m A dc
Total Device Dissipation @ T_A = 25 ºC Derate above 25 ºC	$P_D^{(2)}$	310 2.82	mW mW/ºC
Operating Junction Temperature	$T_J^{(2)}$	135	ºC
Storage Temperature Range	$T_{stg}^{(2)}$	– 65 to + 150	ºC

ELECTRICAL CHARACTERISTICS (T_A = 25 ºC unless otherwise noted)

Characteristic		Symbol	Min	Typ	Max	Unit
OFF CHARACTERISTICS						
Gate-Source Breakdown VoltageBV_{GSS} (I_G = – 10 μAdc, V_{DS} = 0)			25	—	Vdc —	
Gate Reverse Current (V_{GS} = – 15 Vdc, V_{DS} = 0) (V_{GS} = – 15 Vdc, V_{DS} = 0, T_A = 100 ºC)		I_{GSS}	— —	— —	1.0 200	n A dc
Gate-Source Cutoff Voltage (V_{DS} = 15 Vdc, I_D = 10 nAdc)	2N5457 2N5458 2N5459	$V_{GS\ (off)}$	0.5 1.0 2.0	— — —	6.0 7.0 8.0	V dc
Gate-Source Voltage (V_{DS} = 15 Vdc, I_D = 100 μAdc) (V_{DS} = 15 Vdc, I_D = 200 μAdc) (V_{DS} = 15 Vdc, I_D = 400 μAdc)	2N5457 2N5458 2N5459	V_{GS}	— — —	2.5 3.5 4.5	— — —	V dc
ON CHARACTERISTICS						
Zero-Gate-Voltage Drain Current (1) (V_{DS} = 15 Vdc, V_{GS} = 0)	2N5457 2N5458 2N5459	I_{DSS}	1.0 2.0 4.0	3.0 6.0 9.0	5.0 9.0 16	mA dc
DYNAMIC CHARACTERISTICS						
Forward Transfer Admittance (1)$\lvert y_{fs} \rvert$ (V_{DS} = 15 Vdc, V_{GS} = 0, f = 1 kHz)	2N5457 2N5458 2N5459		1000 1500 2000	3000 4000 4500	μ mhos 5000 5500 6000	
Output Admittance (1) (V_{DS} = 15 Vdc, V_{GS} = 0, f = 1 kHz)		$\lvert y_{os} \rvert$	—	10	50	μ mhos
Input Capacitance (V_{DS} = 15 Vdc, V_{GS} = 0, f = 1 MHz)		C_{iss}	—	4.5	7.0	pF
Reverse Transfer Capacitance (V_{DS} = 15 Vdc, V_{GS} = 0, f = 1 MHz)		C_{rss}	—	1.5	pF 3.0	

(1) Pulse Test : Pulse Width ≤ 630 ms, Duty Cycle ≤ 10 %
(2) Continuous package improvements have enhanced these guaranteed Maximum Ratings as follows P_D = 1.0 W @ T_C = 25 ºC, Derate above 25 ºC - 8.0 mW/ºC, T_J = – 65 to + 150 ºC, θ_{JC} = 125 ºC/W

* Courtesy of Motorola, Inc.

MODEL TEST PAPERS

MODEL TEST PAPER–1

ANALOG DEVICES AND CIRCUITS

Time : 3 Hours *Maximum Marks : 60*

Note : *Section-A is compulsory. Attempt any* ***FOUR*** *questions from Section B and* ***TWO*** *questions from Section C.*

SECTION-A **[Marks : 2 Each]**

1. (*a*) What is junction diode ? Draw characteristics of ideal diode.

(*b*) Explain how zener diode maintains constant voltage across the load.

(*c*) Give reasons why common emitter (CE) configuration is widely used in amplifier circuits.

(*d*) What does UJT stand for ? Justify the name UJT.

(*e*) Explain the difference between a voltage and power amplifier.

(*f*) What is crossover distortion ? Explain in brief.

(*g*) How does negative feedback increase bandwidth of an amplifier ?

(*h*) What is an oscillator ? How does it differ from an amplifier ?

(*i*) Explain how phase reversal of signal takes place when it is amplified by a single stage CE voltage amplifier.

(*j*) What are hybrid capacitances ?

SECTION-B **[Marks : 5 Each]**

2. Explain the tunnel diode in details. Give its volt-ampere characteristics.

3. What is LCD ? Discuss its advantages, disadvantages and applications.

4. Define stability factor. Explain the self-bias circuit with diagrams and derive expression for stability factor.

5. The parameters of a crystal oscillator equivalent circuit are L_s = 0.8 H; C_s = 0.08 pF, R_s = 5 kΩ and C_p = 1.0 pF. Determine the resonance frequencies f_s and f_p.

6. A power amplifier working in class A operation has a transformer as a load. If the transformer has a turn ratio of 10 and the secondary load is 100 Ω, find maximum ac power output given that zero-signal collector current is 0.14 A.

SECTION-C **[Marks : 10 Each]**

7. (*a*) Draw the input and output characteristics curves of a CB configuration for transistor. Explain the shape of the curves and clearly indicate cutoff, active and saturation regions.

(*b*) In a CE transistor amplifier when signal changes by 0.02 V, the base current changes by 10 μA and collector current by 1 mA. If the collector resistance R_C = 5 kΩ and R_L = 10 kΩ. Find (*i*) Voltage gain, (*ii*) Current gain, (*iii*) Input impedance, (*iv*) AC load resistance and (*v*) Power gain.

8. Draw and explain the working of R-C phase shift oscillator and also derive an expression for its frequency of oscillations.

An R-C phase shift oscillator has to be designed to provide a sine wave of frequency 1 kHz. If the value of capacitance is 0.016 μF, calculate the value of the resistances.

9. (*a*) Draw the hybrid-pi model of a BJT in common emitter (CE) configuration and discuss about each component in the model.

(*b*) A BJT has h_{ie} = 6 kΩ and h_{fe} = 224 at I_C = 1 mA with f_T = 80 MHz and $C_{b'c}$ = 12 pF. Determine (*i*) g_m (*ii*) $r_{b'e}$ (*iii*) $r_{bb'}$ and (*iv*) $C_{b'e}$ at room temperature and a collector current of 1 mA.

MODEL TEST PAPER–2

ANALOG DEVICES AND CIRCUITS

Time : 3 Hours *Maximum Marks : 60*

Note : *Section-A is compulsory. Attempt any* **FOUR** *questions from Section B and* **TWO** *questions from Section C.*

SECTION-A **[Marks : 2 Each]**

1. (*a*) How do the transition width and contact potential across a P-N junction vary with the applied bias voltage ?

(*b*) State differences between LED and photodiodes.

(*c*) Explain why an ordinary transistor called bipolar.

(*d*) What do you understand by terms channel and drain in JFET ?

(*e*) Why do we use transformer in the output stage of an amplifier ?

(*f*) Although a class-B single ended power amplifier has high efficiency, yet it is less used in practical circuits. Explain why ?

(*g*) Why is negative feedback employed in high gain amplifiers ?

(*h*) How does Hartley oscillator differ from Colpitt's oscillator in construction ?

(*i*) What is an amplifier and what are the characteristics of an ideal voltage amplifier ?

(*j*) What do you understand by the term "virtual base" ?

SECTION-B **[Marks : 5 Each]**

2. Draw the energy band diagram for a P-N junction under open-circuited condition. Clearly indicate various energy levels in P-region, space region and N-region. How will it be modified if P-N Junction is forward biased ?

3. Explain the working of diode as a shunt regulator. For the circuit of Fig. 1, determine the range of R_L and I_L that will result in V_{R_L} being maintained at 10 V.

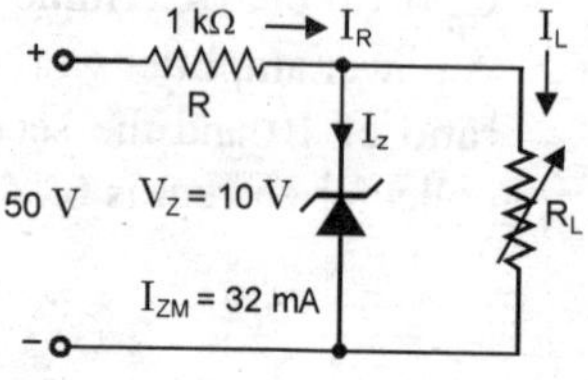

Fig. 1

4. Draw diagram showing constructional details of N-channel MOSFET. Explain the operation in depletion mode.

5. Calculate the gain of a negative feedback amplifier with an internal gain $A_v = 75$ and feedback fraction $m_v = 1/15$. What will be the gain if A_v doubles ?

6. Draw high frequency hybrid π-model for a BJT. Derive an expression for g_m.

SECTION-C **[Marks : 10 Each]**

7. (*a*) Draw and explain V-I characteristics of tunnel diode.

(*b*) What are LCDs ? In what respect LCDs are advantageous over LEDs ? Give their drawbacks in comparison to LEDs and applications.

8. (*a*) What is bias stabilization and what is its need ? Explain.

(*b*) A silicon transistor uses potential divider method of biasing. $V_{CC} = 12$ V, $R_1 = R_2 = 10$ kΩ, $R_E = 3$ kΩ, $R_C = 4$ kΩ. Determine operating point using Thevenin's theorem.

9. With suitable circuit diagram and waveforms, explain the operation of a class B power amplifier using complementary symmetry transistors.

□□□

Index

A

B

C

D

E

G

H

I

T

U

V

W

Z

❑❑❑

Win Prizes

ATTENTION READERS

Win Prizes

We request for your frank assessment regarding various aspects of the book as given below :

ANALOG DEVICES AND CIRCUITS

By : J.B. GUPTA (PTU)

Please Fill up the Sheet in Neat and Clean Writing

(i) What topics of your syllabus, important for your examination point of view, are not covered in this book ?

....

....

....

....

(ii) What are the chapters wherein the treatment of the subject matter is not systematic or organised or updated ?

....

....

....

....

(iii) Have you come across misprints/mistakes/factual inaccuracies in the book ? Please specify the chapters and the page numbers.

....

....

....

....

(iv) What other books on the same subject have you found/heard better than the present book ? Please specify in terms of price and quality.

....

....

....

....

(v) Further suggestions and comments for the improvement of this book.

....

....

....

....

(PS : If need be please attach a separate sheet to write your views)

OTHER DETAILS

(i) Who recommended you this book ? (Please tick (✓) in the box near the option relevant to you.

☐ Teacher ☐ Friends ☐ Bookseller

....

....

....

....

(ii) Name of the recommending teacher, his designation and address:

....

....

....

....

(iii) Name and address of the bookseller you purchased the book from :

....

....

....

....

(iv) Name and address of your institution (Please mention the University or Board as the case may be)

....

....

....

....

(v) Your name and complete address :

....

....

....

....

(vi) Course you are enrolled in :

....

....

....

....

The best assessment will be awarded each month : The award will be in the form of our publications as decided by the Editorial Board.

Please mail the filled up coupon at the vearliest to :

S.K. KATARIA & SONS®

4760-61/23, Ansari Road,

Daryaganj, New Delhi-110002

Phone: +91-11-23243489, 23269324

TeleFax : +91-11-23243489

email : katariabooks@yahoo.com